Methods in Enzymology

Volume 270
HIGH RESOLUTION SEPARATION
AND ANALYSIS OF
BIOLOGICAL MACROMOLECULES
Part A
Fundamentals

METHODS IN ENZYMOLOGY

EDITORS-IN-CHIEF

John N. Abelson   Melvin I. Simon

DIVISION OF BIOLOGY
CALIFORNIA INSTITUTE OF TECHNOLOGY
PASADENA, CALIFORNIA

FOUNDING EDITORS

Sidney P. Colowick and Nathan O. Kaplan

Methods in Enzymology

Volume 270

High Resolution Separation and Analysis of Biological Macromolecules

Part A
Fundamentals

EDITED BY

Barry L. Karger

DEPARTMENT OF CHEMISTRY
BARNETT INSTITUTE
NORTHEASTERN UNIVERSITY
BOSTON, MASSACHUSETTS

William S. Hancock

HEWLETT-PACKARD
PALO ALTO, CALIFORNIA

ACADEMIC PRESS

San Diego   New York   Boston   London   Sydney   Tokyo   Toronto

Academic Press, Inc.
A Division of Harcourt Brace & Company
525 B Street, Suite 1900, San Diego, California 92101-4495

United Kingdom Edition published by
Academic Press Limited
24-28 Oval Road, London NW1 7DX

International Standard Serial Number: 0076-6879

International Standard Book Number: 0-12-182171-4

PRINTED IN THE UNITED STATES OF AMERICA
96   97   98   99   00   01   MM   9   8   7   6   5   4   3   2   1

Table of Contents

Section I. Liquid Chromatography
A. Methods

B. Columns and Instrumentation

v

Contributors to Volume 270

Article numbers are in parentheses following the names of contributors.
Affiliations listed are current.

MARIE-ISABEL AGUILAR (1), *Department of Biochemistry and Centre for Bioprocess Technology, Monash University, Clayton, Victoria 3168, Australia*

J. FRED BANKS, JR. (21), *Analytica of Branford, Inc., Branford, Connecticut 06405*

RONALD C. BEAVIS (22), *Department of Chemistry and Pharmacology, Skirball Institute, New York University, New York, New York 10016*

BRUCE W. BIRREN (11), *Division of Biology, California Institute of Technology, Pasadena, California 91125*

PETR BOČEK (17), *Institute of Analytical Chemistry, Academy of Sciences of the Czech Republic, CZ-611 42 Brno, Czech Republic*

RICHARD M. CAPRIOLI (20), *Analytical Chemistry Center and Department of Biochemistry and Molecular Biology, University of Texas Medical School, Houston, Texas 77030*

BRIAN T. CHAIT (22), *Laboratory for Mass Spectrometry and Gaseous Ion Chemistry, The Rockefeller University, New York, New York 10021*

MARCELLA CHIARI (10), *Institute of Hormone Chemistry, National Research Council, Milan 20133, Italy*

GARGI CHOUDHARY (3), *Department of Chemical Engineering, Yale University, New Haven, Connecticut 06520*

BRUCE JON COMPTON (15), *AutoImmune Inc., Lexington, Massachusetts 02173*

MERCEDES DE FRUTOS (4, 6), *Instituto de Quimica Organica, General y Fermentaciones Industriales (C.S.I.C.), 28006 Madrid, Spain*

GUY DROUIN (12), *Department of Biology, University of Ottawa, Ottawa, Ontario, Canada K1N 6N5*

PETR GEBAUER (17), *Institute of Analytical Chemistry, Academy of Sciences of the Czech Republic, CZ-611 42 Brno, Czech Republic*

CECILIA GELFI (10), *Institute of Advanced Biomedical Technologies, National Research Council, Milan, Italy*

METTE GRØNVALD (15), *Department of Chemistry and Chemical Engineering, The Engineering Academy of Denmark, TK 7058, A 892036 Copenhagen, Denmark*

MILTON T. W. HEARN (1), *Department of Biochemistry and Centre for Bioprocess Technology, Monash University, Clayton, Victoria 3168, Australia*

STELLAN HJERTÉN (13), *Department of Biochemistry, Uppsala University, Uppsala, Sweden*

CSABA HORVÁTH (3), *Department of Chemical Engineering, Yale University, New Haven, Connecticut 06520*

IAN JARDINE (23), *Finnigan MAT, San Jose, California 95134*

JAMES W. JORGENSON (18), *Department of Chemistry, University of North Carolina, Chapel Hill, North Carolina 27599*

LUDMILA KŘIVÁNKOVÁ (17), *Institute of Analytical Chemistry, Academy of Sciences of the Czech Republic, CZ-611 42 Brno, Czech Republic*

BARRY L. KARGER (2), *Department of Chemistry, Barnett Institute, Northeastern University, Boston, Massachusetts 02115*

IRA S. KRULL (8), *Department of Chemistry, Northeastern University, Boston, Massachusetts 02115*

ERIC LAI (11), *Department of Pharmacology, University of North Carolina, Chapel Hill, North Carolina 27599*

JOHN P. LARMANN, JR. (18), *Department of Chemistry, University of North Carolina, Chapel Hill, North Carolina 27599*

THOMAS T. LEE (19), *Department of Chemistry, Stanford University, Stanford, California 95305*

ix

ANTHONY V. LEMMO (18), *Department of Chemistry, University of North Carolina, Chapel Hill, North Carolina 27599*

BARBARA D. LIPES (11), *Department of Pharmacology, University of North Carolina, Chapel Hill, North Carolina 27599*

NORIO MATSUBARA (14), *Faculty of Science, Himeji Institute of Technology, Kamigori, Hyogo 678-12, Japan*

PASCAL MAYER (12), *Department of Biology, University of Ottawa, Ottawa, Ontario, Canada K1N 6N5*

JEFF MAZZEO (8), *Waters Chromatography Division, Millipore Corporation, Milford, Massachusetts 01757*

ROHIN MHATRE (8), *PerSeptive Biosystems, Inc., Framingham, Massachusetts 01701*

STAN MICINSKI (15), *Washington State University, Pullman, Washington 99164*

ALVIN W. MOORE, JR. (18), *Department of Chemistry, University of North Carolina, Chapel Hill, North Carolina 27599*

MILOS V. NOVOTNY (5), *Department of Chemistry, Indiana University, Bloomington, Indiana 47405*

SANDEEP K. PALIWAL (4, 6), *SyStemix Inc., Palo Alto, California 94304*

FRED E. REGNIER (4, 6), *Department of Chemistry, Purdue University, Lafayette, Indiana 47906*

PIER GIORGIO RIGHETTI (10), *Faculty of Pharmacy and Department of Biomedical Sciences and Technologies, University of Milan, Milan 20133, Italy*

ROBERTO RODRIGUEZ-DIAZ (16), *Bio-Rad Laboratories, Hercules, California 94547*

GERARD P. ROZING (9), *Waldbronn Analytical Division, Hewlett Packard GmbH, D76337 Waldbronn, Germany*

JAE C. SCHWARTZ (23), *Finnigan MAT, San Jose, California 95134*

WILLIAM E. SEIFERT, JR. (20), *Analytical Chemistry Center, University of Texas Medical School, Houston, Texas 77030*

GARY W. SLATER (12), *Department of Physics, University of Ottawa, Ottawa, Ontario, Canada K1N 6N5*

LLOYD R. SNYDER (7), *LC Resources, Inc., Orinda, California 94563*

MICHAEL SZULC (8), *Quality Control R&D Laboratory, Biogen Corporation, Cambridge, Massachusetts 02142*

SHIGERU TERABE (14), *Faculty of Science, Himeji Institute of Technology, Kamigori, Hyogo 678-12, Japan*

TIM WEHR (16), *Bio-Rad Laboratories, Hercules, California 94547*

CRAIG M. WHITEHOUSE (21), *Analytica of Branford, Inc., Branford, Connecticut 06405*

JANET C. WRESTLER (11), *Department of Pharmacology, University of North Carolina, Chapel Hill, North Carolina 27599*

SHIAW-LIN WU (2), *Department of Analytical Chemistry, Genentech, Inc., South San Francisco, California 94080*

EDWARD S. YEUNG (19), *Department of Chemistry and Ames Laboratory, Iowa State University, Ames, Iowa 50011*

MINGDE ZHU (16), *Bio-Rad Laboratories, Hercules, California 94547*

Preface

All areas of the biological sciences have become increasingly molecular in the past decade, and this has led to ever greater demands on analytical methodology. Revolutionary changes in quantitative and structure analysis have resulted, with changes continuing to this day. Nowhere has this been seen to a greater extent than in the advances in macromolecular structure elucidation. This advancement toward the exact chemical structure of macromolecules has been essential in our understanding of biological processes. This trend has fueled demands for increased ability to handle vanishingly small quantities of material such as from tissue extracts or single cells. Methods with a high degree of automation and throughput are also being developed.

In the past, the analysis of macromolecules in biological fluids relied on methods that used specific probes to detect small regions of the molecule, often in only partially purified samples. For example, proteins were labeled with radioactivity by *in vivo* incorporation. Another approach has been the detection of a sample separated in a gel electrophoresis by means of blotting with an antibody or with a tagged oligonucleotide probe. Such procedures have the advantages of sensitivity and specificity. The disadvantages of such approaches, however, are many, and range from handling problems of radioactivity, as well as the inability to perform a variety of *in vivo* experiments, to the invisibility of residues out of the contact domain of the tagged region, e.g., epitope regions in antibody-based recognition reactions.

Beyond basic biological research, the advent of biotechnology has also created a need for a higher level of detail in the analysis of macromolecules, which has resulted in protocols that can detect the transformation of a single functional group in a protein of 50,000–100,000 daltons or the presence of a single or modified base change in an oligonucleotide of several hundred or several thousand residues. The discovery of a variety of posttranslational modifications in proteins has further increased the demand for a high degree of specificity in structure analysis. With the arrival of the human genome and other sequencing initiatives, the requirement for a much more rapid method for DNA sequencing has stimulated the need for methods with a high degree of throughput and low degree of error.

The bioanalytical chemist has responded to these challenges in biological measurements with the introduction of new, high resolution separation and detection methods that allow for the rapid analysis and characterization of macromolecules. Also, methods that can determine small differences in many thousands of atoms have been developed. The separation techniques include affinity chromatography, reversed phase liquid chromatography (LC), and capillary electrophoresis. We include mass spectrometry as a high resolution separation method, both given the fact that the method is fundamentally a procedure for separating gaseous ions and because separation–mass spectrometry (LC/MS, CE/MS) is an integral part of modern bioanalysis of macromolecules.

The characterization of complex biopolymers typically involves cleavage of the macromolecule with specific reagents, such as proteases, restriction enzymes, or chemical cleavage substances. The resulting mixture of fragments is then separated to produce a map (e.g., peptide map) that can be related to the original macromolecule from knowledge of the specificity of the reagent used for the cleavage. Such fingerprinting approaches reduce the characterization problem from a single complex substance to a number of smaller and thus simpler units that can be more easily analyzed once separation has been achieved.

Recent advances in mass spectrometry have been invaluable in determining the structure of these smaller units. In addition, differences in the macromolecule relative to a reference molecule can be related to an observable difference in the map. The results of mass spectrometric measurements are frequently complemented by more traditional approaches, e.g., N-terminal sequencing of proteins or the Sanger method for the sequencing of oligonucleotides. Furthermore, a recent trend is to follow kinetically the enzymatic degradation of a macromolecule (e.g., carboxypeptidase). By measuring the molecular weight differences of the degraded molecule as a function of time using mass spectrometry [e.g., matrix-assisted laser desorption ionization–time of flight (MALDI–TOF)], individual residues that have been cleaved (e.g., amino acids) can be determined.

As well as producing detailed chemical information on the macromolecule, many of these methods also have the advantage of a high degree of mass sensitivity since new instrumentation, such as MALDI–TOF or capillary electrophoresis with laser-based fluorescence detection, can handle vanishingly small amounts of material. The low femtomole to attomole sensitivity achieved with many of these systems permits detection more sensitive than that achieved with tritium or ^{14}C isotopes and often equals that achieved with the use of ^{32}P or ^{125}I radioactivity. A trend in mass spectrometry has been the extension of the technology to ever greater mass ranges so that now proteins of molecular weights greater than 200,000 and oligonucleotides of more than 100 residues can be transferred into the gas phase and then measured in a mass analyzer.

The purpose of Volumes 270 and 271 of *Methods in Enzymology* is to provide in one source an overview of the exciting recent advances in the analytical sciences that are of importance in contemporary biology. While core laboratories have greatly expanded the access of many scientists to expensive and sophisticated instruments, a decided trend is the introduction of less expensive, dedicated systems that are installed on a widespread basis, especially as individual workstations. The advancement of technology and chemistry has been such that measurements unheard of a few years ago are now routine, e.g., carbohydrate sequencing of glycoproteins. Such developments require scientists working in biological fields to have a greater understanding and utilization of analytical methodology. The chapters provide an update in recent advances of modern analytical methods that allow the practitioner to extract maximum information from an analysis. Where possible, the chapters also have a practical focus and concentrate on methodological details which are key to a particular method.

The contributions appear in two volumes: Volume 270, High Resolution Separation of Biological Macromolecules, Part A: Fundamentals and Volume 271, High Resolution Separation of Biological Macromolecules, Part B: Applications. Each volume is subdivided into three main areas: liquid chromatography, slab gel and capillary electrophoresis, and mass spectrometry. One important emphasis has been the integration of methods, in particular LC/MS and CE/MS. In many methods, chemical operations are integrated at the front end of the separation and may also be significant in detection. Often in an analysis, a battery of methods are combined to develop a complete picture of the system and to cross-validate the information.

The focus of the LC section is on updating the most significant new approaches to biomolecular analysis. LC has been covered in recent volumes of this series, therefore these volumes concentrate on relevant applications that allow for automation, greater speed of analysis, or higher separation efficiency. In the electrophoresis section, recent work with slab gels which focuses on high resolution analysis is covered. Many applications are being converted from the slab gel into a column format to combine the advantages of electrophoresis with those of chromatography. The field of capillary electrophoresis, which is a recent, significant high resolution method for biopolymers, is fully covered.

The third section contains important methods for the ionization of macromolecules into the gas phase as well as new methods for mass measurements which are currently in use or have great future potential. The integrated or hybrid systems are demonstrated with important applications.

We welcome readers from the biological sciences and feel confident that they will find these volumes of value, particularly those working at the interfaces between analytical/biochemical and molecular biology, as well as the immunological sciences. While new developments constantly occur, we believe these two volumes provide a solid foundation on which researchers can assess the most recent advances. We feel that biologists are working during a truly revolutionary period in which information available for the analysis of biomacromolecular structure and quantitation will provide new insights into fundamental processes. We hope these volumes aid readers in advancing significantly their research capabilities.

WILLIAM S. HANCOCK
BARRY L. KARGER

METHODS IN ENZYMOLOGY

VOLUME LV. Biomembranes (Part F: Bioenergetics)
Edited by SIDNEY FLEISCHER AND LESTER PACKER

VOLUME LVI. Biomembranes (Part G: Bioenergetics)
Edited by SIDNEY FLEISCHER AND LESTER PACKER

VOLUME LVII. Bioluminescence and Chemiluminescence
Edited by MARLENE A. DELUCA

VOLUME LVIII. Cell Culture
Edited by WILLIAM B. JAKOBY AND IRA PASTAN

VOLUME LIX. Nucleic Acids and Protein Synthesis (Part G)
Edited by KIVIE MOLDAVE AND LAWRENCE GROSSMAN

VOLUME LX. Nucleic Acids and Protein Synthesis (Part H)
Edited by KIVIE MOLDAVE AND LAWRENCE GROSSMAN

VOLUME 61. Enzyme Structure (Part H)
Edited by C. H. W. HIRS AND SERGE N. TIMASHEFF

VOLUME 62. Vitamins and Coenzymes (Part D)
Edited by DONALD B. MCCORMICK AND LEMUEL D. WRIGHT

VOLUME 63. Enzyme Kinetics and Mechanism (Part A: Initial Rate and Inhibitor Methods)
Edited by DANIEL L. PURICH

VOLUME 64. Enzyme Kinetics and Mechanism (Part B: Isotopic Probes and Complex Enzyme Systems)
Edited by DANIEL L. PURICH

VOLUME 65. Nucleic Acids (Part I)
Edited by LAWRENCE GROSSMAN AND KIVIE MOLDAVE

VOLUME 66. Vitamins and Coenzymes (Part E)
Edited by DONALD B. MCCORMICK AND LEMUEL D. WRIGHT

VOLUME 67. Vitamins and Coenzymes (Part F)
Edited by DONALD B. MCCORMICK AND LEMUEL D. WRIGHT

VOLUME 68. Recombinant DNA
Edited by RAY WU

VOLUME 69. Photosynthesis and Nitrogen Fixation (Part C)
Edited by ANTHONY SAN PIETRO

VOLUME 70. Immunochemical Techniques (Part A)
Edited by HELEN VAN VUNAKIS AND JOHN J. LANGONE

VOLUME 71. Lipids (Part C)
Edited by JOHN M. LOWENSTEIN

VOLUME 72. Lipids (Part D)
Edited by JOHN M. LOWENSTEIN

VOLUME 73. Immunochemical Techniques (Part B)
Edited by JOHN J. LANGONE AND HELEN VAN VUNAKIS

VOLUME 92. Immunochemical Techniques (Part E: Monoclonal Antibodies and General Immunoassay Methods)
Edited by JOHN J. LANGONE AND HELEN VAN VUNAKIS

VOLUME 93. Immunochemical Techniques (Part F: Conventional Antibodies, Fc Receptors, and Cytotoxicity)
Edited by JOHN J. LANGONE AND HELEN VAN VUNAKIS

VOLUME 94. Polyamines
Edited by HERBERT TABOR AND CELIA WHITE TABOR

VOLUME 95. Cumulative Subject Index Volumes 61–74, 76–80
Edited by EDWARD A. DENNIS AND MARTHA G. DENNIS

VOLUME 96. Biomembranes [Part J: Membrane Biogenesis: Assembly and Targeting (General Methods; Eukaryotes)]
Edited by SIDNEY FLEISCHER AND BECCA FLEISCHER

VOLUME 97. Biomembranes [Part K: Membrane Biogenesis: Assembly and Targeting (Prokaryotes, Mitochondria, and Chloroplasts)]
Edited by SIDNEY FLEISCHER AND BECCA FLEISCHER

VOLUME 98. Biomembranes (Part L: Membrane Biogenesis: Processing and Recycling)
Edited by SIDNEY FLEISCHER AND BECCA FLEISCHER

VOLUME 99. Hormone Action (Part F: Protein Kinases)
Edited by JACKIE D. CORBIN AND JOEL G. HARDMAN

VOLUME 100. Recombinant DNA (Part B)
Edited by RAY WU, LAWRENCE GROSSMAN, AND KIVIE MOLDAVE

VOLUME 101. Recombinant DNA (Part C)
Edited by RAY WU, LAWRENCE GROSSMAN, AND KIVIE MOLDAVE

VOLUME 102. Hormone Action (Part G: Calmodulin and Calcium-Binding Proteins)
Edited by ANTHONY R. MEANS AND BERT W. O'MALLEY

VOLUME 103. Hormone Action (Part H: Neuroendocrine Peptides)
Edited by P. MICHAEL CONN

VOLUME 104. Enzyme Purification and Related Techniques (Part C)
Edited by WILLIAM B. JAKOBY

VOLUME 105. Oxygen Radicals in Biological Systems
Edited by LESTER PACKER

VOLUME 106. Posttranslational Modifications (Part A)
Edited by FINN WOLD AND KIVIE MOLDAVE

VOLUME 107. Posttranslational Modifications (Part B)
Edited by FINN WOLD AND KIVIE MOLDAVE

Section I

Liquid Chromatography

A. Methods
Articles 1 through 4

B. Columns and Instrumentation
Articles 5 through 9

[1] High-Resolution Reversed-Phase High-Performance Liquid Chromatography of Peptides and Proteins

By MARIE-ISABEL AGUILAR and MILTON T. W. HEARN

Introduction

Reversed-phase high-performance liquid chromatography (RP-HPLC) has become a commonly used method for the analysis and purification of peptides and proteins.[1-3] The extraordinary popularity of RP-HPLC can be attributed to a number of factors, including the excellent resolution that can be achieved for closely related as well as structurally disparate substances under a large variety of chromatographic conditions; the experimental ease with which chromatographic selectivity can be manipulated through changes in mobile phase composition; the generally high recoveries, even at ultramicroanalytical levels; the excellent reproducibility of repetitive separations carried out over long periods of time, due in part to the stability of the various sorbents under many mobile phase conditions; the high productivity in terms of cost parameters; and the potential, which is only now being addressed, for the evaluation of different physicochemical aspects of solute–eluent or solute–hydrophobic sorbent interactions and assessment of their structural consequences from chromatographic data.

The RP-HPLC experimental system usually comprises an *n*-alkylsilica-based sorbent from which peptides or proteins are eluted with gradients of increasing concentration of an organic solvent such as acetonitrile containing an ionic modifier, e.g., trifluoroacetic acid (TFA). With modern instrumentation and columns, complex mixtures of peptides and proteins can be separated and low picomolar amounts of resolved components can be collected. Separations can be easily manipulated by changing the gradient slope, temperature, ionic modifier, or the organic solvent composition. The technique is equally applicable to the analysis of enzymatically derived mixtures of peptides and also for the analysis of synthetically derived peptides. An example of the high-resolution analysis of a tryptic digest of bovine growth hormone is shown in Fig. 1. Figure 1 demonstrates the rapid

---

[1] M. T. W. Hearn (ed.), "HPLC of Proteins, Peptides and Polynucleotides—Contemporary Topics and Applications." VCH, Deerfield, FL, 1991.

[2] K. M. Gooding and F. E. Regnier (eds.), "HPLC of Biological Macromolecules: Methods and Applications." Marcel Dekker, New York, 1990.

[3] C. T. Mant and R. S. Hodges (eds.), "HPLC of Peptides and Proteins: Separation, Analysis and Conformation." CRC Press, Boca Raton, FL, 1991.

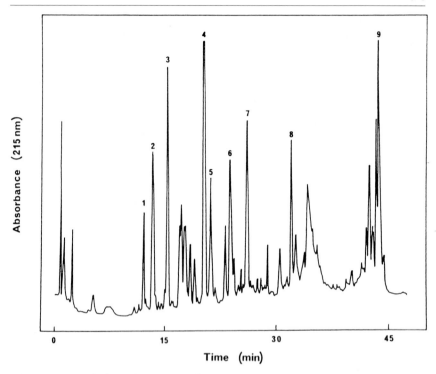

Fig. 1. Reversed-phase chromatographic profile of a tryptic digest of bovine growth hormone on an *n*-octadecylsilica sorbent, particle diameter 5 μm, average pore size 30 nm, packed into a 25 cm × 4.6 mm i.d. column. Gradient elution was carried out from 0 to 50% acetonitrile in 0.1% TFA over 60 min at a flow rate of 1 ml/min. Detection was at 215 nm. (From A. J. Round, M. I. Aguilar, and M. T. W. Hearn, unpublished results, 1995.)

and highly selective separation that can be achieved with tryptic digests of proteins, using RP-HPLC as part of the quality control or structure determination of a recombinant or natural protein. The chromatographic separation shown in Fig. 1 was obtained with an octadecylsilica (C_{18}) stationary phase packed in a column of dimensions 25 cm (length) × 0.46 cm (i.d.). Separated components can be directly subjected to further analysis such as automated Edman sequencing or electrospray mass spectroscopy. For the purification of synthetically derived peptides, the crude synthetic product is typically separated on an analytical scale to assess the complexity of the mixture. This step is usually followed by large-scale purification and collection of the product, with an aliquot of the purified sample then subjected to further chromatography under different RP-HPLC conditions or another HPLC mode to check for homogeneity. Finally, the isolation and analysis of many proteins can also be achieved using high-resolution RP-HPLC techniques. In these cases, the influence of protein conformation, subunit assembly, and extent of microheterogeneity becomes an important consideration in the achievement of a high resolution separation and recovery of the active substance by RP-HPLC techniques. Nevertheless, RP-HPLC methods can form an integral part of the successful isolation of proteins in their native structure, as has been shown, for example, in the purification of transforming growth factor α,[4] inhibin,[5] thyroid-stimulating hormone,[6] growth hormone,[7] and insulin.[8] However, it should be noted that the recovery of more refractory proteins can present a serious problem in RP-HPLC either in terms of recovered mass or the loss of activity.

The success of RP-HPLC, which is illustrated by the selected examples in Table I,[9-60] is also due to the ability of this technique to probe the hydrophobic surface topography of a biopolymer. This specificity arises

[4] F. J. Moy, Y.-C. Li, P. Rauenbuehler, M. E. Winkler, H. A. Scheraga, and G. T. Montelione, *Biochemistry,* **32,** 7334 (1993).

[5] R. G. Forage, J. M. Ring, R. W. Brown, B. V. McInerney, G. S. Cobon, P. Gregson, D. M. Robertson, F. J. Morgan, M. T. W. Hearn, J. K. Findlay, R. E. H. Wettenhall, H. G. Burger, and D. M. de Kretser, *Proc. Natl Acad. Sci. U.S.A.,* **83,** 3091 (1986).

[6] M. A. Chlenov, E. I. Kandyba, L. V. Nagornaya, I. L. Orlova, and Y. V. Volgin, *J. Chromatogr.* **631,** 261 (1993).

[7] B. S. Welinder, H. H. Sorensen, and B. Hansen, *J. Chromatogr.* **398,** 309 (1987).

[8] B. S. Welinder, H. H. Sorensen, and B. Hansen, *J. Chromatogr.* **361,** 357 (1986).

[9] A. T. Jones and N. B. Roberts, *J. Chromatogr.* **599,** 179 (1992).

[10] R. Rosenfeld and K. Benedek, *J. Chromatogr.* **632,** 29 (1993).

[11] A. Calderan, P. Ruzza, O. Marin, M. Secchieri, G. Bovin, and F. Machiori, *J. Chromatogr.* **548,** 329 (1991).

[12] R. H. Buck, M. Cholewinski, and F. Maxl, *J. Chromatogr.* **548,** 335 (1991).

[13] E. Perez-Paya, L. Braco, C. Abad, and J. Dufourck, *J. Chromatogr.* **548,** 351 (1991).

[14] S. Visser, C. J. Slangen, and H. S. Rollema, *J. Chromatogr.* **548,** 361 (1991).

[15] S. Linde, B. S. Welinder, B. Hansen, and O. Sonne, *J. Chromatogr.* **369,** 327 (1986).

[16] D. J. Poll and D. R. K. Harding, *J. Chromatogr.* **469,** 231 (1989).

[17] R. C. Chloupek, R. J. Harris, C. K. Leonard, R. G. Keck, B. A. Keyt, M. W. Spellman, A. J. S. Jones, and W. S. Hancock, *J. Chromatogr.* **463,** 375 (1989).

[18] P. M. Young and T. E. Wheat, *J. Chromatogr.* **512,** 273 (1990).

[19] E. Watson and W. C. Kenney, *J. Chromatogr.* **606,** 165 (1992).

[20] D. L. Crimmins and R. S. Thoma, *J. Chromatogr.* **599,** 51 (1992).

[21] D. Rapaport, G. R. Hague, Y. Pouny, and Y. Shai, *Biochemistry* **32,** 3291 (1993).

[22] J. Tözsér, D. Friedman, I. T. Weber, I. Blaha, and S. Oroszlan, *Biochemistry* **32,** 3347 (1993).

[23] J.-J. Lacapère, J. Gavin, B. Trinnaman, and N. M. Green, *Biochemistry* **32,** 3414 (1993).

[24] E. Gazit and Y. Shai, *Biochemistry* **32,** 3429 (1993).

[25] M. Pacaud and J. Derancourt, *Biochemistry* **32,** 3448 (1993).

[26] T. P. King, M. R. Coscia, and L. Kochoumian, *Biochemistry* **32,** 3506 (1993).

[27] K. Mock, M. Hail, I. Mylchrest, J. Zhou, K. Johnson, and I. Sardine, *J. Chromatogr.* **646,** 169 (1993).

[28] P. A. Grieve, A. Jones, and P. F. Alewood, *J. Chromatogr.* **646,** 175 (1993).

through selective interactions between the immobilized ligand on the surface of the stationary phase and the biopolymer in question. Initially, practical applications made in this field of high-resolution chromatographic methods have greatly exceeded the development of detailed theoretical

[29] P. F. Alewood, A. J. Bailey, R. I. Brinkworth, D. Fairlie, and A. Jones, *J. Chromatogr.* **646,** 185 (1993).

[30] J. J. Gorman and B. J. Shiel, *J. Chromatogr.* **646,** 193 (1993).

[31] A. T. Jones and J. N. Keen, *J. Chromatogr.* **646,** 207 (1993).

[32] L. Fabri, H. Maruta, H. Muramatsu, T. Muramatsu, R. J. Simpson, A. W. Burgess, and E. C. Nice, *J. Chromatogr.* **646,** 213 (1993).

[33] Y. Eswel, Y. Shai, T. Vorhgar, E. Carafoli, and Y. Salomon, *Biochemistry* **32,** 6721 (1993).

[34] N. E. Zhou, C. M. Kay, B. D. Sykes, and R. S. Hodges, *Biochemistry* **32,** 6190 (1993).

[35] X. Liu, S. Magda, Z. Hu, T. Aiuchi, K. Nakaya, and Y. Kurihara, *Eur. J. Biochem.* **211,** 281 (1993).

[36] S. Fulton, M. Meys, J. Protentis, N. B. Afeyan, J. Carlton, and J. Haycock, *Biotechniques* **12,** 742 (1992).

[37] D. Müller, C. Schulz, H. Baumeister, F. Buck, and V. Richter, *Biochemistry* **31,** 11138 (1992).

[38] P. Le Marechal, B. M. C. Hoang, J.-M. Schmitter, A. Van Dorsselaer, and P. Decottignies, *Eur. J. Biochem.* **210,** 421 (1992).

[39] T. Weimbs and W. Stoffel, *Biochemistry* **31,** 12289 (1992).

[40] S. Murao, K. Ohkuni, M. Nagao, K. Hirayama, K. Fukuhara, K. Oda, H. Oyama, and T. Shin, *J. Biol. Chem.* **268,** 349 (1993).

[41] D. L. Lohse and R. J. Linhardt, *J. Biol. Chem.* **267,** 24347 (1992).

[42] D. O. O'Keefe, A. L. Lee, and S. Yamazaki, *J. Chromatogr.* **627,** 137 (1992).

[43] G. Chaga, L. Anderson, and J. Porath, *J. Chromatogr.* **627,** 163 (1992).

[44] G. Teshima and E. Canova-Davis, *J. Chromatogr.* **625,** 207 (1992).

[45] J. Koyama, J. Nomura, Y. Shojima, Y. Ohtsu, and I. Horii, *J. Chromatogr.* **625,** 217 (1992).

[46] S. O. Ugwu and J. Blanchard, *J. Chromatogr.* **584,** 175 (1992).

[47] S. Awasthi, F. Ahmad, R. Sharma, and H. Ahmad, *J. Chromatogr.* **584,** 167 (1992).

[48] F. Honda, H. Honda, and M. Koishi, *J. Chromatogr.* **609,** 49 (1992).

[49] N. Nimura, H. Itoh, T. Kinoshita, N. Nagae, and M. Nomura, *J. Chromatogr.* **585,** 207 (1992).

[50] S. E. Blondelle and R. A. Houghten, *Biochemistry* **30,** 4671 (1991).

[51] K. Asai, K. Nakanishi, I. Isobe, Y. Z. Eksioglu, A. Hirano, K. Hama, T. Miyamoto, and T. Kato, *J. Biol. Chem.* **267,** 20311 (1992).

[52] G. P. Lunstrum, A. M. McDonough, M. P. Marinkovich, D. R. Keene, N. P. Morris, and R. E. Burgeson, *J. Biol. Chem.* **267,** 20087 (1992).

[53] C. J. Rhodes, B. Lincoln, and S. E. Shoelson, *J. Biol. Chem.* **267,** 22719 (1992).

[54] M. H. Sayre, N. T. Schochner, and R. D. Kornberg, *J. Biol. Chem.* **267,** 23383 (1992).

[55] D. L. Rousseau, Jr., C. A. Guyer, A. H. Beth, I. A. Papayannopoulos, B. Wang, R. Wu, B. Mroczkowski, and J. V. Staros, *Biochemistry* **32,** 7893 (1993).

[56] H. Peled and Y. Shai, *Biochemistry* **32,** 7879 (1993).

[57] R. L. Moritz and R. J. Simpson, *J. Chromatogr.* **599,** 119 (1992).

[58] J. Liu, K. J. Volk, E. H. Kerns, S. E. Klohr, M. S. Lee, and I. E. Rosenberg, *J. Chromatogr.* **632,** 45 (1993).

[59] P. R. Griffin, J. A. Coffman, L. E. Hood, and J. R. Yates III, *Int. J. Mass Spectrom. Ion Proc.* **111,** 131 (1991).

[60] J. B. Smith, L. R. Miesbauer, J. Leeds, D. L. Smith, J. A. Loo, R. D. Smith, and C. G. Edmonds, *Int. J. Mass Spectrom. Ion Proc.* **111,** 229 (1991).

descriptions of the molecular basis of the interactions of biological macro-molecules with these hydrophobic chromatographic surfaces.[61] More recently, however, the widespread practical application of RP-HPLC with biomacromolecules has been accompanied by a significant improvement in our understanding of the molecular basis of the retention process and its impact on conformational stability.[62–65] As a consequence, the use of high-resolution chromatographic techniques for the physicochemical character-ization of the interactive phenomena of peptides and proteins is also now providing new insight into the dynamic behavior of biomacromolecules at hydrophobic surfaces.

Parameters That Control Resolution

Theoretical Considerations

The capacity factor k' of a solute can be expressed in terms of the retention time t_r through the relationship

$$k' = (t_r - t_0)/t_0 \tag{1}$$

where t_0 is the retention time of a nonretained solute. The development of high resolution separations of peptides and proteins involves the separa-tion of sample components through manipulation of both retention times and solute peak shape. The practical significance of k' in defining a particu-lar chromatographic separation window therefore resides in the concept of solute selectivity, α, which is defined as the ratio of the capacity factors for adjacent peaks as follows:

$$\alpha = k'_1/k'_2 \tag{2}$$

The second experimental factor that is involved in defining the quality of a separation is the solute peak width. The degree of peak broadening is related to the column efficiency, which is normally expressed in terms of

[61] J. Frenz, W. S. Hancock, W. J. Henzel, and C. Horvath, *in* "HPLC of Biological Macromole-cules: Methods and Applications" (K. M. Gooding and F. E. Regnier, eds.), p. 145. Marcel Dekker, New York, 1990.

[62] W. R. Melander, H.-J. Lin, J. Jacobson, and C. Horvath, *J. Phys. Chem.* **88,** 4527 (1984).

[63] J. Jacobson, W. R. Melander, G. Vaisnys, and C. Horvath, *J. Phys. Chem.* **88,** 4536 (1984).

[64] S. Lin and B. L. Karger, *J. Chromatogr.* **499,** 89 (1990).

[65] S. A. Cohen, K. Benedek, Y. Tapuhi, J. C. Ford, and B. L. Karger, *Anal. Chem.* **144,** 275 (1985).

TABLE I

Peptides and Proteins Separated by RP-HPLC

Peptide/protein	Column	Mobile phase	Ref.
Pepsin isozyme peptide map	Exsil, 300 Å, C_{18}, 5 μm, 15 cm × 4.6 mm i.d.	0.1% Trifluoroacetic acid (TFA), 0–48% acetonitrile (AcCN), 50 min, 1.5 ml/min	9
Brain-derived neuro-trophic factor	Vydac, protein, C_4, 15 cm × 4.6 mm i.d.	0.1% TFA, 18–31% AcCN, 44 min, 1 ml/min	10
Casein kinase-related peptides	ROsil, C_{18}, 3 μm, 10 cm × 4.6 mm i.d.	20 mM Na_2HPO_4, pH 5.6/2 mM tetrabutylammonium hydrogen sulfate, 5–25% AcCN, 35 min, 1 ml/min, 25°	11
Ornipressin	Hypersil ODS, 5 μm, 12.5 cm × 4.6 mm i.d.	20 mM Tetramethylammonium hydroxide, pH 2.5, 5–30% AcCN, 25 min, 1 ml/min, 60°	12
Mellitin	μBondapak C_{18}, 30 cm × 7.8 mm i.d.	0.1% TFA, 30–70% AcCN, 20 min, 1 ml/min	13
Bovine caseins	HiPore RP-318, 25 cm × 4.6 mm i.d.	0.1% TFA, 23–63% AcCN, 38 min, 0.8 ml/min, 30°	14
Inhibin	Ultrapore RPSC-C3, 7.5 cm × 4.6 mm i.d.	0.1% TFA, 0–50% AcCN, 90 min, 1 ml/min	5
Proinsulin	LiChrosorb RP-18, 5 μm, 25 cm × 4.0 mm i.d.	125 mM ammonium sulfate, pH 4, 30–34% AcCN, 60 min, 1 ml/min	15
Glycoprotein hormones	Vydac 214TP, C_4, 10 μm, 25 cm × 4.6 mm i.d.	0.1% TFA, 18–63% AcCN, 30 min, 1 ml/min, 100 mM Na_2HPO_4, pH 6.8, 12.5–50% AcCN, 40 min, 2 ml/min	6
Growth hormone	Nucleosil C_{18}, 7 μm, 25 cm × 4.0 mm i.d.	225 mM $(NH_4)_2HPO_4$/90 mM NaH_2PO_4, pH 2.5, 0–90% AcCN, 60 min, 1 ml/min	7
Insulin	LiChrosorb RP-18, 5 μm, 25 cm × 4.0 mm i.d.	250 mM triethylammonium phosphate, pH 3, 25–30% AcCN, 30 min, 1 ml/min	8
Lysozyme tryptic peptides	Vydac C_4, 25 cm × 4.6 mm i.d.	0.1% formic acid, 5–20% AcCN, 60 min, 0.5 ml/min	16
Tissue plasminogen activator	Novapak C_{18}, 5 μm, 15 cm × 3.9 mm i.d.	0.1% TFA, 0–60% AcCN, 85 min, 1 ml/min	17
Cytochrome c tryptic map	Delta-Pak C_{18}, 300 Å, 5 μm, 15 cm × 3.9 mm i.d.	0.1% TFA or 6 mM HCl or 6 mM HFBA, 0–60% AcCN, 1 ml/min, 35°	18
Platelet-derived growth factor	Vydac C_4, 25 cm × 4.6 mm i.d.	0.1% TFA, 12–15% AcCN, 60 min, 1 ml/min	19
Tropomyosins	Vydac 214TP, C_4, 25 cm × 4.6 mm i.d.	0.1% TFA, 0–90% AcCN, 60 min, 1 ml/min	20
Pardoxin analogs	C_4 sorbent	0.1% TFA, 25–80% AcCN, 40 min, 1 ml/min	21

TABLE I (*continued*)

Peptide/protein	Column	Mobile phase	Ref.
Equine infectious anemia virus protein	Vydac C_4, 300 Å, 25 cm × 4.6 mm i.d.	0.1% TFA, 0–100% AcCN, 28 min, 1 ml/min	22
Sarcoplasmic reticulum CNBr peptides	Zorbax C_{18}, 15 cm × 4.6 mm i.d.	40 mM ammonium acetate, pH 6.0, or 0.1% TFA, 0–90% AcCN, 100 min, 1 ml/min	23
δ-Endotoxin analogs	Vydac C_4, 300 Å, 25 cm × 4.6 mm i.d.	0.1% TFA, 25–80% AcCN, 40 min, 0.6 ml/min	24
Actin-bundling protein peptides	Aquapore C_4, 25 cm × 4.6 mm i.d.	0.1% TFA, 0–60% AcCN	25
Mellitin analogs	Vydac C_{18}, 10 cm × 9 mm i.d.	0.1% TFA, 0–30% 2-propanol, 100 min, 1 ml/min	26
Growth hormone tryptic peptides	Reliasil C_{18}, 5 μm, 15 cm × 1 mm i.d.	0.1% TFA, 0–60% AcCN, 30 min, 0.05 ml/min, 40°	27
Bactenecins	DeltaPak C_{18}, 300 Å, 30 cm × 3.9 mm i.d.	0.1% TFA, 27–45% AcCN, 25 min, 1 ml/min	28
TNF-α-related peptides	DeltaPak C_{18}, 300 Å, 30 cm × 3.9 mm i.d.	0.1% TFA, 4.5–50% AcCN, 50 min, 1 ml/min	29
Ovalbumin tryptic peptides	Vydac 218TP54, 5 μm, 25 cm × 4.6 mm i.d.	0.1% TFA, 0–8% AcCN, 85 min, 1 ml/min, 22°	30
Pepsin-3b peptic peptides	Exsil C_{18}, 300 Å, 5 μm, 15 cm × 4.6 mm i.d.	0.1% TFA, 0–48% AcCN, 70 min, 1.5 ml/min, 22°	31
RA-inducible midkine	Brownlee RP-300, 30 cm × 2.1 mm i.d.	0.1% TFA, 0–60% AcCN, 60 min, 0.1 ml/min, 45°	32
Calmodulin-binding peptides	Vydac C_4, 300 Å, 25 cm × 4.6 mm i.d.	0.1% TFA, 10–60% AcCN, 40 min, 0.9 ml/min	33
NBD-labeled peptides	Vydac C_4, 300 Å, 25 cm × 4.6 mm i.d.	0.1% TFA, 15–60% AcCN, 40 min, 0.9 ml/min	33
Amphipathic α-helical peptides	C_{18}, 300 Å, 6.5 μm, 25 cm × 10 mm i.d.	0.1% TFA, 0–100% AcCN, 100 min, 2 ml/min	34
Mabinlin II	TSK gel TMS-250, 7.5 cm × 4.6 mm i.d.	0.05% TFA, 10–50% AcCN, 25 min, 1 ml/min	35
Mabinlin II-related peptides	Vydac 218TP54, 25 cm × 4.6 mm i.d.	0.05% TFA, 5–60% AcCN, 60 min, 1 ml/min	35
Vasoactive intestinal polypeptide	Poros R/H, 10 μm, 10 cm × 4.6 mm i.d.	12 mM HCl, 0–30% AcCN, 5 min, 5 ml/min	36
Atrial natriuretic peptide	Vydac C_{18}, 300 Å, 5 μm, 10 cm × 2.1 mm i.d.	0.1% TFA, 0–70% AcCN, 60 min, 0.2 ml/min	37
Carboxymethylthioredoxin peptide maps	Vydac C_4, 15 cm × 4.6 mm i.d.	0.1% TFA, 0–70% AcCN, 1 ml/min, 10 mM ammonium formate, pH 7.5, 0–60% AcCN	38
Proteolipid protein thermolytic peptides	C_{18}, 300 Å, 5 μm, 25 cm × 4.0 mm i.d.	0.1% TFA : 10 mM triethylamine, 5–50% AcCN, 45 min, 1 ml/min	39
Insulin B-chain digests	Nucleosil C_{18}, 5 μm, 25 cm × 4.0 mm i.d.	0.05% TFA, 0–35% AcCN	40

(*continued*)

TABLE I (*continued*)

Peptide/protein	Column	Mobile phase	Ref.
Heparin lyase tryptic peptides	Vydac C_{18}	0.1% TFA, 0–80% AcCN, 120 min	41
Transforming growth factor α-*Pseudomonas aeruginosa* exotoxin A	Hy-tach C_{18} (nonporous), 30 cm × 4.6 mm i.d.	0.1% TFA, 34–64% AcCN, 6 min, 1 ml/min, 80°	42
Lactate dehydrogenase CNBr fragments	PEP-RPC HR 5/5, 5 cm × 5 mm i.d.	0.1% TFA, 0–50% AcCN, 30 min, 0.7 ml/min	43
H_2O_2-treated human growth hormone	PLRP-S, 300 Å, 10 μm, 30 cm × 7.5 mm i.d.	25 mM ammonium acetate, pH 7.5, 34–39% 1-propanol, 100 min, 1 ml/min, 40 °	44
H_2O_2-treated human growth hormone tryptic peptides	Vydac C_{18}, 300 Å, 5 μm, 25 cm × 4.6 mm i.d.	0.1% TFA, 57–77% AcCN, 40 min, 0.5 ml/min, 40°	44
Ribonuclease A, cytochrome c, lysozyme, bovine serum albumin, ovalbumin	C_8, 300 Å, 5 μm, 25 cm × 4.6 mm i.d.	0.1% TFA, 15–60% AcCN, 30 min, 1.5 ml/min, 40°	45
Melanotan II	Vydac C_{18}, 5 μm, 15 cm × 2.1 mm i.d.	100 mM Na_2HPO_4/triethylamine, pH 2.5, 21% AcCN, 0.25 ml/min	46
Dinitrophenyl-3-glutathione	PEP-RPC C_{18}, 5 cm × 4 mm i.d.	0.1% TFA, 0–70% AcCN	47
Aprotinin, cytochrome c, bovine serum albumin, fibrinogen, apoferritin	C_{18}-Coated polyethylene, 10 μm, 10 cm × 10 mm i.d.	0.1% TFA, 5–70% AcCN, 15 min, 0.5 ml/min	48
Insulin, cytochrome c, lysozyme, bovine serum albumin, α-lactalbumin	C_{18} nonporous 2, 5, 20 μm, 3 cm × 4.6 mm i.d.	0.1% TFA, 9–90% AcCN, 20 min, 2 ml/min, 0.1% TFA, 20–90% AcCN, 48 sec, 4 ml/min, 40°	49
Mellitin analogs	Vydac C_{18}, 25 cm × 4.6 mm i.d.	0.1% TFA, 10–75% AcCN, 65 min, 1 ml/min	50
Gliostatin and platelet-derived endothelial cell growth factor tryptic peptides	μRPC C_2/C_{18}, 10 cm × 2.1 mm i.d.	0.1% TFA, 0–50% AcCN, 45 min, 0.15 ml/min	51
Collagen type-XII tryptic peptides	Vydac C_4, 25 cm × 4.6 mm i.d.	0.1% TFA, 0–70% AcCN, 200 min, 1 ml/min	52
Iodinated proinsulin peptides	C_{18} sorbent	0.1% TFA, 16–40% AcCN, 90 min	53
RNA polymerase II general initiation factor a	HiPore RP 304 C_4, 25 cm × 4.6 mm i.d.	0.1% TFA, 0–100% AcCN, 90 min, 1 ml/min	54
Murine epidermal growth factor mutant	Aquapore RP-300 C_8, 22 cm × 4.6 mm i.d.	0.1% TFA, 4–40% AcCN, 70 min, 1 ml/min	55

TABLE I (continued)

Peptide/protein	Column	Mobile phase	Ref.
Shaker K⁺ channel-related peptides	Vydac C$_4$, 300 Å, 25 cm × 4.6 mm i.d.	0.1% TFA, 25–80% AcCN, 40 min, 0.6 ml/min	56
Murine interleukin 6 tryptic peptides	Brownlee RP-300 C$_8$, 7 μm, 5 cm × 0.32 mm i.d.	0.1% TFA, 0–60% AcCN, 60 min, 3.6 μl/min	57
Ribonuclease B and α$_1$-acid protein tryptic peptides	Vydac C$_{18}$, 300 Å, 5 μm, 32 cm × 0.25 mm i.d.	0.1% TFA, 0–48% AcCN, 120 min, 3.0 μl/min	58
β-Lactoglobulin B, α-s1-phosphocasein, myoglobin tryptic peptides	C$_{18}$, 5 μm, 15 cm × 0.32 mm i.d.	0.1% TFA, 0–100% AcCN, 2.0 μl/min	59
β-Crystallins	SynChropak C$_8$, 300 Å, 5 μm, 25 cm × 1.0 mm i.d.	0.1% TFA, 20–60% AcCN, 40 min, 0.05 ml/min	60
β-Crystallin tryptic peptides	Aquapore RP C$_{18}$, 5 cm × 1.0 mm i.d.	0.1% TFA, 3% glycerol, 3% thioglycerol, 10–40% AcCN, 30 min, 0.05 ml/min	60

the number of theoretical plates, N, as follows:

$$N = (t_r)^2/\sigma_t^2 \tag{3}$$

N can also be expressed in terms of the reduced plate height equivalent h, the column length L, and the particle diameter of the stationary phase d_p as

$$N = hL/d_p \tag{4}$$

Resolution between components of a mixture depends on both selectivity and band width, according to

$$R_s = \tfrac{1}{4}(N)^{1/2}(\alpha - 1)[1/(1 + k')] \tag{5}$$

This relationship therefore describes the interdependence of the quality of a separation on the relative retention, relative selectivity, and peak width. The objective in the development of a high resolution separation is the choice of experimental conditions that maximize R_s by thorough and systematic modulation of k' and α. To obtain high resolution separations, R_s values > 1 are required. Thus three strategies are available for improving resolution: (1) increase α; (2) vary k' over a predefined range, e.g., 1 < k' < 10; or (3) increase N, typically by using very small particles (about 2–5 μm) in microbore columns. This chapter focuses on the steps that can be taken for enhancing the high resolution separation of peptides and proteins using RP-HPLC.

Retention Relationships of Peptides and Proteins in RP-HPLC

The rapid growth in the number of applications of RP-HPLC in peptide and protein analysis or purification has greatly exceeded the development of physically relevant, mechanistic models that adequately detail the thermodynamic and kinetic processes that are involved in the interaction of peptides or proteins with nonpolar sorbents. In the absence of rigorous models that predict the effect of experimental parameters on retention and band width in terms of the detailed structural hierarchy of the ligand–peptide or the ligand–protein interaction, investigators often resort to arbitrary changes in experimental parameters to effect improved peptide separations. However, a number of predictive nonmechanistic optimization models have been reported and effectively applied to the RP-HPLC elution of peptides or proteins.[66-70] For example, k' for a peptide separated under linear elution conditions with isocratic RP-HPLC can be expressed as a linear function of the organic volume fraction ψ according to

$$\log k' = \log k_0 - S\psi \tag{6}$$

For gradient elution separation of peptides in RP-HPLC, an analogous relationship between the median capacity factor, $\bar{k}$, and the median organic mole fraction, $\bar{\psi}$, can be used:

$$\log \bar{k} = \log k_0 - S\bar{\psi} \tag{7}$$

where S is the slope of the plot of $\log \bar{k}$ versus $\bar{\psi}$ and $\log k_0$ is the intercept of these plots. Depending on the magnitude of the S and $\log k_0$ values and how these parameters change with variations in temperature, eluant pH, etc., a variety of dependencies of k' on ψ can be specified as depicted in Fig. 2. These scenarios provide direct insight into the relationship between solute structure and retention behavior and how improved high resolution separations can be achieved. For example, cases (c) and (d) in Fig. 2 are representative of typical behavior for the RP-HPLC behavior of strongly hydrophobic polypeptides and proteins, while cases (a) and (b) demonstrate a typical dependency of retention on ψ of polar peptides and small polar proteins. The S and $\log k_0$ values for polypeptides and proteins are usually large when compared to the corresponding values for small organic molecules.[70,71] This feature of polypeptide and protein retention behavior is

[66] X. Geng and F. E. Regnier, *J. Chromatogr.* **296,** 15 (1984).

[67] L. R. Snyder, *in* "HPLC—Advances and Perspectives" (C. Horvath, ed.), Vol. 1, p. 208. Academic Press, New York, 1983.

[68] J. L. Glajch, M. A. Quarry, J. F. Vaster, and L. R. Snyder, *Anal. Chem.* **58,** 280 (1986).

[69] M. T. W. Hearn and M. I. Aguilar, *J. Chromatogr.* **359,** 33 (1986).

[70] M. T. W. Hearn and M. I. Aguilar, *J. Chromatogr.* **397,** 47 (1987).

[71] M. A. Stadalius, H. S. Gold, and L. R. Snyder, *J. Chromatogr.* **296,** 31 (1984).

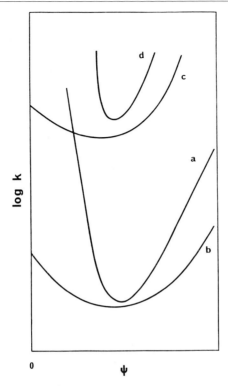

Fig. 2. Schematic representation of the retention dependencies for peptides or proteins chromatographed on RP-HPLC sorbents. Illustrated here are four scenarios for the dependence of log k' versus ψ. As the contact area increases, the slope of the plots increases, which results in a narrowing of the elution window over which the solute will elute. (Reprinted from M. T. W. Hearn et al. Reversed phase high performance liquid chromatography of peptides and proteins, in "Modern Physical Methods in Biochemistry" (A. Neuberger and L. L. M. Van Deenan, eds.), p. 113, Copyright 1989 with kind permission of Elsevier Science–NL, Sara Burgerhartstraat 25, 1055 KV Amsterdam, The Netherlands.)

believed to be a consequence of multisite peptide–ligand interactions. A practical consequence of this behavior is that high-resolution isocratic elution of polypeptides or proteins can rarely be carried out, as the experimental window of solvent concentration required for peptide elution is narrow. Complex mixtures of peptides or proteins are therefore routinely resolved by gradient elution methods when high resolution is mandatory.

Evaluation of the S and log k_0 values is important for several reasons. First, this information can be directly applied to the enhancement of resolution via optimization procedures through the determination of changes in selectivity and resolution as a function of chromatographic parameters such as flow rate, solvent strength, temperature, particle diameter, and column length.[67,70] Second, analysis of these chromatographic variables also pro-

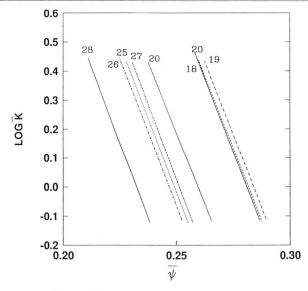

FIG. 3. Plots of log $\bar{k}$ versus $\bar{\psi}$ for D-amino acid-substituted analogs of neuropeptide Y (NPY)[18–38] separated on a C_4 sorbent with acetonitrile as the organic modifier at 25°. The plots were derived from best fit analysis to the data points (which have been excluded for clarity). The amino acid sequence of NPY[18–36] = ARYYSALRHYINLITRQRY-NH$_2$. The retention plot for each D-substituted analog is designated by the residue position of the D-amino acid substitution. (Data derived from results in Ref. 23.)

vides quantitative guidelines for the preparation of improved hydrophobic stationary phases through the characterization of different stationary-phase topographies and the effect of different column configurations.

Third, knowledge of the S and log k_0 values greatly simplifies the determination of physicochemical relationships between solute structure and chromatographic selectivity. Subtle differences in the experimentally observed S values in response to changes in operating parameters such as column temperature and surface hydrophobicity for several classes of peptide analogs related to β-endorphin,[70,72] myosin light chain,[73] luteinizing hormone-releasing hormone,[69] interleukin 2,[74] and neuropeptide Y (NPY)[75] have been reported that enable conformationally dependent differences in the interactive sites on the peptide solutes to be visualized. Figure 3 shows

[72] M. I. Aguilar, A. N. Hodder, and M. T. W. Hearn, *J. Chromatogr.* **327,** 115 (1985).
[73] M. T. W. Hearn and M. I. Aguilar, *J. Chromatogr.* **392,** 33 (1987).
[74] M. Kunitani, D. Johnson, and L. R. Snyder, *J. Chromatogr.* **371,** 313 (1986).
[75] M. I. Aguilar, S. Mougos, J. Boublik, J. Rivier, and M. T. W. Hearn, *J. Chromatogr.* **646,** 53 (1993).

the plots of log $\overline{k}$ versus $\overline{\psi}$ for a series of NPY analogs differing in sequence only by the substitution of a single D-amino acid residue. These plots clearly demonstrate the sensitivity of RP-HPLC to resolve small differences in peptide structure. More specifically, the ability of these high-resolution RP-HPLC procedures to discriminate between these analogs indicates that the stationary-phase ligands can act as a molecular probe of peptide surface topography.

The mechanism by which peptide or protein solutes are retained in RP-HPLC depends on the hydrophobic expulsion of the peptide from a polar mobile phase and concomitant adsorption onto the nonpolar sorbent.[76,77] Under these conditions, peptides or proteins are retained to different extents depending on their intrinsic hydrophobicities, the eluotropicity of the mobile phase, and the nature of the sorbent ligands. Experimental data with species variants of proteins, as well as recombinant mutants, indicate that proteins interact with the chromatographic surface in an orientation-specific manner.[78–80] Their chromatographic retention behavior in terms of their affinity and kinetics of the interaction is therefore determined by the molecular composition of the specific contact region(s). The contact region for small peptides has been shown to involve the contribution from all or a large proportion of the molecular surface of the solute. As a result, the retention time of small peptides in RP-HPLC can be predicted with reasonable accuracy by summating the retention coefficients for all constituent amino acid residues.[81,82] For larger polypeptides or proteins, the chromatographic retention data indicate that the contact region represents a relatively small portion of the total solute surface. Although the hydrophobic surface area of a protein may increase with increasing molecular weight, it is not the molecular weight per se but rather the polarity and spatial disposition of the surface amino acid residues involved in the interaction with the stationary phase that ultimately control the mechanistic pathway of the binding process. Since the magnitude of log k_0 is a measure of the free energy changes associated with the binding of the solute to the stationary phase under initial elution conditions, it can also be anticipated that log k_0 values should progressively increase with incremental increases in solute hydrophobicity. However, if a peptide assumes any degree of pre-

[76] W. R. Melander, D. Corradini, and C. Horvath, *J. Chromatogr.* **317,** 67 (1984).

[77] C. Horvath, W. Melander, and I. Molnar, *J. Chromatogr.* **125,** 129 (1976).

[78] J. F. Pollit, G. Thévenon, L. Janis, and F. E. Regnier, *J. Chromatogr.* **443,** 221 (1988).

[79] R. M. Chicz and F. E. Regnier, *J. Chromatogr.* **500,** 503 (1990).

[80] F. E. Regnier, *Science* **238,** 319 (1987).

[81] M. C. J. Wilce, M. I. Aguilar, and M. T. W. Hearn, *J. Chromatogr.* **632,** 11 (1993).

[82] D. Guo, C. T. Mant, A. K. Taneja, and R. S. Hodges, *J. Chromatogr.* **359,** 519 (1986).

ferred secondary structure or preferred folding, no simple relationship will exist between the retention time and the summated retention coefficients.

Stationary Phases

The choice of sorbent material is one of the first decisions to be made in the design of a high resolution RP-HPLC separation of a peptide or protein. The chromatographic packing materials that are generally used in RP-HPLC are commonly based on microparticulate porous silica that is chemically modified by a derivatized silane containing an *n*-alkyl hydrophobic ligand.[83,84] The most commonly used ligands are *n*-butyl, *n*-octyl, and *n*-octadecyl, while phenyl and cyanopropyl ligands can also provide alternative selectivity.[85] During the immobilization of the ligands, only about half of the original surface silanol hydroxyl groups react, as a result of steric crowding of the ligands. The sorbents can then be subjected to further silanization with a small reactive silane to yield a so-called end-capped packing material. The nature of the *n*-alkyl chain is an important factor that can be used to change selectivity of peptide or protein mixtures. While the specific molecular basis of these differences in selectivity is not yet established, the relative hydrophobicity and molecular flexibility of the ligands together with the degree of exposure of the surface silanol groups are known to play an important role in the interactive process.[86,87] An example of the effect of ligand chain length on the resolution of tryptic peptides of porcine growth hormone is shown in Fig. 4. It can be seen that the peaks labeled T_3 (sequence, EFER) and T_{13} (sequence, ELEDGSPR) are fully resolved with the C_4 sorbent yet cannot be separated on the C_{18} sorbent. Conversely, peptides T_5 (sequence, YSIQNAQAAFCFSETI-PAPTG) and T_{18} (sequence, NYGLLSCFK) elute as a single peak with the C_4 sorbent but are fully resolved on the C_{18} sorbent. Moreover, the choice of the chain length of the *n*-alkyl ligand can have a significant impact on the recovery, as well as the conformational integrity of a protein. While higher protein recoveries have been reported with the shorter and less hydrophobic *n*-butyl or cyanopropyl sorbents, proteins have also been isolated in high yield using the *n*-octadecyl sorbent.[4–6,15] In an attempt to control the denaturation of proteins by RP-HPLC sorbents, porous and nonporous silica supports also can be coated with polymethacrylate-based

[83] K. K. Unger, B. Anspach, R. Janzen, G. Jilge, and K. D. Lork, *in* "HPLC—Advances and Perspectives" (C. Horvath, ed.), Vol. 5, p. 2, Academic Press, New York, 1988.
[84] M. Henry, *J. Chromatogr.* **544,** 413 (1991).
[85] N. E. Zhou, C. T. Mant, J. J. Kirkland, and R. S. Hodges, *J. Chromatogr.* **548,** 179 (1991).
[86] I. Yarovsky, M. I. Aguilar, and M. T. W. Hearn, *Anal. Chem.* **67,** 2145 (1995).
[87] K. Albert and E. Bayer, *J. Chromatogr.* **544,** 345 (1991).

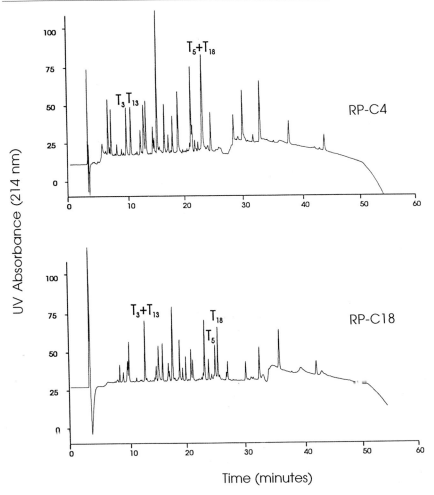

FIG. 4. The influence of *n*-alkyl chain length on the separation of an identical mixture of tryptic peptides derived from porcine growth hormone. *Top:* Bakerbond (J. T. Baker, Phillipsburg, NJ) RP-C_4, 25 cm × 4.6 mm i.d., 5-μm particle size, 30-nm pore size. *Bottom:* Bakerbond RP-C_{18}, 25 cm × 4.6 mm i.d., 5-μm particle size, 30-nm pore size. Conditions, linear gradient from 0 to 90% acetonitrile with 0.1% TFA over 60 min, flow rate of 1 ml/min, 25°. (From A. J. Round, M. I. Aguilar, and M. T. W. Hearn, unpublished results, 1995.)

polymers to produce a series of sorbents with varying surface hydrophobicity in which the underlying silanol groups also have been masked.[88,89]
The use of these sorbents allows peptide and protein selectivity to be
manipulated through changes in the solute conformation.

Silica-based packings are susceptible to hydrolytic cleavage of the siloxane backbone, particularly when using mobile-phase pH values greater
than pH 7, even when coated with a layer of polymer such as polybutadiene.
In these cases, where high-pH separations are needed, alternative stationary-phase materials have been developed such as cross-linked polystyrene–
divinylbenzene,[90,91] porous graphitized carbon,[92] and porous zirconia,[93]
which all offer superior stability at alkaline pH values and different options
for high resolution separations. However, only the polymeric-based sorbents have been used for the RP-HPLC analysis of peptides and proteins.

The geometry of the sorbent particle is also an important factor that
requires consideration. The pore size of the RP-HPLC sorbent generally
ranges between 100 and 300 Å, depending on the size of the peptide solutes, while porous materials of 300- to 4000-Å pore size should be used
for proteins. The selection of an optimal pore size for a particular sorbent
is made on the basis that the solute molecular diameter must be at least
one-tenth the size of the pore diameter of the packing material to avoid
restricted diffusion of the solute and also to allow the total surface area of
the sorbent material to be accessible. The other important variable of the
reversed-phase material is the particle diameter, d_p. As is evident from
Eq. (4), resolution improves as the particle diameter decreases. The most
commonly used range of particle diameters with high-resolution RP-HPLC
sorbents is 3–5 μm. However, there are examples of the use of nonporous
particles with smaller particle diameter.[94]

Mobile Phases

The ability to manipulate solute resolution through changes in the composition of the mobile phase represents a powerful characteristic of RP-

[88] M. Hanson, K. K. Unger, C. T. Mant, and R. S. Hodges, *J. Chromatogr.* **599,** 65 (1992).

[89] M. Hanson, K. K. Unger, C. T. Mant, and R. S. Hodges, *J. Chromatogr.* **599,** 77 (1992).

[90] N. Tanaka, K. Kimata, Y. Mikawa, K. Hosoya, T. Araki, Y. Ohtsu, Y. Shiojima, R. Tsuboi, and H. Tsuchiya, *J. Chromatogr.* **535,** 13 (1990).

[91] B. S. Welinder, *J. Chromatogr.* **542,** 83 (1991).

[92] F. Belliardo, O. Chiantore, D. Berek, I. Novak, and C. Lucarelli, *J. Chromatogr.* **506,** 371 (1990).

[93] H.-J. Wirth, K.-O. Eriksson, P. Holt, M. Aguilar, and M. T. W. Hearn, *J. Chromatogr.* **646,** 129 (1993).

[94] G. Jilge, R. Janzen, H. Giesche, K. K. Unger, J. N. Kinkel, and M. T. W. Hearn, *J. Chromatogr.* **397,** 71 (1987).

HPLC systems. RP-HPLC is usually carried out on n-alkyl-bonded silicas or other reversed-phase sorbents with an acidic mobile phase and elution of the peptides or proteins is achieved by the application of a gradient of increasing organic solvent concentration. The most commonly used mobile-phase additives are 10 mM trifluoroacetic acid (TFA), phosphoric acid, perchloric acid, or heptafluorobutyric acid.[18] At low pH values, silica-based sorbents are chemically stable and the surface silanols are fully protonated. TFA is the most popular of the acidic additives owing to its volatility, while significant changes in solute selectivity can be obtained with phosphoric acid. Formic acid, hydrochloric acid, and acetic acid can also be utilized.[16,95] Other mobile-phase additives such as nonionic detergents can be used in the isolation of more hydrophobic proteins such as membrane proteins.[96]

The three most common organic solvent modifiers are acetonitrile, methanol, or 2-propanol, which all exhibit high optical transparency in the detection wavelengths used in the RP-HPLC of peptides and proteins. While acetonitrile provides lower viscosity solvent mixtures, 2-propanol is a stronger eluent. An example of the influence of organic solvent on the separation of peptides is shown in Fig. 5. Changes in selectivity are clearly evident for peaks 9–12, 13–15, 17, and 18. The nature of the organic solvent can also influence the conformation of protein samples[97] and therefore may have a significant impact on the level of recovery of biologically active sample.

Operating Parameters

Several operating parameters will also influence the resolution of peptides and proteins in RP-HPLC. These parameters include the gradient time, the gradient shape, the mobile-phase flow rate, and the operating temperature. Typically, linear gradients with conventional analytical columns are applied from 5% organic solvent up to between 50 and 100% solvent over the time range of 20–120 min while flow rates between 0.5 and 2 ml/min are commonly used. With microbore columns, flow rates in the range 50–250 μl/min can be employed. The choice of the gradient conditions will depend on the selectivity between the solutes of interest. The influence of gradient time on the separation of growth hormone tryptic peptides is shown in Fig. 6. While longer retention times are generally observed with longer gradient times, improved resolution can also be obtained, as is evident for peaks T_3 and T_{13} and also T_5 and T_{18}. Variation

[95] G. Thévenon and F. E. Regnier, *J. Chromatogr.* **476**, 499 (1989).
[96] G. W. Welling, R. Van der Zee, and S. Welling-Wester, *J. Chromatogr.* **418**, 223 (1987).
[97] P. Oroszlan, S. Wicar, G. Tashima, S.-L. Wu, W. S. Hancock, and B. L. Karger, *Anal. Chem.* **64**, 1623 (1993).

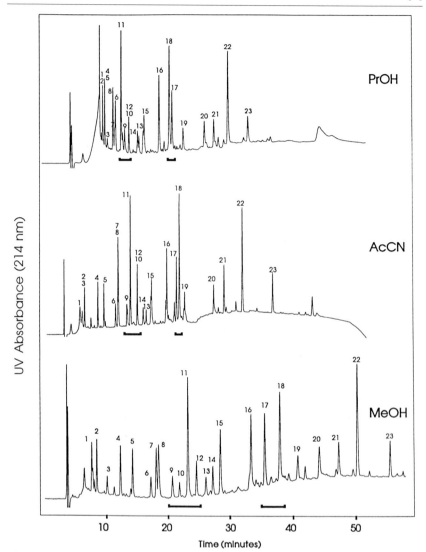

Fig. 5. Effect of organic modifier on the reversed-phase separation of tryptic peptides derived from porcine growth hormone. Column, Bakerbond RP-C$_4$, 25 cm × 4.6 mm i.d., 5-μm particle size, 30-nm pore size; conditions, linear gradient from 0 to 90% 2-propanol (top), acetonitrile (middle), or methanol (bottom) with 0.1% TFA over 60 min, flow rate of 1 ml/min, 37°. (From A. J. Round, M. I. Aguilar, and M. T. W. Hearn, unpublished results, 1995.)

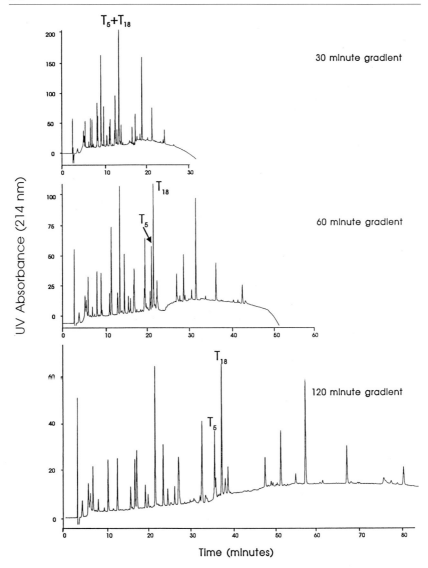

FIG. 6. Effect of gradient time on the separation of tryptic peptides of porcine growth hormone in RP-HPLC. Column, Bakerbond RP-C₄, 25 cm × 4.6 mm i.d., 5-μm particle size, 30-nm pore size; conditions, linear gradient from 0 to 90% acetonitrile over 30 min (top), 60 min (middle), and 120 min (bottom). (From A. J. Round, M. I. Aguilar, and M. T. W. Hearn, unpublished results, 1995.)

in the operating parameters relates to the differences in retention depicted in plots of log $\bar{k}$ versus $\bar{\psi}$, such as those shown in Fig. 3, and the conditions that maximize the differences between these retention plots. However, the use of longer gradient times or lower flow rates increases the residence time of the protein solute on the sorbent, which may then result in an increase in the degree of denaturation.

Peptide and protein separations are generally carried out with the operating temperature controlled slightly above ambient for improved reproducibility. However, the operating temperature is an additional parameter that can be used to modulate peptide and protein resolution. Solute retention in RP-HPLC is influenced by temperature through changes in solvent viscosity. While the conformation of proteins can be disrupted under conditions employed in RP-HPLC,[10,98,99] peptides have been shown to adopt significant secondary structure with reversed-phase sorbents.[100,101] Thus the use of temperature to deliberately manipulate the secondary and tertiary structure of peptides and proteins can improve separations, as shown in Fig. 7, where the resolution between peptides T_5 and T_{18} is achieved only at elevated temperature.

Inverse gradient elution chromatography has also been utilized for the micropreparative isolation of proteins from sodium dodecyl sulfate polyacrylamide gel electrophoresis (SDS–PAGE) electroeluates.[102] This approach takes advantage of the U-shaped or bimodal dependency of protein retention on organic solvent composition that is depicted in Fig. 2. In the inverse gradient procedure, proteins are bound to the reversed-phase sorbent under conditions of high organic solvent concentration and then eluted with a gradient of decreasing organic solvent concentration. As a consequence, large amounts of SDS, buffer salts, and acrylamide-related contaminants that interfere with the Edman sequencing procedure can be readily removed and the proteins are recovered in a form suitable for amino acid sequence analysis.

The overall resolution of complex peptide or protein mixtures also can be increased by the use of two-dimensional chromatography (2D-HPLC), whereby a reversed-phase column is coupled to an ion-exchange column.[103] Hence, the components of a mixture are separated on the basis of electrostatic charge differences in the first dimension followed by selectivity based on hydrophobicity in the second dimension. While 2D-HPLC can be per-

[98] X. M. Lu, K. Benedek, and B. L. Karger, *J. Chromatogr.* **359**, 19 (1986).
[99] M. T. W. Hearn, A. N. Hodder, and M. I. Aguilar, *J. Chromatogr.* **327**, 47 (1985).
[100] A. W. Purcell, M. I. Aguilar, and M. T. W. Hearn, *J. Chromatogr.* **593**, 103 (1992).
[101] N. E. Zhou, C. T. Mant, and R. S. Hodges, *Peptide Res.* **3**, 8 (1990).
[102] R. J. Simpson, R. L. Moritz, E. C. Nice, and B. Grego, *Eur. J. Biochem.* **165**, 21 (1987).
[103] N. Takahashi, N. Ishioka, Y. Takahashi, and F. W. Putnam, *J. Chromatogr.* **326**, 407 (1985).

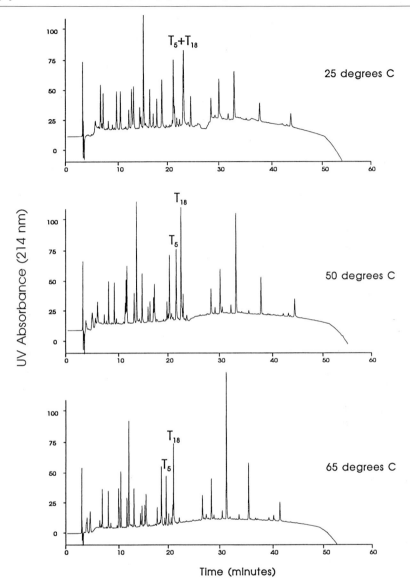

FIG. 7. Effect of temperature on the separation of tryptic peptides of porcine growth hormone. Column, Bakerbond RP-C$_4$, 25 cm × 4.6 mm i.d., 5-μm particle size, 30-nm pore size; conditions, linear gradient from 0 to 90% acetonitrile over 60 min at 25° (top), 50° (middle), and 65° (bottom). (From A. J. Round, M. I. Aguilar, and M. T. W. Hearn, unpublished results, 1995.)

formed manually without the need for special equipment, there have been several approaches to the automation of 2D-HPLC.[104,105]

The most commonly used mode of detection in RP-HPLC of peptides and proteins involves on-line ultraviolet detection. Elution is typically monitored at 210–220 nm, which is specific for the absorbance of the peptide bond, and detection is often performed at 280 nm, which corresponds to the aromatic amino acids tryptophan and tyrosine. The advent of photodiode array detectors has expanded the detection capabilities by allowing complete solute spectral data to be accumulated on-line. The spectra can be used to identify peaks more specifically on the basis of spectral characteristics and for the assessment of peak purity.[106] In addition, second derivative spectroscopy can also yield information on peak identity[107] and on the conformational integrity of proteins following elution.[98,108]

A number of precolumn and postcolumn derivatization procedures have also been developed to increase the sensitivity and specificity of detection of peptides and proteins in RP-HPLC. For example, precolumn derivatization of peptides with phenyl isothiocyanate allows ultraviolet (UV) detection at the picomolar level.[109] Alternatively, automated precolumn derivatization of peptides with fluorescamine[110] or of glycopeptides with 1,2-diamino-4,5-dimethoxybenzene[111] allows fluorescence detection of peptides at the femtomolar level.

Column Geometry

The selection of column dimensions is made on the basis of the desired level of efficiency and the sample loading capacity. For small peptides and proteins up to 10 kDa, resolution can be improved with increases in column length. Thus, for systems such as tryptic mapping, column lengths between 15 and 25 cm with standard 4.6-mm i.d. columns are commonly used. However, for larger proteins increased column length may adversely affect the protein mass recovery and maintenance of biological activity owing to denaturation and/or irreversible binding to the sorbent. In these cases, column lengths between 2 and 10 cm can be used.

[104] N. Takahashi, Y. Takahashi, N. Ishioka, B. S. Blumberg, and F. W. Putnam, *J. Chromatogr.* **359**, 181 (1986).
[105] K. Matsuoka, M. Taoka, T. Isobe, T. Okuyama, and Y. Kato, *J. Chromatogr.* **506**, 371 (1990).
[106] J. Frank, A. Braat, and J. A. Duine, *Anal. Biochem.* **162**, 65 (1987).
[107] F. Nyberg, C. Pernow, U. Moberg, and R. B. Eriksson, *J. Chromatogr.* **359**, 541 (1985).
[108] M. T. W. Hearn, M. I. Aguilar, T. Nguyen, and M. Fridman, *J. Chromatogr.* **435**, 271 (1988).
[109] F. J. Collida, S. P. Yadav, K. Brew, and E. Mendez, *J. Chromatogr.* **548**, 303 (1991).
[110] V. K. Boppana, C. Miller-Stein, J. F. Politowski, and G. R. Rhodes, *J. Chromatogr.* **548**, 319 (1991).
[111] O. Shirota, D. Rice, and M. Novotny, *Anal. Biochem.* **205**, 189 (1992).

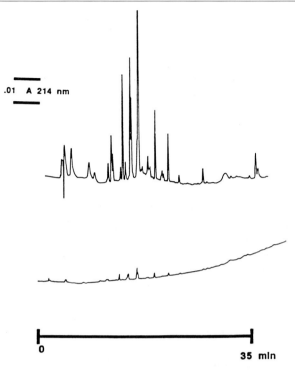

FIG. 8. Effect of column diameter on the sensitivity of peptide separations in RP-HPLC. Columns: lower chromatogram, Aquapore RP300 C$_8$, 22 cm × 4.6 mm i.d., 7-μm particle size, 30-nm pore size; upper chromatogram, Aquapore RP300 C$_8$, 25 cm × 1 mm i.d., 7-μm particle size, 30-nm pore size. Conditions: lower chromatogram, linear gradient from eluent A to B [eluent A = 0.1% TFA; eluent B = 0.088% TFA in acetonitrile–water (70:30)] from 0 to 60% acetonitrile in 45 min at a flow rate of 1 ml/min; upper chromatogram, same conditions as above, except for a flow rate of 50 μl/min. The sample size was 100 pmol on both columns. (Reprinted with permission from C. T. Mant and R. S. Hodges (eds.), "HPLC of Peptides and Proteins: Separation, Analysis and Conformation." CRC Press, Boca Raton, FL, 1991. Copyright CRC Press, Boca Raton, Florida.)

The choice of column internal diameter can then be based on the sample capacity required. Most analytical applications in the microgram range are generally carried out with columns of 4.6-mm i.d. However, an additional major requirement in the high-resolution purification of peptides or proteins generated from previously uncharacterized proteins is to maximize the detection sensitivity, because the supply of the protein is limited. RP-HPLC separations of 5–25 pmol of sample are now carried out with narrow-bore columns of 1- to 2-mm i.d. The flow rate of the mobile phase is also reduced to maintain the same linear flow velocity. These so-called narrow-bore (2-mm i.d.) or microbore (1-mm i.d.) columns allow samples to be eluted in smaller volumes than is possible with 4-mm i.d. columns, which results in higher solute concentration, thereby increasing mass sensitivity. The use of these columns also avoids the need for further concentration steps, thereby minimizing possible sample losses. An example of the different sensitivity obtained with the same sample loaded onto a 4-mm i.d. column and a 1-mm i.d. column is shown in Fig. 8. Further miniaturization of RP-HPLC systems has also been reported with the use of capillary columns with internal diameters between 0.25 and 0.32 mm.[57,58]

Future Directions

RP-HPLC is now firmly established as the central tool in the analysis and isolation of peptides and proteins, particularly in the field of analytical biotechnology. The widespread opportunities offered by RP-HPLC have been greatly expanded through coupling of chromatographic systems with electrospray mass spectroscopic (MS) analysis. An ever-increasing demand of investigators is for the availability of faster separations. Thus the so-called coupled procedures based on LC-MS, LC-CE (capillary electrophoresis), or LC-biosensor methods will allow more rapid on-line analysis through immediate identification of solute molecular weight, purity level, and activity characteristics. Identification of the nature of the contact region established between the peptide or protein and the reversed-phase ligand is also a crucial step for the full experimental validation of the predictions of the solvophobic model and also to advance our understanding of the mechanism of peptide and protein interactions with chromatographic surfaces. Definition of the precise molecular characteristics of the hydrophobic contact region established between proteins and RP-HPLC sorbents is a major focus of current studies on the conformational analysis of peptides and proteins, the development of new optimization procedures, and the design of new ligands. There is no doubt that the demands of the biotechnology industry will drive the nature of the applications of RP-HPLC, which will generally be related to the establishment of compositional and conformational purity of recombinantly derived proteins. However, the ultimate challenge resides in the attainment of detection sensitivity at the low femtomolar/high attomolar level. Currently, this should be feasible with miniaturized systems coupled with fluorescence derivatization methods and laser-induced fluorescence detection. As we gain further insight into the molecular basis of the interaction between peptides and proteins and the factors that control their orientation and conformation at hydrophobic surfaces, new approaches to the significant enhancement of resolution can also be determined.

[2] Hydrophobic Interaction Chromatography of Proteins

By Shiaw-Lin Wu and Barry L. Karger

Introduction

Hydrophobic interaction chromatography (HIC) is a useful tool for purifying proteins with maintenance of biological activity. In this method, hydrophobic ligands (e.g., *n*-alkyl or phenyl groups) are chemically bound to matrices (polymer or silica gels), and protein components interact with these ligands through the application of a high concentration of an antichaotropic salt [e.g., $(NH_4)_2SO_4$)]. This interaction is diminished by decreasing salt concentration, leading to protein elution, as shown in Fig. 1.

Hydrophobic interaction chromatography, first reported in the 1950s under the name "salting out chromatography,"[1-3] is a technique in which a high concentration of salt (e.g., ammonium sulfate or sodium chloride) is used to adsorb proteins, with desorption a result of decreasing salt concentration. Soon after its initial description, researchers reported nonspecific hydrophobic interaction in size-exclusion chromatography[4,5] and affinity chromatography,[6] and even used this interaction for separation.[7,8] Reports published almost simultaneously in 1973[9-12] further characterized this interaction by demonstrating that hydrophobic interaction (or retention) can be enhanced by increasing the concentration of antichaotropic salts or by increasing the length of the *n*-alkyl chain on the support matrices; these are two main features of HIC. Since that time, high-performance HIC has been developed for rapid separations with high efficiency.[13-16]

[1] A. Tiselius, *Ark. Kem. Min. Geol.* **26B** (1948).
[2] C. C. Shepard and A. Tiselius, *in* "Chromatographic Analysis," p. 275. *Discussions of the Faraday Society*, No. 7. Hazell, Watson and Winey, London, 1949.
[3] J. Porath, *Biochim. Biophys. Acta* **39,** 193 (1960).
[4] B. Gelotte, *J. Chromatogr.* **3,** 330 (1960).
[5] N. V. B. Marsden, *Ann. N.Y. Acad. Sci.* **125,** 428 (1965).
[6] G. J. Doellgast and G. Kohlaw, *Fed. Proc.* **31,** 424 (1972).
[7] P. Cuatrecasas and C. B. Afinsen, *Annu. Rev. Biochem.* **40,** 259 (1971).
[8] R. J. Yon, *Biochem. J.* **126,** 765 (1972).
[9] B. H. J. Hofstee, *Anal. Biochem.* **52,** 430 (1973).
[10] J. Porath, L. Sundberg, N. Fornstedt, and L. Olsson, *Nature* (*London*) **245,** 465 (1973).
[11] S. Hjerten, *J. Chromatogr.* **87,** 325 (1973).
[12] R. A. Rimerman and G. W. Hatfield, *Science* **182,** 1268 (1973).
[13] Y. Kato, T. Kitamura, and T. Hashimoto, *J. Chromatogr.* **292,** 418 (1984).
[14] N. T. Miller, B. Feibush, and B. L. Karger, *J. Chromatogr.* **316,** 519 (1984).

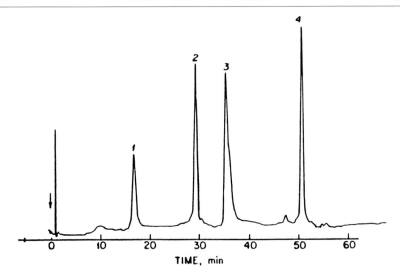

Fɪɢ. 1. Separation of a standard protein mixture by HIC. Peak identification: (1) cytochrome *c*, (2) ribonuclease A, (3) lysozyme, (4) α-chymotrypsinogen A. Column: Methyl polyether, 0.46 × 10 cm. Linear gradient from 3 *M* ammonium sulfate + 0.5 *M* ammonium acetate, pH 6.0, to 0.5 *M* ammonium acetate, pH 6.0, in 60 min. Flow rate, 1 ml/min; temperature, 25°. (Reprinted from *J. Chromatogr.*, **326,** N. T. Miller and B. L. Karger, p. 45, Copyright 1985 with kind permission of Elsevier Science–NL, Sara Burgerhartstraat 25, 1055 KV Amsterdam, The Netherlands.)

Reversed-phase high-performance liquid chromatography (RP-HPLC), a powerful tool for peptide separation,[17] was developed for the analysis of proteins in this period (see [1]).[17a] A low recovery of biological activity and mass in protein separations was, however, often encountered in RP-HLPC,[18-20] and the elution order could be significantly different than in HIC.[21] More detailed studies showed that the low recovery in RP-HPLC was due to a change in protein conformation as a consequence of the high-

[15] J.-P. Chang, Z. El Rassi, and C. Horvath, *J. Chromatogr.* **319,** 396 (1985).

[16] Y. Kato, T. Kitamura, and T. Hashimoto, *J. Chromatogr.* **360,** 260 (1986).

[17] M. T. W. Hearn, *Methods Enzymol.* **104,** 190 (1984).

[17a] M.-I. Aguilar and M. T. W. Hearn, *Methods Enzymol.* **270,** Chap. 1, 1996 (this volume).

[18] S. Y. M. Lau, A. K. Tanija, and R. S. Hodges, *J. Chromatogr.* **317,** 129 (1984).

[19] A. J. Salder, R. Micanovic, G. E. Katzenstein, R. V. Lewis, and C. R. Middaugh, *J. Chromatogr.* **317,** 93 (1984).

[20] S. A. Cohen, K. P. Benedek, S. Dong, Y. Tapuhi, and B. L. Karger, *Anal. Chem.* **56,** 217 (1984).

[21] J. L. Fausnaugh, L. A. Kennedy, and F. E. Regnier, *J. Chromatogr.* **317,** 141 (1984).

density n-alkyl surface and the harsh mobile-phase conditions (e.g., low pH and the use of organic modifier for elution).[22,23]

The comparison between HIC and RP-HPLC is interesting because the retention mechanism is based on the same fundamental principle, i.e., hydrophobicity. However, because the structure of the protein is significantly altered in RP-HPLC, the two methods are, in effect, often separating different conformational species. The weaker hydrophobic stationary phase in HIC (i.e., shorter alkyl chain length or lower ligand density on the support matrices) with milder elution conditions (i.e., aqueous solution near neutral pH) leads to more suitable conditions for protein separation of active forms.[24–28] Today, HIC has been refined and is widely used for the separation of biopolymers based on differences in hydrophobicity.[27–30]

Protein Structure and Chromatographic Surfaces

In HIC, a portion of the outer surface of the protein is in direct contact with the chromatographic surface. The amino acid residues that are on the exterior of the protein are determined by the basic structural characteristics of the protein and by the chromatographic conditions (e.g., column temperature, ligand type, and mobile phase). Each protein will have a specific stability, and hence its conformation can be significantly altered by variation in chromatographic conditions.[22,31,32] Therefore, it is difficult to formulate a general strategy for the separation of proteins. There are, however, some practical guidelines that can be given. First, it is important to have information about specific properties of the protein. The size, chemical nature, function, amino acid sequence, helicity, and three-dimensional structure are important parameters to consider in selecting chromatographic conditions. For example, since membrane proteins are strongly hydrophobic in nature, a mild chromatographic surface (e.g., ether ligand) with a nonionic

[22] K. Benedek, S. Dong, and B. L. Karger, J. Chromatogr. 317, 227 (1984).

[23] R. H. Ingraham, S. Y. M. Lau, A. K. Taneja, and R. S. Hodges, J. Chromatogr. 327, 77 (1985).

[24] S. Shaltiel, Methods. Enzymol. 104, 69 (1984).

[25] S. Hjerten, Methods Biochem. Anal. 27, 89 (1981).

[26] Y. Kato, T. Kitamura, and T. Hashimoto, J. Chromatogr. 266, 49 (1983).

[27] J. L. Fausnaugh, E. Pfannkoch, S. Gupta, and F. E. Regnier, Anal. Biochem. 137, 464 (1984).

[28] N. T. Miller, B. Feibush, K. Corina, S. Powers-Lee, and B. L. Karger, Anal. Biochem. 148, 510 (1985).

[29] D. L. Gooding, M. N. Schmuck, and K. M. Gooding, J. Chromatogr. 296, 107 (1984).

[30] S. C. Goheen and S. C. Engelhorn, J. Chromatogr. 317, 55 (1984).

[31] S. Lin, P. Oroszlan, and B. L. Karger, J. Chromatogr. 536, 17 (1991).

[32] P. Oroszlan, S. Wicar, G. Teshima, S.-L. Wu, W. S. Hancock, and B. L. Karger, Anal. Chem. 64, 1623 (1992).

detergent (e.g., Triton X-100) or a zwitterionic detergent [e.g., 3-[(3-cholamidopropyl)dimethylammonio]-1-propane sulfonate (CHAPS), a derivative of cholic acid] added to the mobile phase to enhance solubility are good choices for such hydrophobic species.[33,34] For glycoproteins, which are generally less hydrophobic in solution, a stronger hydrophobic surface (e.g., C_4 or phenyl ligands) with a higher concentration of antichaotropic salts to facilitate protein binding may be effective.[35,36]

For the separation of protein variants (e.g., single or multiple amino acid residue substitution through mutation, oxidation, deamidation, or cleavage), unfortunately, no simple rule can be given. Initially, a purely empirical approach was followed to separate protein variants by HIC. Nonetheless, several reports showed that HIC is indeed capable of discriminating subtle structural differences[37–39] and suggested that in successful cases, the altered amino acid positions were in or near the contact surface region of the protein. As an example, in Fig. 2, a minor impurity peak (near 16 min at 10°) was well separated at subambient temperatures but not at high temperature (50°) with a weakly hydrophobic methyl polyether column.[40a,b] However, if the protein variation were buried inside the molecule, then some alteration of protein structure would be needed to expose the variation to the contact region. Such conformational manipulation can be probed with specific chromatographic conditions (e.g., column type, mobile-phase pH, temperature, and additives). For example, in Fig. 3, methionyl human growth hormone (in which an extra methionine residue is on the N-terminal sequence of native growth hormone) can be better separated from native human growth hormone by using a more hydrophobic phenyl stationary phase than an ether column.[38] In this study, the temperature and pH can also affect the separation. At subambient temperatures, neutral pH, and with a weakly hydrophobic surface, the separation of methionyl and native growth hormone was not possible since they have similar conformations.[32,38] Therefore, manipulation of structure was important for separation, e.g., mildly elevated temperature, strongly hydrophobic column, or denatured additives.[32,38] It should be pointed out

[33] Y. Kato, T. Kitamura, K. Nakamura, A. Mitsui, Y. Yamasaki, and T. Hashimoto, *J. Chromatogr.* **391,** 395 (1987).

[34] D. Josic, W. Hofmann, and W. Reutter, *J. Chromatogr.* **371,** 43 (1986).

[35] S.-L. Wu, *LC-GC* **10,** 430 (1992).

[36] A. Alpert, *J. Chromatogr.* **444,** 269 (1988).

[37] J. L. Fausnaugh and F. E. Regnier, *J. Chromatogr.* **359,** 131 (1986).

[38] S.-L. Wu, W. S. Hancock, B. Pavlu, and P. Gellerfors, *J. Chromatogr.* **500,** 595 (1990).

[39] A. M. Jespersen, T. Christensen, N. K. Klausen, P. F. Nielsen, and H. H. Sørensen, *Eur. J. Biochem.* **219,** 365 (1994).

[40a] S.-L. Wu, K. Benedek, and B. L. Karger, *J. Chromatogr.* **359,** 3 (1986).

[40b] S.-L. Wu and B. L. Karger, *in* "HPLC of Peptides and Proteins: Separation, Analysis, and Conformation" (R. S. Hodges, ed.), p. 613. CRC Press, Boca Raton, FL, 1991.

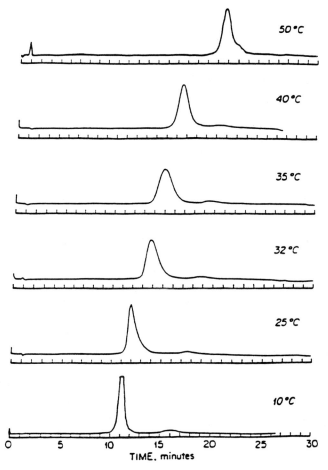

FIG. 2. Effect of temperature on the elution behavior of Ca^{2+}-depleted α-lactalbumin. Column: Methyl polyether, 0.46×10 cm. Linear gradient from 2 M ammonium sulfate + 0.5 M ammonium acetate, pH 6.0, to 0.5 M ammonium acetate, pH 6.0, in 30 min. Flow rate, 1 ml/min. (Reprinted from *J. Chromatogr.*, **359**, S.-L. Wu, K. Benedek, and B. L. Karger, p. 3, Copyright 1986 with kind permission of Elsevier Science–NL, Sara Burgerhartstraat 25, 1055 KV Amsterdam, The Netherlands.)

that these conformational changes in HIC are often subtle and reversible, such that the native state of the protein can often be recovered after separation.[28,33,41,42]

In 1986, the first report on the use of an on-line photodiode array

[41] T. Arakawa and S. N. Timasheff, *Methods Enzymol.* **144**, 49 (1985).
[42] R. E. Shansky, S.-L. Wu, A. Figueroa, and B. L. Karger, *in* "HPLC of Biological Macromolecules" (K. M. Gooding and F. E. Regnier, eds.), p. 95. Marcel Dekker, New York, 1990.

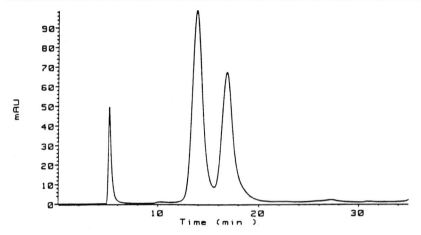

Fig. 3. Separation of premixed Genentropin rhGH (14 min) and Somatonorm Met-hGH (17 min). Column: TSK-phenyl-5PW, 0.75 × 7.5 cm. Linear gradient from mobile phase A [0.5 *M* sodium sulfate, 0.03 *M* Tris-HCl, 0.5% v/v acetonitrile (pH 8)] to mobile phase B {0.03 *M* Tris-HCl, 5% v/v acetonitrile, 0.07% w/v Brij 35 [polyoxethylene (23) lauryl ether or C12E23], pH 8} in 60 min. Flow rate, 0.5 ml/min; temperature, 25°. (Reprinted from *J. Chromatogr.*, **500**, S.-L. Wu, W. S. Hancock, B. Pavlu, and P. Gellerfors, p. 595, Copyright 1990 with kind permission of Elsevier Science–NL, Sara Burgerhartstraat 25, 1055 KV Amsterdam, The Netherlands.)

detector to determine the conformational changes in HIC was published.[40a,b] Those changes were shown to be indeed reversible and to correlate well with the earlier observation of chromatographic behavior in RP-HPLC.[20,32] More detailed characterization was subsequently developed by studying the structural changes on the surface by an on-column fluorescence spectroscopic method.[31,32,43] An experienced protein chemist with an understanding of how to manipulate protein conformation can often meet the challenge of separating protein variants.[35,38,39,44–46]

Practice of Hydrophobic Interaction Chromatography

Equipment and Buffers

Any standard HPLC commercial equipment is suitable for HIC. A dual pumping system for gradient elution is desirable for protein separations.

[43] X. M. Lu, A. Figueroa, and B. L. Karger, *J. Am. Chem. Soc.* **110**, 1978 (1988).
[44] R. M. Riggin, G. K. Dorulla, and D. J. Miner, *Anal. Biochem.* **167**, 199 (1987).
[45] M. G. Kunitani, R. L. Cunico, and S. J. Staats, *J. Chromatogr.* **443**, 205 (1988).
[46] E. Canova-Davis, G. M. Teshima, J. T. Kessler, P. J. Lee, A. Guzzeta, and W. S. Hancock, *in* "Analytical Biotechnology" (C. Horvath and J. G. Nikelly, eds.), p. 90. ACS Symposium Series No. 434. American Chemical Society, Washington, D.C., 1990.

With the use of high salt concentrations in HIC, pumps with stainless steel components must be thoroughly flushed with water when not in use. A 2-hr water wash at a flow rate of 1 ml/min is a minimum requirement in such cases, and an overnight wash is highly recommended to avoid any salt deposits that can block microcolumn connections or corrode metal surfaces (the presence of halides, e.g., sodium chloride, can cause corrosion). To avoid the problems associated with stainless steel, metal-free or "biocompatible" HPLC systems can be selected. In these designs, wetted parts consist only of glass, titanium, or fluoroplastics. Care should also be taken in the preparation of salt and buffer solutions. The concentration of mobile-phase components should be at least 10% less than their solubility limit (e.g., no more than 3.5 M for ammonium sulfate). Note that sodium sulfate is much less soluble ($\sim$1.5 M) than ammonium sulfate. Filtration of all mobile phases through 0.22-μm or smaller pore size membranes prior to use is mandatory.

Sample Preparation and Collection

Before beginning an HIC run, it is important to consider the sample concentration and injection volume. An HIC method generally starts from a high concentration of antichaotropic salt with a gradient of decreasing salt concentration. Thus, water (low ionic strength) can be considered as a strong solvent for elution in HIC, similar to the role played by the organic solvent in RP-HPLC.

The injection of large volumes of aqueous samples can cause distortion of peak shapes or even retention of sample components.[38,42] Figure 4 shows the effect of sample concentration and injection volume on the resultant chromatography. In Fig. 4A, recombinant human growth hormone (rhGH) is dissolved in water at two different concentrations, 1 and 10 mg/ml. To inject the same mass of rhGH, the injection volume is 25 μl for the 10-mg/ml solution and 250 μl for the 1-mg/ml solution, respectively. A distorted peak (fronted) is observed for the 250-μl injection. The distortion disappears if the rhGH concentration is diluted 10 times to 0.1 mg/ml (i.e., rhGH is dissolved in water and 250 μl is injected), as shown in Fig. 4B. This distortion also disappears if the rhGH is dissolved in 1 M ammonium sulfate (a salt concentration below that required for protein precipitation) instead of water, and 250 μl (1 mg/ml) is injected. Because the sample is dissolved in a low ionic strength buffer (e.g., water) and the column is equilibrated in a high ionic strength mobile phase A (e.g., antichaotropic salt), the incomplete mixing of these two buffers during injection (especially a large-volume injection) could cause the distorted peaks in the chromatogram. Therefore, it is useful to have some antichaotropic salt present in the sample

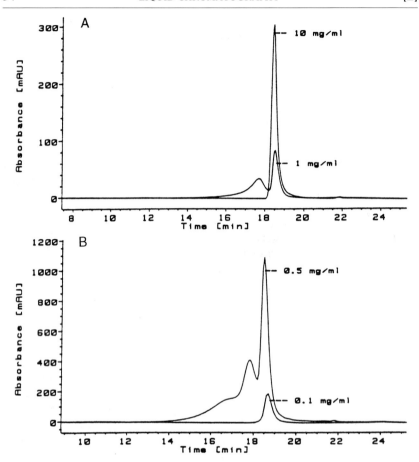

FIG. 4. Effect of sample concentration and injection volume on the elution behavior of recombinant human growth hormone. Column: TSK-ether-5PW, 0.75×7.5 cm; temperature, $50°$; 30-min linear gradient from 2 M ammonium sulfate + 0.1 M potassium phosphate, pH 7, to 0.1 M potassium phosphate, pH 7. (A) An overlay of a 10-mg/ml sample concentration and 25-μl injection volume with a 1-mg/ml sample concentration and 250-μl injection volume. (B) An overlay of a 0.5-mg/ml sample concentration and 250-μl injection volume with a 0.1-mg/ml and 250-μl injection volume. (Reprinted from *J. Chromatogr.*, **500**, S.-L. Wu, W. S. Hancock, B. Pavlu, and P. Gellerfors, p. 595, Copyright 1990 with kind permission of Elsevier Science–NL, Sara Burgerhartstraat 25, 1055 KV Amsterdam, The Netherlands.)

buffer. As a practical guide, a 3:1 dilution of sample buffer with mobile phase A is effective for a relatively large-volume injection (e.g., injection of 250 μl in a typical 7.5×75 mm analytical column), where mobile phase A is the starting concentration of antichaotropic salt. An alternative approach for large-volume injection is to load the sample in a series of small-volume injections.[38]

After separation in HIC, a collected fraction often contains a large amount of salt. It may be necessary to desalt the collected sample for subsequent experiments. Dialysis (e.g., membranes with different cutoff sizes), desalting through a size-exclusion column [e.g., Pharmacia (Piscataway, NJ) PD-10 column], or desalting through a reversed-phase column (e.g., Waters Sep-Pak, or any C_4 reversed-phase column) are methods most often used. Aggregation and precipitation can occur if the concentrated sample is eluted from the HIC column with high salt concentration at a pH close to the pI of the protein, or if the sample remains in the high salt for a long period of time.[47,48] In this case, it is important either to desalt or dilute the collected sample quickly with a solution in which the protein will be maintained in its native state.

Column

The heart of any chromatographic system is the column. From the previous discussion, the design of an appropriate chromatographic surface is important for protein separations in HIC. Initially, researchers prepared their own chromatographic surfaces and packed the particles into high-performance columns. This approach is not in practice today, and researchers are dependent on the commercial availability of high-quality packed columns that have been specifically designed for protein separation. Table I lists a number of commercially available columns with their suppliers. As suggested earlier, the choice of packing surfaces must be balanced against the liability of the proteins to be separated.[35,40a,44,49] Columns made from the packings in Table I are stable at high flow rates and resist swelling and shrinkage under a variety of mobile-phase conditions. Common features of analytical columns are small particle size (10 μm or less) and large pore size (300 Å or more). Large-scale preparative packings generally have larger particle diameters (typical 20–40 μm) with the same pore diameters as analytical scale.

It is well known that the slow diffusion rates of proteins in and out of pores of the packing can yield broad bands and result in low efficiency. Nonporous particles (1.5–2.5 μm) eliminate protein diffusion into pores and, as a result, high efficiency can be generated at high velocity, leading

[47] N. Grinberg, R. Blanco, D. M. Yarmush, and B. L. Karger, *Anal. Chem.* **61,** 514 (1989).
[48] I. S. Krull, H. H. Stuting, and S. Krzysko, *J. Chromatogr.* **252,** 29 (1988).
[49] S.-L. Wu, A. Figueroa, and B. L. Karger, *J. Chromatogr.* **371,** 3 (1986).

TABLE I
COMMERCIAL ANALYTICAL HIC COLUMNS

Manufacturer (location)/name	Particle/pore size	Phase/support
J. T. Baker (Phillipsburg, NJ)		
BakerBond HI-propyl	5 μm/300 Å	Propyl/silica
BakerBond PREPSCALE	40 μm/275 Å	HI-propyl/silica
Beckman (Fullerton, CA)		
Spherogel CAA-HIC5	5 μm/300 Å	Methyl polyether/silica
Pharmacia (Piscataway, NJ)		
Phenyl-Superose	10 μm	Phenyl/cross-linked agarose
Alkyl-Superose	10 μm	Branched C_5/cross-linked agarose
PerSeptive Biosystems (Cambridge, MA)		
PH	10 μm/6000 Å	Phenyl/polymer
PE	10 μm/6000 Å	Phenyl ether/polymer
BU	10 μm/6000 Å	Butyl/polymer
ET	10 μm/6000 Å	Ether/polymer
PolyLC (Columbia, MD)		
Polyethyl A	5 μm/330 Å	Polyalkylaspartamide/silica
Polypropyl A	5 μm/330 Å	Polyalkylaspartamide/silica
Supelco (Bellefonte, PA)		
LC-HINT	5 μm	"Polar" bonded phase/silica
SynChrom (Lafayette, IN)		
SynChropak Methyl, Propyl, Pentyl, Benzyl, Hydroxypropyl	6.5 μm/100 Å, 300 Å, 500 Å, 1000 Å	Polyamide coating with ligand indicated/silica
SynChroprep	30 μm/300 Å	
Toso-Haas (Philadelphia, PA)		
TSK-phenyl-5PW	10 μm/1000 Å	Phenyl/polymer
TSK-ether-5PW	10 μm/1000 Å	Oligopolyethylene glycol/polymer
Tsk-*n*-butyl-PW	2.5 μm, nonporous	*n*-Butyl/polymer

to high-speed protein separations.[35,50–52] Figure 5 illustrates one example of the separation possibilities with such particles. The separation time for recombinant tissue plasminogen activator (rtPA) variants is decreased from 2.5 hr to 15 min by using a nonporous column. However, because of the small sample loading capacity with nonporous particles, it is difficult to collect sufficient sample for further characterization. Rather, nonporous

[50] R. Janzen, K. K. Unger, H. Giesche, J. N. Kinkel, and M. T. W. Hearn, *J. Chromatogr.* **397**, 91 (1987).
[51] K. Kalhgatgi and C. Horvath, *J. Chromatogr.* **398**, 335 (1987).
[52] Y. Kato, S. Nakatani, T. Kitamura, Y. Yamasaki, and T. Hashimoto, *J. Chromatogr.* **502**, 416 (1990).

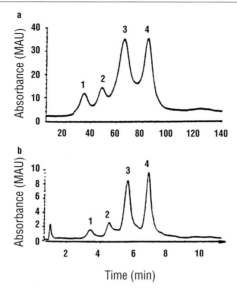

FIG. 5. Separation of recombinant tissue plasminogen activator (rtPA) variants using (a) porous and (b) nonporous resin columns. The variants include type I (three carbohydrate sites) and type II (two carbohydrate sites) forms of the two-chain (plasmin cleavage between Arg-275 and Ile-276) and one-chain molecules. Conditions in (a): column, TSK-phenyl-5PW, 0.75 × 7.5 cm, 10-μm particle size, and 1000-Å pore diameter; 120-min linear gradient from 0.6 M ammonium sulfate + 0.1 M potassium phosphate + 20% glycerol + 0.005% C12E8 (octaethylene glycol monododecyl ether), pH 7, to 0.1 M potassium phosphate + 20% glycerol + 0.04% C12E8, pH 7; flow, 0.8 ml/min; temperature, 60°. Conditions in (b): column, TSK-butyl-NPR HIC, 0.46 × 3.5 cm, 2.5-μm particle size; buffer conditions are the same as in (a), except that the elution from 1.2 M ammonium sulfate to 0 M ammonium sulfate is in 10 min; flow, 1 ml/min; temperature, 60°. Peak identification: 1, type I (one chain); 2, type II (two chain); 3, type I (one chain); 4, type II (one chain). (From Ref. 35, with permission of LC-GC magazine.)

particles are typically used in the analytical mode as a second method to help cross-validate a higher capacity separation method.[35]

Perfusive packings are another approach to minimize the slow diffusion rates of proteins in and out of pores.[53] In this approach, in contrast to the nonporous particle, the porosity is greatly increased (4000–6000 Å) to overcome the diffusion restriction of the protein by convective mass transfer or flow within or through the particle. Efficiency in theory is not reduced at higher flow rates, and run times can thus be reduced simply by increasing the flow rate. Not only does this approach save time, but the amount of sample that can be loaded is much higher than for nonporous resins.[35,53]

[53] N. Afeyan, S. Fulton, and F. E. Regnier, *J. Chromatogr.* **544**, 267 (1991).

Such columns can be readily scaled for preparative operation. However, it is to be noted that perfusive particle columns in many cases are like other HPLC columns—the separation works for a number of proteins but not for all.[35,54a] Nevertheless, these columns are increasing in popularity.

For any of the above-described columns, the use of a small guard column after the injection port and an occasional cleaning of the HIC column with a reversed-phase eluent (e.g., a gradient of acetonitrile in 0.1% trifluoroacetic acid) are recommended. For polymer-type supports, a stronger wash with 0.1 N sodium hydroxide for 20 to 30 column volumes can be used. In general, silica-based supports have a lower column lifetime than polymer supports. In our experience, more than 200 runs can be made with a polymer support column without impairing the separation efficiency. The retention time and peak area in HIC are generally as reproducible as with any other HPLC mode [e.g., approximately 0.5% of relative standard deviation (rsd) in retention time and 2–3% rsd in peak area].

Mobile Phase

Antichaotropic salts, buffers, and additives (e.g., metal ions, organic solvents, or detergents) are the main components of HIC mobile phases. The choice of salt generally follows the same order as for the salting out of proteins. Table II[54b,c] lists some typical salts with their molal surface tension increment, σ, which is proportional to the retention of a protein in HIC, for a given salt concentration.[42] Salts with a positive σ raise the surface tension of water and are antichaotropic ("structure making," i.e., water molecules have increased structure or order) and reduce the solubility of a protein, i.e., salt-out. Salts of many organic acids and bases (not listed) have lower values of σ, thereby enhancing solubility, i.e., salt-in. The molal surface tension increment thus provides a measure of the effect of salt type and concentration on HIC retention. Solubility is another important factor in the selection of the salt for HIC. Sodium citrate in this respect would seem to be a good choice because of its high σ value and high solubility. However, this salt exhibits high ultraviolet (UV) absorbency in the low-wavelength UV region and is prone to microbial growth. Also, sodium sulfate would appear to be a good choice, but, as already noted, it is not highly soluble (limit approximately 1.5 M). Ammonium sulfate has generally been found to be the most useful salt for HIC. It has a high σ value and is highly soluble (about 4 M) even at subambient temperatures. In addition, concentrated solutions of this salt resist microbial growth and

[54a] K. Nugent and K. Olson, *Biochromatogr.* **5,** 101 (1990).
[54b] W. R. Melander and C. Horvath, *Arch. Biochim. Biophys.* **183,** 200 (1977).
[54c] W. R. Melander, D. Coradini, and C. Horvath, *J. Chromatogr.* **317,** 67 (1984).

TABLE II
SALTS FOR HIC[a]

Salt	Molal surface tension increment $\sigma \times 10^3$ (dyn · g/cm · mol)
Tripotassium citrate	3.12
Na$_2$SO$_4$	2.73
K$_2$SO$_4$	2.58
(NH$_4$)$_2$SO$_4$	2.16
Na$_2$HPO$_4$	2.02
Dipotassium tartrate	1.96
NaCl	1.64
KClO$_4$	1.40
NH$_4$Cl	1.39
NaBr	1.32
NaNO$_3$	1.06
NaClO$_3$	0.55
KSCN	0.45
MgCl$_2$	3.16[b]
CaCl$_2$	3.66[b]

[a] $\gamma = \gamma^0 + \sigma m$. γ is the surface tension of the salt; γ^0 is the surface tension of pure water (72 dyn/cm); σ is the molal surface tension increment; m is the molality of the salt (see Refs. 42 and 54b,c).
[b] Exhibits specific binding effects (see Ref. 41).

have a high UV transparency. In our experience, a 2 M ammonium sulfate solution can be used even after 2 months of storage. A highly purified grade of ammonium sulfate is recommended for use in HIC. Sodium chloride has a somewhat lower σ value but can be employed because of its high solubility and low cost. However, the chloride ion may corrode wetted parts of stainless steel.

The choice of buffer is simple in HIC. High buffering capacity in the neutral pH region and high UV transparency are the two main considerations. Sodium phosphate, potassium phosphate, and Tris at 50 to 100 mM are typical buffer choices.

Mobile-Phase Additives

Buffer additives can often change the relative retention of proteins and thus dramatically affect separation. There are many examples of the use of buffer additives in practice. Metal ions (e.g., calcium or magnesium having specific binding effects),[31,42,43] organic solvents (e.g., polyhydric alcohol or acetonitrile),[33,42] and detergents (e.g., Tween, Triton X-100, or CHAPS)[33,35,38,42] are typical additives used for protein separations. Their general and specific effects are discussed as follows.

For proteins that are strongly adsorbed even at low salt concentration (e.g., membrane proteins), additives to mobile phase B (decrease in the σ value) become important to increase the elution strength and to enlarge the separation window. Figure 6 shows that a detergent must be used in mobile phase B in order to elute enkephalinase (Enk), a large membrane protein with a molecular mass of 98,000 Da. No matter what HIC columns are selected, Enk cannot be eluted unless a detergent is added to mobile phase B. The recommended detergents are octaethylene glycol monodode-cyl ether (C12E8) and octylglucoside. It is interesting to note that C12E8 was found to yield much better separation than other detergents in this case. It is generally believed that the micelle size of the detergent and its

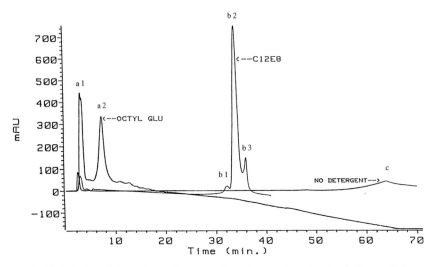

FIG. 6. Effect of detergents on the separation of recombinant enkephalinase. Column: TSK-phenyl-5PW, 0.75 × 7.5 cm. There are three sets of buffer conditions: (1) A1 = 1.2 M ammonium sulfate + 0.1 M potassium phosphate + 0.8% octylglucoside + 20% glycerol, pH 7, and B1 = 0.1 M potassium phosphate + 0.8% octylglucoside + 20% glycerol, pH 7; (2) A2 = 1.2 M ammonium sulfate + 0.1 M potassium phosphate + 0.005% C12E8 + 20% glycerol, pH 7, and B2 = 0.1 M potassium phosphate + 0.04% octaethylene glycol + 20% glycerol, pH 7; (3) A3 = 1.2 M ammonium sulfate + 0.1 M potassium phosphate + 20% glycerol, pH 7, and B3 = 0.1 M potassium phosphate + 20% glycerol, pH 7. Elution is the same for these three sets of buffers, from 100% A to 100% B in 60 min with an additional 10 min at 100% B; temperature, 25°; flow, 0.8 ml/min. Peak identification: a1, high molecular weight impurity; a2, enkephalinase; b1, b2, and b3, enkephalinase and its variants; c, enkepha-linase.

A

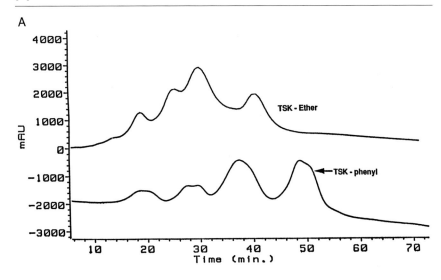

B

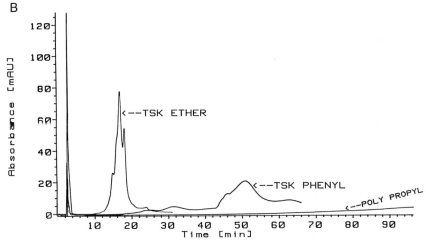

FIG. 7. (A) Effect of detergent on the separation of recombinant tissue plasminogen activator (rtPA) variants. All conditions as Fig. 5a. (B) Shown is an overlay of the separation of rtPA variants by three type of HIC columns. Column A, TSK-ether-5PW; column B, TSK-phenyl-5PW; column C, polypropyl A. All other conditions as in (A) except for the absence of detergent additives in mobile phase B.

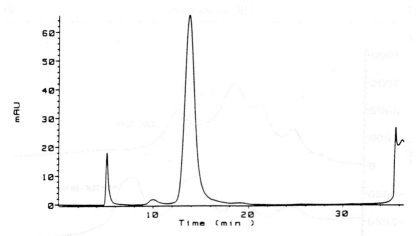

FIG. 8. Effect of detergent and organic modifier on the separation of recombinant human growth hormone variants. Column: TSK-phenyl-5PW, 0.75 × 7.5 cm; 60-min linear gradient from 0.5 M sodium sulfate + 0.03 M Tris-HCl, 0.5% acetonitrile, pH 8, to 0.03 M Tris-HCl, 5% acetonitrile + 0.07% Brij 35 [polyoxyethylene (23) lauryl ether; C12E23]; flow, 0.5 ml/min; temperature, 25°. (Reprinted from *J. Chromatogr.*, **500**, S.-L. Wu, W. S. Hancock, B. Pavlu, and P. Gellerfors, p. 595, Copyright 1990 with kind permission of Elsevier Science–NL, Sara Burgerhartstraat 25, 1055 KV Amsterdam, The Netherlands.)

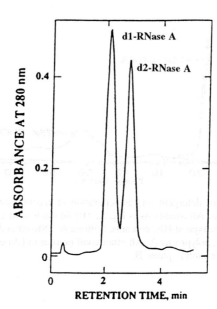

FIG. 9. Separation of two deamidated ribonuclease A variants labeled as d1-RNase A (isoaspartic acid form) and d2-RNase A (aspartic acid form) by HIC. Column, Spherogel HIC-CAA; 10-min linear gradient from 2.1 M ammonium sulfate + 0.5 M ammonium acetate, pH 6, to 0.9 M ammonium sulfate + 0.5 M ammonium acetate, pH 6; flow, 1 ml/min; temperature, 25°. (Reprinted with permission from Ref. 59. Copyright the American Society for Biochemistry and Molecular Biology.)

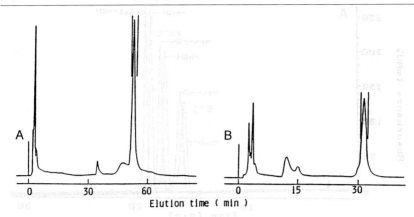

FIG. 10. Separation of monoclonal antibody from ascites fluid by two types of HIC column. Column A, TSK-phenyl-5PW, 0.75 × 7.5 cm; column B, TSK-ether-5PW, 0.75 × 7.5 cm; 60-min linear gradient from 1.5 M ammonium sulfate + 0.1 M sodium phosphate, pH 7, to 0.1 M sodium phosphate, pH 7; flow, 1 ml/min; temperature, 25°; sample, anti-chicken 14K lectin, diluted ascites fluid (1.5 mg in 100 μl). (From Ref. 68a.)

particular binding to the protein are important for protein separations.[33,42,55–58] Figure 7A (page 41) shows that this detergent (C12E8) can even elute recombinant tissue plasminogen activator (rtPA), a glycoprotein with a molecular mass of 66,000 Da, much better than without detergent in the mobile phase (Fig. 7B). A poor recovery and low resolution are shown in phenyl, ether, and polypropyl columns without C12E8 detergent (Fig. 7B), while a much higher recovery and resolution are obtained with the C12E8 detergent (Fig. 7A). There are two advantages in the use of C12E8: (1) the separation window is widened by adding the detergent; and (2) the micelle size (approximately 120 molecules of C12E8; 60,000 Da) is not too large and thus does not fully encapsulate rtPA, which could result in loss in resolution.

Figure 8 presents another example of the separation of recombinant human growth hormone (rhGH) and its variants. The peaks eluting at 10, 14, and 19 min have been shown to be (1) a desPhe variant in which the N-terminal phenylalanine is lacking, (2) intact rhGH, and (3) a two-chain form in which the peptide bond is cleaved between residue 142 and 143 (Thr-142 and Tyr-143), respectively.[38] The HIC column is the same as used for the rtPA separation in Fig. 7 (TSK-phenyl-5PW). However, the mobile-

[55] L. M. Hjelmeland and A. Chrambach, *Methods Enzymol.* **104**, 305 (1984).
[56] L. M. Hjelmeland and A. Chrambach, *in* "Membranes, Detergents, and Receptor Solubilization," p. 35. Alan R. Liss, New York, 1984.
[57] D. Josic, W. Hofmann, and W. Reutter, *J. Chromatogr.* **371**, 43 (1986).
[58] K. L. Kadam, *Enzyme Microb. Technol.* **8**, 266 (1986).

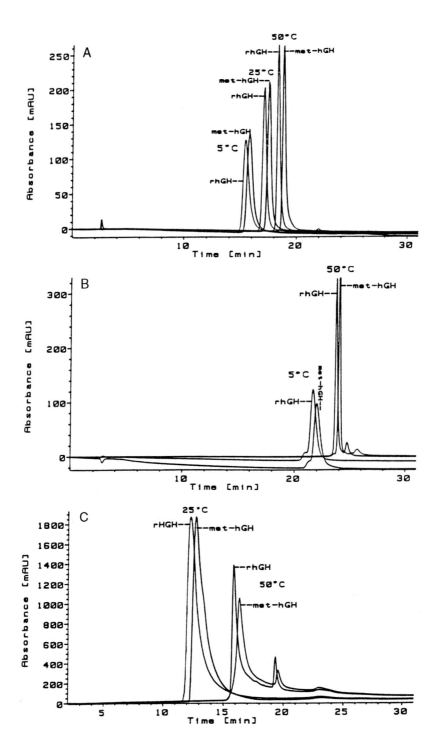

phase conditions are somewhat different, with the addition of small amounts of acetonitrile and Brij 35 detergent. This HIC system can also easily differentiate the two forms of human growth hormone, methionyl (Met)-hGH and rhGH, as shown previously in Fig. 3. Other additives such as ethylene glycol, glycerol, metal ions (e.g., calcium or magnesium ions), urea, and guanidine hydrochloride have been utilized in HIC for various protein separations.[31,33,42] One point that needs to be emphasized is that detergents and several other additives can adhere to HIC surfaces. It is important to wash with pure water or 0.1 N sodium hydroxide for 10 to 15 min, depending on the extent of adherence, between each run for reproducible retention results.

As another example of the power of HIC, Fig. 9 (page 42) shows the separation of different forms of deamidated ribonuclease A.[59] One (d1-RNase A) has an isoaspartic acid and the other has an aspartic acid (d2-RNase A) at residue 67, as a substitution for asparagine. Moreover, in this work, the slight difference in peptide backbone between aspartic acid and isoaspartic acid was differentiated by HIC. It has been further shown that the relevant residues are located on the outer surface of the protein, and these correlate with biological binding activity. Thus, in this example, HIC not only can separate the variants but can also quantitate biological activity.[59]

A number of applications have appeared on the use of HIC to purify receptors,[60a,b] labile enzymes,[61] hormones,[62-64] and antibodies.[65-68] Figure

[59] A. D. Donato, M. A. Ciardiello, M. de Nigris, R. Piccoli, L. Mazzarella, and G. D'Alessio, *J. Biol. Chem.* **268**, 4745 (1993).

[60a] S. M. Hyder, R. D. Wiehle, D. W. Brandt, and J. L. Whitcliff, *J. Chromatogr.* **327**, 237 (1985).

[60b] S. L. Hyder, N. Sato, and J. L. Whitcliff, *J. Chromatogr.* **397**, 251 (1987).

[61] V. Stocchi, P. Cardoni, P. Ceccaroli, G. Piccoli, L. Cucchiarini, R. D. Bellisand, and M. Dacha, *J. Chromatogr.* **676**, 51 (1994).

[62] N. Sakihama, M. Shin, and H. Toda, *J. Biochem.* **100**, 43 (1986).

[63] J. Hiyama, A. Surus, and A. G. C. Renwick, *J. Endocrinol.* **125**, 493 (1990).

[64] M. A. Chlenov, E. I. Kandyba, L. V. Nagornaya, I. L. Orlova, and Y. V. Volgin, *J. Chromatogr.* **631**, 261 (1993).

[65] S. C. Goheen and S. C. Engelhorn, *J. Chromatogr.* **326**, 235 (1985).

[66] S. A. Berkowitz and M. P. Henry, *J. Chromatogr.* **389**, 317 (1987).

[67] E. Harlow and D. Lane, "Antibodies: A Laboratory Manual." Cold Spring Harbor Laboratory, Cold Spring Harbor, NY, 1988.

FIG. 11. Effect of temperature on the separation of Met-hGH and rhGH on the TSK-ether-5PW, 0.75 × 7.5 cm (A), polypropyl A, 0.46 × 15 cm (B), and Spherogel HIC-CAA, 0.46 × 15 cm (C) columns. Met-hGH and rhGH were injected separately and overlaid together in the same chromatogram. All buffer and elution conditions as in Fig. 4. (Reprinted from *J. Chromatogr.*, **500**, S.-L. Wu, W. S. Hancock, B. Pavlu, and P. Gellerfors, p. 595, Copyright 1990 with kind permission of Elsevier Science–NL, Sara Burgerhartstraat 25, 1055 KV Amsterdam, The Netherlands.)

10^{68a} (page 43) shows the separation of a monoclonal antibody from ascites fluid by means of a hydrophobic phenyl column (Fig. 10A) and a less hydrophobic ether column (Fig. 10B). The ether column appears better suited for the monoclonal antibody separation. The large molecular mass of antibody (approximately 150 kDa) possesses a large hydrophobic surface (or force). Therefore it has a stronger interaction with a more hydrophobic column (e.g., phenyl column), which resulted in a broader or tailing peak (slower desorption rate) and less recovery than a less hydrophobic column (e.g., ether column).

As discussed earlier, a change in column temperature or pH can alter protein conformation and potentially change chromatographic behavior. One can thus use conformational manipulation to optimize the separation pattern. In general, higher temperature will yield higher efficiency (sharper peaks) and lower back pressure of the column (see Fig. 11A, page 44). A temperature study is highly recommended for the development of a separation protocol (e.g., 10 through 60°). One must, of course, be cautious not to use too high a temperature to denature the protein, which would produce a broad peak or multiple peaks with low protein recovery (see Fig. 11C). pH, in theory, should not significantly influence hydrophobic interactions since only the surface charges of the protein change with pH. However, the ionization of specific amino acids can have a significant effect on retention if they are located in the chromatographic contact region. Retention in general is expected to be changed most dramatically with pH near the isoelectric point of the protein. This effect is also observed in the salting out of proteins.[69] However, owing to solubility considerations, the recovery of a protein is often low at or near its pI.[42,47] In general, HIC can be performed over a wide range of pH values.[21,70–73] Furthermore, the elution program in HIC, as in other modes of HPLC, can affect separation. In general, a longer (i.e., shallower) gradient will resolve a mixture better than a shorter (sharper) one. However, the price to be paid is broader peaks and longer retention times when using a shallower gradient. A smaller particle size (e.g., 5 μm) will yield a higher peak efficiency than a larger particle (e.g., 20 μm) with a tradeoff in higher back pressure and cost. A longer column length (e.g., 250 mm) will have a slight advantage in separa-

[68] A. H. Guse, A. D. Milton, H. Schulze-Koops, B. Muller, E. Roth, B. Simmer, H. Wachter, E. Weiss, and F. Emmrich, *J. Chromatogr.* **661**, 13 (1994).
[68a] Y. Yamasaki, T. Kitamura, and Y. Kato, Poster 420, 9th International Symposium on HPLC of Proteins, Peptides and Polynucleotides, Philadelphia, November 6–9, 1989.
[69] M. Dixon and E. C. Webb, *Adv. Protein Chem.* **16**, 197 (1961).
[70] P. Strop, D. Lechova, and V. Tomesek, *J. Chromatogr.* **259**, 255 (1983).
[71] N. T. Miller and B. L. Karger, *J. Chromatogr.* **326**, 45 (1985).
[72] H. Engelhardt and U. Schon, *J. Liquid Chromatogr.* **9**(15), 3225 (1986).
[73] Y. Kato, T. Kitamura, and T. Hashimoto, *J. Chromatogr.* **298**, 407 (1984).

tion relative to a shorter one (e.g., 150 mm), with a tradeoff in a slightly broader peak and longer retention time. A linear relationship exists between the gradient time and retention, peak capacity, and peak height.[71] In this study, the effect of column length and particle diameter can also be understood for separation.

Conclusion

In this chapter, general guidelines concerning separation conditions in HIC are discussed with a variety of examples (e.g., *Escherichia coli*-expressed proteins, glycosylated proteins, membrane proteins, and antibodies). Protein structure and conformation play a central role in determining chromatographic retention and behavior. Conformational effects (e.g., choices of column temperature and type vs protein structure and hydrophobicity) are important tools with which to manipulate protein structure for variant separation. The introduction of new matrix supports (e.g., perfusive and nonporous particle) has enhanced the speed of separation dramatically (from hours to a few minutes). To enlarge the separation window and increase selectivity, mobile-phase additives (e.g., detergents, metal ions, urea, guanidine hydrochloride, ethylene glycol, glycerol, and organic solvents) are particularly important and are widely applied.

[3] Ion-Exchange Chromatography

By GARGI CHOUDHARY and CSABA HORVÁTH

Introduction

Ion-exchange chromatography has been the most widely used technique for the isolation and purification of biological macromolecules since the introduction of cellulosic ion exchangers in the 1950s. These hydrophilic sorbents with fixed ionogenic functions facilitated the separation of proteins and other biopolymers without denaturation and with relatively high selectivity and resolution. The availability of pure proteins was essential for structural and other studies that laid down the foundation for substantial advances in biochemistry, biophysics, and molecular biology in the ensuing years. Conventional anion-exchange resins were already being used in the late 1940s for the separation of small nucleic acid fragments. Later, partition chromatography was employed in the separation of large nucleic acids by ion exchange.

The first dedicated liquid chromatograph, the amino acid analyzer, was introduced in 1957. This forerunner of the present high-performance liquid chromatography (HPLC) instrument was used to analyze the amino acid composition of protein hydrolysates and the composition of peptide mixtures by ion-exchange chromatography, using columns packed with microparticulate cation-exchange resins. The construction of the amino acid analyzer, the first "automated" instrument for determining the amino acid composition of the proteins and peptides, was an important step in the history of chromatography and life sciences in general.

The mid-1960s brought about the development of high-performance liquid chromatography, employing sophisticated instrumentation with high-sensitivity detectors and high-efficiency columns for the separation of non-volatile substances with the speed, resolution, sensitivity, and convenience of gas chromatography. During the 1970s, the main driving force to the development of HPLC was the need for the analysis of small drug molecules mainly in the pharmaceutical industry. In the 1980s, advances in biotechnology and the concomitant need for high bioanalytical performance were responsible for the establishment of HPLC as a widely used tool for the analysis of biopolymers in science and technology.

In the 1980s, a specialized technique of HPLC dedicated to protein chromatography was introduced: fast protein liquid chromatography (FPLC; Pharmacia, Uppsala, Sweden). Since then the instrument has catered to the specific needs of the protein chemists, and these liquid chromatographs have become popular and have greatly contributed to the acceptance of HPLC as a high-resolution technique in protein chemistry.

In the 1990s, we are again witnessing the introduction of new types of high-performance liquid chromatographs especially designed for fast protein chromatography. These instruments employ novel columns, but their striking feature is the level of computerization that serves not only as a means to control the analytical system but also to facilitate rapid method development. Ion-exchange chromatography has contributed to the progress in life sciences and biotechnology in a major way, serving in pre-HPLC times mainly as a preparative method. It is a highly versatile and selective separation method and is not only a high-resolution analytical tool but also a superior (micro) preparative technique.

Over the years electrophoretic techniques were the leading bioanalytical tools for biopolymers. Slab gel electrophoresis has yielded spectacular results in resolving closely related nucleic acids of high molecular mass. Isoelectric focusing and sodium dodecyl sulfate-polyacrylamide gel electrophoresis (SDS–PAGE) have served well the analytical needs of life scientists and biotechnologists in protein analysis. Electrophoresis has undergone the same metamorphosis as did liquid chromatography at the advent of

HPLC. As a result, capillary zone electrophoresis (CZE) has emerged as an instrumental bioanalytical technique of formidable potential. In the future we may expect growing competition between high-performance ion-exchange chromatography also in the form of capillary electrochromatography and capillary electrophoresis.

This chapter is devoted to the employment of ion-exchange chromatography in the analysis of biological macromolecules. A brief review of the fundamentals is followed by a discussion of the various modes of ion-exchange chromatography. Then the stationary phases employed in the ion-exchange columns, which is the heart of the chromatographic system, are described with particular attention given to the novel ion exchangers introduced for high-speed protein separations. Details regarding methodology and method development are interspersed throughout the text.

Retention in Ion-Exchange Chromatography

Effect of Salt

Salt is the most commonly used modulator in ion-exchange chromatography. Boardman and Partridge[1] used the mass action law to describe the reversible binding of proteins in the presence of salt in the mobile phase to the fixed ionic groups at the surface of the ion-exchanger stationary phase. According to their stoichiometric displacement model, the binding of a charged protein molecule to the oppositely charged surface of the stationary phase is accompanied by a simultaneous expulsion of an equivalent number of counterions. Therefore, the binding equilibrium in the absence of specific salt effects can be described[2,3] as follows:

$$PC_{N_C} + N_S\overline{S} \rightleftharpoons \overline{P} + N_S S + N_C C \tag{1}$$

where P is the protein in mobile phase, C is the monovalent coion to the protein, S is the salt counterion, and N_C and N_S are the number of coions and counterions involved in the exchange process. Species that are bound to the stationary phase are denoted by overbars.

The chromatographic retention factor k' is defined as

$$k' = \phi K \tag{2}$$

[1] N. K. Boardman and S. M. Partridge, *Biochem. J.* **59,** 543 (1955).
[2] W. Kopaciewicz, M. A. Rounds, J. Fausnaugh, and F. E. Regnier, *J. Chromatogr.* **366,** 3 (1983).
[3] A. Velayudhan and Cs. Horváth, *J. Chromatogr.* **367,** 160 (1986).

where ϕ is the stationary-to-mobile phase ratio and K is the equilibrium constant for the displacement process according to Eq. (1). The dependence of retention factor in ion-exchange chromatography on the salt concentration within the hermeneutics of the stoichiometric displacement model has been expressed[3] as

$$\log k' = \text{const.} - (Z_P/Z_S) \log[S] \tag{3}$$

where Z_P is the characteristic charge on the protein and is defined as the number of bound monovalent counterions that are freed when one protein molecule binds to the surface. The number of salt counterions of the valence, Z_S, that are expelled on the binding of one molecule of protein according to the stoichiometric exchange equals Z_P/Z_S, and it need not be an integer. The slope of the plot of $\log k'$ versus the logarithm of salt concentration yields the characteristic charge, Z_P, that is not the net charge on the protein, although the slope has been found to reflect the net charge on the eluite molecules for small ionic species.[4] Such plots offer a useful means to organize and analyze experimental data.

This theory is based on the assumption that the protein is bound to the ion exchanger at discrete sites and such a simple stoichiometric law is familiar and appealing to chemists. However, coulombic interactions have a long range and are more appropriately treated by the rules of physics. A theoretical framework elucidating the effect of salt concentration on the retention factor of proteins, based on the electrostatic interactions, has been developed.[5] The theory predicts a linear relation between the logarithmic retention factor and reciprocal square root of the ionic strength of the eluting salt. Figure 1 illustrates the two kind of plots used for data analysis according to the stoichiometric and the electrostatic models.

The above-described theories do not explicitly consider the effect on protein retention of hydrophobic interactions that may occur concomitantly with the electrostatic interactions. However, an analysis of the physicochemical phenomena accounting for the effect of hydrophobic interactions was provided[6] by a combination of the salient features of the counterion condensation[7] and solvophobic theories.[8] The result of this approach is a three-parameter equation that describes the effect of salt on the retention of proteins in ion-exchange chromatography as

$$\log k' = A - B \log m_S + Cm_S \tag{4}$$

[4] J. Ståhlberg, *Anal. Chem.* **66**, 440 (1994).
[5] J. Ståhlberg, B. Jönnson, and Cs. Horváth, *Anal. Chem.* **63**, 1867 (1991).
[6] W. R. Melander, Z. El Rassi, and Cs. Horváth, *J. Chromatogr.* **469**, 3 (1989).
[7] G. S. Manning, *Q. Rev. Biophys.* **11**, 179 (1978).
[8] W. R. Melander and Cs. Horváth, *Arch. Biochem. Biophys.* **183**, 200 (1977).

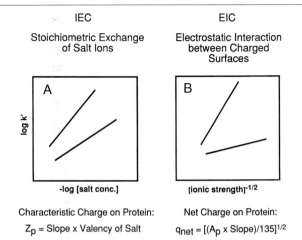

FIG. 1. Schematic illustration of the linear plots for the dependence of the retention factor on the salt concentration according to the two major models for ion-exchange chromatography. (A) Stoichiometric displacement model; characteristic charge on the protein, Z_P, is given by the product of the slope of this plot and the valency of the eluting salt. (B) Nonstoichiometric model based on electrostatic interaction between charged surfaces; the net charge on protein is given by $[(A_P \times \text{slope})/135]^{1/2}$, where A_P is the area of the protein surface that interacts with the stationary phase.

where m_S is the molality of the eluting salt. Parameter B, called the "electrostatic interaction parameter," is given by Z_r/Z_s, so that it depends on the characteristic charge of the protein and the salt counterion. In essence, parameter B, governs the change in retention with the salt concentration in ion-exchange chromatography. Parameter C, termed the "hydrophobic interaction parameter," is determined by the contact area on protein binding at the chromatographic surface and the properties of the eluting salt. It is given by the limiting slope of the log k' versus m_S plot at sufficiently high salt concentrations.

As seen in Fig. 2A, a plot of log k' against log m_S yields U-shaped plots. The limiting slope of the plots at low salt concentrations gives the electrostatic interaction parameter. It can be seen that at low salt concentration the retention decreases with increase in the ionic strength because of a decrease in electrostatic interactions; however, if the ionic strength is increased beyond a certain value, the hydrophobic interactions become operative and the retention begins to increase. Another approach to the treatment of the combined effect of electrostatic and hydrophobic interactions is based on the combination of the nonstoichiometric theories for

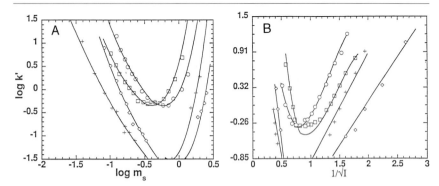

FIG. 2. The dependence of the retention factor on the salt concentration under the combined effect of electrostatic and hydrophobic interactions. Experimental conditions: Column, 80 × 6.2 mm, packed with Zorbax Bioseries WAX-300; flow rate, 1.5 ml/min; isocratic elution with 20 mM phosphate buffer containing ammonium sulfate, pH 6.0; temperature, 25°. (A) Combined effect of electrostatic and hydrophobic interactions[6]; the theoretical parameters A, B, C for the sample proteins are as follows: (+) ribonuclease, −3.99, 3.46, 2.69; ($\diamond$) cytochrome c, −3.83, 4.13, 2.20; ($\square$) α-pchymotrypsinogen A, −3.58, 3.94, 4.16; (○) lysozyme, −3.17, 4.25, 3.12, respectively. (B) Nonstoichiometric theory for the combined effect of electrostatic and van der Waals effect[9]; the column parameters are $A_S/V_0 = 3.4 \times 10^6$ m^2/m^3 and $s_S = -0.220$ C/m^2; the respective parameters s_p, A_p, and H, for sample proteins are as follows: (+) ribonuclease, 0.0387 C/m^2, 4160 Å^2, and 1.24 × 10^{-21} J, ($\diamond$) cytochrome c, 0.0430 C/m^2, 4410 Å^2, 0.94 × 10^{-21} J, ($\square$) α-chymotrypsinogen A, 0.0430 C/m^2, 4834 Å^2, 2.11 × 10^{-21} J, (○) lysozyme, 0.0500 C/m^2, 4410 Å^2, 1.55 × 10^{-21} J.

electrostatic and van der Waals interactions.[9] Figure 2B shows that the plot of log k' against the reciprocal square root of salt concentration also yields U-shaped graphs. Thus, both the stoichiometric and nonstoichiometric theories explain the retention dependence of proteins on salt concentration over a wide range satisfactorily. Nevertheless, ion-exchange chromatography remains a rather complex process and we may keep Helfferich's[10] adage in mind that "the mechanism of separation may be as far removed from the exchange of ions as chromatography is from color."

Effect of pH

In the first approximation the retention of a protein on an ion exchanger depends on its net charge. For this reason the operational pH range is selected according to the isoelectric point of the protein as well as the nature of functional groups present on the support matrix.

[9] J. Ståhlberg, B. Jönnson, and Cs. Horváth, *Anal.Chem.* **64**, 3118 (1992).
[10] F. G. Helfferich, *in* "Advances in Chromatography" (J. C. Giddings and R. A. Keller, eds.), Vol. 1, p. 4. Marcel Dekker, New York, 1965.

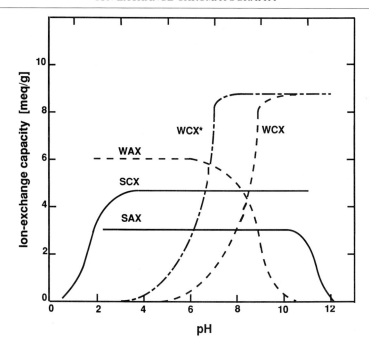

FIG. 3. Schematic illustration of the effect of pH on the ion-exchange capacity of various types of ion exchanger. SAX, Titration behavior of a strong anion exchanger; WAX, titration behavior of a weak anion exchanger; SCX, titration behavior of a strong cation exchanger; WCX, titration behavior of a weak cation exchanger at low salt concentration; WCX*, titration behavior of a weak cation exchanger at high salt concentration.

Ion exchangers are insoluble solid matrices containing fixed ionogenic groups. The two major classes of ion exchangers are cation exchangers and anion exchangers, having negatively and positively charged functional groups, respectively. Strong cation exchangers (SCX) usually contain sulfonic acid groups and strong anion exchangers (SAX) often have quaternary ammonium functions. The ionogenic functions in weak anion exchangers (WAX) are primary, secondary, or tertiary amines, whereas weak cation exchangers (WCX) usually contain carboxylic groups. The terms *strong* and *weak* (borrowed from "strong and weak electrolytes") do not refer to binding strength but rather reflect that "weak" ion-exchanger groups are ionized only in a narrower pH range than the strong ion-exchanger groups. Columns packed with mixed anion and cation exchangers (mixed-bed ion-exchange columns) are also used for the HPLC of proteins.[11]

Figure 3 shows the titration curves for four types of ion exchangers.

[11] Y. F. Maa, F. D. Antia, Z. El Rassi, and Cs. Horváth, *J. Chromatogr.* **443,** 31 (1988).

The titration curve of a sulfonated polystyrene-based cation exchanger (SCX) illustrates that above pH 3, sulfonic acid groups are completely ionized and exhibit maximal ion-exchange capacity. The titration curve of a strongly basic styrenic anion exchanger (SAX) shows that below pH 10 the quaternary ammonium groups are completely ionized and the capacity of the anion exchanger is thus maximal. The titration behavior of a weak ion-exchanger depends on the ionic strength also. The WCX plot represents the titration curve in water at low ionic strength whereas the WCX* plot is typical for the titration at higher ionic strength, e.g., in the presence of 0.1 M NaCl. It is seen that at lower ionic strength the curve is displaced toward higher pH. The comparison of WCX and SCX also illustrates that the exchange capacity of a weak ion exchanger varies considerably with the pH. The WAX curve is the titration curve of a weak anion exchanger that is completely ionized only below pH 6. The charge on weak ion exchangers depends on the pK_a of the fixed ionogenic functions, the mobile-phase pH, and the salt concentration. As the charge densities on both the column and proteins are pH dependent, retention is less predictable on weak ion-exchange columns.

Figure 4 shows the dependence of the protein net charge and the protein retention in ion-exchange chromatography on the pH of the mobile-phase relative to the pI value of the protein. It follows from the titration curve of a protein in Fig. 4A that the net charge of a protein is zero at its isoelectric point. Therefore, when the eluent pH equals the pI of the protein, theoretically there should be no electrostatic interaction between the neutral protein and the chromatographic surface. When the mobile-phase pH decreases below the pI of the protein, the protonation of the accessible amino groups increasingly imparts a cationic character to the protein. This increase in the (positive) net charge results in a stronger retention of the protein on a cation exchanger. Similarly, when the pH increases, the protein acquires more anionic character due to ionization of the carboxyl groups. In this case, both the (negative) net charge of the protein and its retention on an anion exchanger increases. Manipulation of the mobile-phase pH therefore represents an attractive method for regulating the degree of protein interaction with a particular stationary phase in ion-exchange chromatography. The solubility of many proteins drastically diminishes in the vicinity of their isoelectric points; therefore, care must be exercised to avoid isoelectric precipitation on the column. The solubility of the sample components at the pH and salt concentrations to be used during the separation should always be tested in advance. However, contrary to the "ideal behavior" illustrated in Fig. 4B some degree of retention may occur at a mobile phase pH equal to the pI value of the protein in anion- and cation-exchange columns. There are two reasons for this. First, the pH in the

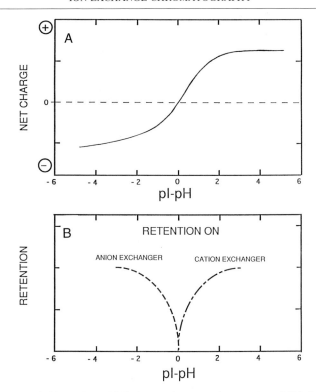

Fig. 4. Schematic illustration of (A) the net charge on the protein and (B) the chromatographic retention factor as a function of the pH of the mobile phase and p*I* of the protein.

microenvironment of an ion exchanger is not the same as that of the bulk
eluting buffer because of the Donnan exclusion in the ion exchanger. The
pH inside the resin is usually about 1 unit higher and lower than that of
the eluting buffer in an anion and a cation exchanger, respectively.[12] Such
differences in the microenvironmental pH suffice for significant protein
retention to occur even though the bulk pH of the mobile phase equals
the p*I* of the protein. Second, nonuniform distribution of the charged groups
on the protein surface may also cause the characteristic charge to deviate
from the net charge.[2] In preparative chromatography it may be advisable
to carry out simple test tube experiments by equilibrating a small quantity
of the ion exchanger with the starting buffer chosen and then with the
protein sample in the same buffer. After centrifugation, the supernatant is

[12] R. K. Scopes, in "Protein Purification: Principles and Practice" (C. R. Cantor, ed.), 2nd
Ed., p. 152. Springer-Verlag, New York, 1994.

assayed for the protein. Thus, the fraction of protein adsorbed can be determined in order to assess the strength of binding and from that the operating conditions can be chosen accordingly. Modern instrumentation facilitates such experiments by allowing the rapid chromatographic measurement of the retention of all proteins in the sample under different mobile phase conditions.

Elution Chromatography

Elution chromatography is used most widely in both analytical and preparative ion-exchange chromatography in the laboratory. The sample mixture is introduced into the column as a pulse, and the components are swept by the flowing mobile phase through the column packed with the stationary phase containing fixed ionogenic functions. The components are separated as a consequence of their unequal partitioning between the mobile and the stationary phases and appear at the column outlet as individual peaks. Elution chromatography is termed *isocratic* when the eluent strength, i.e., the salt concentration or the pH of the mobile phase, do not change during the chromatographic run. Most biological molecules of interest are polyelectrolytes and at a given mobile-phase composition will either adsorb strongly to the surface of the ion exchanger or show no tendency to adsorb at all, following the "all or nothing principle" espoused by Tiselius.[13] When the elution windows of the sample components do not overlap, as is often the case with proteins, separation does not occur by isocratic elution. However, by gradually increasing the eluent strength during the chromatographic run, i.e., by using gradient elution, they can be readily separated. The change in eluent strength, i.e., the gradients, can be either continuous or stepwise.

Stepwise elution is tantamount to the consecutive application of several isocratic runs with the eluent strength increased at each step, so that the sample components are eluted sequentially. It is used mainly for recovering a specific component in concentrated form. In this case three elution steps suffice. In the first step, all the components that are bound weakly to the chromatographic surface (light end) are eluted; in the second step, the desired product is recovered; and finally, the stronger binding components (heavy end) are removed from the column. The results of preliminary experiments by gradient elution are useful for development of the stepwise gradient that is preferentially used in macropreparative/process chromatography. When the protein to be captured represents a small fraction of the total protein in the feed solution, the conditions are preferably chosen such

[13] A. Tiselius, *Angew. Chem.* **245,** 67 (1955).

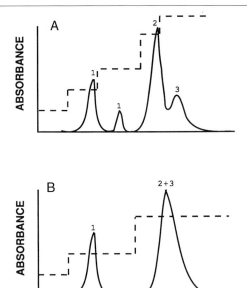

FIG. 5. Pitfalls in the chromatographic separation of three proteins (1, 2, and 3) by stepwise elution. (A) Peak splitting of peak 1; (B) incomplete separation of peaks 2 and 3.

that this component stays bound to the column while contaminants pass through. Thereafter, the thermodynamic conditions, i.e., the salt concentration or pH value, is changed so that the protein of interest is obtained as a peak at the breakthrough of the eluting buffer. When a separation method involving stepwise elution is developed, the choice of the right conditions is particularly important because the efficiency is determined by the selectivity of the system. Figure 5 serves as a caveat by illustrating the results of improper stepwise elution: peak splitting of component 1 or poor resolution of components 2 and 3.

For analytical separations by ion-exchange chromatography linear gradient elution is normally used. Figure 6[13a] shows the effect of gradient shape as measured by the composition (salt concentration or pH) of the eluent entering the column as a function of time on the separation of a multicomponent mixture. Elution by linear salt gradient is the most popular technique in HPLC and more complicated gradient schemes are used for elution only in preparative chromatography. In gradient elution the migra-

---

[13a] L. R. Snyder and J. J. Kirkland, in "Introduction to Modern Liquid Chromatography," 2nd Ed., p. 677. John Wiley & Sons, New York, 1979.

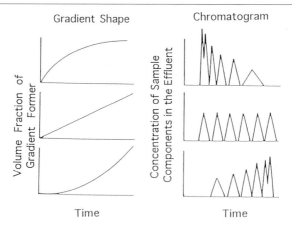

FIG. 6. Schematic illustration of the effect of the gradient shape on the chromatographic separation. (Adapted from Ref. 13a.)

tion velocities of the eluites increase with the eluent strength until they move with the velocity of the mobile phase. The rear of the eluite peak is exposed to a higher eluent strength than the front and therefore the front moves somewhat slower than the rear. This is tantamount to a focusing effect of the gradient elution that results in a reduction of peak tailing and in enhanced separation efficiency.

Salt Gradients

Gradient elution with increasing salt concentration is most commonly used in the ion-exchange chromatography of macromolecules. Figure 7 illustrates plots of the logarithmic retention factor, k', versus the salt concentration for pairs of two small and two large polyelectrolyte molecules, like proteins. It can be seen that for macromolecules, a small change in salt concentration results in an abrupt change in the retention. Isocratic elution, therefore, is not of much use in protein separation and the employment of gradient elution is imperative. Figure 8 shows the separation of a mixture of β-lactoglobulins A and B under isocratic and gradient elution conditions. It can be seen that the column loading capacity is greater and peak splitting is smaller with gradient than with isocratic elution. The average retention factor of an eluite under conditions of gradient elution can be evaluated from the gradient steepness parameter, b, as

$$\bar{k} = 1/1.15b \qquad (5)$$

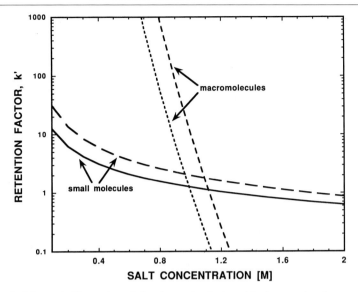

FIG. 7. Schematic illustration of the change in the logarithmic retention factor of small ionized molecules and large polyelectrolytic eluites with the change in the concentration of the eluting salt.

The gradient steepness parameter depends on the characteristic charge of the protein, the flow velocity, the gradient time, the column volume, and the initial and final salt concentration. It has been shown[14] that $\bar{k}$ is the same as the k' value of the eluite band halfway in the column. To optimize a separation it is important to consider the objectives of the experiment, since the desired features of a separation, i.e., speed, resolution, and column loading capacity, are often mutually exclusive. In ion-exchange separations using salt gradients the speed of separation is determined by the gradient steepness parameter. Shallow gradients (small b) will give maximum resolution but the separation time and band broadening are concomitantly increased. However, although steep gradients facilitate faster chromatography and may give sharper peaks, their resolution is usually inferior. In practice, the gradient volume is kept constant and an increase in resolution is obtained by using more efficient columns. Gradients with salt concentration varying from 0.05 to 1 M over 10–20 column volumes at the flow rate recommended for the ion-exchanger stationary phase are used for initial investigative experiments. In gradient elution, the total analysis time is the sum of the retention time of the last peak and the column regeneration

[14] R. W. Stout, S. I. Sivakoff, R. D. Ricker, and L. R. Snyder, *J. Chromatogr.* **353,** 439 (1986).

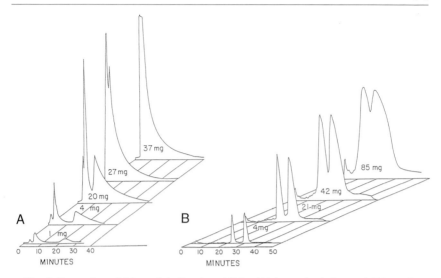

FIG. 8. Separation of β-lactoglobulins A and B by (A) isocratic elution and (B) gradient elution. Column, 75 × 7.5 mm, packed with 10-μm porous TSKgel DEAE 5-PW; flow rate, 1 ml/min. Isocratic elution with 130 mM phosphate buffer, pH 7.0. Gradient elution from 25 to 263 mM phosphate buffer, pH 7.0, in 40 min followed by isocratic elution for 5 min; sample loop, 0.5 ml; temperature, 22°. (Reproduced with permission from Ref. 20.)

time, i.e., the time it takes after the gradient run is completed to bring the column back to its initial conditions. Regeneration requirements vary with the support and the functional groups of the ion-exchanger stationary phase. Manufacturers usually recommend the appropriate regeneration procedure. It normally consists of washing with 1–5 column volumes of 0.1–0.5 M NaOH followed by rinsing with 5 column volumes of a concentrated salt solution, e.g., 2 M NaCl. The removal of lipoproteins and membrane lipids may require the additional use of nonionic detergents or ethanol. In any cleanup method, reversal of flow direction can be employed to flush out particulates.

pH Gradients

The use of a pH gradient offers another way to elute proteins in ion-exchange chromatography. It can be formed like salt gradients, by mixing two or more buffers so that the pH of the eluent entering the column changes with time as desired. The generation of such "external" pH gradients in a reliable and reproducible manner is considered to be rather difficult. For this reason the introduction of "internal" pH gradients generated in an

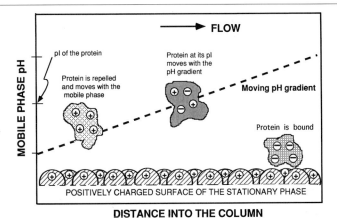

FIG. 9. Schematic illustration of the principle of chromatofocusing.

ampholytic mobile phase led to chromatofocusing,[15,16] a particular branch of ion-exchange chromatography. The development of this technique was inspired by isoelectric focusing. However, in chromatofocusing the combination of a flow field and an ion-exchanger stationary phase replaces the electric field of isoelectric focusing. The main feature of this technique is that the internal pH gradient is formed in the mobile phase by suitable ampholytes having high buffering power and low ionic strength. The internal gradient moves down the column with a velocity significantly lower than that of the mobile phase.

Figure 9 schematically illustrates the process of chromatofocusing. Initially the positively charged ion-exchanger column is equilibrated with an ampholyte buffer of high pH. Thereafter, a buffer having a lower pH is introduced and, as it passes through the column, an axial pH gradient is formed due to titration of the fixed groups on the ion exchanger. The pH gradient moves down the column at a velocity lower than that of the mobile phase flow. The proteins are selectively adsorbed when the pH is greater than their pI and they are desorbed when the pH is lower or equal to their pI. It is seen in Fig. 9 that the protein molecule moving down the chromatofocusing column encounters at some axial distance a mobile phase pH that is lower than the pI of the protein. At this point, the protein becomes positively charged and is repelled by the stationary phase. As a result it moves with the mobile phase velocity. Since this velocity is higher than that of the pH gradient, the protein reaches a point where the pH of

[15] L. A. A. Sluyterman and O. Elgersma, *J. Chromatogr.* **150,** 17 (1978).
[16] L. A. A. Sluyterman and O. Elgersma, *J. Chromatogr.* **150,** 31 (1978).

the eluent equals its pI value. The net charge on the protein at this stage is zero, so that it moves with the pH gradient. Protein molecules, which diffuse to a region of higher pH, will acquire a net negative charge and will be adsorbed on the positively charged stationary phase. As the pH gradient moves down the column, to the region where the protein is adsorbed, the local mobile phase pH will be gradually lower and at some point in time it becomes equal to the pI. As a result the protein molecule is desorbed and moves again with the mobile phase to the column outlet. Evidently proteins having different pI values will elute from the column at different times and thus will be separated.

The separation is enhanced by a focusing effect: random molecular diffusion of eluites downstream from the position, where they are neutral, is arrested by their binding to the oppositely charged stationary phase, while diffusion upstream results in the protein moving with the mobile phase velocity, which is faster than the velocity of pH gradient. In both cases the protein molecules that "try to escape" from the band will be returned and thus the band will be sharper than it would be without the focusing effect.

Figure 10[16a] shows the separation of human very low density lipoproteins (VLDLs) by chromatofocusing on a 50 × 5 mm prepacked Mono P column using the fast protein liquid chromatography (FPLC) system from Pharmacia (Uppsala, Sweden). The column was equilibrated with 25 mM Bis–Tris buffer, pH 6.3, containing 6 M urea. The pretreated and dried mixture of VLDL apolipoproteins was dissolved in the buffer overnight. The insoluble part was centrifuged and the supernatant containing about 10 mg of protein or less was applied to the column. The pH gradient was developed with 3 ml of polybuffer 74, pH 4 (Pharmacia) diluted to 30 ml with 6 M urea solution. The chromatographic run was carried out at a flow rate of 1 ml/min and seven fractions were collected in 30 min. From each fraction urea and polybuffer were removed by dialysis and subsequent gel filtration on a Sephadex G-50 column (Pharmacia) with the FPLC system.

Other Modes of Chromatography

Frontal Chromatography

In the frontal mode, the mixture to be separated is fed continuously into the column under conditions that favor the binding of all but one component. This component is obtained in a pure form in the column effluent until the stationary phase is saturated and the other sample compo-

---

[16a] P. Weisweiler, C. Friedl, and P. Schwandt, *Biochim. Biophys. Acta* **875,** 48 (1986).

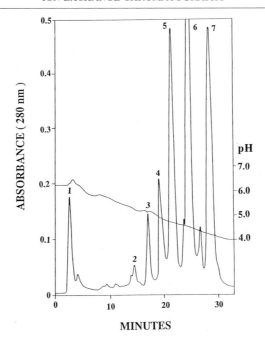

FIG. 10. Separation of human very low density lipoproteins (VLDLs) by chromatofocusing. Column, 50 × 5 mm, packed with 10-μm porous Mono P, equilibrated with 0.025 mM Bis–tris buffer containing 6 M urea, pH 6.3; flow rate, 1 ml/min; pH gradient with 3 ml of Polybuffer 74, pH 4.0, diluted to 30 ml with 6 M urea solution; sample, 8 mg of protein in 6 M urea. (Reprinted from *Biochim. Biophys. Acta*, **875**, P. Weisweiler, C. Friedl, and P. Schwandt, p. 48, Copyright 1986 with kind permission of Elsevier Science–NL, Sara Burgerhartstraat 25, 1055 KV Amsterdam, The Netherlands.)

nents begin to break through. In essence, frontal chromatography is a binary separation process in which only the least retained component can be separated from the others. In preparative chromatography, where the sample occupies a finite length of the column, the process of sample introduction itself is frontal chromatography. The technique can be used not only to recover pure fractions of the least retained component, but also to measure single and multicomponent isotherms[17] that are needed in preparative/ process chromatography. An example of the application of frontal chromatography for protein separation is a technique called isoelectric chromatography.[18] In this case, ion-exchange chromatography is performed consecutively on tandem anion- and cation-exchange columns at a pH where the protein to be purified is not retained. As mentioned before, this pH is not necessarily the isoelectric pH of the protein. The chromatographic process

[17] J. M. Jacobson, J. H. Frenz, and Cs. Horváth, *Ind. Eng. Chem. Res.* **26**, 43 (1987).
[18] P. Petrilli, G. Sannia, and G. Marino, *J. Chromatogr.* **135**, 511 (1977).

results in the adsorption of acidic and basic proteins whereas the protein of interest remains unadsorbed.

Displacement Chromatography

Unlike linear elution chromatography used in analytical work, displacement is a nonlinear chromatographic technique that employs sample load so high that the retention time and peak shape are not independent of the composition and the amount of the sample. This is because displacement chromatography exploits the competition between the sample components for the binding site on the ion-exchange stationary phase to bring about their separation. Although the principles of displacement have been known for more than 50 years, owing to more recent theoretical and practical developments this chromatographic mode has received a new impetus. In displacement chromatography, the sample is loaded onto the column under conditions that allow strong binding of the components to be separated by the stationary phase. Thereafter a displacer, which binds more strongly to the stationary phase than to any of the components, is passed through the column. As the displacer front progresses, the sample components are forced to compete for the adsorption sites and are finally separated into adjacent rectangular bands. When the separation is completed, an isotachic state is reached, and all bands migrate with the velocity of the displacer. In the displacement train thus obtained, the concentration of each component zone is determined by the adsorption isotherm of the corresponding component as well as the isotherm and concentration of the displacer. Because this technique allows an increase in the concentration of a component with respect to its concentration in the feed, displacement chromatography can be used for the enrichment of certain components in trace analysis[19] and in preparative/process applications. The effect of feed and displacer concentration on the separation of β-lactoglobulins A and B is depicted in Fig. 11. The effect of protein loading on the displacement profile is illustrated by comparing Fig. 11a and Fig. 11b. As can be seen, the final pattern was not reached in Fig. 11a, and it suggests that a longer column would be needed for the full development of the displacement train under the operating conditions employed. Figure 11c and Fig. 11d show the effect of displacer concentration. A decrease in displacer concentration from 20 mg/ml of chondroitin sulfate in Fig. 11c to 3 mg/ml in Fig. 11d effected a concomitant decrease in the plateau concentration of the displacement bands. The capacity of the column for the separation of β-lactoglobulins A and B was fivefold greater in the displacement mode than with gradient

[19] J. Frenz, J. Bourell, and W. S. Hancock, *J. Chromatogr.* **512**, 299 (1990).

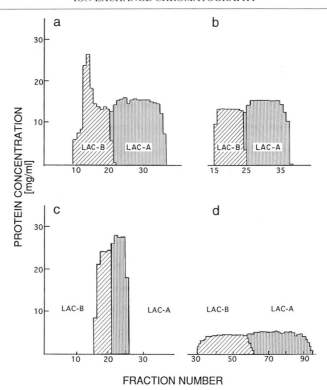

FRACTION NUMBER

FIG. 11. Displacement chromatograms of β-lactoglobulin A and B. Column, 75 × 7.5 mm, packed with 10-μm porous TSK DEAE 5-PW; flow rate, 0.1 ml/min; carrier, 25 mM sodium phosphate, pH 7.0. (a) Feed, 100 mg of mixture; displacer, chondroitin sulfate (10 mg/ml); (b) feed, 78 mg of mixture; displacer, chondroitin sulfate (10 mg/ml); (c) feed, 62 mg of mixture; displacer, chondroitin sulfate (20 mg/ml); (d) feed, 70 mg of mixture; displacer, chondroitin sulfate (3 mg/ml). Sample loop, 4.0 ml; temperature, 22°. (Reproduced with permission from Ref. 20.)

elution.[20] Whereas the theoretical and practical aspects of protein displacement are under active investigation, at present displacement chromatography is mainly used for peptide purification.

Problems associated with displacement chromatography originate from the nonlinear nature of the process, which is much more complex than linear elution both in theory and practice and scarcity of the isotherm data. For ion-exchange displacement chromatography the steric mass action model appears to yield good results.[21] Selecting a proper displacer can also

[20] A. W. Liao, Z. El Rassi, D. M. LeMaster, and Cs. Horváth, *Chromatographia* **24**, 881 (1987).
[21] C. A. Brooks and S. M. Cramer, *AIChE J.* **38**(12), 1969 (1992).

present a problem, particularly when biopolymers must be separated. A large number of displacers for proteins, e.g., carboxymethyldextrans, chondroitin sulfate, carboxymethylstarch, and polyethyleneimine are commercially available, whereas other displacers such as certain low molecular weight dendritic polymers[22] for protein purification by ion-exchange displacement chromatography have not yet been produced commercially. Column regeneration must also be considered when selecting the displacer; for instance, polyethyleneimine binds tenaciously to certain negatively charged stationary phases.

Stationary Phases

Three main types of support materials, i.e., polysaccharide based, silica based, and polymer based, are available for ion-exchange chromatography with packed columns. The choice of support has many aspects, for instance, required particle and pore size, stability, as well as acceptable hydrophobicity and swelling characteristics at the operational range of pH. Because ion-exchange chromatography is a nondenaturing technique and is mainly used for micropreparative and preparative separation, the capacity of the ion exchanger is of major importance. The capacity of the ion exchanger at the operational pH depends on the nature of accessible functional groups and their charge density as well as on the pore size and the molecular weight of the protein. Large protein molecules require matrices with larger pore sizes than those that work well with small proteins. Exclusion limits of ion-exchange matrices are usually given by the manufacturers. Ion-exchange capacity is a dynamic variable and usually decreases when the flow rate is increased above a certain value. Table I (pages 68–71) lists some of the commercially available cation and anion exchangers. Ion exchangers should always be equilibrated with the mobile phase before use. Prepacked high-performance ion-exchange columns are readily equilibrated in the column. If protocols are not provided by the manufacturer, the following equilibration procedure could be adopted. The column is first rinsed with water; thereafter it is washed with 1–5 column volumes of 0.1 M NaOH in the case of an anion exchanger and with 0.1 M HCl in the case of a cation exchanger to titrate all groups. The ion exchanger is then transferred to the correct counterion by appropriate acid or base. Last, the ion exchanger is equilibrated by washing with 1–5 column volumes of the mobile phase.

[22] G. Jayaraman, Yu-Fei Li, J. A. Moore, and S. M. Cramer, *J. Chromatogr.* **102,** 143 (1995).

Polysaccharide-Based Ion Exchangers

In 1956, Sober and Peterson[23] introduced a number of cellulosic ion exchangers for the chromatographic separation of proteins. Since then the weak anion exchanger diethylaminoethyl (DEAE)-derivatized cellulose and the weak cation exchanger carboxymethyl-derivatized cellulose have been most widely used stationary phases. Cellulose-based ion exchangers opened the way to protein chromatography providing excellent separations, high protein recovery, and good loading capacity by the standard of the time. Furthermore, they exhibited satisfactory chemical stability in a wide pH range. Later, ion exchangers based on cross-linked dextran and agarose having spherical shape and uniform particle size distribution as well as relatively high column loading capacity were introduced. A great advantage of these polysaccharide-based ion exchangers is that the integrity of the native protein structure is maintained during the chromatographic separation process. Because the poor mechanical strength of these sorbents has precluded their use at high mobile phase velocities, polysaccharide-based ion exchangers with a relatively high degree of cross-linking have been introduced. Such stationary phases exhibit greatly improved mechanical stability and enhanced flow properties. Moreover, such particles do not significantly swell or shrink with change in ionic strength and pH. It should be noted that in most cases an increase in hydrophobic properties of the support is concomitant to the increase of the degree of cross-linking. The agarose-based ion exchangers generally have larger pores than do dextran ion exchangers. For instance, the exclusion limit for globular proteins is 4 $\times$ 10^6 D[24] for Sepharose Fast Flow, whereas it ranges from 1 to 5 $\times$ 10^5 D for Sephadex-based ion exchangers.

Silica-Based Ion Exchangers

Most silica-based "bonded" stationary phases presently used in HPLC have the organic moieties containing the appropriate ionogenic groups bound to the silica gel surface by siloxane bridges. Silica-based supports have several advantages. The particles are rigid and have favorable pore structure. The technology for their manufacturing is well developed and there are numerous processes for the production of silica particles with a narrow size distribution and with a desired pore structure. Particles having pore sizes greater than 250 Å are used for the separation of proteins. The

[23] H. A. Sober and E. A. Peterson, *J. Am. Chem. Soc.* **78,** 751 (1956).
[24] "Technical Report: Sepharose Fast Flow Ion-Exchangers." Pharmacia LKB Biotechnology, Uppsala, Sweden, 1989.

TABLE I
COMMERCIAL ION EXCHANGERS

Name	Support	Functional group	Particle (μm)	Pore (Å)	Supplier (location)
Strong anion exchangers					
Mono Q	Cross-linked PS/DVB	Quaternary amine	10	—	Pharmacia (Piscataway, NJ)
Source 15Q	Cross-linked PS/DVB	Quaternary amine	15	—	Pharmacia
Q-Sepharose High Performance	Cross-linked agarose (6%)	Quaternary amine	24–44	—	Pharmacia
Fast Flow	Cross-linked agarose (6%)	Quaternary amine	45–165	—	Pharmacia
QAE Sephadex	Cross-linked dextran	Quaternary aminomethyl	40–125	—	Pharmacia
Macro-Prep high Q	Methacrylate copolymer	Quaternary amine	50	1000	Bio-Rad (Hercules, CA)
Macro-Prep Q	Methacrylate copolymer	Quaternary amine	50	1000	Bio-Rad
Poros HQ	Gigaporous PS/DVB	Quaternary amine	10, 20, 50	7000/700[a]	PerSeptive Biosystems (Farmingham, MA)
Q-HyperD	Gel-in-the-cage	Quaternary amine	35–60	—	Sepracore (Marlborough, MA)
QMA Spherosil M	Silica	Quaternary aminomethyl	40–100	1000	Sepracore
QA Trisacryl M	Acrylic polymer	Quaternary amine	40–80	—	Sepracore
PL-SAX 1000 Å	Gigaporous PS/DVB	Quaternary amine	8, 10	1000	Polymer Laboratories (Amherst, MA)
PL-SAX 4000 Å	Gigaporous PS/DVB	Quaternary amine	8, 10	4000	Polymer Laboratories

Name	Matrix	Functional group	Particle size	Pore size	Manufacturer
TSKgel Sugar AX	Cross-linked PS/DVB	Quaternary amine	8, 10	—	TosoHaas (Montgomeryville, PA)
TSKgel SCX	Cross-linked PS/DVB	Quaternary amine	5	—	TosoHaas
SynChropak Q300	Silica	Quaternary amine	6	300	SynChrom (Lafayette, IN)
AminoPac PA1	Cross-linked PS/DVB	Quaternary amine	10	Pellicular[b]	Dionex (Sunnyvale, CA)
CarboPac PA1	Cross-linked PS/DVB	Quaternary amine	10	Pellicular[b]	Dionex
CarboPac PA-100	Cross-linked PS/DVB	Quaternary amine	8, 5	Pellicular[b]	Dionex
NucleoPac PA-100	Cross-linked PS/DVB	Quaternary amine	13	Pellicular[b]	Dionex
ProPac PA1	Cross-linked PS/DVB	Quaternary amine	10	Pellicular[b]	Dionex
IonPac AS10	Cross-linked PS/DVB	Quaternary amine	8, 5	2000	Dionex
Weak anion exchangers					
DEAE-Sepharose Fast Flow	Cross-linked agarose (6%)	Diethylaminoethyl	45–165	—	Pharmacia
DEAE-Sephacel	Cross-linked cellulose	Diethylaminoethyl	40–160	—	Pharmacia
DEAE-Sephacel	Cross-linked dextran	Diethylaminoethyl	40–125	—	Pharmacia
DEAE Bio-Gel A	Cross-linked agarose (4%)	Diethylaminoethyl	80–150	—	Bio-Rad
Macro-Prep DEAE	Methacrylate copolymer	Diethylaminoethyl	50	1000	Bio-Rad
Poros PI	Gigaporous PS/DVB	Polyethyleneimine	20	7000/700[a]	PerSeptive
DEAE Spherodex M	Silica	Diethylaminoethyl	40–80	—	Sepracore
DEA Spherosil M	Silica	Diethylaminoethyl	40–100	1000	Sepracore
DEAE Trisacryl M	Acrylic polymer	Diethylaminoethyl	40–80	—	Sepracore
DEAE Trisacryl Plus M	Acrylic polymer	Diethylaminoethyl	40–80	—	Sepracore
D Zephyr	Silica	Diethylaminoethyl	20	1000	Sepracore
TSKgel DEAE-5PW	Methacrylate copolymer	Diethylaminoethyl	10	1000	TosoHaas
TSKgel DEAE-3SW	Silica	Diethylaminoethyl	10	250	TosoHaas

(continued)

TABLE I (continued)

Name	Support	Functional group	Size Particle (μm)	Size Pore (Å)	Supplier (location)
TSKgelDEAE-2SW	Silica	Diethylaminoethyl	5	125	TosoHaas
TSKgel DEAE-NPR	Methacrylate copolymer	Diethylaminoethyl	2, 5	Pellicular	TosoHaas
SynChropak AX100	Silica	Polyethyleneimine	5	100	SynChrom
SynChropak AX300	Silica	Polyethyleneimine	6	300	SynChrom
SynChropak AX1000	Silica	Polyethyleneimine	7	1000	SynChrom
Strong cation exchangers					
Mono S	Cross-linked PS/DVB	Sulfonic acid	10	—	Pharmacia
Source 15S	Cross-linked PS/DVB	Sulfonic acid	15	—	Pharmacia
SP-Sepharose					
High Performance	Cross-linked agarose (6%)	Sulfopropyl	24–44	—	Pharmacia
Fast Flow	Cross-linked agarose (6%)	Sulfopropyl	45–165	—	Pharmacia
SP-Sephadex	Cross-linked dextran	Sulfopropyl	40–125	—	Pharmacia
Macro-Prep high S	Methacrylate copolymer	Sulfonic acid	50	1000	Bio-Rad
Macro-Prep S	Methacrylate copolymer	Sulfonic acid	50	1000	Bio-Rad
Poros S	Gigaporous PS/DVB	Sulfoethyl	10, 20	7000/700[a]	PerSeptive
Poros HS	Gigaporous PS/DVB	Sulfopropyl	10, 20, 50	7000/700[a]	PerSeptive
PL-SCX 1000 Å	Gigaporous PS/DVB	Sulfonic acid	8, 10	1000	Polymer Laboratories
PL-SCX 4000 Å	Gigaporous PS/DVB	Sulfonic acid	8, 10	4000	Polymer Laboratories
S-HyperD	Gel-in-the-cage	Sulfopropyl	35–60	—	Sepracore
SP Spherodex M	Silica	Sulfopropyl	40–80	—	Sepracore

SP Trisacryl M	Acrylic polymer	Sulfopropyl	40–80	—	Sepracore
SP Trisacryl Plus M	Acrylic polymer	Sulfopropyl	40–80	—	Sepracore
S Zephyr	Silica	Sulfonic acid	20	1000	Sepracore
TSKgel SP-5PW	Methacrylate copolymer	Sulfopropyl	10	1000	TosoHaas
TSKgel SP-NPR	Methacrylate copolymer	Sulfopropyl	2, 5	Pellicular	TosoHaas
TSKgel Aminopak	Cross-linked PS/DVB	Sulfonic acid	5	—	TosoHaas
TSKgel SCX	Cross-linked PS/DVB	Sulfonic acid	5	—	TosoHaas
SynChropak S300	Silica	Sulfonic acid	6	300	SynChrom
SynChropak S1000	Silica	Sulfonic acid	7	1000	SynChrom
IonPac CS10	Cross-linked PS/DVB	Sulfonic acid	8, 5	Pellicular[b]	Dionex
IonPac Fast Cation I	Cross-linked PS/DVB	Sulfonic acid	13	Pellicular[b]	Dionex
Omni PACX 100	Cross-linked PS/DVB	Sulfonic acid	8, 5	Pellicular[b]	Dionex
Omni PACX 500	Cross-linked PS/DVB	Sulfonic acid	8, 5	60	Dionex
Weak cation exchangers					
CM-Sepharose Fast Flow	Cross-linked agarose (6%)	Carboxymethyl	45–165	—	Pharmacia
CM-Sephadex	Cross-linked dextran	Carboxymethyl	40–125	—	Pharmacia
CM Bio-Gel A	Cross-linked agarose (4%)	Carboxymethyl	80–300	—	Bio-Rad
Macro-Prep CM	Methacrylate copolymer	Carboxymethyl	50	1000	Bio-Rad
Poros CM	Gigaporous PS/DVB	Carboxymethyl	20	7000/700[a]	PerSeptive
CM Spherodex M	Silica	Carboxymethyl	40–80	—	Sepracore
CM Trisacryl M	Acrylic polymer	Carboxymethyl	40–80	—	Sepracore
TSKgel CM-5PW	Methacrylate copolymer	Carboxymethyl	10	1000	TosoHaas
TSKgel CM-3SW	Silica	Carboxymethyl	10	250	TosoHaas
SynChropak CM100	Silica	Carboxymethyl	5	100	SynChrom
SynChropak CM300	Silica	Carboxymethyl	6	300	SynChrom

[a] Bidisperse pore size distribution; convection in large pores, diffusion in small pores.
[b] Retentive layer is formed from latex particles carrying ionic functions.

thermodynamic and kinetic properties of silica-supported bonded stationary phases are favorable to obtain high column efficiencies, to control selectivity, and to attain reproducible results.[25] Unlike traditional polymeric ion-exchange resins, siliceous bonded ion exchangers do not swell or shrink on changing ionic strength. Furthermore, they can be used with hydroorganic mobile phases or even with neat organic eluents. They are eminently suitable for the separation of large molecules with relatively high efficiencies but their ion-exchange capacity is comparatively low. The use of silica-based ion exchangers is limited to neutral or acidic eluents, i.e., from pH 2 to pH 8, because the stationary phase disintegrates in contact with alkaline solvents. Table II[25a] illustrates the characteristic properties of typical anion and cation exchangers.

Ion Exchangers with Styrenic or Acrylic Support

We distinguish between two types of polymeric ion exchangers; "soft" gel ion exchangers and composite sorbents with rigid, macroporous support. Soft ion-exchange resins have been used for the ion-exchange chromatography of small molecules since the 1940s. Polystyrene/divinylbenzene copolymers are used as supports with sulfonic acid and with quaternary ammonium functions as cation and anion exchangers, respectively. The particles are compressible, and cannot be used at high column pressures. Moreover, soft gel resins have the disadvantage of swelling and shrinking with changing salt concentration in the eluent. The small pore size precludes the employment of conventional ion-exchange resins for the ion-exchange chromatography of proteins. Ion exchangers with porous and nonporous polymethacrylate support are also available.

Rigid macroporous polymeric supports are prepared by cross-linking copolymerization process in the presence of a porogen. For protein chromatography the hydrophobic macroreticular polymer matrix thus obtained is subsequently coated with a thin hydrophilic layer to which the ionogenic groups are attached. Because of the high degree of cross-linking, such support particles are rigid, do not swell, and the column can withstand high pressure gradients. Polymer-based ion exchangers are chemically stable from pH 1 to pH 13. Advances in the synthesis of such materials led to significant improvements in the properties of polymeric type ion exchangers; that in turn has facilitated the preparation of spherical, microparticulate ion exchangers of a narrow particle size range.

In addition to the microarchitecture of the support material that deter-

[25] I. Halász and I. Sebastien, *Adv. Chromatogr.* **14,** 75 (1976).
[25a] Y. Kato, "Study: Comparison of Ion-Exchange Columns." Tosoh Corporation, Montgomeryville, PA, 1993.

TABLE II
PROPERTIES OF TYPICAL ANION AND CATION EXCHANGERS USED IN PROTEIN CHROMATOGRAPHY[a]

Property	Anion exchangers			Cation exchangers		
	TSKgel DEAE-5PW[b] (Toyo Soda)	Mono Q[c] (Pharmacia)	SynChropak AX-300[b] (SynChrom)	TSKgel SP-5PW[d] (ToyoSoda)	Mono S[d] (Pharmacia)	Synchropak CM-300[e] (SynChrom)
Support	Macroreticular polymer with hydrophilic surface	Macroreticular polymer with hydrophilic surface	Silica	Macroreticular polymer with hydrophilic surface	Macroreticular polymer with hydrophilic surface	Silica
Particle size (μm)	10	10	10	10	10	10
Ionogenic functions	$-CH_2CH_2N(C_2H_5)_2$	$-CH_2N^+(CH_3)_3$	$-NHCH_2CH_2-$	$-CH_2CH_2CH_2SO_3^-$	$-CH_2SO_3^-$	$-CH_2COO^-$
Operational pH range	2–12	2–12	2–8	2–12	2–12	2–8
pK_a	11.3	11.4	NA	2.5	2.6	NA
Binding capacity (mg/ml)	30[f]	65[f]	30[f]	40[g]	75[h]	NA
Exclusion limit for polyethylene glycol (Da)	1,000,000	500,000	150,000	1,000,000	500,000	150,000

[a] Adapted with permission from Ref. 25a.
[b] Weak anion exchanger.
[c] Strong anion exchanger.
[d] Strong cation exchanger.
[e] Weak cation exchanger.
[f] Bovine serum albumin (BSA).
[g] Hemoglobin.
[h] Immunoglobulin G (IgG).

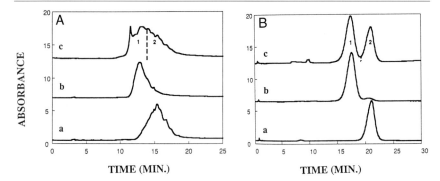

Fig. 12. Chromatography on two strong cation exchangers of (a) rhDNase, (b) its deaminated variant, and (c) an admixture of the two. Columns: (A) 75 × 7.5 mm, packed with 10-
μm TSK SP-5PW, (B) 50 × 4.6 mm packed with 5-μm LiChrosphere SO_3^-. Flow rate, 1.0
ml/min; gradient elution from 0 to 580 mM NaCl and 1 mM $CaCl_2$ in 10 mM sodium acetate,
pH 4.5, in 26 min. (Reprinted from *J. Chromatogr.*, **634**, J. Frenz *et al.*, p. 229, 1993, with
kind permission of Elsevier Science–NL, Sara Burgerhartstraat 25, 1055 KV Amsterdam,
The Netherlands.)

mines the mechanical and mass transfer properties of the ion exchangers,
the molecular architecture of the fixed ionogenic moieties at the chromatographic surface also plays a significant role in determining the efficacy of
separation. This is exemplified in Fig. 12 by the separation of recombinant
human deoxyribonuclease I (rhDNase) and its deaminated variant on two
different ion exchangers. Whereas a conventional cation-exchange column
(Fig. 12A) could not resolve these closely related protein variants, they
could be separated on a column packed with a tentacle-type cation exchanger (Fig. 12B). It has been assumed that the longer anionic ligates of
the tentacle ion exchanger could gain access to the interior portion of
the protein.[26]

Special Configurations of Stationary Phase

As the sample molecules move through the ion-exchanger column, band
spreading occurs as a result of longitudinal diffusion, flow maldistribution,
resistance to mass transport, and slowness of sorption kinetics. Slowness
of diffusion inside the anfractuous pore system of stationary phase particles
is the major source of bands spreading for macromolecules because of
their small diffusion coefficients. The intraparticulate diffusion resistance
is reduced by decreasing the particle diameter and there are various tech-

[26] J. Frenz, J. Cacia, C. P. Quan, M. B. Sliwkowski, and M. Vasser, *J. Chromatogr.* **634**,
229 (1993).

nologies available to produce particles with uniform size distribution in the 3- to 20-μm diameter range. However, columns packed with small particles have high flow resistance due to their low permeability and thus require high column inlet pressures, despite the relatively short column lengths.

Pellicular Configuration

An early approach to reduce stationary phase mass transfer resistances was to form thin ion-exchange shells on the surface of fluid impervious core, e.g., glass beads. Such pellicular anion and cation exchangers[27] with an average particle size of 40 μm were first introduced in the 1960s. At the beginning, pellicular stationary phases dominated HPLC; however, with the availability of methods to obtain uniform 5- to 10-μm particles, they were largely replaced by totally porous microparticulate bonded phases for the separation of small molecules. Micropellicular stationary phases having a small particle diameter ($d_p = 1.5–5$ μm) regained interest in the 1980s for rapid separation of biological macromolecules.[28,29] The main advantage of pellicular ion-exchange resins is the rapid mass transfer in the actual ion-exchange stationary phase having the configuration of a spherical annulus, because the diffusional path is limited to the thin exterior retentive shell. In addition, the impervious core imparts high mechanical strength to the stationary phase to withstand high pressures. However, pellicular sorbents have relatively low capacity for small molecules, although this disadvantage is moderated with increasing molecular weight of the eluite. Micropellicular stationary phases facilitate rapid separations as seen by comparing the chromatograms depicted in Fig. 13A[29a] and Fig. 13B.[29b] Figure 13A shows the separation of peptide mixture containing eight components on a cation-exchange column packed with 5-μm porous polymeric particles, whereas Fig. 13B shows the separation of a similar mixture on a cation-exchange column packed with 2.5-μm pellicular particles.

At present the major application of pellicular anion exchangers is in the chromatography of carbohydrates at a pH higher than pH 12, where most carbohydrates become anionic and can be separated on a column packed with a strong anion exchanger. Because carbohydrates undergo chemical changes on prolonged exposure to strong alkali, the analysis time must be short. Micropellicular anion exchangers with a thin layer of latex

[27] Cs. Horváth, B. Preiss, and S. R. Lipsky, *Anal. Chem.* **39,** 1422 (1967).

[28] K. Kalghatgi, *J. Chromatogr.* **499,** 267 (1990).

[29] Y. H. Maa, S. Lin, Cs. Horváth, U. Yang, and D. M. Crothers, *J. Chromatogr.* **508,** 61 (1990).

[29a] A. J. Alpert and P. C. Andrew, *J. Chromatogr.* **443,** 85 (1988).

[29b] Y. Kato, S. Nakatani, S. Kitamura, T. Onaka, and A. Hashimoto, *J. Chromatogr.* **513,** 384 (1990).

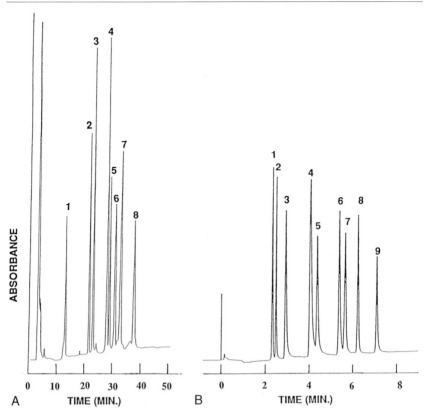

Fɪɢ. 13. Peptide separation on strong cation exchangers. (A) Column, 200 × 4.6 mm, packed with 5-μm porous PolySULFOETHYL Aspartamide; flow rate, 0.7 ml/min; linear gradient from 0 to 250 mM KCl in 5 mM KH$_2$PO$_4$, pH 3.0, containing 25% (v/v) acetonitrile in 42 min; sample, 5 mg of each peptide in 50 ml. Peaks: (1) Oxytocin, (2) [Arg^8]vasopressin, (3) somatostatin, (4) substance P, free acid, (5) Substance P, (6) bovine pancreatic polypeptide, (7) anglerfish peptide Y, (8) human neuropeptide Y. [(A) reprinted from *J. Chromatogr.*, **443**, A. J. Alpert and P. C. Andrew, p. 85, 1988, with kind permission of Elsevier Science–NL, Sara Burgerhartstraat 25, 1055 KV Amsterdam, The Netherlands.] (B) Column, 35 × 4.5 mm, packed with 2.5-μm pellicular TSKgel SP-NPR; flow rate, 1.5 ml/min; linear gradient from 0 to 250 mM Na$_2$SO$_4$ in 20 mM acetate buffer, pH 3.5, containing 60% (v/v) acetonitrile in 30 min; sample, 0.25 mg for components 2, 3, and 5 and 0.5 mg for others. Peaks: (1) γ-Endorphin, (2) bombesin, (3) luteinizing hormone-releasing hormone, (4) somatostatin, (5) α-melanocyte-stimulating hormone, (6) substance P, (7) glucagon, (8) insulin, (9) β-endorphin. [(B) reprinted from *J. Chromatogr.*, **513**, Y. Kato *et al.*, p. 384, 1990, with kind permission of Elsevier Science–NL, Sara Burgerhartstraat 25, 1055 KV Amsterdam, The Netherlands.]

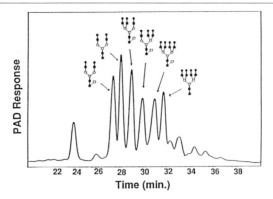

FIG. 14. Separation of glycoprotein-derived oligosaccharides on an AS6 pellicular anion-exchanger column by using a model BioLC chromatographic system equipped with a model PAD-2 pulsed amperometric detector, all from Dionex (Sunnyvale, CA). The column was preequilibrated in 100 mM NaOH; after a 5-min hold oligosaccharides were eluted at a flow rate of 1 ml/min with a linear gradient from 0 to 100 mM sodium acetate in 100 mM NaOH over 40 min. Symbols used in the shorthand structural representation are as follows: (■) GlcNAc; (○) mannose; (●) galactose; and (□) fucose. (Reprinted with permission from Ref. 30a. Copyright 1990 American Chemical Society.)

microspheres with quaternized amino groups coated on superficially sulfonated rigid polystyrene/divinylbenzene beads of 5- or 10-μm diameter are eminently suitable for such demanding analytical applications.[30] Figure 14[30a] shows the high pH separation of glycoprotein-derived oligosaccharides using a Dionex BioLC system equipped with a pellicular anion-exchange column (AS6) and pulsed amperometric detector.[31] The separation is carried out at a flow rate of 1 ml/min with linear gradient from 0 to 100 mM sodium acetate containing 100 mM NaOH in 40 min. Figure 14 illustrates several characteristics of high-pH anion-exchange (HPAE) chromatography. HPAE is sensitive to molecular size, sugar composition, and linkages between the monosaccharide units.[32] The chromatogram exemplifies typical oligosaccharide separation. This technique gives near-baseline resolution between isomeric triantennary oligosaccharides (the third labeled peak is a 2,4-branched triantennary structure while the fourth is a 2,6-branched triantennary structure).

[30] R. D. Rocklin and C. A. Pohl, *J. Liq. Chromatogr.* **6,** 1577 (1983).
[30a] M. W. Spellman, *Anal. Chem.* **62,** 1714 (1990).
[31] L. J. Basa and M. W. Spellman, *J. Chromatogr.* **499,** 205 (1990).
[32] M. R. Hardy and R. R. Townsend, *Proc. Natl. Acad. Sci. U.S.A.* **85,** 3289 (1988).

Gigapore with
convection

Diffusive pore

FIG. 15. A schematic illustration of the gigaporous bidisperse stationary phase configuration. (Adapted from Ref. 33.)

Gigaporous Sorbents

Another way to enhance intraparticular mass transfer is by the use of rigid gigaporous support particles that have a bidisperse pore size distribution. In the larger pores, which have diameters greater than one-hundredth of the particle diameter, intraparticular convective mixing may take place at sufficiently high reduced velocities so that the rate of mass transfer inside the particles is enhanced and thus fast protein chromatography is facilitated.[33,34] When convective transport in the gigapores is operative, the use of such columns is termed perfusion chromatography.[35] Figure 15 illustrates schematically such an ion-exchange particle with convection in the gigapores and diffusion in the smaller pores, which are the loci of most ionogenic groups for protein binding. Columns packed with gigaporous ion exchanger can be operated at high flow velocities, but not without loss of efficiency. Furthermore, the loading capacity of such columns is generally lower than those packed with most other types of conventional ion exchangers.

"Gel-in-the-Cage" Configuration

Another composite stationary phase of novel configuration represented by the so-called "gel-in-the-cage" (commercially "gel-in-a-shell") structure, in which the interior of a rigid framework of the support particles is filled with a retentive hydrogel carrying fixed ionogenic functional groups,[36] is

[33] N. B. Afeyan, N. F. Gordon, I. Mazsaroff, L. Varady, S. P. Fulton, Y. B. Yang, and F. E. Regnier, *J. Chromatogr.* **519**(1), 1 (1990).

[34] L. L. Loyd and F. P. Warner, *J. Chromatogr.* **512**, 365 (1990).

[35] F. E. Regnier, *Nature* (*London*) **350**, 634 (1991).

[36] E. Boschetti, *J.Chromatogr.* **658**, 207 (1994).

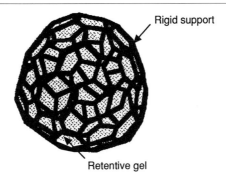

Rigid support

Retentive gel

FIG. 16. Artist's rendition of the so-called "gel-in-the-cage" stationary phase configuration. (Reprinted from *J. Chromatogr.*, **705**, H. Chen and Cs. Horváth, p. 3, 1995, with kind permission of Elsevier Science–NL, Sara Burgerhartstraat 25, 1055 KV Amsterdam, The Netherlands.)

illustrated in Fig. 16.[36a] The current trend in HPLC is to achieve high speed and resolution without loss of column loading capacity in the separation of macromolecules, which is tantamount to high throughput rate and high product recovery. As discussed in the preceding section, soft gels have a high concentration of ion-exchange sites accessible to the protein by diffusion and therefore have high protein-binding capacity. To exploit the advantages of soft hydrogel-type ion exchangers they are placed into gigaporous support particles of high pore volume that serve as a mechanically stable cage. This can be a useful combination of the mechanical stability of the "cage" particles with the high binding capacity and high recovery offered by the hydrogel filling. The latter can have a low level of cross-linking so that high flow velocities can be used and fast separations obtained. Figure 17[36b] shows separation of six standard proteins obtained on an S-HyperD column.

Comparison of Three Stationary Phases

The pellicular, gigaporous, and gel-in-the-cage types of stationary phase, which have enhanced mass transfer properties with respect to the conventional ion exchanger, exhibit certain differences in terms of the conditions under which their potential can be best utilized.

Columns packed with bidisperse gigaporous particles offer the possibility of high-speed separations in applications that do not require high resolving power.[37] Figure 18 illustrates the time of separation as a function of

[36a] H. Chen and Cs. Horváth, *J. Chromatogr.* **705**, 3 (1995).
[36b] J. Horvath, E. Boschetti, L. Guerrier, and N. Cooke, *J. Chromatogr.* **679**, 10 (1994).
[37] D. D. Frey, E. Schweinheim, and Cs. Horváth, *Biotechnol. Prog.* **9**, 273 (1993).

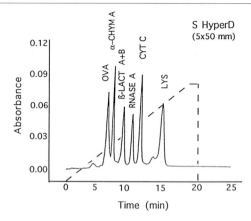

FIG. 17. Separation of six standard proteins on a novel strong cation-exchanger column. Column, 50 × 5 mm, packed with 10-μm S-HyperD; flow rate, 1.25 ml/min; linear gradient from 0 to 750 mM LiCl in 20 mM sodium formate, pH 4, over 18.8 min. (Reprinted from *J. Chromatogr.*, 679, J. Horváth *et al.*, p. 10, 1994, with kind permission of Elsevier Science–NL, Sara Burgerhartstraat 25, 1055 KV Amsterdam, The Netherlands.)

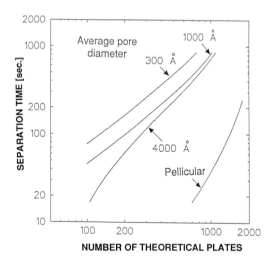

FIG. 18. Plots of separation time versus plate number calculated for 5-cm columns packed with 8-μm particles having mean pore diameters of 300, 1000, and 4000 Å. Conditions correspond to a tortuosity factor of 4, a molecular diffusivity of 10^{-6} cm^2/sec for an eluite having a 40-Å molecular diameter and a retention factor of 3.3. Convection was considered to be zero in the particles having 300- and 1000-Å pores but it was assumed that the convective velocity of the mobile phase through the particles was 1% of the interstitial flow velocity in the 4000-Å gigaporous sorbent. (Reprinted with permission from Ref. 37. Copyright 1993 American Chemical Society.)

the theoretical number of plates when columns packed with particles of usual dimensions are employed. It is seen that columns packed with particles having 4000-Å gigapores give faster separation than columns packed with particles with 1000-Å pore material, but only when the number of theoretical plates required to bring about the separation is moderate, i.e., less than 200 in this particular case. However, when high column efficiency, i.e., when a large number of theoretical plates is required for separation, the mass transfer advantages of the bidisperse gigaporous adsorbents vanish, as is seen in Fig. 18. Thus, the advantages of such columns are manifest in preparative-scale purifications or rapid laboratory separations with no demand for high column efficiency. Furthermore, columns packed with this kind of stationary phase particle have relatively low loading capacities. When rapid separation of complex mixtures (which require high plate numbers) is planned the use of a column with pellicular packing offers the greatest promise. This is also shown in Fig. 18 and the conclusion is supported by the chromatogram in Fig. 13B.

The results noted previously demonstrate that columns packed with pellicular sorbents are most suitable for rapid analytical separation of large molecules with high resolution because their relatively low loading capacity makes them less suitable for large-scale preparative separations. Nevertheless, pellicular ion exchangers are eminently suitable for micropreparative applications with high speed, efficiency, and recovery due to lack of the cavernous interior of many conventional rigid support particles.

In contrast, columns packed with sorbents having a gel-in-the-cage configuration have relatively high loading capacity, but their efficiency is likely to diminish much faster on increasing the flow velocity beyond a practical value than that of the bidisperse gigaporous sorbents.

Hydroxyapatite Chromatography

Hydroxyapatite chromatography was introduced in 1956[38] for the separation of proteins and nucleic acids having subtle structural differences. Native proteins can bind strongly to hydroxyapatite, but the retention decreases drastically on the denaturation of the protein. Hydroxyapatite retains double-stranded DNA, while allowing single-stranded DNA to pass through the column.[39]

Hydroxyapatite binds both acidic and basic proteins; however, no simple ion-exchange mechanism accounts for the observed chromatographic behavior. The surface of hydroxyapatite has two kinds of adsorbing sites. One

[38] A. Tiselius, S. Hjertén, and Ö. Levin, *Arch. Biochem. Biophys.* **65,** 132 (1956).
[39] G. Bernardi, *Methods Enzymol.* **21,** 95 (1971).

is the positively charged domain where the adsorption of anionic sample components occurs. It consists of calcium ions and serves as an anion exchanger. The other domain entails negatively charged phosphate ions and can be considered as a cation exchanger. In contact with the mobile phase at neutral and alkaline pH, the hydroxyapatite surface is negatively charged as a result of phosphate groups. Hydroxyapatite chromatography is carried out by using a linear gradient of phosphate buffer, pH 6.8, which is an equimolar mixture of mono- and dibasic sodium or potassium phosphates. Acidic pH of the eluent should be avoided because crystalline hydroxyapatite is stable only above pH 5.

For a long time the mechanical and chemical instability of hydroxyapatite was a serious impediment to the wide use of this stationary phase. The introduction of ceramic hydroxyl- and fluoroapatite, which have favorable pore structure and mechanical strength, has renewed interest in hydroxyapatite chromatography. Appropriate chromatographic columns are commercially available from various sources for both analytical and preparative purposes. The chromatographic behavior of proteins on a fluoroapatite column is similar to that on hydroxyapatite column.

Acknowledgments

This work is supported by Grant No. GM20993 from the National Institute of Health, U.S. Public Health Service and by a grant from the National Foundation for Cancer Research.

[4] Analytical Immunology

By MERCEDES DE FRUTOS, SANDEEP K. PALIWAL, and FRED E. REGNIER

Introduction

Immunoassays are based on the specific recognition by antibodies of a small number of amino acids (epitopes) in an antigen. This immunological discrimination, which can be used to determine either antigen or the antibody concentration, is the basis for modern analytical immunology.

The human immune system has the potential to generate 10 million different combinations of immunoglobulin specificities,[1] each of which may complex with an antigen in a different way. The discussion here focuses on the general aspects of antigen–antibody binding that are common to

[1] P. M. Colman, Adv. Immunol. **43**, 99 (1988).

most immunocomplexes. The lock-and-key model proposed in 1900[2] to represent the antibody–antigen binding is still accepted, although it is important to consider the surface topology of each molecular species involved in describing the specific chemical interactions. Knowledge of the steric aspects of antibody–antigen interaction at the molecular level can be obtained by crystallographic studies in some cases. In others, it must be combined with additional biochemical and immunological information.

Getzoff et al.[3] divide antibody–antigen complex formation into recognition and binding. They consider that for recognition and initial complex formation, relatively long-range electrostatic and polar interactions play the most important role, while hydrogen bonds are a dominant force for interaction of the final complex. The above-mentioned forces determine the rate of initial complex formation, whereas the dissociation rate is a function of short-range forces, e.g., hydrophobic and van der Waals interactions.

According to Braun et al.[4] a main focus in immunoassay research through the years has been the effort to find alternative labeling techniques. On the basis of labeling techniques, immunoassays can be classified as radioactive, chromogenic, fluorescent, chemiluminescent, and enzymatic. Among these, the radioactive and or enzymatic methods of detection have been the most frequently used.

Traditionally, immunological assays have been performed in polystyrene tubes or microtiter wells. This involves the binding of either the antibody or the antigen to the polystyrene plate. After antibody–antigen binding, a detection step is required to quantitate the extent of binding. Figure 1 depicts a scheme for the procedure to perform an ELISA (enzyme-linked immunosorbent assay) in a "sandwich" format. Briefly, a sample containing the antigen is incubated in the microtiter well containing the immobilized antibody. Following a wash step to remove the nonimmunologically complexed species, a second antibody labeled with the enzyme is added to identify the antigen specifically and to provide a means for detection. After washing, to eliminate the excess antibody, a substrate for the enzyme is added (e.g., p-nitrophenyl phosphate for alkaline phosphatase) and the amount of the product formed during the enzymatic reaction is measured. The amount of product formed is proportional to the amount of the antigen present.

Several modifications to improve and simplify immunological assays have been introduced. These include kinetic and thermodynamic aspects

[2] P. Ehrlich, Proc. R. Soc. London B 66, 424 (1900).
[3] E. D. Getzoff, J. A. Tainer, R. A. Lerner, and H. M. Geysen, Adv. Immunol. 43, 1 (1988).
[4] T. Braun, A. Klein, and S. Zsindely, Trends Anal. Chem. 11, 5 (1992).

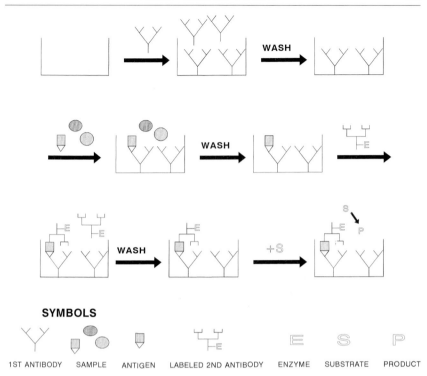

SYMBOLS

1ST ANTIBODY   SAMPLE   ANTIGEN   LABELED 2ND ANTIBODY   ENZYME   SUBSTRATE   PRODUCT

Fig. 1. A general scheme depicting an ELISA in sandwich format.

of antigen–antibody binding,[5–8] comparison of the performance of immuno-sorbents prepared by different coupling methods,[9–11] and use of diverse antibody fragments.[12–14] New techniques, such as capillary electrophoresis, have also been employed to study and quantitate antibody–label conjugates

[5] J. R. Sportsman and G. S. Wilson, *Anal. Chem.* **52,** 2013 (1980).

[6] J. R. Sportsman, J. D. Liddil, and G. S. Wilson, *Anal. Chem.* **55,** 771 (1983).

[7] J. Renard, C. Vidal-Madjar, B. Sebille, and C. Lapresle, *J. Mol. Recognit.* 1996 (in press).

[8] D. S. Hage, D. H. Thomas, and M. S. Beck, *Anal. Chem.* **65,** 1622 (1993).

[9] C. L. Orthner, F. A. Highsmith, J. Tharakan, R. D. Madurawe, T. Morcol, and W. H. Velander, *J. Chromatogr.* **558,** 55 (1991).

[10] J.-N. Lin, I.-N. Chang, J. D. Andrade, J. H. Herron, and D. A. Christensen, *J. Chromatogr.* **542,** 41 (1991).

[11] L. J. Janis and F. E. Regnier, *J. Chromatogr.* **444,** 1 (1988).

[12] M. J. Berry, J. Davies, C. G. Smith, and I. Smith, *J. Chromatogr.* **587,** 161 (1991).

[13] M. J. Berry and J. Davies, *J. Chromatogr.* **597,** 239 (1992).

[14] M. J. Berry and J. J. Pierce, *J. Chromatogr.* **629,** 161 (1993).

and antibody–antigen complexes.[15–17] Figure 2 shows the usefulness of capillary electrophoresis for determining the labeling of immunoglobulin G (IgG) with alkaline phosphatase (AP), and to verify the effectiveness of purification of the conjugate from unreacted components. Thiolated IgG and maleimidated AP were reacted to prepare the conjugate. Comparison of electropherograms of a standard mixture of IgG and AP (Fig. 2A), of the conjugate before purification by fast protein liquid chromatography (FPLC; Pharmacia, Piscataway, NJ) (Fig. 2B), and of the conjugate after FPLC purification (Fig. 2C) demonstrates (1) formation of the immunoglobulin–enzyme conjugate and (2) purification of the conjugate from unreacted material. Attempts to minimize interference have been reviewed by Miller and Valdes.[18] Current trends are directed toward automation and nonisotopic assays, such as those that use enzyme, fluorescent, or bioluminescent labels.[19]

Enzyme amplification techniques have allowed a marked decrease in the detection limits of immunoassays.[20–22] Still higher sensitivities are being achieved with fluorescence, chemiluminescence, or electrochemical detection in combination with enzyme amplification. A wide description of amplification techniques is beyond the scope of this chapter, but it can be found elsewhere.[23] It has even been proposed that future detection systems would be based on amplification of the signal generated by a marker molecule and/or exponential amplification of the target molecule by PCR (polymerase chain reaction) techniques.[24]

Techniques for clinical assays have dominated analytical immunology. Amid the drawbacks of classic immunoassays are that they are time consuming and labor intensive. The use of immunomagnetic beads[25–27] has been

[15] S. J. Harrington, R. Varro, and T. M. Li, *J. Chromatogr.* **559,** 385 (1991).

[16] R. G. Nielsen, E. C. Rickard, P. F. Santa, D. A. Sharknas, and G. S. Sittampalam, *J. Chromatogr.* **539,** 177 (1991).

[17] D. E. Hughes and P. Richberg, *J. Chromatogr.* **635,** 313 (1993).

[18] J. J. Miller and R. Valdes, Jr., *Clin. Chem.* **37,** 144 (1991).

[19] D. S. Hage, *Anal. Chem.* **65,** 420R (1993).

[20] A. Johansson, D. H. Ellis, D. L. Bates, A. M. Plumb, and C. J. Stanley, *J. Immunol. Methods* **87,** 7 (1986).

[21] R. Lejeune, L. Thunus, F. Gomez, F. Frankenne, J.-L. Cloux, and G. Hennen, *Anal. Biochem.* **189,** 217 (1990).

[22] F. J. Dhahir, D. B. Cook, and C. H. Self, *Clin. Chem.* **38,** 227 (1992).

[23] S. Avrameas, *J. Immunol. Methods* **150,** 23 (1992).

[24] J.-L. Guesdon, *J. Immunol. Methods* **150,** 33 (1992).

[25] S. G. Gundersen, I. Haagensen, T. O. Jonassen, K. J. Figenschau, N. de Jonge, and A. M. Deelder, *J. Immunol. Methods* **148,** 1 (1992).

[26] C. H. Pollema, J. Ruzicka, G. D. Christian, and A. Lernmark, *Anal. Chem.* **64,** 1356 (1992).

[27] C. Loliger, E. Ruhlmann, and P. Kuhnl, *J. Immunol. Methods* **158,** 197 (1993).

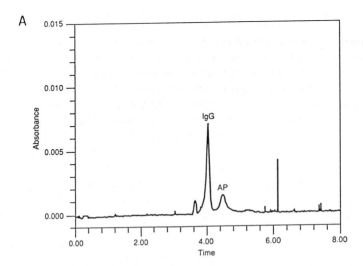

A

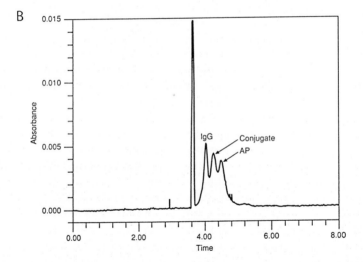

B

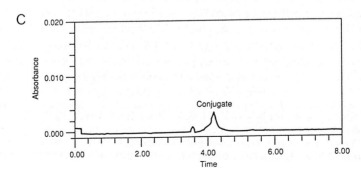

C

seen as a convenient approach to perform less tedious and time-consuming assays. One source of increased time in assays performed in microtiter wells is the slow diffusion of sample components and reagents toward the polystyrene surface. To diminish the time, the diffusion distance has been shortened by immobilizing the antibody on a membrane and placing it in a flow cuvette that is used as a reaction vessel.[28] Use of flow-based techniques [usually flow injection analysis (FIA) or high-performance liquid chromatography (HPLC)][29–32] leads to faster mass transfer, acceleration of binding, and increased reproducibility. Flow-based assays are also easier to automate. A typical ELISA in "sandwich" format performed on an HPLC column can be carried out according to the following method.

1. A primary antibody (Ab) against the analyte of interest [antigen (Ag)] is physically or chemically attached to the column stationary phase.
2. Sample containing the analyte is injected onto the column. Antigen–antibody binding takes place, with the analyte being retained and the rest of the sample components eluted. An enzyme-labeled secondary antibody is injected, and thus a sandwich is formed between the Ab–Ag–Ab–label.
3. A substrate of the enzyme is introduced onto the column while the appropriate mobile phase is used. If necessary, flow is stopped to allow for enzyme–substrate incubation.
4. Mobile phase flow is resumed and the product of the enzymatic reaction is measured downstream, using the visible-light detector of the HPLC equipment.
5. A desorption agent is introduced onto the column to disrupt the antigen–antibody binding, which allows for repeated column use.

Biotechnology is another field in which use of analytical immunology is becoming increasingly important. At present, biotechnology encompasses

[28] B. M. Gorovits, A. P. Osipov, and A. M. Egorov, *J. Immunol. Methods* **157,** 11 (1993).
[29] P. C. Gunaratna and G. S. Wilson, *Anal. Chem.* **65,** 1152 (1993).
[30] N. B. Afeyan, N. F. Gordon, and F. E. Regnier, *Nature (London)* **358,** 603 (1992).
[31] D. S. Hage and P. C. Kao, *Anal. Chem.* **63,** 586 (1991).
[32] U. de Alwis and G. S. Wilson, *Anal. Chem.* **59,** 2786 (1987).

FIG. 2. Capillary zone electropherograms of (A) standard mixture of AP (alkaline phosphatase) + IgG (1:1), (B) unpurified AP–IgG conjugate containing the conjugate and unreacted AP and IgG, and (C) sample from FPLC (fast protein liquid chromatography) purification of AP–IgG conjugate corresponding to the fraction with the highest concentration of conjugate. (Reprinted from *J. Chromatogr.,* **559,** S. J. Harrington, R. Varro, and T. M. Li, p. 385, Copyright 1991 with kind permission of Elsevier Science–NL, Sara Burgerhartstraat 25, 1055 KV Amsterdam, The Netherlands.)

not only the fermentation industry but also recombinant DNA and monoclonal antibody technologies[33] for the manufacture of commercial products.[34] By their very nature, biotechnological processes often give rise to contaminants of the target products. These contaminants include proteins from the host organism, DNA, residual cellular proteins from the medium, and variants of the recombinant product, which originate by translational errors, posttranslational modifications, degradation, etc. Thus, to determine the purity of the target product and the nature and concentration of contaminants, highly selective and sensitive procedures are needed. It is in this context that analytical immunology has proved to be useful in biotechnology. When the high selectivity and affinity of antibodies for their target antigens are coupled with antibody-labeling techniques that allow detection at subpicogram levels, analytical immunology becomes an indispensable tool for biotechnology. As mentioned previously, a product obtained by genetic engineering techniques can be contaminated with proteins whose structures are slightly different from that of the target recombinant protein. These variants need to be detected and eliminated because they can be immunogenic or have diminished biological activity. A combination of chromatographic and immunological methods allows discrimination between proteins with shared epitopes. As an example, avian lysozyme variants in a sample were bound to an anti-lysozyme-containing affinity column and then released onto a cation-exchange column. On the cation-exchange column, the protein variants were separated and detected by ultraviolet (UV) absorption.[35] One of the disadvantages of performing an assay in this fashion is the lack of sensitivity. In those cases in which direct UV detection of analytes is not sufficiently sensitive, sensitivity can be increased by introducing a labeled antigen or antibody prior to the detection step. As an example, anti-human growth hormone antibodies in the sera of patients were determined with enhanced sensitivity using fluorescently labeled human growth hormone.[36] Capillary electrophoresis (CE) has also been used in this context for analysis of immunocomplexes.[37–39] Besides the above-mentioned advantages, flow techniques usually employ lower sample volumes than conventional assays.

[33] R. L. Garnick, N. J. Solli, and P. A. Papa, *Anal. Chem.* **60,** 2546 (1988).
[34] J. W. G. Smith, Aspects of regulating medicinal products of biotechnology. *In* "The World Biotech Report 1984, Volume 1: Europe." Online Publications, Pinner, United Kingdom, 1984.
[35] L. J. Janis, A. Grott, F. E. Regnier, and S. J. Smith-Gill, *J. Chromatogr.* **476,** 235 (1989).
[36] A. Riggin, F. E. Regnier, and J. R. Sportsman, *Anal. Chim. Acta* **249,** 185 (1991).
[37] N. M. Schultz and R. T. Kennedy, *Anal. Chem.* **65,** 3161 (1993).
[38] K. Shimura and B. L. Karger, *Anal. Chem.* **66,** 9 (1994).
[39] F.-T. A. Chen and J. C. Sternberg, *Electrophoresis* **15,** 13 (1994).

Another advance in immunological assays is the use of surface plasmon resonance (SPR) to monitor the association of antigen and antibodies at the surface in real time.[40] Surface plasmon resonance measures the increase in protein layer thickness at the surface as the immunological complex forms. Because association and subsequently dissociation are determined in real time, SPR provides a powerful tool for the estimation of kinetics and binding constants in addition to quantitative information on either antibody or antigen concentration in samples. A limitation of the technique is that the lower limit of detection is in the range of 0.1 to 1 mg/ml. Techniques for immobilizing antibodies are also an important variable in SPR.[41]

A wide variety of methods have been explored for antibody immobilization. Concern is that during immobilization, antibody activity may be compromised by either steric or chemical effects. As an alternative to chemical methods, biospecific adsorption has been used. Protein A or protein G adsorbed or bound to the surface have been used to capture antibodies or antibody–antigen complexes.[42,43] Association of antibodies with these proteins occurs through the Fc region, thus orienting the antigen-binding region of the antibodies away from the sorbent surface. As an example, in assays performed in plates,[43] the method is performed as follows: A solution of protein A in appropriate buffer is incubated on the plate overnight at 4° or at 37° for 1 hr. Unoccupied binding sites on the plate are blocked overnight with blocking buffer. Samples of preincubated antibody and biotynylated antigen are transferred to the wells of the protein A-coated plate and incubated for 1 hr at 37°. Streptavidin–alkaline phosphatase conjugate is added and incubated for 1 hr at 37°. p-Nitrophenyl phosphate is then added and absorption at 405 nm is measured using an ELISA reader.

Other novel methods have been developed for the analysis of immunological and biochemical properties of cells in suspension. Three groups of in vitro cellular enzyme-amplified techniques can be considered.[44] In the first group, a cell-ELISA (CELISA) is used for the analysis of cell surface antigens. Rapid analysis of monoclonal antibody specificities is achieved in this way. The second group consists of cellular immunomagnetic techniques, which are used for the analysis of secreted products of cells at the single-cell level. These assays have been referred to as ELISPOT, spot-ELISA,

[40] M. Brigham-Burke and D. J. O'Shannessy, *Chromatographia* **35**, 45 (1993).
[41] M. C. Millot, F. Martin, D. Bousquet, B. Sebille, and Y. Levy, *Sensors Actuators* (*B. Chem.*) 1996 (in press).
[42] L. J. Janis and F. E. Regnier, *Anal. Chem.* **61**, 1901 (1989).
[43] P. K. M. Ngai, F. Ackermann, H. Wendt, R. Savoca, and H. R. Bosshard, *J. Immunol. Methods* **158**, 267 (1993).
[44] J. D. Sedwick and C. Czerkinsky, *J. Immunol. Methods* **150**, 159 (1992).

and ELISA-plaque, and are used for detection and enumeration of cells secreting immunoglobulins or cytokines. The third group of assays has been developed for the analysis of cellular proliferation *in vitro*.

Recombinant DNA techniques are also being utilized to produce immunological assay reagents. Several attempts to use fusion proteins between a peptide and an enzyme as tracers in competitive binding assays have been reported. By taking advantage of the flexibility of genetic engineering, Witkowski *et al.*[45] prepared a synthetic octapeptide epitope combined with bacterial alkaline phosphatase. They found that the genetically prepared conjugate performs significantly better than one prepared by conventional methods. The validation of a recombinant conjugate between a larger protein and alkaline phosphatase (AP) was also assessed in a competitive enzyme immunoassay for rat prolactin (rPrl).[46] The use of rPrl–AP hybrid protein expressed in *Escherichia coli* did not modify the specificity of the assay. Both rPrl and AP domains retained antigenicity. This system allows the production of a remarkably stable hybrid, using a fast and inexpensive fermentation process. The cloned enzyme donor immunoassay (CEDIA)[47] is another example and is considered the most sensitive of the commercially available homogeneous enzyme immunoassays.[48] Recombinant DNA techniques allowed the production of a homogeneous population of conjugates and enabled exact control of the site of enzyme modification. Genetically engineered conjugates also increase the long-term assay reproducibility by avoiding the batch-to-batch variability.

Immunological assays can be grouped according to several different criteria: assay format, detection method, or type of sample analyzed. It should also be pointed out that the concentration of the molecules of interest can vary in a wide range from sample to sample and thus has a large impact on the assay method chosen. We have mentioned a number of approaches that have been used to increase the sensitivity of immunoassays. This enhanced sensitivity is accompanied by either increased cost or difficulty of analysis. Often, high-sensitivity assays are employed when the concentration of analyte is orders of magnitude above the detection limit of the assay. For this reason we discuss immunological assays on the basis of analyte concentration.

[45] A. Witkowski, S. Daunert, M. S. Kindy, and L. G. Bachas, *Anal. Chem.* **65,** 1147 (1993).
[46] D. Gillet, E. Ezan, F. Ducancel, C. Gaillard, T. Ardouin, M. Istin, A. Menez, J.-C. Boulain, and J.-M. Grognet, *Anal. Chem.* **65,** 1779 (1993).
[47] D. R. Henderson, S. B. Friedman, J. B. Harris, W. B. Manning, and M. A. Zoccoli, *Clin. Chem.* **32,** 1637 (1986).
[48] S. H. Jenkins, *J. Immunol. Methods* **150,** 91 (1992).

Assays Performed to Detect Analytes at High Levels

Various immunoassay formats have been applied to determine antigens or antibodies present in microgram levels or higher. In the traditional plate format, this requires numerous incubation and washing steps, which renders assays slow and labor intensive. Advantages of performing the assays in flow systems are clearly shown in papers in which antigens or antibodies are measured by using immunoaffinity chromatography.[42,49] The analyte captured on the immunoaffinity column is released using a desorption buffer. The effect of desorption buffer on recovery of antibody from protein G columns has been studied.[42] Comparison of 0.1 M glycine (pH 2.2), 0.01 M glycine in 0.15 M NaCl (pH 2.2), 2% (v/v) acetic acid in 0.15 M NaCl (pH 2.4), 2% (v/v) acetic acid in 0.25 M NaCl (pH 2.4), 20% (v/v) acetic acid (pH 2.1), 0.05 M sodium citrate (pH 2.2), and 0.1 M glycine in 2% (v/v) acetic acid (pH 2.9) has shown that desorption is dependent on ionic strength and that among these desorption buffers, the last one gave the highest recovery of rabbit IgG from protein G columns. Ultraviolet detection of the effluent at the column outlet is sensitive enough to detect microgram amounts of antibodies or antigens in the sample. This chromatographic format of immunological assay also permits easy automation. Sensitivity is limited in this method by the fact that the desorbing agent may cause changes in the absorbance (due to a change in the refraction index) and interfere with the analyte detection.

Both the traditional plate format assays and the immunoaffinity chromatographic-based assays work on the basis of either confirming the presence or absence of some epitopes in the analyte. Thus, for this reason, in both formats, molecules cross-reacting with the same antibody are, in principle, not discriminated. Frontal immunoaffinity chromatography (FIC) has been applied successfully to distinguish protein variants using only an immunochromatography column.[50] In this method, polyclonal antibodies against an antigen are immobilized on a protein A column that is operated in the frontal mode. This means that when a high volume of a solution containing a constant concentration of antigen is introduced into the antibody column, antigen is initially retained by the column and no signal is detected. When the amount of antigen introduced exceeds the loading capacity of the column, the nonretained antigen is detected as a step of constant height. Alteration or deletion of epitopes in the variants causes a decrease in the column loading capacity. Frontal immunoaffinity chroma-

[49] A. Riggin, F. E. Regnier, and J. R. Sportsman, *Anal. Chem.* **63,** 468 (1991).
[50] C. Xu and F. E. Regnier, *Anal. Methods Instr.* **3,** 145 (1993).

tography can even differentiate between primary structure and conformational variants of antigens, as shown using polyclonal antibodies against bovine serum albumin (BSA) to discriminate native and reduced bovine serum albumin, and albumins from bovine, ovine, porcine, human, and guinea pig serum.

Another method that allows cross-reacting species to be differentiated without eluting the retained fraction from the affinity column is the tandem column immunosorbent subtraction method.[51] This method was first used by combining a protein G column containing antibodies against human growth hormone (hGH) with a size-exclusion chromatography (SEC) column. This approach was used to investigate the cross-reactivity of hGH-related compounds (hGH dimer, N-terminal methionyl hGH, deamido hGH, etc.) with anti-hGH antibodies. The assay was performed by first introducing the sample through the protein G column, without the antibodies, and then through the SEC column. Afterward, anti-hGH antibodies were immobilized onto the protein G column and a second chromatogram obtained by passing the sample through the protein G–anti-hGH column and the SEC column. The protein G column with immobilized antibodies subtracts hGH from the sample. Figure 3 shows the first and the second chromatogram and also the subtraction chromatogram used to examine the cross-reactivity of hGH-related compounds with anti-hGH antibodies.

A similar method has been used for monitoring the production of recombinant proteins in mammalian cell culture.[52] By using subtractive immunoassays, glycosylated and nonglycosylated forms of γ-interferon produced by recombinant Chinese hamster ovary (CHO) cells were quantitated at the microgram level every 15 min.

As reported,[53] more rapid methods are needed for process monitoring. Through the use of high-speed immunosorbent columns and subtractive chromatographic methods, it is possible to determine antigen concentration at greater than 1% in a sample in 5–10 sec. Automation of this subtractive immunoassay process has allowed antigen concentration and purity to be monitored in close to real time. When data are fed into a computer, real-time quality control and process validation are achieved.[30]

Dual-column systems are also valuable in cases in which the concentration of antigen or antibodies is determined after the disruption of the antigen–antibody complex from the immunoaffinity column. The immunoaffinity column generally eliminates interferences caused by other sub-

[51] A. Riggin, J. R. Sportsman, and F. E. Regnier, *J. Chromatogr.* **632,** 37 (1993).
[52] T. K. Nadler, S. K. Paliwal, F. E. Regnier, R. Singhvi, and D. I. C. Wang, *J. Chromatogr.* **659,** 317 (1994).
[53] S. K. Paliwal, T. K. Nadler, and F. E. Regnier, *Trends Biotech.* **11,** 95 (1993).

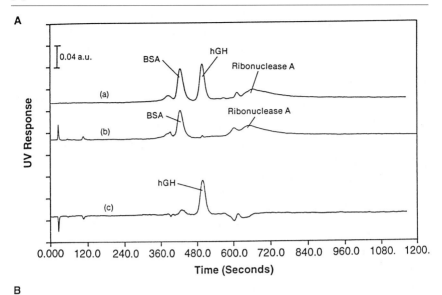

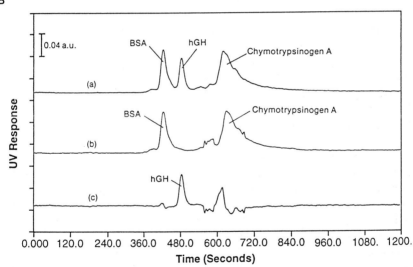

FIG. 3. Recognition of hGH by (A) a monoclonal antibody and (B) anti-hGH in rabbit serum, (a) before and (b) after immobilization of antibodies onto a protein G column. The difference chromatogram (c) was obtained by subtracting each data point of (b) from (a). (Reprinted from *J. Chromatogr.*, **632,** A. Riggin, J. R. Sportsman, and F. E. Regnier, p. 37, Copyright 1993 with kind permission of Elsevier Science–NL, Sara Burgerhartstraat 25, 1055 KV Amsterdam, The Netherlands.)

stances present, while the second HPLC column negates the refractive index problems caused by the desorption buffer and slow desorption kinetics. An anti-interferon column coupled to a reversed-phase (RP) column has been used to assay recombinant leukocyte interferon from *Saccharomyces cerevisiae*.[54] The method is capable of detecting microgram amounts of interferon and provides information about the monomer and oligomer contents in crude mixtures.

Assays Performed to Detect Analytes at Low Levels

The ELISA sensitivity in microtiter plates has been increased through the introduction of affinity-purified antibodies, careful choice of the medium for Ab–Ag reaction, and reduction of nonspecific binding. To detect 4 pg of protein A in therapeutic antibodies purified on a protein A column, IgYs used in a noncompetitive ELISA were isolated from egg yolk and affinity selected, and sensitivity was improved by increasing the sample incubation time, although it has the cost of reducing the maximum trace contamination detectable.[55] The importance of antibody selection has been demonstrated in the case of interleukin 8. By selecting the best monoclonal antibody out of 18 clones and using it with polyclonal antibody conjugate, the detection limit of interleukin 8 was decreased to 0.1 pg, far more sensitive than in previous assays.[56]

Although the lack of a separation step in homogeneous immunoassays can limit the sensitivity of the method owing to interference from other sample contaminants, the coupling of immunoassays with recombinant DNA technologies has led to a sensitive homogeneous enzyme immunoassay, capable of detecting picogram levels of antigens.[48]

Modifications aimed at increasing the signal detection and simultaneously decreasing the background have allowed the detection of about 1 pg of IgE.[57] The background was reduced by substituting polystyrene microspheres for the conventional nitrocellulose membrane used in the Ab–Ag solid phase reaction. In this assay, the use of ultrafine gold particles as labeling material and quantitative laser-induced photothermal beam deflection clearly increased the detection signal.

An increase in sensitivity has also been obtained through the use of sequential enzyme amplification in the detection step. This was done by employing a kinetic technique to measure color development.[22] Femtogram

[54] L. Rybacek, M. D'Andrea, and S. J. Tarnowski, *J. Chromatogr.* **397,** 355 (1987).
[55] M. A. J. Godfrey, P. Kwasowski, R. Clift, and V. Marks, *J. Immunol. Methods* **149,** 21 (1992).
[56] N. Ida, S. Sakurai, K. Hosoi, and T. Kunitomo, *J. Immunol. Methods* **156,** 27 (1992).
[57] C.-Y. Tu, T. Kitamori, T. Sawada, H. Kimura, and S. Matsuzawa, *Anal. Chem.* **65,** 3631 (1993).

levels of alkaline phosphatase (AP) could be detected when using AP as a label, initiating two sequential catalytic reactions.[20] Using such techniques, it has also been possible to detect picogram amounts of insulin. This is 10 times more sensitive than existing methods and also has a wider dynamic range.

Flow-based methods have also been used to increase the sensitivity of detection. Detection techniques, such as amplification in a flow-based format,[21] have allowed detection of human growth hormone in the range of femtograms. This is 100 to 1000 times more sensitive than a classic sandwich assay. Flow-based chemiluminescent detection has been applied to analyze parathyroid hormone (PTH) as a way to enhance sensitivity. The method is carried out in a sandwich format, using a second antibody labeled with an acridinium ester. The first antibody–PTH complex is then dissociated, and PTH eluted from the column in association with the labeled antibody is combined with an alkaline peroxide postcolumn reagent. Measurement of the light produced at 430 nm allows detection of 0.1 pg of PTH.[31] Another modification of flow-based immunoassays is to carry out the assays in capillary columns, following the method indicated in Fig. 4. In this case, protein G was covalently bound to the wall of a 100-μm i.d. capillary column. A buffer containing 0.1% (w/v) bovine serum albumin, 0.1% (w/v) Brij 35, and 0.1 M NaCl added to 50 mM Tris (pH 7.4) markedly decreased nonspecific adsorption without altering specific binding when used as loading buffer. By using loading buffer as mobile phase at 2 μl/min, hIgG and the Fab fraction of anti-hIgG labeled with alkaline phosphatase were sequentially or simultaneously injected onto the protein G capillary. Mobile phase was changed to detection buffer (50 mM Tris, pH 9) and p-nitrophenyl phosphate, used as substrate, injected at 10 μl/min. Flow through the column was stopped to incubate the substrate inside the column. When the incubation was complete, flow was increased to 100 μl/min and the peak height measured on-column at 405 nm. To prepare the column for the next analysis, desorbing solution [0.3 M MgCl$_2$ in 2% (v/v) acetic acid] was injected. Optimization of the method results in rapid (20 min) analysis, with a detection limit of 300 zmol of IgG. In addition to the speed and sensitivity of the method, small volumes (500 nl) of sample and antibody are consumed.[58]

A similar detection limit (280 zmol) has been obtained for the Fab fragment of anti-insulin by an immunoassay using capillary zone electrophoresis with fluorescence detection.[37] In this assay, the preformed and labeled complex of antibody and antigen is introduced into a capillary and detected by laser-induced fluorescence, after separation. Capillary isoelectric focus-

[58] M. de Frutos, S. K. Paliwal, and F. E. Regnier, *Anal. Chem.* **65,** 2159 (1993).

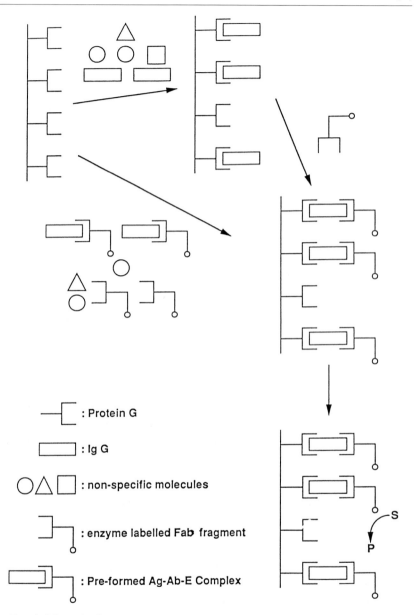

: Protein G

: Ig G

: non-specific molecules

: enzyme labelled **Fab** fragment

: Pre-formed Ag-Ab-E Complex

FIG. 4. Schematic diagram showing the protocol used to determine hIgG concentration by capillary HPLC ELISA. (Reprinted with permission from Ref. 58. Copyright 1993 American Chemical Society.)

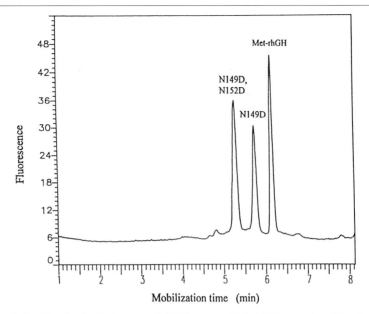

FIG. 5. Capillary isoelectric focusing of rhGH variants: Met-rhGH, monodeamidated variant (N149D), and dideamidated variant (N149D, N152D) detection by laser-induced fluorescence using tetramethylrhodamine iodoacetamide-labeled antibody as an affinity probe. (Reprinted with permission from Ref. 38. Copyright 1994 American Chemical Society.)

ing in combination with laser-induced fluorescence detection has been used to detect methionyl recombinant human growth hormone (Met-rhGH) and its mono- and dideamidated variants simultaneously by using a fluorescently labeled Fab' fragment of anti-hGH (Fig. 5).[38]

Besides the above-mentioned methods, coupling of the separation and detection steps has been achieved by other means. The immunoassay may be performed in a microtiter plate or in a column and then directed to a chromatographic or flow injection analysis (FIA) system. Tandem chromatographic–immunological analyses have been reviewed.[59] Thompson *et al.*[60] performed a microtiter plate assay for IgG–alkaline phosphatase in which the product of the enzymatic reaction was injected in an FIA system. The analyte was detected amperometrically at the level of about 1 pg. Fluorescence-based detection systems allow the detection of picograms of digoxin in serum with a system in which an immunoaffinity column extracts

[59] M. de Frutos and F. E. Regnier, *Anal. Chem.* **65,** 17A (1993).
[60] R. Q. Thompson, M. Porter, C. Stuver, H. B. Halsall, W. R. Heineman, E. Buckley, and M. R. Smyth, *Anal. Chim. Acta* **271,** 223 (1993).

the analyte from the sample and a reversed-phase column is then used to purify the product further.[61]

A reversal of columns described previously has also been used to detect and quantitate digoxin and its metabolites at picogram amounts. This was done by performing the assay with a postcolumn derivatization system. Components of the sample separated in a reversed-phase column are reacted with fluorescein-labeled anti-digoxigenin antibodies. Free antibodies are removed before detection, by using an immunoaffinity column containing digoxin.[62]

Assays Performed to Detect Analytes at Intermediate Levels

The methods that have been described up to this point have been used to detect analytes present in a sample at either high (microgram) or low (picogram) amounts. There are a large number of cases in which the analyte of interest exists in the sample at an intermediate concentration, i.e., in the nanogram range.

Modifications used for detecting high or low levels of analytes make it possible to apply the methods for analysis at intermediate levels of antigen. For instance, the automated real-time immunoassay described by Afeyan et al.,[30] which combines immunodetection and perfusion chromatography,[62a] can be used with direct UV detection to measure hundreds of nanograms of analyte, while the same system is capable of detecting picogram quantities if used in the flow-through ELISA format.

Venembre et al.[63] have examined the influence of detection technique on sensitivity and performance of analysis. Enhanced chemiluminescence and colorimetric techniques were compared as visualization steps on a Western blot analysis for human α_1-antitrypsin (AAT). Results indicate that chemiluminescence detection is more sensitive (detection limit, 50 pg) but not suitable for quantitation. In contrast, colorimetric methods can be used as quantitative technique but have a detection limit of 2 ng. By combining both techniques, it was possible to show that glycosylation of AAT in the monocytic THP-1 cell line is different from that of serum, and also allows discrimination and quantitation of four AAT forms in the monocyte supernatant.

---

[61] E. Reh, *J. Chromatogr.* **433**, 119 (1988).

[62] H. Irth, A. J. Oosterkamp, W. van der Welle, U. R. Tjaden, and J. van der Greef, *J. Chromatogr.* **633**, 65 (1993).

[62a] L. R. Snyder, Methods in Enzymology, **270**, Chapter 7 (this volume).

[63] P. Venembre, N. Seta, A. Boutten, M. Dehoux, M. Aubier, and G. Durand, *Clin. Chim. Acta* **227**, 175 (1994).

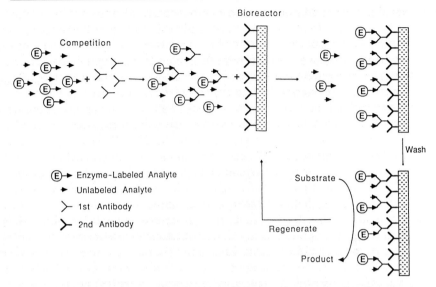

FIG. 6. Schematic diagram of the protocol used to perform a competitive immunoassay for insulin on a flow injection system. (Reprinted from Ref. 65, with permission.)

of analytes without disrupting the immunocomplex has been proposed.[64] The procedure, called "kinetic immunochromatographic sequential addition," makes use of a protein A column with a predetermined antibody capacity. On this column, a sample containing an unknown amount of antigen and then a known amount, called label, are sequentially injected. The nonretained label, proportional to the amount of antigen in the sample, is quantitated by UV detection. A complete analysis of human transferrin can be performed in less than 3 min with a detection limit of 50 ng.

Hundred-nanogram levels of insulin can be detected with a competitive assay proposed by Lee and Meyerhoff,[65] which is performed in a flow system according to the scheme shown in Fig. 6. This assay photometrically detects the product of the enzymatic reaction of horseradish peroxidase, which is used as a label.

Lower detection limits, in the range of 3 ng, have been achieved through amperometric or fluorescent detection. Amperometric detection is used to quantitate α-fetoprotein (AFP) through the measurement of 4-aminophenol, the product of the enzymatic reaction of alkaline phosphatase, which

[64] S. A. Cassidy, L. J. Janis, and F. E. Regnier, Anal. Chem. 64, 1973 (1992).
[65] I. H. Lee and M. E. Meyerhoff, Mikrochim. Acta (Wien) III, 207 (1988).

nol, the product of the enzymatic reaction of alkaline phosphatase, which is used as a label.[66] The assay of AFP is performed in a sandwich format in polystyrene plates; the product of the enzymatic reaction is injected into a flow-injection analyzer–electrochemical detection system. Fluorescent detection is used to analyze IgG in a competitive assay carried out in a flow system utilizing immunomagnetic beads that are covalently bound to the antibody.[26] Besides combining high speed and sensitivity, it is claimed that this assay shows high reproducibility because the magnetic beads are discharged after each assay.

Tandem chromatography–immunology has been shown to increase sensitivity compared to a single affinity column by at least threefold. Protein G affinity–reversed-phase tandem column chromatography has been used to determine antibodies to human growth hormone (hGH) in sera from patients under treatment with recombinant human growth hormone. By using fluorescence detection, the minimum detectable concentration for anti-hGH in serum with this method is 50 ng/ml. This corresponds to about 5 ng of antibody injected onto the column.[36] The reversed-phase step increases the sensitivity by avoiding the serum background interference.

Similar to the tandem immunochromatographic methods mentioned previously, a dual microcolumn system has been developed that allows 10 ng of human transferrin[67] to be detected by direct UV analysis. The first column (250-μm i.d.) contains an antibody that is hydrophobically adsorbed onto an octylsilica-based stationary phase. Antigens desorbed from this column are concentrated onto a reversed-phase column (250-μm i.d.), where quantitation is performed. Although in this case the system is used for the analysis of only one antigen, it should be possible to quantitate antigens that show cross-reactivity with the antibody, as has been discussed previously with larger diameter columns. One advantage of this system is the small amount of sample and antibody required for analysis.

Conclusion

Improvements in antibody selectivity, antibody and tracer labeling, antibody immobilization, detection, and automation of immunoassays together with the enhanced selectivity and sensitivity provided by integrating separation systems into the assay have enormously increased the value of immunoassays. The large number of variations on this technique have proved to be quite useful in quantitating an analyte in a sample, determining its purity,

[66] Y. Xu, H. B. Halsall, and W. R. Heineman, *Clin. Chem.* **36,** 1941 (1990).
[67] C. L. Flurer and M. Novotny, *Anal. Chem.* **65,** 817 (1993).

discriminating variants of the target substance formed during its production, and in performing real-time monitoring of biomolecules. Further work should concentrate on decreasing costs and making assays less cumbersome and laborious.

[5] Microcolumn Liquid Chromatography in Biochemical Analysis

By Milos V. Novotny

Introduction

Biochemical methodologies based on various forms of liquid chromatography (LC) have been extremely important to progress in both structural investigations and physicochemical measurements performed on biological macromolecules.[1] The high-performance LC (HPLC) instruments featuring analytical-scale (4.6-mm i.d.) and preparative columns have now become standard equipment in biochemical laboratories. While such instrumentation performs admirably[1a] in the day-to-day tasks of a biochemist (ranging from isolation work to routine sequencing methods to various other determinations), there is an increasing need to work at the microscale and nanoscale. Miniaturization of chromatographic separations and the associated small-scale manipulations is certainly one way of dealing with the problem of analyzing minute quantities of important biomolecules isolated from complex biological materials. Increasing sophistication of chromatographic detectors in terms of sensitivity and structural information provides further incentive for the development of microcolumn separation methods.

Historically, serious interest in small-bore LC columns started in the mid-1970s. Three different research groups who pioneered the area were pursuing different objectives: expansion of the effectiveness of size-exclusion chromatography for small molecules,[2] miniaturization of LC systems as a general trend in analytical chemistry,[3] and development of columns with dramatically improved analytical capabilities.[4,5] In subsequent years,

[1] M.-I. Aguilar and M. T. W. Hearn, *Methods Enzymol.* **270,** Chap. 1, 1996 (this volume); S.-L. Wu and B. L. Karger, *Methods Enzymol.* **270,** Chap. 2, 1996 (this volume).
[1a] M. T. W. Hearn, *J. Chromatogr.* **418,** 3 (1987).
[2] R. P. W. Scott and P. Kucera, *J. Chromatogr.* **125,** 251 (1976).
[3] D. Ishii, K. Asai, K. Hibi, T. Jonokuchi, and M. Nagaya, *J. Chromatogr.* **144,** 157 (1977).
[4] T. Tsuda and M. Novotny, *Anal. Chem.* **50,** 271 (1978).
[5] T. Tsuda and M. Novotny, *Anal. Chem.* **50,** 632 (1978).

the miniaturized LC columns have provided various advantages to chemical analysis. These advantages have been felt either directly as extended capabilities of HPLC, or indirectly through the utilization of miniaturized instrumental components in the fields of capillary zone electrophoresis,[6,7] microcolumn supercritical-fluid chromatography,[8] and in coupling of the microcolumn systems to fast atom bombardment mass spectrometry[9-11] and electrospray mass spectrometry.[12-15]

The efforts leading to miniaturized LC can be roughly categorized into two types of investigation: (1) studies aiming at a moderate degree of miniaturization, i.e., a gradual scale-down of the well-proven conventional HPLC, and emphasizing the benefits of reduced mobile phase consumption (for economy and environmental reasons); and (2) investigations seeking significant departure from the analytical norm of a typical LC experiment and emphasizing superior separation performance, new detection techniques, and capabilities of working with small samples. For the obvious reasons, only the first direction generated immediate interest from the instrument companies, who started to advocate "nonradical miniaturization" by developing 2.1-mm i.d. (and, to a lesser degree, 1-mm i.d.) columns similar to those used in the pioneering studies of Scott and Kucera.[2] The lack of interest in small-size microcolumns had initially been due to instrumental difficulties in the design of reliable sampling and gradient mixing devices.

Earlier detectors in micro-LC were predominantly the miniaturized versions of the conventional HPLC detectors [ultraviolet (UV) absorbance, electrochemical, and fluorometric devices]. This was dictated by the necessity of coping with reduced volumetric flow rates. Because more unique and powerful detection devices, such as microelectrodes, laser-based detectors, and novel mass spectrometers (which preferably use microliter-per-minute flow rates) have become attractive in bioanalysis, the interest in

[6] J. W. Jorgenson and K. D. Lukacs, *Anal. Chem.* **53,** 1298 (1981).
[7] J. W. Jorgenson and K. D. Lukacs, *Science* **222,** 266 (1983).
[8] M. Novotny, S. R. Springston, P. A. Peaden, J. C. Fjeldsted, and M. L. Lee, *Anal. Chem.* **53,** 407A (1981).
[9] Y. Ito, T. Takeuchi, D. Ishii, M. Goto, and T. Mizuno, *J. Chromatogr.* **358,** 201 (1985).
[10] M. A. Moseley, L. J. Deterding, J. S. M. de Wit, K. B. Tomer, R. T. Kennedy, N. Bragg, and J. W. Jorgenson, *Anal. Chem.* **61,** 1577 (1989).
[11] M. A. Moseley, L. J. Deterding, K. B. Tomer, and J. W. Jorgenson, *Anal. Chem.* **63,** 1467 (1991).
[12] M. Hail, S. Lewis, I. Jardine, J. Liu, and M. Novotny, *J. Microcol. Sep.* **2,** 285 (1990).
[13] E. C. Huang and J. D. Henion, *Anal. Chem.* **63,** 732 (1991).
[14] D. F. Hunt, J. E. Alexander, A. L. McCormack, P. A. Martino, H. Michel, J. Shabanowitz, N. Sherman, M. A. Moseley, J. W. Jorgenson, and K. B. Tomer, *in* "Techniques in Protein Chemistry II" (J. J. Villafranca, ed.). Academic Press, San Diego, CA, 1991.
[15] M. A. Moseley, J. Shabanowitz, D. F. Hunt, K. B. Tomer, and J. W. Jorgenson, *J. Am. Soc. Mass Spectrom.* **3,** 289 (1992).

smaller microcolumns has rapidly increased. Perhaps this is best exemplified by the use of microcolumns in electrospray mass spectrometry. Interest in working with small biological samples has also been on the rise in the areas of impurity characterization in biotechnology, drug metabolism research, and biochemical investigations in general. The extent of this interest is best represented by the exciting opportunities for investigations of the chemical composition of single biological cells.[16-18] Clearly, microcolumn separation techniques seem key to the best investigations of this kind, where separating power must be combined with highly sensitive detection tools.

The primary aim of this chapter on microcolumn LC and its related techniques is to provide a perspective on its value in biochemical research. Whereas much developmental work in micro-LC employed separation of small molecules, the primary emphasis of our discussion is on biological macromolecules and their degradation products. After a brief discussion of microcolumn types and their technology, the instrumental aspects pertinent to the field of biomolecular analysis are described. Selected applications further demonstrate certain unique capabilities of microcolumn separation systems.

Types of Miniaturized Columns

When microcolumn LC was being explored in the late 1970s, three major types of microcolumns were under development in different laboratories[2-5]: (1) packed microcolumns, often referred to as "microbore columns;" (2) semipermeable, loosely packed capillaries; and (3) open tubular microcolumns. Their structural characteristics are depicted in Fig. 1,[19] emphasizing typical dimensions of their inner diameter and particle size. The 1.0-mm i.d. columns explored in the initial studies[2,20,21] featured low volumetric flow rates (typically 50–100 μl/min) and enhanced mass sensitivities of some detectors. The same features were also emphasized by Ishii and co-workers in Japan in a report[3] that clearly demonstrated the instrumental feasibility and relative simplicity of a drastically miniaturized LC system. Overall, 1-mm i.d. columns have become a logical extension of the HPLC

[16] R. T. Kennedy and J. W. Jorgenson, *Anal. Chem.* **61,** 436 (1989).
[17] R. T. Kennedy, M. D. Oates, B. R. Cooper, B. Nickerson, and J. W. Jorgenson, *Science* **246,** 57 (1989).
[18] B. R. Cooper, J. A. Jankowski, D. L. Leszczyszyn, R. M. Wightman, and J. W. Jorgenson, *Anal. Chem.* **64,** 691 (1992).
[19] M. Novotny, *Anal. Chem.* **60,** 500A (1988).
[20] R. P. W. Scott and P. Kucera, *J. Chromatogr.* **169,** 51 (1978).
[21] R. P. W. Scott (ed.), "Small-Bore Liquid Chromatography Columns." Wiley-Interscience, New York, 1984.

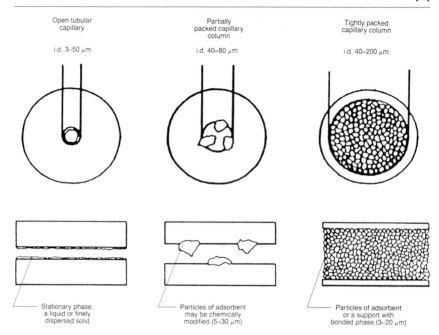

FIG. 1. Different geometries and typical dimensions of LC microcolumns.

column technology that needed relatively minor adjustment in the then available instrumentation. Open tubular LC columns represent an extreme of technology seeking a breakthrough performance through a drastic decrease in column inner diameters. According to the Golay equation[22] derived for the benefits of gas phase high-efficiency chromatographic analysis, the sorption–desorption kinetics of the column is largely controlled by the mobile phase solute diffusion. Semipermeable, loosely packed capillaries for LC were first described by Tsuda and Novotny.[4] In spite of their theoretical potential and a good compromise between high theoretical plate numbers and peak capacity, these microcolumns did not become popular. During the last several years, small-bore, totally packed microcolumns have become central to the field of microcolumn HPLC. A crucial step in column technology was the development of packed fused silica capillaries by Takeuchi and Ishii,[23] Yang,[24] and our laboratory[25] in the early 1980s. Flexible

[22] M. J. E. Golay, in "Gas Chromatography" (V. J. Coates, H. J. Noebels, and I. S. Fagerson, eds.), p. 1. Academic Press, New York, 1958.
[23] T. Takeuchi and D. Ishii, J. Chromatogr. 238, 409 (1982).
[24] F. J. Yang, J. Chromatogr. 236, 265 (1982).
[25] J. Gluckman, A. Hirose, V. L. McGuffin, and M. Novotny, Chromatographia 17, 303 (1983).

fused-silica tubing, originally developed for capillary gas chromatography, provides a nearly ideal envelope for a great variety of chromatographic packing materials. Its flexibility is further essential in interfacing the columns to specialized equipment. Packed capillaries of less than 200-μm i.d. have now become almost standard in the practice of LC coupled with electrospray mass spectrometry. Further improvements in technology of packed capillary columns were reported in 1988 by Karlsson and Novotny[26] and in 1989 by Kennedy and Jorgenson[27]: Highly efficient columns with inner diameters between 20 and 50 μm could be prepared using particle slurries.

Throughout the 1990s, biochemical researchers have begun to explore columns of reduced diameters for the reasons of trace enrichment and micropreparative capabilities. The microcolumns most typically used were 1-mm i.d. columns reminiscent of those pioneered by Scott and Kucera.[2] The benefits of reducing the inner diameter from 4.6 to 1.0 mm were substantial: (1) recording of the recovered proteins in nanograms per milliliter rather than micrograms per milliliter, (2) increased mass sensitivity permitting effective coupling to a mass spectrometer, and (3) better compatibility with proteolytic digest systems,[28] microsequencing,[29–31] and a previous separation in sodium dodecyl sulfate–polyacrylamide slab gel electrophoresis (SDS–PAGE). The introduction of packed fused silica capillaries in the protein analysis area[32,33] has led to further improvements, bringing analytical capabilities down to the nanogram-to-picogram range.

Microcolumn Characteristics

Following the initial success of Scott and Kucera[2,20,21] in reducing the column inner diameter of packed stainless steel columns down to 1 mm, "microbore columns" became of interest, particularly to researchers interested in a moderate sample scale-down. However, the experimental data indicated[34] that small particles were difficult to pack, and the resulting columns were actually less efficient for a length unit than the conventional

[26] K.-E. Karlsson and M. Novotny, *Anal. Chem.* **60**, 1662 (1988).
[27] R. T. Kennedy and J. W. Jorgenson, *Anal. Chem.* **61**, 1128 (1989).
[28] R. M. Caprioli, W. T. Moore, B. Da Gue, and M. Martin, *J. Chromatogr.* **443**, 355 (1988).
[29] E. C. Nice, B. Grego, and R. J. Simpson, *Biochem. Int.* **11**, 187 (1985).
[30] S. Yuen, M. W. Hunkapiller, K. J. Wilson, and P. M. Yuan, *Anal. Biochem.* **168**, 5 (1988).
[31] R. J. Simpson, R. L. Moritz, E. C. Nice, and B. Grego, *Eur. J. Biochem.* **165**, 21 (1987).
[32] C. L. Flurer, C. Borra, S. Beale, and M. Novotny, *Anal. Chem.* **60**, 1826 (1988).
[33] C. L. Flurer, C. Borra, F. Andreolini, and M. Novotny, *J. Chromatogr.* **448**, 73 (1988).
[34] P. Kucera, *in* "Microcolumn High-Performance Liquid Chromatography" (P. Kucera, ed.), p. 39. Elsevier, Amsterdam, 1984.

4.6-mm i.d. columns. Serendipitously, the use of fused silica capillaries with small column diameters[23-27] was an important advance because (1) 100- to 300-μm i.d. columns, typically packed in 0.5- to 1.0-m lengths, were found to be at least as efficient as the conventional columns; (2) reproducible packing procedures, presumably owing to the smoothness of the fused silica wall, could be fairly easily developed for a number of packing materials with relatively small (3–10 μm) particles; (3) such microcolumns exhibited the characteristic advantages expected from this column type, with the use of relatively easy-to-assemble instrumentation, and (4) convenient flexibility of fused silica tubing permitted easy formation of the on-column optical cells.

While the current interest of biochemical researchers in microcolumn LC stems primarily from their practical microisolation capabilities and compatibility with electrospray mass spectrometry, an important point that is not yet widely appreciated is the fact that the fused silica capillaries filled with small particles perhaps represent the best of modern HPLC in terms of approaching its theoretical potential. With 200- to 300-μm i.d. tubing, packed with 5-μm spherical siliceous particles (e.g., those employed for the reversed-phase LC), it has been common to prepare reproducible columns with theoretical efficiencies.

The studies with smaller inner diameters (below 100 μm) resulted in a fairly surprising conclusion: Using appropriate packing techniques, it is actually easier to pack small-diameter columns than the large ones! In addition, it is now well established that with decreasing the column diameter for the same particle size, the separation performance increases.[26,27] The best column performance observed in the study of Karlsson and Novotny[26] was with a 1.95 m × 44-μm i.d. microcolumn (packed with 5-μm particles) delivering 226,000 theoretical plates in 33 min. Kennedy and Jorgenson,[27] who extended the packing technology down to 20-μm i.d. columns, further explained that this improved performance due to the column diameter decrease was the result of a decreasing range of flow velocities in the column and a more uniform packing density caused by the desirable interactions between the particles and the column wall.

Following a discussion on the effect of column inner diameter, particle size and shape effects should be reviewed briefly. In all forms of liquid chromatography, microcolumn LC included, a decrease in particle size is the most straightforward way to enhance the separation performance of a system. In the contemporary practice of HPLC, spherical (totally porous) particles are most commonly employed. Not surprisingly, packing fused silica capillaries with 5-μm particles is now a most common procedure in microcolumn LC. During the earlier work with 3-μm particles, problems of packing reproducibility were initially encountered.[25] These can, however,

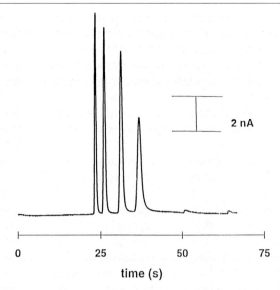

FIG. 2. Fast chromatographic separation of electrochemically active compounds. Elution order (from left to right): ascorbic acid, methylhydroquinone, 4-methylcatechol, and 4-hydroxy-3-methoxyphenylacetic acid. (From L. J. Cole, N. M. Schultz, and R. T. Kennedy, *J. Microcol. Sep.,* Copyright © 1993 John Wiley & Sons, Inc. Reprinted by permission of John Wiley & Sons, Inc.)

be overcome in time, as demonstrated by Hoffmann and Blomberg,[35] who were able to achieve reduced plate–height values around 2.5 with the microcolumns packed with 3-μm particles, but somewhat less satisfactory results with 2-μm materials. Yet another variation in microcolumn design involves the use of particles that have significant inner-channeling effects (perfused particles[36]) or particles featuring only superficially porous (pellicular) sorption geometries.[37] These particle types have been evaluated by Cole *et al.*[38] for work with packed capillaries. They showed that 8-μm pellicular particles, packed into 50-μm columns, generated nearly 7000 theoretical plates in 10 sec. The analytical merits of such a system are demonstrated in Fig. 2, where certain electrochemically active compounds were rapidly separated.[38]

Further studies on the combination of smaller particles and small column diameters clearly have considerable merit. Besides the lack of suitable packing technology, at present, the limitations of high pressure drop with

[35] S. Hoffmann and L. Blomberg, *Chromatographia* **24,** 416 (1987).
[36] S. Fulton, N. Afeyan, N. Gordon, and F. Regnier, *J. Chromatogr.* **547,** 452 (1991).
[37] K. Kalghatgi and C. Horváth, *J. Chromatogr.* **398,** 335 (1987).
[38] L. J. Cole, N. M. Schultz, and R. T. Kennedy, *J. Microcol. Sep.* **5,** 433 (1993).

small-particle beds may make progress in the area slow. An interesting variation in this regard is electroosmotically driven chromatography (electrochromatography) in packed columns.[39,40] Encouraging results attained using this approach clearly emphasize the need for small columns, so that the Joule heat generated during the process can easily be conducted away.

Column Preparation Techniques

The emphasis on separation of biomolecules in miniaturized columns has increased substantially. Because separation selectivity is the principal means of isolation and purification in biomolecular chromatography, in general, there is a need for microcolumns to be available for a variety of separation modes[41]: size-exclusion and ion-exchange principles, reversed-phase and hydrophobic interaction, immobilized metal affinity, borate complexation, lectin selectivity, and immunoaffinity, among others. Some of the packing materials employed in these separations will undoubtedly be siliceous, while others may be organic beads. Packing these materials into suitable microcolumn tubing may still necessitate the development of procedures beyond those reported to date. Some of the microcolumn technologies may also overlap with the development of microreactors[42,43] (off- or on-line) intended for the use of microscale enzymatic degradation and sample preparation procedures for samples available in minute quantities.

Although microcolumn LC has gained in popularity, there is a considerable lag in the availability of commercial microcolumns when compared to the standard (4.6-mm i.d.) columns. Whereas 1.0-mm metal columns have also been available for some time, the fused silica microcolumns that most researchers are currently interested in are rarely available commercially. Numerous laboratories seem to have acquired the know-how of packing the fused silica tubes with particles according to the previously published procedures. The packing procedures are relatively straightforward, while the fused silica tubing material can be purchased routinely. Such packing procedures are briefly described as follows.

Besides their chromatographic performance, the fused silica packed microcolumns possess a number of practical advantages over the conventional LC columns: (1) flexibility and ease of handling in instrumental modifications, (2) low cost of production, (3) minimal consumption of the

[39] J. H. Knox and I. H. Grant, *Chromatographia* **32**, 317 (1991).
[40] N. W. Smith and M. B. Evans, *Chromatographia* **38**, 649 (1994).
[41] M. Novotny, *J. Microcol. Sep.* **2**, 7 (1990).
[42] K. A. Cobb and M. Novotny, *Anal. Chem.* **61**, 2226 (1989).
[43] K. A. Cobb and M. Novotny, *Anal. Chem.* **64**, 879 (1992).

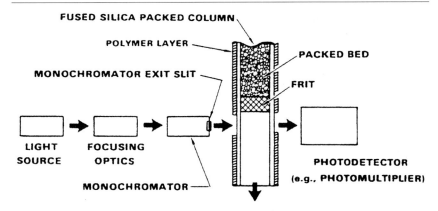

FIG. 3. Schematic diagram for on-column detection with a packed microcolumn. (Reprinted from *J. Chromatogr.*, **236**, F. J. Yang, p. 265, Copyright 1982 with kind permission of Elsevier Science–NL, Sara Burgerhartstraat 25, 1055 KV Amsterdam, The Netherlands.)

mobile phase, and (4) easy generation of the "optical window" for spectroscopic detection directly on-column, as exemplified in Fig. 3.[24] In a small section at the end of the tube, the polymeric overcoat is removed (e.g., through heat or the action of a strong acid), so that an optical beam can be focused onto this capillary section. For fluorescence measurements, the light beam can actually be directed straight to the column packing,[44] achieving minimum band broadening and high-sensitivity measurements. For electrochemical detection, a carbon fiber electrode can be directed inside the capillary.[45]

Yet another important part of microcolumn technology is placement of a small frit that retains the packing material inside the capillary.[44] To insert such a frit inside the capillary, microscopic manipulation is desirable (as it is, in fact, for a number of observations pertaining to construction and maintenance of microcolumns). In earlier work,[25] a piece of quartz wool was used as a column end piece, but this simple procedure was often found to increase the pneumatic resistance of the microcolumns and irreversible adsorption of the solutes. A procedure utilizing a small, porous Teflon end frit was subsequently developed,[44] which provided considerable improvement over the previous technique. In this procedure, a small disk is cut from the porous Teflon sheet and, using a simple guide device, inserted inside a capillary tube and cemented into place with a specialty glue. This procedure can easily be used for 200- to 500-μm i.d. capillaries, but becomes difficult with smaller microcolumns. Some alternative procedures involve

[44] D. C. Shelly, J. C. Gluckman, and M. V. Novotny, *Anal. Chem.* **56**, 2990 (1984).
[45] L. A. Knecht, E. J. Guthrie, and J. W. Jorgenson, *Anal. Chem.* **56**, 479 (1984).

a special ceramic frit support,[46] a section of coarser particles held in place by another epoxy-glued fused silica capillary,[26] or a sintered glass bead section.[27] The last procedure, developed by Kennedy and Jorgenson,[27] is particularly appealing because of its simplicity. Briefly, the frits are made by gently tapping one end of the capillary into a small pile of glass beads (until a 0.1- to 0.2-mm section is filled), pushing the plug of beads, under a stereomicroscope, into the desirable position in a future microcolumn using a 25-μm tungsten wire and, finally, sintering the particles in place by short exposure to heat.

Fused silica tubing up to 2 m in length can be slurry packed for preparation of efficient microcolumns; however, smaller pieces are most frequently utilized for various practical applications. The packing conditions may differ in terms of type of sorbent and particle size, as well as desired column diameter.[25–27,35,47,48] It is commonly agreed that packing of reversed-phase materials is the easiest. Slurries of sorption materials suspended in an appropriate solvent are forced into microcolumns, which are terminated with an end frit, by use of a small-volume, high-pressure bomb (packing reservoir) which may be magnetically stirred. The packing conditions, such as the slurry ratio (milliliters of solvent divided by grams of sorbent), the use of additives, and packing pressures and flows, vary somewhat in different reports. Optimization of the slurry ratio for packing of reversed-phase materials had already been reported in an early paper[25]; however, more dilute slurries appear necessary when packing columns with inner diameters of 50 μm, or less.[27] A brief slurry sonication prior to packing,[26,27,38] or even during the packing procedure,[48] has been advocated. The use of nonionic detergents, such as Nonidet P-40 (NP-40)[47,48] or Triton X-100,[27] can be beneficial in aiding a uniform dispersion of slurry components.

The packing flows and pressures are obviously important in the preparation of efficient microcolumns. Yet it is difficult to recognize a clear "recipe for success" from the studies published, presumably owing to the variation in microcolumn types, particle size, and inner diameter. Until the industry brings some degree of standardization into the area, individual researchers are advised to follow the literature procedures that "work" in their hands and for their specific applications. To highlight some important directions, it seems essential that different slurry composition and packing conditions pertain to the preparation of columns with very small inner diameters,[26,27,38] and that "slow" or "fast" packing conditions may be needed for materials

[46] H. J. Cortes, C. D. Pfeiffer, B. E. Richter, and T. S. Stevens, *HRC-CC, J. High Resolut. Chromatogr. Chromatogr. Commun.* **10,** 446 (1987).
[47] C. Borra, S. M. Han, and M. Novotny, *J. Chromatogr.* **385,** 75 (1987).
[48] F. Andreolini, C. Borra, and M. Novotny, *Anal. Chem.* **59,** 2428 (1987).

with different surface chemistry.[48] A stepwise change in pressure during the packing procedure has also been advocated.[38] Packing pressures of no more than several 1000 psi (typical for most procedures) and the relative simplicity of a packing apparatus, together with the low price of packing materials per column, make it still attractive for researchers to prepare their own slurry-packed microcolumns.

For most applications, microcolumns packed with 5-μm particles seem to provide the best combination of the desired efficiencies and time of analysis. Columns packed with reversed-phase materials may typically generate 100,000 theoretical plates/m and, indeed, 1-m long columns can be beneficial in separating complex mixtures of low molecular mass components.[49–51] The microcolumns needed for protein separation (microisolation)[32,33] are usually shorter (10–20 cm). It has been demonstrated that the use of pellicular packings can result in a significant improvement in performance and speed over conventional packings.[38] Employing a smaller particle size (1–3 μm) is yet another route to improvement in efficiency and speed, albeit at the expense of higher inlet pressures and the difficulties associated with it. Documentation of the use of very small particles in the micro-LC literature is limited. Gluckman et al.[25] prepared microcolumns with 3-μm reversed-phase packing, yielding around 130,000 plates/m, but the column-to-column reproducibility was lower than for the columns packed with larger particles. Hoffmann and Blomberg[35] achieved reduced plate–height values around 2.5 with 3-μm particles, but experienced packing difficulties with a 2-μm material.

As far as the microcolumns with different selectivities are concerned, some polar packings were reported earlier,[48] in addition to the commonly used reversed-phase systems. In relation to the separation of proteins and other biomolecules, additional packings featuring alkyl chains of different length (C_1, C_4, and C_8) are needed for the reversed-phase and hydrophobic interaction modes of separation. The column-packing procedures for these may not differ significantly from those used for C_{18} materials.[52] It has also been demonstrated[53] that C_8 materials can be used as a suitable matrix for the preparation of immunoadsorbent microcolumns. Following the example of Janis and Regnier[54] with conventional columns, Flurer and Novotny[53] demonstrated successful immobilization of various antibodies on C_8 microcolumns through applying small volumes of appropriate immunoglobulin

[49] M. Novotny, K.-E. Karlsson, M. Konishi, and M. Alasandro, J. Chromatogr. 292, 159 (1984).
[50] M. Novotny, M. Alasandro, and M. Konishi, Anal. Chem. 55, 2375 (1983).
[51] J. F. Banks, Jr., and M. V. Novotny, J. Chromatogr. 475, 13 (1989).
[52] C. L. Flurer, Doctoral dissertation, Indiana University, Bloomington, 1992.
[53] C. L. Flurer and M. Novotny, Anal. Chem. 65, 817 (1993).
[54] J. L. Janis and F. E. Regnier, J. Chromatogr. 444, 1 (1988).

G (IgG) solutions from a miniature reservoir. Through the use of frontal analysis, it was determined that 600–700 μg of IgG was needed to fully saturate 30 cm $\times$ 250-μm i.d. microcolumns. Exposed nonspecific binding sites on the column additionally were covered with conalbumin or lysozyme. The use of microcolumns in the immunoadsorptive mode is attractive from the point of view of minimum consumption of expensive immunoreagents through such a procedure. The use of such microcolumns in protein micro-isolation efforts is described in Microisolation and Single-Cell Analysis (as follows).

The wide utilization of affinity-based separations in biochemistry will undoubtedly lead to its parallels in microcolumn chromatography. Owing to the slow commercialization of the general method and its applications to microisolation, little activity has thus far been seen in this direction. Among the potentially useful forms of selectivity toward proteins is the so-called immobilized-metal affinity chromatography (IMAC), explored initially by Porath and co-workers[55] and adapted to HPLC by many others. The principle of this separation is well known to biochemists: Various metals are required in electron-transfer reactions, enzymatic action, transport processes, protein synthesis, etc. The biological selectivity toward metal ions, such as Zn^{2+}, Cu^{2+}, Fe^{3+}, and Al^{3+}, can thus be utilized in designing the chromatographic materials with the appropriate ligated structures that can bind selectively certain proteins. Appropriately derivatized siliceous materials,[56,57] compact organic resins,[58] and cross-linked agarose[59,60] were thus utilized in the art of protein HPLC. Flurer and Novotny[61] described preparation of slurry-packed microcolumns using a commercial chelate (Cu^{2+}) resin and studied the analytical advantages of this microcolumn mode, such as increased mass sensitivity, compatibility with the imidazole gradient, and optical detection at low wavelength. IMAC can also be utilized in probing topology of the protein surface and some other tertiary structural aspects.[62]

Instrumental Aspects

The separation column is the heart of an analytical system to which all other components must be adjusted. In microcolumn LC, this primarily

[55] J. Porath, J. Carlsson, I. Olsson, and G. Belfrage, *Nature* (*London*) **258**, 598 (1975).

[56] Z. El Rassi and C. Horváth, *J. Chromatogr.* **359**, 241 (1986).

[57] A. Figueroa, C. Corradini, B. Feibush, and B. L. Karger, *J. Chromatogr.* **371**, 335 (1986).

[58] Y. Kato, K. Nakamura, and T. Hashimoto, *J. Chromatogr.* **354**, 511 (1986).

[59] M. Belew, T.-T. Yip, L. Andersson, and R. Ehrnström, *Anal. Biochem.* **164**, 457 (1987).

[60] M. Belew, T.-T. Yip, L. Andersson, and J. Porath, *J. Chromatogr.* **403**, 20 (1987).

[61] C. L. Flurer and M. Novotny, *J. Microcol. Sep.* **4**, 497 (1992).

[62] E. Sulkowski, *Trends Biotechnol.* **3**, 1 (1985).

means a substantial decrease in the mobile phase flow rates: typically, 40–100 μl/min for 1.0-mm i.d. columns, several microliters per minute for 0.2- to 0.3-mm fused silica, slurry-packed capillaries, and only nanoliters per minute for open tubular capillaries. The crucial parts of miniaturized LC equipment with regard to such low flow rates are (1) the solvent delivery system (pumps and gradient elution), (2) sample introduction, and (3) detectors. With the HPLC routine equipment designed for flow rates on the order of milliliters per minute, the extracolumn band dispersion due to large volumes of injection valves and detectors would be a serious problem if such equipment were used with microcolumns. Following relatively minor instrument modifications,[2,21] typical HPLC instruments can accommodate 1-mm i.d. columns; however, specialized equipment must be used in work with smaller microcolumns. The design of reliable micropumps, high-pressure valves dispensing small sample volumes, small-sized fittings, and miniaturized detectors have all become crucial to the successful practice of microcolumn LC.

Reproducible flows of a few microliters per minute (or less) are needed to achieve reliable microcolumn LC. First, precision and accuracy of sampling volumes depend on the stability of flow rate. Also, reproducibility of retention times in long chromatographic runs is strongly influenced by the solvent delivery system. It has been shown by numerous laboratories that various older pump designs can be satisfactorily modified to serve microcolumn LC (using a split-flow operation), but the more recent systems, designed specifically as micropumps, tend to perform more reliably. The syringe type pumps, rather than reciprocating cylinder designs, seem generally preferable. Some commercially available hydraulic syringes are capable of metering accurately a flow of microliters per minute, and slightly below, which is satisfactory for 100- to 300-μm i.d. packed columns. Microcolumns with smaller diameters still need flow splitting.

The problems with generating solvent gradients in microcolumn LC were recognized in much of the earlier work. Obviously, reliably mixing the flows of liquids in microliter-per-minute flow regimes is a major engineering problem. Owing to the obvious value of gradient elution for the analysis of complex mixtures and isolation of proteins, microcolumn LC systems must fully adopt these powerful means of retentive manipulation. Although numerous "workable" systems for gradient elution in micro-LC have been proposed, the sought-for capabilities should be equivalent to those described in the theory and practice of conventional LC.

Various techniques of generating gradients for microcolumn LC were reviewed by Berry and Schwartz,[63] who emphasized the use of "slowed-

[63] V. V. Berry and H. E. Schwartz, in "Microbore Column Chromatography: A Unified Approach to Chromatography" (F. J. Yang, ed.), p. 67. Marcel Dekker, New York, 1988.

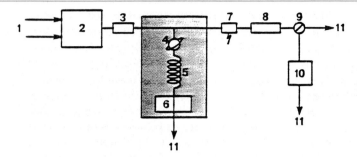

Fig. 4. Schematic of a micro-LC pumping system that utilizes flow splitting. 1, Solvent reservoirs; 2, high-pressure pump; 3, static mixer; 4, injection valve; 5, microcolumn; 6, detector; 7, pressure transducer; 8, pulse dampener; 9, purge valve; 10, microflow controlling device; and 11, waste. (Reproduced from Ref. 64 with permission of Dr. Alfred Huethig Publishers.)

down" conventional pumps through a simple pump modification and the flow-splitting approach. The flow-splitting procedure, described for the first time with microcolumn LC by van der Wal and Yang,[64] is probably the most commonly used procedure. Such a procedure is exemplified by the Varian 500 LC pump (Varian Instruments, Walnut Creek, CA) (Fig. 4), although it is clear that several other pumping systems are feasible to use. As seen from Fig. 4, the flow delivering the eluent with a continuously changing composition is split before the sample introduction point, while the major part of the eluent is permitted to bypass the analytical system. This simple solution is applicable to low-pressure as well as high-pressure gradient mixer systems, although an excessive volume between the point of mixing and the column inlet could be critical to the gradient compositional precision and accuracy.

   Introducing small sample volumes and amounts into microcolumns is mandatory for preventing column overloading and minimizing the width of input functions with respect to the column variance. The maximum permissible sample volumes can be estimated[65] from the general characteristics of a particular microcolumn or from empirical measurements. The volumes may range from small fractions of a microliter for 100- to 300-μm i.d. slurry-packed capillaries[66] to as little as a few picoliters for the small-bore open tubular columns. Direct sampling procedures are rarely used

[64] S. van der Wal and F. J. Yang, HRC-CC, J. High Resolut. Chromatogr. Chromatogr. Commun. 6, 216 (1983).
[65] M. Novotny, in "Microcolumn High-Performance Liquid Chromatography" (P. Kucera, ed.), p. 194. Elsevier, Amsterdam, 1984.
[66] J. C. Gluckman and M. Novotny, in "Microcolumn Separations" (M. Novotny and D. Ishii, eds.), p. 57. Elsevier, Amsterdam, 1985.

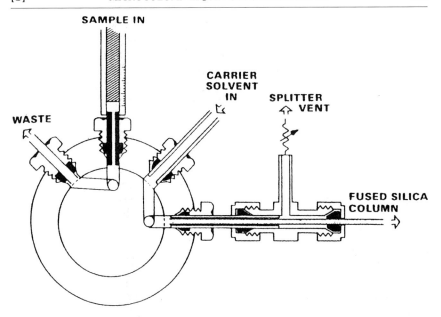

FIG. 5. A 0.5-μl internal loop sampling valve modified for split injection. (Reprinted from *J. Chromatogr.*, **236**, F. J. Yang, p. 265, Copyright 1982 with kind permission of Elsevier Science–NL, Sara Burgerhartstraat 25, 1055 KV Amsterdam, The Netherlands.)

to deliver a desired sample size to the microcolumn. The split injection techniques[4,24,67] employ a large injection volume, which is quantitatively divided between the column and a venting restrictor according to their relative flow restriction. Various splitting ratios can be utilized for different types of microcolumns or experimental conditions. An example of a splitting injection device is given in Fig. 5,[24] showing a modified internal loop sampling valve. Split injection is convenient and relatively precise. The disadvantages of this approach are the difficulties in injecting dilute samples (problems in detection sensitivity) and the fact that it is wasteful of precious samples. Using miniaturized sampling valves[47] has become feasible for slurry-packed capillaries with internal diameters above approximately 200 μm. To avoid band broadening associated with sample dilution and viscous drag while emptying the loop of a sampling valve, the use of the "moving-injection" technique[68] with electrically activated injection is highly recommended. Employing this procedure with valve loops 50–100 nl in volume is clearly satisfactory[47] for the commonly used fused silica microcolumns.

[67] T. Tsuda and G. Nakagawa, *J. Chromatogr.* **199**, 249 (1980).
[68] M. C. Harvey and S. D. Stearns, *J. Chromatogr. Sci.* **21**, 473 (1983).

There is still a need for the development of generally applicable sampling technologies in which the solutes originally present in dilute solutions can be reliably "focused" at the column inlet prior to the actual chromatographic analysis. The use of a solvent modifier[24] at the time of sampling, or an effective utilization of the solute solubility changes during a gradient elution, can be helpful in such a solute-focusing step. In a high-sensitivity analysis of sialic acid fluorescent derivatives,[69] we were able to focus effectively on a reversed-phase material as dilute as 10^{-9}–10^{-10} M. During preconcentration, the dilute sample was first placed in a small vessel (volumes ranging from 20 to 200 μl) and transported with pure water to the C_8 precolumn, in which the sample was first recovered, and after an appropriate dissolution (mobilization), the solute is, once again, refocused at the beginning of the C_{18} analytical column. Thereafter, a steep gradient was initiated to result in elution of the concentrated sample as a sharp peak.

Volumes and geometries of detector cells can also be expressed as extracolumn variances[66] and compared with the column variance. For as long as microcolumn LC has existed, significant efforts have been directed to the miniaturization of conventional detectors (based on UV absorbance, fluorescence, refractive index, electrochemical principles, etc.). Many LC detectors have now been drastically reduced in size (from microliter to nanoliter volumes and below). The key point in these investigations has been to make a sensible compromise between detection performance and extracolumn volumetric variance. Designs have been reported in which the solute band is probed directly on the column packing or immediately at the point of elution. Unfortunately, postcolumn chemical derivatization, which is commonly used in the practice of conventional LC, encounters serious limitations in microcolumn work. However, some interesting ideas on miniaturization in the postcolumn mixing arrangement in capillary electrophoresis[70,71] could potentially be utilized in future designs in microcolumn LC.

Detectors and Ancillary Techniques

At the beginning, miniaturization of LC detectors was largely viewed as a nuisance born of the necessity for a concomitant variance decrease with the reduced flows in microcolumns. It has been gradually recognized that such miniaturization can have far-reaching consequences for detection performance: (1) enhanced mass sensitivity of the concentration-sensitive detectors, (2) optimum performance of some modern mass spectrometric

[69] O. Shirota, D. Rice, and M. Novotny, *Anal. Biochem.* **205,** 189 (1992).
[70] D. J. Rose and J. W. Jorgenson, *J.Chromatogr.* **447,** 117 (1988).
[71] S. L. Pentoney, Jr., X. Huang, D. S. Burgi, and R. N. Zare, *Anal. Chem.* **60,** 2625 (1988).

devices, and (3) opportunities to design new detector types. Neither the separation step nor the detection performance need be compromised seriously due to miniaturization.

In the biochemical practice of microcolumn LC, and similarly in high-performance capillary electrophoresis, the UV absorbance detector remains the "workhorse." This is because in the separation and analysis of proteins and nucleotides, as well as in peptide mapping, the measurement of absorbance remains the most useful and universal means of detection. Improvements in the design of this detector thus remain a high priority. It often surprises those unfamiliar with microcolumn techniques how sensitive the miniaturized UV detectors actually are. On-column UV detection, in which a part of the capillary column is used as the optical window, often yields subnanogram detection levels[50,72] in spite of extremely short path lengths. The widely used cross-flow detectors are typically an order of magnitude less sensitive than more elaborate UV-absorbance cells of the Z-type configuration.[51,72] The extra bonus of using microcolumns in LC is the possibility of working with less transparent solvents and gradient elution solvent mixtures. An example of this is the already-mentioned case of immobilized-metal affinity chromatography, in which the use of microcolumns permitted gradient elution with imidazole as the buffer component.[61]

Spectroscopic consequences of detector cell miniaturization have their continuation with the photodiode-array devices[73,74] and fluorescence detection. In the latter case, there is no penalty for miniaturization, as best evidenced by the popular laser-induced fluorescence (LIF). Extremely low detection limits (low femtogram range) obtained with LIF are a reflection of the enhanced mass sensitivity possible with miniaturized columns.

The literature of the last decade is replete with reports of applying laser technologies to LC detection. Laser technologies combine well with microcolumns because of the highly collimated nature of laser beams, so that nanoliter or even picoliter volumes are easily probed at the outlet of the column. The use of lasers for enhancing detection capabilities in both microcolumn LC and high-performance capillary electrophoresis (HPCE) is likely in future to be an active area at the cutting edge of technological innovations. A variety of detection principles have been explored,[75,76] including thermal lens detection, refractive index, light scattering, photoioni-

[72] C. Borra, F. Andreolini, and M. Novotny, *Anal. Chem.* **59**, 2428 (1987).

[73] T. Takeuchi and D. Ishii, *HRC-CC, J. High Resolut. Chromatogr. Chromatogr. Commun.* **7**, 151 (1984).

[74] J. Gluckman, D. Shelly, and M. Novotny, *Anal. Chem.* **57**, 1546 (1985).

[75] E. S. Yeung (ed.), "Detectors for Liquid Chromatography." Wiley-Interscience, New York, 1986.

[76] E. S. Yeung, *in* "Microbore Column Chromatography: A Unified Approach to Chromatography" (F. J. Yang, ed.), p. 117. Marcel Dekker, New York, 1989.

zation, indirect polarimetry, circular dichroism, etc., for 1-mm i.d. columns, and smaller capillaries. However, the most straightforward and useful utilization of laser in detection is LIF. This measurement mode for highly fluorescent molecules is phenomenally sensitive, leading in a few isolated cases to detection limits below the attomole (10^{-18} mol) and zeptomole (10^{-21} mol) levels.[77] Although more practical and realistic cases are much less sensitive than indicated by the best-case detection at single-molecule levels, LIF is slowly becoming responsible for major methodological breakthroughs in biochemical analysis. Although commercialization of LIF in conjunction with microcolumn LC has not yet occurred at this stage, much has been described in terms of LIF instrumentation, including the designs of various microcells.[76–80] Simple fiber optic designs for collection of the light emitted from the laser-illuminated capillary (on-column detection) is probably the simplest and best arrangement. Only a few biochemically important compounds exhibit native fluorescence at the wavelength of "practical" lasers, such as the helium/cadmium laser, or the argon ion laser. The native fluorescence of proteins observed by the LIF approach necessitates expensive lasers[81] owing to the low-excitation wavelength. The use of fluorescence labeling in connection with LIF detection is perhaps the most direct way of dealing with nonfluorescent solutes. Some fluorescence-tagging applications are dealt with in the next section.

Some forms of electrochemical detection are extremely suitable for miniaturization. Unlike with UV absorbance (the limitation through Beer's law), there are no obvious drawbacks of small volumes with most electrochemical techniques. In addition, electrochemical techniques provide some of the most sensitive and selective measurements known. The general advances in the design and technologies of microelectrodes are likely to provide support to investigations of novel LC detectors in future. The earlier detectors designed for micro-LC[82,83] had volumes around 200–300 nl, depending on a particular electrode design, and were typically amperometric in their response mechanism. It was subsequently demonstrated by Slais and Krejci,[84] Manz and Simon,[85] and Knecht et al.[45] that much smaller detection volumes were feasible. Their detectors were primar-

[77] N. Dovichi, J. Martin, J. Jett, and R. Keller, *Science* **219,** 845 (1983).
[78] S. Folestad, L. Johnson, B. Josefsson, and B. Galle, *Anal. Chem.* **54,** 925 (1982).
[79] L. W. Hershberger, J. B. Callis, and G. D. Christian, *Anal. Chem.* **51,** 1444 (1979).
[80] J. Gluckman, D. Shelly, and M. Novotny, *J. Chromatogr.* **317,** 443 (1984).
[81] T. T. Lee and E. S. Yeung, *Anal. Chem.* **64,** 3045 (1992).
[82] Y. Hirata, P. T. Lin, M. Novotny, and R. M. Wightman, *J. Chromatogr.* **181,** 787 (1980).
[83] M. Goto, Y. Koyanagi, and D. Ishii, *J. Chromatogr.* **208,** 261 (1981).
[84] K. Slais and M. Krejci, *J. Chromatogr.* **235,** 21 (1982).
[85] A. Manz and W. Simon, *Anal. Chem.* **59,** 74 (1987).

ily designed for open tubular microcolumns with extremely small diameters, utilizing a tip consisting of an even smaller microelectrode positioned at the column outlet. An approximate estimation[45] indicated an effective detector volume of a few picoliters.

Knecht *et al.*[45] constructed a unique carbon fiber electrode system in which the tiny electrode is inserted directly in the column outlet. At such a reduced volume, the detector response actually becomes coulometric and is limited only by the number of electrochemically active molecules. Nearly 100% coulometric efficiencies were achieved with the design. Yet another useful attribute of such a microelectrochemical detector is the capability of recording current–voltage characteristics for the solutes emerging from the microcolumn outlet.[86] The remarkable sensitivity of this type of detection has made it feasible to record chromatograms of electrochemically active substances from single biological cells, such as snail neurons[17] and adrenomedullary cells.[18]

Undoubtedly, the most exciting detection method, which is becoming routinely combined with microcolumn LC, is mass spectrometry. The LC/MS combination (in its various forms) has been extensively developed since the early 1970s, but only recently has become truly revolutionized through new ionization techniques, such as dynamic fast atom bombardment (FAB) and electrospray ionization. Importantly, miniaturized LC columns have played a crucial role in these advances. It is beyond the scope of this chapter to review the developments in this field.

Analysis of Complex Mixtures

Mixtures of compounds derived from biological materials (i.e., plant extracts, physiological fluids, or even the digests of biomacromolecules) can be exceedingly complex. This inherent complexity has long been the primary incentive for the development of various chromatographic and electrophoretic techniques. Among various approaches to the separation and analytical measurements, HPLC has assumed considerable scope in modern biochemical applications. Microcolumn LC, being an extension of the very versatile HPLC toward analyzing smaller quantities of biological materials, is one of the most attractive options among the so-called "nanoscale separations." Another important, yet not so widely appreciated, fact is that microcolumn LC yields, under the right experimental conditions, efficiencies (plate numbers) that are superior to those of conventional HPLC. Numerous systems are now known to readily yield 50,000–100,000

[86] J. G. White, R. L. St. Claire, and J. W. Jorgenson, *Anal. Chem.* **58,** 293 (1986).

theoretical plates in the analysis of nonvolatile mixtures of steroids, bile acids, prostaglandins, carbohydrates, etc. In addition, microcolumn LC as a peptide-mapping technique is highly complementary to HPCE. In the following section, some applications of microcolumn LC to complex mixtures are reviewed. In addition, the role of "hyphenated techniques" in dealing with extraordinarily complex mixtures is briefly mentioned.

The determination of different metabolites simultaneously is an important problem from the biochemical and biomedical points of view. Multicomponent analyses, serving the concept of "metabolic profiles"[87–89] and distinguishing between normal and pathological conditions, are best performed by high-resolution separation methodologies that can simultaneously determine a multitude of metabolically related components of biological samples. While gas chromatography was initially explored for the purpose,[87–90] LC and HPCE will become in the future the major methods of choice in dealing with the nonvolatile constituents of physiological fluids.

In principle, metabolic profiling can involve any class of compounds where a special pattern of metabolites (qualitative and/or quantitative changes) is indicative of a metabolic abnormality. Steroids, bile acids, polyols, and prostaglandins are among the common examples of successful attempts at metabolic profiling[49,50,80,91] by microcolumn LC. Microcolumn methodologies of this kind are further demonstrated with steroids, which are among the most versatile molecules that nature has designed. They are known to exist in various conjugation forms and affect a variety of metabolic processes. Owing to the inherent complexity of these metabolic processes, a great number of metabolites exist in human urine or blood. Because it is difficult to measure such metabolites directly by means of common LC detectors, precolumn derivatization procedures have been developed[49,50] to convert the steroids of interest to spectroscopically active derivatives. For the levels of metabolites detected in aliquots of urinary samples, a simple benzoylation of hydroxysteroids is sufficient for UV detection. The general hydrophobicity of such compounds is conveniently utilized in reversed-phase separations with gradient elution.[50] To demonstrate the applicability of this approach to the investigation of biomedically interesting materials, Fig. 6 shows the urinary steroid profiles of a normal human male and a diabetic male patient.[59] Several of the profile constituents are

[87] C. Dalgliesh, E. Horning, M. Horning, K. Knox, and K. Yarger, *Biochem. J.* **101,** 792 (1966).
[88] E. C. Horning and M. G. Horning, *Methods Med. Res.* **12,** 369 (1970).
[89] E. Jellum, *J. Chromatogr.* **143,** 427 (1977).
[90] M. Novotny and D. Wiesler, *in* "New Comprehensive Biochemistry" (Z. Deyl, ed.), Vol. 8. Elsevier, Amsterdam, 1984.
[91] F. Andreolini, S. Beale, and M. Novotny, *HRC-CC, J. High Resolut. Chromatogr. Chromatogr. Commun.* **11,** 20 (1988).

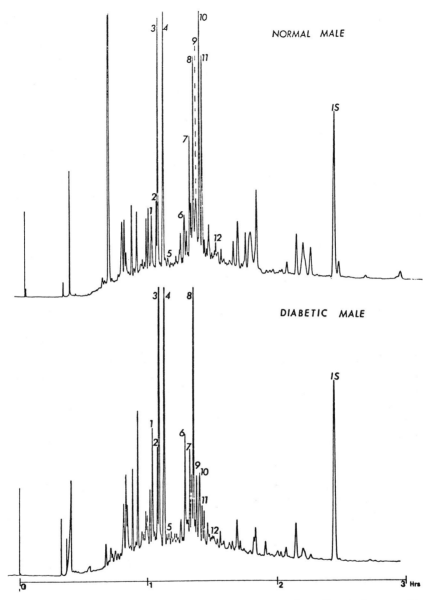

Fig. 6. Comparison of representative chromatographic metabolic profiles of urinary steroids from normal and diabetic human males. Reversed-phase chromatography with stepwise gradient conditions: 80% (v/v) acetonitrile (AcN)/H_2O (15 min); 85% AcN/H_2O (14 min); 90% AcN/H_2O (15 min); 95% AcN/H_2O (18 min); 100% AcN (∞). Peaks: (1) 11-Hydroxyandrosterone, (2) 11-hydroxyetiocholanolone, (3) allotetrahydrocortisol, (4) tetrahydrocortisol, (5) tetrahydrocortisone, (6) β-cortolone, (7) β-cortol, (8) α-cortolone, (9) α-cortol, (10) etiocholanolone, (11) androsterone, (12) dihydroepiandrosterone. (Reprinted from Ref. 50 with permission. Copyright 1983 American Chemical Society.)

characteristically altered in the diabetic sample, as is consistent with the result using a different analytical methodology.[92] Although the steroid metabolites were enzymatically deconjugated prior to derivatization with benzoyl chloride, it is also feasible to run them, under different experimental conditions, in their unconjugated form[49] as dansyl derivatives.

The analysis of steroid hormones and their metabolites in blood is more methodologically involved than the determination of steroid urinary profiles. This is due to (1) low circulation levels, (2) greater amounts of interfering compounds, and (3) smaller volumes of sample available for analysis. Consequently, fluorescence detection after microcolumn LC is a more appropriate choice than UV detection. Introduction of an appropriate fluorophore into the molecules of interest becomes a necessity. Toward this goal, we have introduced 7-(chlorocarboxylmethoxy)-4-methylcoumarin[93] as a unique fluorescence-tagging reagent for various hydroxy compounds. The advantage of this derivatization is that the resulting derivative has its excitation maximum near the 325-nm output of a helium/cadmium laser light source, so that LIF can be used as the detection technique at very high sensitivity.[80] Figure 7 shows the chromatogram of a solvolyzed steroid fraction from plasma[80]; the detected steroid peaks have been estimated at picogram levels. In a preliminary fashion, the same reagent was shown to be applicable for tagging hydroxy groups of prostaglandins.[49]

Identification and quantitative measurements of various bile acids and their conjugates in different physiological fluids and tissues are of great value in the study of various hepatobiliary diseases. Knowledge of the degree of conjugation or of the ratios between metabolites in physiological fluids can also provide clinically useful information. While different bile acid groups (according to their type of conjugation) can be separated from each other as based on selectivity, microcolumn LC can be used for each group as a highly effective, final determination method.[91] To introduce a fluorophore into the structures of bile acids, 4-(bromomethyl)-7-methoxycoumarin was employed[91] with a crown ether catalyst, yielding the derivatives excitable by the 325-nm line of the helium/cadmium laser. Detection by LIF easily permits visualization of low picomole-per-milliliter quantities. The derivatization with bromomethylcoumarin is also applicable to fluorescent tagging of prostaglandin molecules.[94]

To analyze some exceedingly complex mixtures in future studies, it will be important to combine the best methodologies that the area has to offer: (1) the utmost in separation efficiency and selectivity, (2) wide utilization

[92] M. Alasandro, D. Wiesler, G. Rhodes, and M. Novotny, *Clin. Chim. Acta* **126,** 243 (1982).
[93] K.-E. Karlsson, D. Wiesler, M. Alasandro, and M. Novotny, *Anal. Chem.* **57,** 229 (1985).
[94] V. L. McGuffin and R. N. Zare, *Proc. Natl. Acad. Sci. U.S.A.* **82,** 8315 (1985).

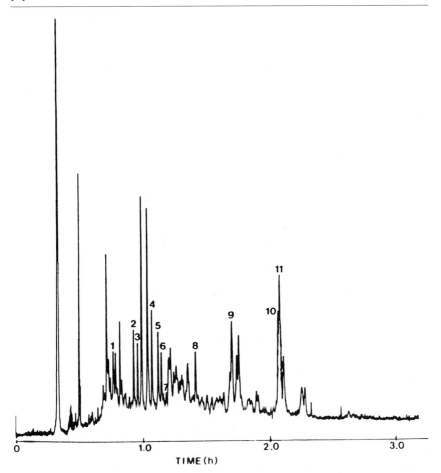

Fɪɢ. 7. Chromatogram of solvolyzed and derivatized plasma steroids. Tentatively identified components: (1) 5α-androstane-3α,11β-diol-17-one; (2) 5-β-androstane-3α,11β-diol-17-one; (3) 5β-pregnane-3α,-11β,17α,20β,21-tetrol-20-one; (4) 5β-pregnane-3α,17α,20β,21-tetrol-11-one; (5) 5β-pregnane-3α,11β,17α20β,21-pentol; (6) 5β-pregnane-3α,17α,20α,21-tetrol-11-one; (7) 5β-pregnane-3α,11β,17α,20α,21-pentol; (8) 5α-androstan-3α-ol-17-one; (9) 5-androsten-3β-ol-17-one; (10) 5β-pregnane-3α,20α,21-triol; and (11) 5β-androstane-3α,17β-diol. (Reprinted from *J. Chromatogr.*, **317**, J. Gluckman, D. Shelly, and M. Novotny, p. 443, Copyright 1984 with kind permission of Elsevier Science–NL, Sara Burgerhartstraat 25, 1055 KV Amsterdam, The Netherlands.)

of the best detection technologies (LIF, electrochemistry, and mass spectrometry, in particular), and (3) development of separation "hyphenated" techniques with orthogonal separating capabilities. Short of some unforeseen developments in separation science, dramatic future improvements in the efficiency of LC microcolumns are not immediately obvious. However, a wider employment of selectivity in group separations within complex mixtures prior to the utilization of "brute force" (high-plate columns) has a significant rationale. The detection techniques of high sensitivity and selectivity will need to be fully compatible with various gradient elution techniques. Numerous exciting possibilities are envisioned in the area of combined separation techniques: liquid chromatography/liquid chromatography; liquid chromatography/capillary gas chromatography; supercritical fluid extraction and/or chromatography/liquid chromatography; and, particularly, liquid chromatography/capillary electrophoresis. The developments in combining various multidimensional separation techniques have been reviewed by Cortes.[95] Owing to its various instrumental attributes, microcolumn LC lends itself uniquely to combining with additional techniques.

The rationale behind combining two microcolumn LC runs is in the complementary utilization of different phase systems, such as, for example, an ion-exchange separation of proteins in the first dimension, followed by size-exclusion chromatography. When the complementary techniques operate at roughly the same separation speeds, the "fraction storage" in the valve loops may present some difficulties. Obviously, the most attractive two-dimensional techniques are those in which the first run is slower than the second run, so that a continuous "on the fly" sampling from the first column into the second column is readily feasible. This condition is easily fulfilled in the combination of microcolumn LC and capillary electrophoresis, as shown initially by Bushey and Jorgenson[96,97] in 1990, and refined considerably by the same research group in the following years.[98–100] For example, in peptide-mapping efforts, reversed-phase LC and HPCE are truly orthogonal techniques[101,102]: one separating on the basis of sample hydrophobicity, and the other according to the charge-to-mass ratio of the solutes. Two-dimensional (2D) separations of peptides from larger proteins[100] provide a glimpse of future exciting possibilities in the rapid investigation of structurally related proteins, and in studies of complex biological samples in general.

[95] H. J. Cortes, *J. Chromatogr.* **626,** 3 (1992).

[96] M. M. Bushey and J. W. Jorgenson, *Anal. Chem.* **62,** 978 (1990).

[97] M. M. Bushey and J. W. Jorgenson, *J. Microcol. Sep.* **2,** 293 (1990).

[98] A. V. Lemmo and J. W. Jorgenson, *J. Chromatogr.* **663,** 213 (1993).

[99] J. P. Larmann, Jr., A. V. Lemmo, A. W. Moore, Jr., and J. W. Jorgenson, *Electrophoresis* **14,** 439 (1993).

[100] A. V. Lemmo and J. W. Jorgenson, *Anal. Chem.* **65,** 1576 (1993).

[101] J. Frenz, S.-L. Wu, and W. S. Hancock, *J. Chromatogr.* **480,** 379 (1989).

[102] R. G. Nielsen, G. S. Sittampalam, and E. C. Rickard, *Anal. Biochem.* **177,** 20 (1989).

Microisolation and Single-Cell Analysis

There are several useful attributes of microcolumn LC for small-scale isolation: (1) a relatively high mass sensitivity in detecting ever-smaller fractions of biomolecules, such as trace-level peptides and proteins; (2) less dilution during sample introduction and fraction collection; and (3) perhaps most important, significantly reduced irreversible adsorption due to the drastically decreased quantities of sorption materials. Using a microcolumn system, chromatographic manipulations can be done with a few microliters of solution, rather than the milliliters required by a conventional HPLC system. It should be acknowledged that the general trend of reducing the contact of protein solutes with sorption materials (with the simultaneous increase in sample recovery) was practiced previously through both a gradual decrease in column diameter[28–31] and the use of shorter columns. The use of small fused silica microcolumns for this task[32,33,53,61] adds a significant improvement.

In isolating small quantities of different proteins through microcolumn LC, different retention modes can be explored. Flurer et al. developed microcolumn LC systems involving the reversed-phase and size-exclusion principles,[32,33] ion-exchange chromatography and hydrophobic interaction,[52] immobilized-metal affinity,[61] and immunoaffinity.[53] An example of subnanogram handling of standard proteins is shown in Fig. 8[32] with size-exclusion chromatography. Obviously, a miniaturized UV detector is capable of "seeing" such small quantities, but an important question arises concerning further structural information on these fractions. Low-picomole quantities of proteins may still be compatible with the sensitivity of modern methods of amino acid analysis and state-of-the-art gas phase sequencers and, most certainly, the modern mass spectrometric techniques. It appears that significant improvements toward greater sensitivity are not currently nearly as limited by the availabilityof sensitive measurement technologies as they are by our capabilities to carry out chemical reactions (e.g., hydrolysis, specific fragmentations, compound tagging) at such low levels. Some inroads into the general problems are illustrated as follows.

In terms of amino acid analysis, laser-based detection and electrochemical measurements seem most attractive at present. Figure 9 displays a reversed-phase chromatogram[32] of the amino acids obtained from 430 fmol of bovine insulin and subjected to acid hydrolysis and fluorogenic derivatization with 3-benzoyl-2-quinoline-carboxaldehyde, which converts the primary amines to their fluorescent derivatives. Reproducible amino acid analyses at similar levels were also demonstrated with electrochemical detection, after derivatization with naphthalene-2,3-dicarboxaldehyde.[103]

[103] M. D. Oates and J. W. Jorgenson, Anal. Chem. 61, 432 (1989).

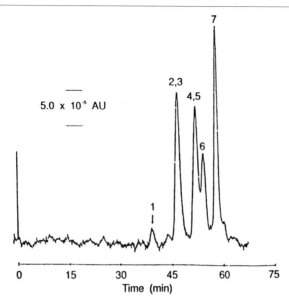

FIG. 8. Chromatogram of standard proteins on a 48 cm × 250-μm i.d. PC 300 microcolumn. Proteins in order of elution: (1) thyroglobulin, (2) transferrin, (3) bovine albumin, (4) β-lactoglobulin A, (5) carbonate dehydratase, (6) ribonuclease A, all approximately 200 pg each; and (7) 400 pg of tetraglycine. (Reprinted with permission from Ref. 32. Copyright 1988 American Chemical Society.)

Both microcolumn LC and HPCE can be utilized for determinations at these and perhaps lower levels.

    Site-specific fragmentation is one of the most informative techniques in protein characterization studies. A peptide map is a characteristic "signature" of a protein that contains a wealth of structural information. Not only can different site-specific enzymatic and chemical cleavages be utilized in the general procedures to yield complementary maps, but the differing reactivities of the formed peptides or their different spectroscopic properties can potentially be utilized to reveal accurately the subtle differences between individual proteins. The value of comparative studies is particularly evident in the detection of posttranslational amino acid modifications, the identification and localization of genetic variants, and the quality control and monitoring of genetically engineered protein products. Trends in peptide-mapping methodology have endorsed the value of HPCE, microcolumn LC, their combination with each other (LC/HPCE), and with electrospray mass spectrometry.

    Once again, the general point to recognize in conjunction with high-

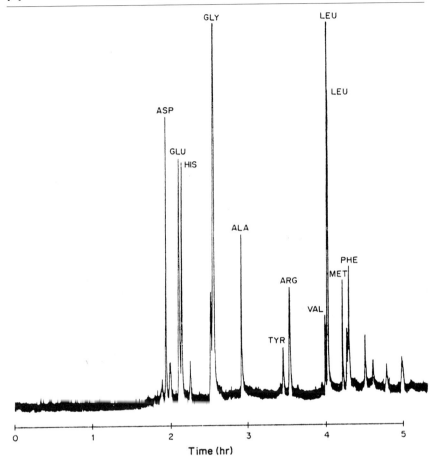

FIG. 9. Chromatogram of amino acids from 430 fmol of hydrolyzed bovine insulin obtained with a laser fluorescence detector. (Reprinted with permission from Ref. 32. Copyright 1988 American Chemical Society.)

sensitivity peptide mapping is the crucial nature of procedures for small-scale sample preparation. Cobb and Novotny[42,104] have reported procedures for the treatment of picomole-to-femtomole total protein quantities that involve the use of small-scale denaturation, reduction of disulfide bonds, alkylation, and protein fragmentation with immobilized enzymes. Ultraviolet detection of peptide maps generated by microcolumn LC is illustrated in Fig. 10, which compares phosphorylated and dephosphorylated forms

[104] K. A. Cobb and M. Novotny, *Anal. Biochem.* **200**, 149 (1992).

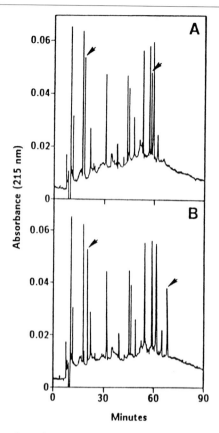

Fɪɢ. 10. Comparison of tryptic digests from (A) phosphorylated and (B) dephosphorylated forms of β-casein, separated by microcolumn HPLC. Arrows point to the two peaks that exhibit different retention times between the two forms. Each chromatogram was obtained from a 300-ng protein digest sample. (Reprinted with permission from Ref. 42. Copyright 1989 American Chemical Society.)

of β-casein.[42] The chromatogram was obtained from a 300-ng protein digest sample, but much smaller quantities of protein can be investigated through the use of HPCE and laser-induced fluorescence of fluorogenically tagged peptides.[43,104]

A good compatibility of microcolumn LC protein isolation with a high-sensitivity peptide-mapping procedure is shown in Fig. 11 with the analysis of a Q-Sepharose fraction of proteins isolated from mouse mucus.[52] In a somewhat similar approach, we have been able to identify glycopeptides containing sialic acid residues in the peptide map of bovine fetuin[69] and

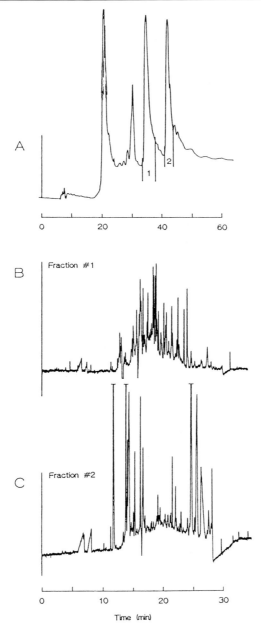

FIG. 11. Microisolation and characterization of proteins from a Q-Sepharose fraction of male mouse mucus: (A) a reversed-phase microcolumn separation of the fraction; (B and C) tryptic maps of fractions 1 and 2, obtained by capillary electrophoresis. (Reproduced from Ref. 52 with permission of the author.)

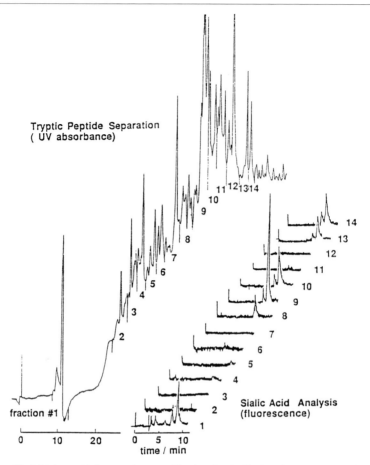

FIG. 12. A high-sensitivity tryptic map of bovine fetuin displayed together with the chromatograms corresponding to the sugar analysis of fractions 1–14. (Reproduced from Ref. 69 with permission of Academic Press.)

other proteins.[105] An example is shown in Fig. 12, in which the tryptic map of bovine fetuin is displayed together with the result of analyzing 14 fractions of the map for sialic acid content.

Isolation of proteins from complex biological mixtures often requires powerful separation principles such as highly selective immunoaffinity, lectin affinity (specific for certain sugar residues in glycoproteins), or pseudoaffinity offered by certain synthetic dyes, or immobilized metals. Besides the

[105] O. Shirota, Doctoral thesis, Indiana University, Bloomington, 1992.

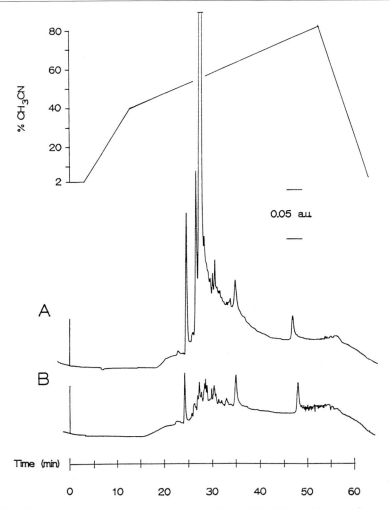

FIG. 13. Reversed-phase chromatographic profiles of (A) 600 nl of human plasma and (B) a 400-nl plasma fraction after elution from an anti-HSA immunoaffinity microcolumn. (Reprinted with permission from Ref. 53. Copyright 1993 American Chemical Society.)

already-mentioned advantages of small columns in microisolation, preparation of microcolumns on the basis of affinity principles is attractive because of cost considerations. Only microgram-to-milligram quantities of expensive reagents such as antibodies are consumed in such procedures. The preparation of an immunoaffinity microcolumn (IAC) has been described by Flurer and Novotny,[53] who utilized the simple procedure of immobilizing IgG on

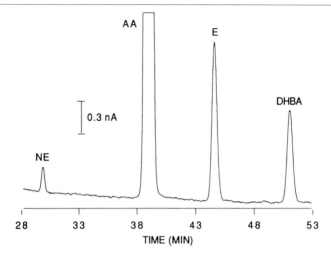

FIG. 14. Recording of the components of a single bovine adrenomedullary cell, using a reversed-phase microcolumn. NE, Norepinephrine; AA, ascorbic acid; E, epinephrine; DHBA, 3,4-dihydroxybenzylamine (internal standard). (Reprinted with permission from Ref. 18. Copyright 1992 American Chemical Society.)

a hydrophobic LC phase described by Janis and Regnier.[54] A description of the dual-column apparatus used in their studies on plasma proteins is given in Ref. 53.

The use of a miniaturized IAC system reduces considerably the sample volumes: quantification of human transferrin and α_1-antitrypsin was easily accomplished with 40 and 200 nl of plasma, respectively.[52] In addition, the system could be utilized for selective removal of a large interfering protein fraction, as was shown with an anti-albumin affinity column. Figure 13 demonstrates this procedure, comparing a 600-nl plasma protein profile (a reversed-phase separation) with a 400-nl plasma aliquot after the removal of albumin through an immunoaffinity.[53]

An increasingly important application of microcolumn separation and micromanipulation systems lies in the analysis of single cells and other small biological entities. In various biological investigations, it is often interesting to know to what extent morphologically different biological objects are similar or dissimilar in their chemical composition. Until recently, various hypotheses on the individual cellular composition were purely speculative, as they were difficult to prove experimentally. The case in point is the area of neurobiology, in which cellular heterogeneity is well established but seldom documented by the parallel knowledge of chemical attributes going hand in hand with physiological observations. Major advances in microelectrodes and microprobe technologies have been joined by the utilization of microcolumn separations[16–18] that are uniquely suited to distinguish closely related organic constituents of the individual cells. Kennedy, Jorgenson, and co-workers have shown, in a series of studies,[16–18] various suitable combinations of microscopic sample manipulation and reversed-phase microcolumn LC with electrochemical and LIF detection. Although open tubular columns were used in their earlier work because of volumetric requirements, small-diameter, slurry-packed capillaries were later adopted. Perhaps the most illustrative example of this type of work is their work on individual adrenomedullary cells[18] (with diameters of approximately 16 μm), in which they accurately evaluated variations in epinephrine and norepinephrine concentration (femtomole levels). A typical chromatogram of the electrochemically active constituents in a bovine adrenomedullary cell is seen in Fig. 14. These first demonstrations of the feasibility of doing meaningful analyses on such small biological objects will undoubtedly stimulate additional applications and efforts to elucidate biochemical processes at the cellular level.

Conclusion

After an apparently slow start and tedious instrumental developments during the 1980s, microcolumn LC has started to attract considerable attention. Working with biological samples at a reduced scale has been increasingly interesting to biochemical scientists. The benefits of microscale isolation techniques also add significantly to the already available separation and detection methodologies. Finally, LC microcolumns with their characteristically low flow rates are highly amenable to coupling with mass spectrometry and other ancillary techniques.

[6] Rapid Separations of Proteins
by Liquid Chromatography

By SANDEEP K. PALIWAL, MERCEDES DE FRUTOS, and FRED E. REGNIER

Introduction

If asked to give the most desirable attributes of a chromatographic system, or in fact any modern instrument, few life scientists would list "speed" or "analysis time." This is surprising in view of the fact that high-performance chromatographic separations of 1 hr or less have revolutionized biochemistry and biotechnology. Few biochemists who worked in laboratories before the 1980s will forget the hours spent doing paper and thin-layer chromatography of peptides to confirm purity or the days spent in cold rooms separating proteins on soft gel columns. Purification of a protein or peptide could consume the bulk of a student's time in graduate school during that era.

High-performance liquid chromatography (HPLC) is now a preferred alternative to soft gels for many biological applications. We are entering a new era in which HPLC materials of the 1980s will be challenged by new chromatographic materials that allow even faster separations. It will be shown here that new technologies allow chromatographic separations of macromolecules to be achieved more rapidly than by HPLC and perhaps two orders of magnitude faster than with soft gel materials. This discussion focuses on the need for greater separation speed, phenomena that limit the speed of chromatographic separations, various techniques for overcoming these limitations, and applications of high-speed separations.

Use of High-Speed Separations

One reason for using high-speed separations is to increase throughput. In many laboratories it is becoming necessary to be able to analyze by chromatography about 50 samples a day. There are a number of examples in the research and development environment, clinical laboratories, pharmaceutical manufacturing, quality control laboratories, and regulatory agencies where the rate of sample analysis is too slow. One example is in monitoring the elution profile from preparative columns. It is calculated that with an existing HPLC system, slightly more than an 8-hr day would be required to examine 60 fractions from a single preparative separation. By increasing the separation speed an order of magnitude, the task could be accomplished in 1 hr. The ability to analyze 500 or more samples a day would dramatically increase the utility of liquid chromatography in peptide sequencing, large-scale diagnostic screening programs, etc.

A second advantage of speed is in the optimization of purification methods. There are generally four or more variables (e.g., pH and ionic strength) that control resolution and selectivity in a purification process. Searching for the right combination can require the examination of 50–100 sets of conditions. This task was so formidable in the time of soft gel columns that chromatographic purifications were seldom optimized. With rapid chromatographic separations, a different set of conditions can be examined every 1–2 min. This makes it possible to optimize a chromatographic step in a day or less. When coupled with a computer-driven system that assists in experiment planning, reagent preparation, data acquisition, and data analysis, optimization of even complex separations can be achieved in a few hours.

A third advantage of high-speed systems is to improve preparative/ process-scale separations. It has been the experience of many biochemists that protein recovery increases with faster separations. Protein denaturation is diminished by shortening the time a protein spends on the surface of a column, and in some cases reduces proteolysis as well. Furthermore, large-scale chromatographic purification of proteins has generally been achieved with a single chromatographic column. High-speed preparative separations in the range of 10 min would allow smaller, higher resolution columns to produce more protein of higher purity when used in a repetitive cycling mode. This would be especially true with an accompanying immunologically based monitoring system in which an antibody-based column is used to purify proteins in a single pass across a column. A smaller column at high flow rates can be (1) used to purify proteins rapidly, allowing the column to be reused for multiple cycles to purify greater amounts, or (2) used as a sensor to analyze fractions rapidly from a process stream.

Finally, a fourth advantage of rapid separation technology is that it allows feedback control of processes. Data collected and analyzed in a time frame equivalent to that of an ongoing process are frequently used to alter or control the direction of that process. When this is done with a chromatographic system, the instrument, in effect, becomes a sensor. "Real-time" monitoring is of utility in the biotechnology industry.[1] In fermentors, control of reagent addition and directing product harvesting through product analysis during fermentation allows both product quality and yield to be increased.[2] The use of high-speed analytical systems to operate "smart" fraction collectors, assure product quality, and validate processes in real time are equally important examples of the impact of sensors on the production of therapeutic proteins by recombinant DNA technology.

Speed Limitations of Chromatographic Supports

Soft gel separation media suffer from two basic problems: severe limitations to intraparticle mass transport and poor mechanical strength. The mass transfer limitations stem from the fact that soft gels are of large particle diameter, generally greater than 74 μm (200 mesh), and of low porosity. The time required for intraparticle diffusion in the stagnant mobile phase results in a severe speed limitation. Although this problem could be diminished by using smaller particles, the mechanical strength of soft gels

[1] S. K. Paliwal, T. K. Nadler, and F. E. Regnier, *Trends Biotechnol.* **11,** 95 (1993).
[2] S. K. Paliwal, T. K. Nadler, D. I. C. Wang, and F. E. Regnier, *Anal. Chem.* **65,** 3363 (1993).

is sufficiently low that they compress when smaller particles are used at mobile phase velocities in excess of 60-cm/hr linear velocity.

These problems were substantially reduced with the introduction of high-performance liquid chromatography (HPLC). The rigidity of microparticulate packings 3–10 μm in diameter with pores up to 1000 Å in mean diameter allowed them to be used at mobile phase velocities of 600–700 cm/hr. Small particles of this porosity gave superior separations in less than one-tenth the time. It became possible to think of HPLC purity determinations, enzyme assays, analysis of tryptic digests, and quality control measurements in less than 1 hr.[3]

Unfortunately, decreasing particle diameter to reduce the time required for intraparticle mass transfer has a price: higher operating pressure and poorer column longevity. When the particle diameter is reduced to 2–3 μm, operating pressures in excess of 300 atm are common with aqueous mobile phases at 1000- to 2000-cm/hr linear velocity.[4,5] Beds of very small particles are also excellent filters. Particulate contamination of samples with dust or precipitated proteins will cause columns to plug quickly. Lower particle size limits in routine separations are generally 3 μm for analytical separations and 20 μm in preparative work.

Several conclusions may be drawn from the preceding discussion. First, the slow rate at which large molecules diffuse into the intraparticulate cavities of chromatographic supports provides the dominant limitation for speed in liquid chromatography. Second, although the use of smaller porous particles in conventional HPLC columns diminishes this problem, intraparticle diffusion is still the dominant mass transfer variable. Third, the practical lower particle size limits for analytical and preparative liquid chromatography columns used for biological samples are 3 and 20 μm, respectively.

Enhanced Mass Transport Media

Three different routes have been taken during the past few years to enhance mass transfer: (1) eliminating the support pores, (2) filling the core of the support to reduce diffusion path length, and (3) introducing particle transacting pores of 6000–8000 Å that cause the liquid to flow through the particle. It is important to note that these new approaches to liquid chromatography do not alter the separation mechanism, only the means of transferring from the mobile phase to the surface of the column.

---

[3] R. J. Garnick, N. J. Solli, and P. A. Papa, *Anal. Chem.* **60,** 2546 (1988).
[4] C. Horváth, B. A. Preiss, and S. R. Lipsky, *Anal. Chem.* **39,** 1422 (1967).
[5] N. Nimura, H. Itoh, T. Kinoshita, N. Nagae, and M. Nomura, *J. Chromatogr.* **585,** 207 (1991).

Nonporous Particle Supports

Elimination of the support pores is an obvious way to circumvent intra-particle diffusion limitations. Through the use of nonporous particles, the only remaining particle-based diffusion limitations are in a shallow layer of stagnant liquid at the surface of the particle. Because this layer is less than 1000 Å deep, it does not pose a serious mass transfer problem. The only limitation of nonporous particle sorbents is that they have approximately 1/100 the surface area of porous supports and therefore are of limited loading capacity. Attempts to reduce this problem have taken the form of reduced support particle size, roughening of the surface of the particle, and creation of a fimbriated layer of stationary phase.[6] The surface area of a nonporous particle bed increases with the inverse square of particle diameter. This is the reason that 3-μm particles are generally used in nonporous particle columns. The creation of surfaces that are physically rough is yet another way to increase surface area. The rugosity of these particles has been likened to the surface of a golf ball. This approach probably doubles the loading capacity. Yet another way to increase the surface area of a sorbent is to use a stationary phase layer that is hairlike (fimbriated).[6]

The use of 2- to 3-μm nonporous silica particles for protein separations was introduced by Unger and colleagues in the early 1980s.[7] It was shown that separation times of a few minutes could be routine, as the hydrophobic interaction liquid chromatographic separation of eight proteins in Fig. 1 demonstrates. Equally rapid separations were achieved in the reversed-phase mode with 2- to 3-μm silica-based materials.[8] In a direct comparison of nonporous and porous immobilized metal affinity chromatography sorbents, four model proteins were separated in 4 min on a nonporous particle column as opposed to 45 min on a porous particle HPLC column.[9] An interesting aspect of this paper was that phosphorylated and nonphosphorylated peptides from erythrocyte membrane proteins were resolved by this mode of chromatography.

Nonporous organic resins have also been used for rapid protein separations. Development of adsorbed polymer coating technology[10] enabled polyethyleneimine to be grafted to the surface of a 7-μm nonporous poly-

---

[6] L. Varady, N. Mu, Y.-B. Yang, S. E. Cook, N. Afeyan, and F. E. Regnier, *J. Chromatogr.* **631**, 107 (1993).

[7] R. Janzen, K. K. Unger, H. Giesche, J. N. Kinkel, and M. T. W. Hearn, *J. Chromatogr.* **397**, 91 (1987).

[8] K. K. Unger, G. Jilge, J. N. Kinkel, and M. T. W. Hearn, *J. Chromatogr.* **359**, 61 (1986).

[9] G. K. Bonn, K. Kalghatgi, W. C. Horne, and C. Horváth, *Chromatographia* **30**, 484 (1990).

[10] A. Alpert and F. E. Regnier, *J. Chromatogr.* **185**, 375 (1979).

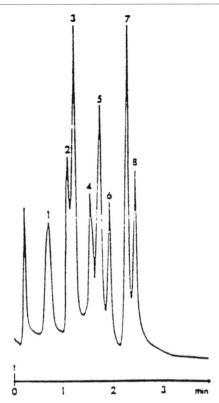

FIG. 1. Fast HIC separation of eight standard proteins on an amide column (36 × 8-mm i.d.); 2.5 min descending gradient from 2.5 to 0 M ammonium sulfate in 0.1 M phosphate buffer (pH 7.0); flow rate, 4.0 ml/min. Elution order: V_0 peak (breakthrough of part of the cytochrome c due to the low ionic strength of the injected solvent and low peak dispersion), cytochrome c (1), myoglobin (2), ribonuclease (3), lysozyme (4), ovalbumin (5), lactate dehydrogenase (6), catalase (7), and ferritin (8). (From Ref. 7.)

methacrylate bead to create a fimbriated anion-exchange sorbent[11] that could separate model proteins in 90 sec (Fig. 2). Similarly, by using a polyvinylimidazole-adsorbed polymer on a silica surface it has been possible to achieve rapid protein separations.[12] With still another approach, deformed nonporous agarose beads were used in the boronate affinity mode

[11] D. J. Burke, J. K. Duncan, L. C. Dunn, L. Cummings, C. J. Siebert, and G. S. Ott, *J. Chromatogr.* **353**, 425 (1986).
[12] G. Jilge, B. Sebille, C. Vidal-Madjar, R. Lemque, and K. K. Unger, *Chromatographia* **37**, 603 (1993).

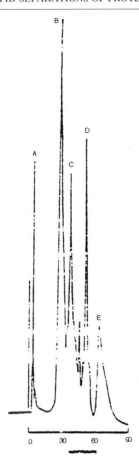

Fig. 2. High-speed separation of a mixture of horse myoglobin (A), human myoglobin (B), conalbumin (C), ovalbumin (D), and bovine serum albumin (E). Buffer A, 20 mM Tris (pH 8.5); buffer B, 20 mM Tris–0.5 M NaCl (pH 8.5); gradient, 0 to 100% B in 42 sec; flow rate, 4.5 ml/min. (From Ref. 11.)

to separate glycated proteins.[13] After hemolysis, the glycated hemoglobins were separated directly from the blood sample in 2 min without removal of cell debris.

As noted previously, the most negative feature of nonporous particle chromatography is the low sample loading capacity and high operating pressure of columns packed with very small particles. When loading capacity is not a problem, it appears that larger particles may be used without

[13] S. Hjerten and J. P. Li, *J. Chromatogr.* **500,** 543 (1990).

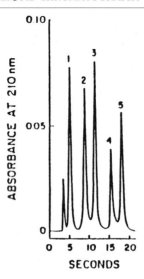

Fig. 3. Chromatogram of a protein mixture. Column: 30 × 4.6 mm packed with 2-μm fluid-impervious spherical silica particles having covalently bound *n*-octyl functionalities at the surface. Linear gradient in 48 sec from 15 to 95% acetonitrile in water containing 0.1% TFA. Flow rate, 4 ml/min. Temperature: 80°. Column inlet pressure, 30 MPa. Sample components: (1) ribonuclease A, 50 ng; (2) cytochrome *c*, 25 ng; (3) lysozyme, 25 ng; (4) L-asparaginase, 50 ng; (5) β-lactoglobulin A, 25 ng. Detection at 210 nm. (From Ref. 14.)

significant loss in resolution. This is particularly true in the ion-exchange, hydrophobic interaction, immobilized metal affinity, and bioaffinity modes. In contrast, large nonporous particles are almost never used with reversed-phase separations.

The pressure associated with the use of 2- to 3-μm particles at high mobile phase velocity has also been addressed through the use of elevated column temperature. The viscosity of an aqueous mobile phase, and therefore operating pressure, is substantially reduced when columns are operated in the range of 70–90°.[14] Using elevated temperature, it has been possible to separate standard proteins in 6 sec with a nonporous reversed-phase column (Fig. 3). High temperature has also been used to perform peptide mapping[15] (Fig. 4). It can be seen that the entire tryptic digest of recombinant tissue plasminogen activator (rtPA) can be resolved in 16 min.[15] The problems with the use of high temperature are obvious: Proteins are denatured when separated at elevated temperature with an acidic mobile phase containing organic solvent. Reversed-phase separation of proteins at ele-

[14] K. Kalghatgi and C. Horváth, *J. Chromatogr.* **398,** 335 (1987).
[15] K. Kalghatgi and C. Horváth, *J. Chromatogr.* **443,** 343 (1988).

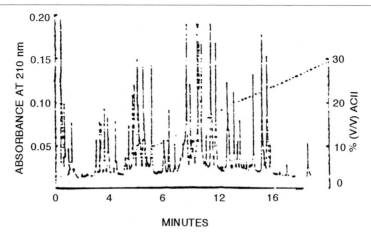

FIG. 4. Tryptic map of rtPA. Column: 75 × 4.6 mm micropellicular C_{18} silica. Eluent A, 5 mM hexylsodium sulfate and 50 mM sodium dihydrogen phosphate in water, adjusted to pH 2.8 with orthophosphoric acid; eluent B, 50 mM sodium dihydrogen phosphate in 60% (v/v) acetonitrile, adjusted to pH 2.8 with orthophosphoric acid; flow rate, 2 ml/min; temperature, 80°; sample, 5 μg of reduced and S-carboxymethylated rtPA digest in 20 μl. (From Ref. 15.)

vated temperature may be used only in analytical studies in which denaturation and dissociation of multimeric proteins are acceptable. Denaturation can also lead to low recoveries and column fouling. But it should also be pointed out that higher temperature increases the speed of analysis, which in turn minimizes protein denaturation.

The manner in which columns are heated has been shown to be an important issue.[16] When only the column is heated, serious temperature equilibration problems can be encountered. At high mobile phase velocity, both axial and radial temperature gradients are created when cold mobile phase enters the column, leading to peak distortion. Furthermore, temperature gradients are created at the column exit. As the mobile phase passes from the column to the detector, the liquid begins to cool, thus creating refractive index gradients in the detector, which increases the noise in absorbance detectors. It is either necessary to use high-efficiency heat exchangers at both the column inlet and outlet or to use a preinlet heat exchanger and operate the column and detector flow cell at elevated temperature. Both approaches require instrument modification.

Operation of columns at moderately elevated temperature with apparent retention of biological activity has been achieved in the bioaffinity mode. Fractionation of human immunoglobulin G (IgG) from contaminat-

[16] G. P. Rozing and H. Goetz, *J. Chromatogr.* **476,** 3 (1989).

ing proteins on a protein A column has been achieved in 3 min at 45°.[17] Further increases in temperature were not useful in efforts to decrease the separation time and increase resolution.

Column lifetime is a serious problem when organosilane-derivatized silica-based columns are operated at elevated temperature. The lifetime of organosilane-bonded phases is generally cut in half by each 10° increase in temperature. This problem has been addressed through the use of organic polymer-based columns. Three-micrometer poly(styrene-divinylbenzene) (PS-DVB) particles have been found to give reversed-phase separations of proteins equivalent to those achieved on alkylsilane-derivatized silica columns.[18,19] The advantage of PS-DVB particles is that the natural hydrophobicity of the matrix is sufficient for reversed-phase separations without derivatization. In addition, the PS-DVB matrix is stable from pH 1 to 14.

Nonporous particle columns are available from three suppliers. Reversed-phase, ion-exchange, and hydrophobic interaction columns may be obtained from TosoHaas (Montgomeryville, PA). Glycotech (New Haven, CT) supplies a poly(styrene-divinylbenzene)-based reversed-phase column and Bio-Rad (Richmond, CA) supplies methacrylate-based resins with fimbriated polyethyleneimine anion-exchange stationary phases.

Core-Filled Supports

Another approach to the intraparticle mass transfer problem is to control the depth of the pores. This has been done by attaching a 1-μm thick layer of 300-Å pore diameter particles to the exterior surface of a larger nonporous particle.[20] These "poreshell" composites have a nonporous (or filled) inner core with a thin, porous external layer. Because the pore depth is 1 μm or less, mass transfer limitations are substantially diminished. It is interesting that small core-filled particles can be of relatively high surface area and capacity. For example, a 5-μm core-filled particle with a 1-μm surface layer of sorbent has 64% of the capacity of a totally porous particle. In contrast, a 10-μm particle with a 1-μm surface layer has only 19% the loading capacity of a totally porous particle. Figure 5 shows a reversed-phase separation of five model proteins in less than 90 sec on this support. It should be noted that the operating pressure of this column was 150 atm at 5 ml/min. This column probably could not be operated at greater than 10 ml/min. Core-filled particle columns are available from Rockland Technologies (Rockland, PN).

[17] L. Varaday, K. Kalghatgi, and C. Horváth, *J. Chromatogr.* **458,** 207 (1988).
[18] Y.-F. Maa and C. Horváth, *J. Chromatogr.* **445,** 71 (1988).
[19] J. K. Swadesh, *BioTechniques* **9,** 626 (1990).
[20] J. J. Kirkland, *Anal. Chem.* **64,** 1239 (1992).

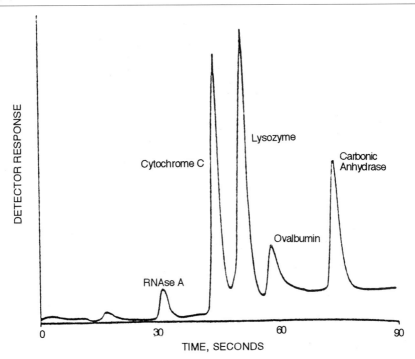

Fig. 5. High-speed separation of proteins with poreshell column. Flow rate, 5 ml/min; back pressure, 150 atm. (From Ref. 20.)

Perfusion-Based Supports

A third approach to the problem of stagnant mobile phase mass transfer is to cause the liquid to flow, or "perfuse," through the particles. Introduction of 6000- to 8000-Å particle transacting pores into sorbents allows a small portion of the mobile phase flowing around the outside of a support to perfuse through the pore matrix.[21,22] Perfusion of mobile phase through chromatographic supports has been shown to accelerate intraparticle mass transfer of macromolecules by one to two orders of magnitude.[23] A 15-sec gradient elution separation of proteins in the reversed-phase mode has been reported using a perfusion chromatography column with a mobile

[21] N. B. Afeyan, N. F. Gordon, I. Mazsaroff, L. Varady, S. P. Fulton, Y. B. Yang, and F. E. Regnier, *J. Chromatogr.* **519,** 1 (1990).
[22] F. E. Regnier, *Nature (London)* **350,** 634 (1991).
[23] S. P. Fulton, N. B. Afeyan, N. F. Gordon, and F. E. Regnier, *J. Chromatogr.* **547,** 452 (1991).

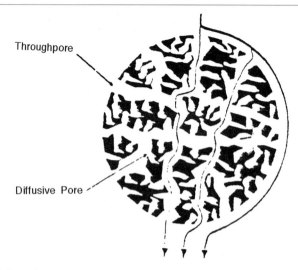

Throughpore

Diffusive Pore

Fig. 6. Schematic diagram of perfusion chromatography packing materials, showing throughpores for fast convective intraparticle mass transport and diffusive pores for high surface area and binding capacity. (From Ref. 24.)

phase velocity of 9000 cm/hr (Fig. 6). At this flow rate the residence time of mobile phase in the column was slightly more than 500 msec.

Unfortunately, pores of this size are of low surface area. The surface area of perfusion chromatography supports is increased by a network of smaller, 500- to 1500-Å pores[24] of 1 μm or less in length that interconnect the large throughpores (Fig. 7). This increases the surface area of perfusion chromatography supports to 30–40% of that of high-capacity soft gel and HPLC supports. Perfusion chromatography sorbents with capacities comparable to those of other sorbents are obtained by applying fimbriated stationary phases to perfusable supports.[6]

Because perfusion chromatography provides a fluid mechanics solution to the mass transfer problem, it is important to understand the fluid dynamics of the system. Liquid flows both through and around the sorbent particles in perfusion chromatography. Mathematical models reveal that when only 2–3% of the mobile phase flows through the particle, there is a substantial increase in performance. Although small, this amount of liquid perfusion is sufficient to sweep macromolecules rapidly through the pore matrix of the particle. It is also important to understand that even though the split ratio of flow, "through" as opposed to "around" the particle, remains

[24] N. B. Afeyan, S. P. Fulton, and F. E. Regnier, "Applications of Enzyme Biotechnology" (J. W. Kelly and T. O. Baldwin, eds.), p. 221. Plenum Press, New York, 1991.

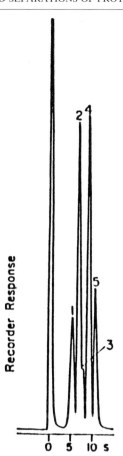

Fig. 7. A high-speed reversed-phase separation of proteins with a 24-sec gradient on a POROS R/M column packed with 20-μm particles and operated at a mobile phase linear velocity of 8700 cm/hr. Peak 1, ribonuclease A: peak 2, cytochrome c; peak 3, lysozyme; peak 4, β-lactoglobulin; peak 5, ovalbumin. (From Ref. 23.)

constant when the flow rate is altered, the absolute flow velocity through the particle changes in direct proportion to the total flow rate through the column. Thus, the faster the mobile phase is pumped through the column, the faster macromolecules are swept through the pore network of the particle.

Enhancement in intraparticle mass transfer is also related to the rate of protein diffusion relative to flow rate through the particle. At low mobile phase velocity, i.e., less than 300-cm/hr linear velocity, proteins are transported inside sorbents faster by diffusion than by perfusion. This means that perfusion chromatography columns provide no advantage over conventional soft gel and HPLC columns at low flow rate. However, at high mobile phase velocity the opposite is true, i.e., the transport rate by perfusion greatly exceeds that by diffusion. The relative rate of transport by convection versus diffusion is known as the Peclet number. Obviously, the Peclet number will be greater for large molecules with small diffusion coefficients. It is easily seen that perfusion provides a greater advantage for large molecules than for small molecules that diffuse 100 times faster.[21]

Perfusion chromatography sorbents are fabricated from poly(styrene-divinylbenzene) (PS-DVB). As noted previously, the PS-DVB matrix can be operated from pH 1 to 14 and is of much greater stability than silica at elevated temperature. Because the surface area of perfusion chromatography sorbents is lower than that of other porous media, fimbriated stationary phases have been used to increase the loading capacity.[6] The loading capacity of fimbriated media approaches those of conventional soft gel and silica-based materials, allowing fimbriated perfusion chromatography media to be used in preparative and process scale separations. Figure 8 is an example of a 600-fold scale-up on perfusive supports.

Loading capacity is a critical issue in preparative/process chromatography and may be specified in terms of either static or dynamic loading capacity. Static loading capacity is the amount of a protein such as bovine serum albumin that a unit volume of sorbent will bind in the batch mode. Dynamic loading capacity is the capacity at a specific flow rate, i.e., under

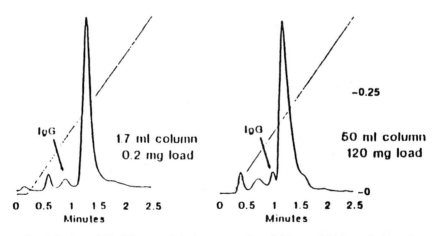

FIG. 8. Scale-up (600×) from analytical to preparative of high-speed, high-resolution anion-exchange purification of IgG from hybridoma cell culture supernatant on POROS perfusion columns. Note that sample load per milliliter column volume was increased 20-fold and the column volume increased 30-fold. (From Ref. 23.)

end use conditions. The static and dynamic loading capacities of a support may vary substantially. At mobile phase linear velocities of 300–3000 cm/hr, the dynamic loading capacity of conventional gel and 300-Å pore diameter silica sorbents will be less than 10% of that of the static loading capacity. In contrast, the loading capacities of enhanced mass transfer media are virtually unchanged over a broad range of flow rates.The dynamic loading capacity of analytical perfusion chromatography sorbents has been reported to be unchanged at up to 9000 cm/hr while the 50-μm particle diameter preparative materials are of constant dynamic capacity up to 1000 cm/hr.[24] Perfusion chromatography support materials can be obtained from PerSeptive Biosystems (Cambridge, MA).

Miscellaneous Types of Rapid Chromatography

Flash chromatography has been widely used for rapid separations.[25] Through gas pressurization of the solvent containers, it is possible to force liquid through chromatography columns more rapidly than by gravity. This type of chromatography might be referred to as "a poor man's HPLC." It has no inherent resolution or speed advantage over HPLC. These materials are available from Crosfield Chemicals (Warrington, UK).

Another approach to rapid chromatography is the use of disks or membranes for protein separations. For such separations, either a packed bed or membranes derivatized with stationary phase may be used. With tightly bound species, it is possible to selectively desorb substances from this column disk at high mobile phase velocity. Because these systems typically have fewer than 100 theoretical plates, band spreading caused by poor mass transfer within the bed is seldom an issue. Membrane or disk chromatography is a crude form of chromatography used in the initial stages of a purification when there are large differences between the target protein and impurities, and resolution is not an issue. It has been reported (Fig. 9) that affinity-based separations may be achieved in less than 20 sec with a very short column.[26] Similarly, rapid assays have been reported using short columns filled with perfusive protein A supports.[2] Short columns have also been used in the reversed-phase[27] and gel-permeation mode.[28]

Rapid two-dimensional separations utilizing column switching techniques have also been reported. By using short analytical columns (3.5 cm in length) packed with nonporous particles, Matsuoka *et al.* were successful

[25] I. Chappell and P. E. Baines, *BioTechniques* **10**, 236 (1991).
[26] R. R. Walters, *Anal. Chem.* **55**, 1395 (1983).
[27] N. Nagae, H. Itoh, N. Nimura, T. Kinoshita, and T. Takeuchi, *J. Microcol. Sep.* **3**, 5 (1991).
[28] H. Tojo, K. Horiike, T. Ishida, T. Kobayashi, M. Nozaki, and M. Okamoto, *J. Chromatogr.* **605**, 205 (1992).

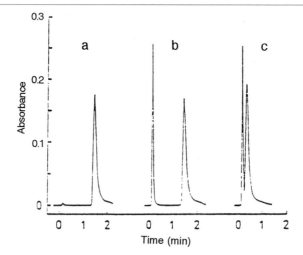

Fig. 9. Separation of 50 μg of concanavalin A (ConA) from 15 μg of BSA at 1 ml/min on a 6.35 mm × 4.6 mm column containing immobilized glucosamine. In (a), only ConA was applied. In (a) and (b), a step change in mobile phase approximately 1 min after injection was used to elute ConA. The time of the step change was adjusted in (c) to meet the resolution requirements. (From Ref. 26.)

in resolving crude peptide fractions from a brain extract into 150 peaks in 80 min.[29] Column switching techniques were also used to develop a rapid assay to quantitate aggregates of human IgG from monomers.[30] This assay utilizes size-exclusion chromatography in the first dimension and a fast affinity chromatographic method (30 sec) in the second dimension. It allows quantitation of aggregates to 1% and takes less than 2 hr.

Methods

It is well known that the pH of the mobile phase buffer plays a significant part in protein retention. But the effect of pH on separation is difficult to predict. It needs to be determined experimentally. By using high-speed separations that utilize any of the above-mentioned methods, it is possible to examine this in more detail in a short time. For example, purification of IgM antibody from the culture supernatant was examined at different pH values on an anion-exchange column (POROS Q/M; 4.6 × 100 mm). The gradient was from 20 mM Tris/Tris–Bis propane (pH 6.0 to 9.0) to

[29] K. Matsuoka, M. Taoka, T. Isobe, T. Okuyama, and Y. Kato, *J. Chromatogr.* **515,** 313 (1990).
[30] T. K. Nadler, S. K. Paliwal, and F. E. Regnier, *J. Chromatogr.* (1996).

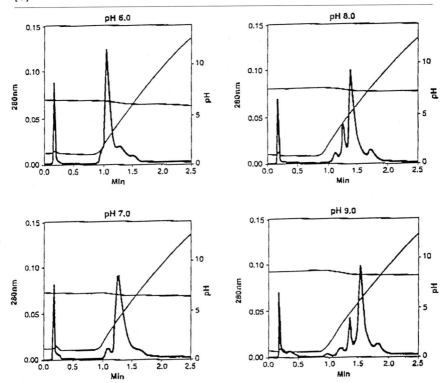

FIG. 10. Purification of IgM antibody from culture supernatant at different pH values. The details of the experiment are discussed in text.

1.0 M NaCl in 3 min. This was done at a flow rate of 10 ml/min (3600 cm/ hr) and the detection was at 280 nm. It can be seen in Fig. 10 that optimal separation with good resolution is obtained at pH 9.0 in less than 3 min. It is also important to note that the entire experiment to examine four different pH values took less than 10 min.

Similarly, pH mobile phase effects can also be studied by using high temperature.[31] By operating a reversed-phase column (30 × 4.6 mm) at 5 ml/min, at 80°, it was possible to examine the effect of acidic and alkaline mobile phases on protein separations. A gradient was run in 1.5 min from either (1) 0.1% trifluoroacetic acid (TFA), pH 2.2, or (2) 1.5 mM Na$_3$PO$_4$, pH 11.0, to 70% acetonitrile. It can be seen from Fig. 11 that in this particular case there is a tremendous advantage in using pH 11.0 buffer, as the resolu-

[31] Y.-F. Maa and C. Horváth, J. Chromatogr. **445**, 71 (1988).

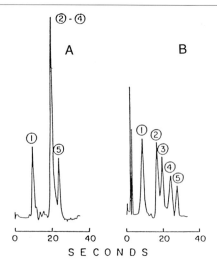

Fɪɢ. 11. Effect of eluent pH on the separation of proteins. The five proteins are (1) ribonuclease A, (2) fetuin, (3) β-lactoglobulin A, (4) α-chymotrypsinogen A, and (5) ovalbumin. (A) pH 2.2 buffer; (B) pH 11 buffer.

tion is far superior to the one at pH 2.2. All five proteins are well separated and the entire separation took less than 1 min. Thus it is possible to look at various pH values and optimize the separation in a short time.

Other applications using fast protein liquid chromatography already have been discussed.

Conclusion

In this chapter we have tried to trace the history of fast separations for proteins. It has been shown that there are multiple methods to achieve faster separations. Use of small (1 to 3 μm) nonporous particles increases the speed of separations but suffers from high back pressure. This problem was solved by operating the columns at high temperature but it has been shown that this leads to denaturation of proteins, and in the case of silica-based supports a loss of bonded phase from the support material. Core-filled supports are another means of performing faster separations. Still another method is the use of perfusion chromatography, in which the pore geometry enables the use of high flow rates with a modest back pressure.

Fast separations are of great importance in routine analytical testing, maximizing production, process optimization, collecting data for feedback control in production, and peptide mapping. Speed of separation is also useful in the process environment, where it plays an important role in increasing throughput and productivity.

Acknowledgments

M.F. acknowledges the support of Fulbright and M. E. C. (Spain) for a scholarship. This work was funded by NIH Grant GM 25431.

[7] Automated Method Development in High-Performance Liquid Chromatography

By LLOYD R. SNYDER

Introduction

Prior to 1980, the separation of large biomolecules by means of liquid chromatography was limited by our knowledge of separation as a function of experimental conditions. At that time it was not possible to anticipate the "automation" of method development, which requires both a theoretical and an empirical understanding of these chromatographic systems. Major advances have now occurred in our understanding of this area,[1-5] so that a consensus on how best to proceed in given cases is becoming apparent. It is also possible to use a few initial experiments to make quantitative predictions as to what will happen when certain separation variables are changed.

It is useful to distinguish between two types of "method automation," both of which involve the use of a computer. Heuristic knowledge that leads to qualitative choices regarding separation conditions is best presented in an *expert system* format.[6] In this case, the user initiates a dialog with the computer concerning the nature of the sample and the goal of separation.

[1] O. Mikes, "High-Performance Liquid Chromatography of Biopolymers and Biooligomers," Parts A and b. Elsevier, Amsterdam, The Netherlands, 1988.

[2] K. M. Gooding and F. E. Regnier (eds.), "HPLC of Biological Macromolecules." Marcel Dekker, New York, 1990.

[3] W. S. Hancock (ed.), "High Performance Liquid Chromatography." John Wiley & Sons, New York, 1990.

[4] C. T. Mant and R. S. Hodges (eds.), "High-Performance Liquid Chromatography of Peptides and Proteins." CRC Press, Boca Raton, FL, 1991.

[5] M. T. W. Hearn (ed.), "HPLC of Proteins, Peptides and Polynucleotides." VCH, New York, 1991.

[6] J. L. Glajch and L. R. Snyder, *J. Chromatogr.* **485,** (1989).

At some point in this process, one or more initial experiments will be recommended by the expert system. The computer may then interrogate the user as to the results of these initial experiments, following which additional *qualitative* recommendations are made by the expert system, further experiments are carried out, and so on, until a successful result is achieved. A good example of an expert system for the separation of biomacromolecules is the SMART Assistant software of Pharmacia (Piscataway, NJ).[7,8]

A second kind of method automation is referred to as *system optimization.* This implies a logical, *quantitative* procedure for efficiently selecting preferred conditions with respect to sample resolution. Several approaches to system optimization have been described for samples of lower molecular weight, especially those of nonbiological origin.[6] In the case of large biomolecules, published work on system optimization is confined mainly to the technique of *computer simulation,* which has been used for both reversed-phase[9] and ion-exchange[10] separations. Computer simulation employs software for the prediction of separation as a function of conditions, based on some theoretical model and a small number of initial experimental runs. Once data for the initial experiments have been entered into the computer, further (very fast) experiments can be carried out using the computer in place of a high-performance liquid chromatography (HPLC) system. The remainder of this chapter describes this new technique, provides several examples of its application, and discusses certain potential problems in its use.

Theory

The chromatographic separation of large biomolecules by either isocratic or gradient procedures is now believed to occur in the same general way as for small molecules.[11] Sample retention is governed by an equilibrium process within the column, so that during its elution each sample band undergoes many partitions between the mobile and stationary phase.

[7] H. Eriksson, K. Sandahl, G. Forslund, and B. R. Osterlund, *Chemometrics Intell. Lab. Systems Lab. Inform. Manage.* **13,** 173 (1991).

[8] H. Eriksson, K. Sandahl, J. Brewer, and B. R. Osterlund, *Chemometrics Intell. Lab. Systems Lab. Inform. Manage.* **13,** 185 (1991).

[9] B. F. D. Ghrist, L. R. Snyder, and B. S. Cooperman, *in* "HPLC of Biological Macromolecules" (K. M. Gooding and F. E. Regnier, eds.), p. 403. Marcel Dekker, Inc., New York, 1990.

[10] T. Sasagawa, Y. Sakamoto, T. Hirose, T. Yoshida, Y. Kobayashi, and Y. Sato, *J. Chromatogr.* **485,** 533 (1989).

[11] L. R. Snyder and M. A. Stadalius, *in* "High-Performance Liquid Chromatography. Advances and Perspectives" (C. Horváth, ed.), Vol. 4, p. 195. Academic Press, New York, 1986.

Gradient elution can be described as the sum of a large number of (approximately) isocratic steps. Although the same fundamental separation process is involved for large and small molecules, this can lead to certain differences in separation as a function of experimental conditions.[11a,b] In general, optimized conditions for large molecules require longer run times than for small molecules.

Isocratic retention as a function of mobile-phase composition can be approximated by certain simple relationships.

Reversed phase: $\log k' = \log k_w - S\phi$ (1)
Ion exchange: $\log k' = A - Z \log(C)$ (2)

Here, k' is the capacity factor of a given sample component, k_w is the value of k' for water as mobile phase, S is a parameter that increases with solute molecular weight M [see Eq. (6)], ϕ is the volume fraction of organic in the mobile phase (% B/100), A and Z are constants, and C is the concentration of salt plus buffer in the mobile phase. In this chapter we are concerned only with reversed-phase chromatography and Eq. (1). However, related principles apply equally well to ion exchange.[11]

Predictions for Change in Experimental Conditions

Because two unknown constants (k_w and S) determine the retention of a given compound as a function of mobile-phase composition ϕ, two isocratic or gradient experiments suffice to obtain values for k_w and S. Retention times can then be predicted for isocratic elution as a function of ϕ, or for gradient elution as a function of gradient conditions (initial % B, gradient time or steepness, gradient shape). Gradient elution can provide the same resolution as with isocratic separation, provided that gradient conditions are adjusted to give an equivalent value of k', referred to as k^* in gradient elution:

$$k^* = 0.85 t_G F/(V_m \Delta\phi S)$$ (3)

Here, t_G is gradient time, F is flow rate, V_m is the column dead volume (volume of mobile phase within the column), and $\Delta\phi$ is the change in ϕ during the gradient (change in % B divided by 100). Equation (3) assumes a linear gradient, which implies so-called linear solvent strength conditions.[11] All chromatographic runs described here are based either on linear gradients or segmented linear gradients.

[11a] L. R. Snyder, M. A. Stadalius, and M. A. Quarry, *Anal. Chem.* **55**, 1412A (1983).
[11b] M. A. Stadalius, M. A. Quarry, T. H. Mourey, and L. R. Snyder, *J. Chromatogr.* **358**, 17 (1986).

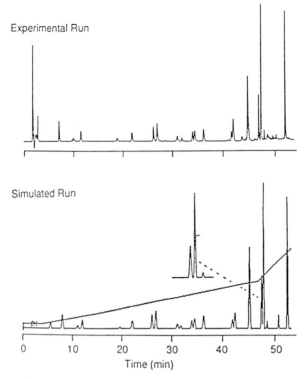

FIG. 1. Comparison of experimental and predicted chromatograms for the separation of a tryptic digest of rhGH. Conditions: 15 × 0.46 cm Nucleosil C_{18} column; 2/32/47% B in 0/48/53 min; solvent A is water and solvent B is acetonitrile, each containing 0.1% trifluoroacetic acid. DryLab G/*plus* (LC Resources) was used for simulation. (Reprinted from *J. Chromatogr.* **594,** R. C. Chloupek, W. S. Hancock, and R. L. Snyder, p. 65, 1992, with kind permission of Elsevier Science–NL, Sara Burgerhartstraat 25, 1055 KV Amsterdam, The Netherlands.)

It is also possible to predict bandwidth as a function of experimental conditions,[11] which with two initial experimental runs allows the quantitative estimation of separation (resolution) as a function of conditions. This is illustrated in Fig. 1, where experimental and simulated chromatograms are compared for the reversed-phase separation of a tryptic digest of recombinant human growth hormone (rhGH).[12] Two experimental runs were carried out initially: 15 × 0.460 cm C_{18} column; 0–60% (v/v) acetonitrile–water gradients (0.1% trifluoroacetic acid in both solvents) in 60 and 240 min. Computer simulation was then used to evaluate separation as a function of gradient conditions; i.e., chromatograms were generated by the

[12] R. C. Chloupek, W. S. Hancock, and L. R. Snyder, *J. Chromatogr.* **594,** 65 (1992).

computer for different values of initial and final % B, gradient time, and gradient shape. The final (optimized) separation obtained by trial-and-error simulations (Fig. 1) yielded a resolution of $R_s > 1.3$ for every band in the sample; by use of a two-segment gradient (see Fig. 1, bottom), it was possible to achieve this separation in a run time of less than 1 hr. The computer simulations shown in Fig. 1 and later simulations were carried out using DryLab software from LC Resources (Walnut Creek, CA).

The preceding model of large-molecule chromatography is occasionally complicated by "nonideal" effects. Large biomolecules such as proteins and nucleic acids have a three-dimensional (3D) structure that can vary as a function of separation conditions. This can lead to a number of problems in their separation,[13] and the relationship between isocratic and gradient separations of the same sample may also be affected.[14] It has been shown that the selection of favored separation conditions can minimize problems of the first type,[13] while a breakdown of the isocratic–gradient relationship now seems fairly rare and seldom has implications for computer simulation because gradient data are generally used for gradient predictions.

Predictions Based on Sample Structure

Proteins and nucleic acids are composed of a relatively small number of constituent amino acids or bases, respectively. This has led several workers to attempt to predict the retention of these compounds on the basis of additivity relationships of the form

$$\log k' = A + \Sigma x_i \tag{4}$$

where A is a constant and x_i is a contribution to $\log k'$ from various structural groups i within the sample molecule. Thus i might refer to each amino acid residue that constitutes a peptide or protein. By determining values of x_i for the 20 or so amino acids that comprise the protein molecule, Eq. (4) allows the prediction of retention (k') for any peptide or protein.

On the basis of the previous discussion, software for the prediction of chromatograms of peptide digests as a function of experimental conditions and sample structure has been described.[15] The advantage of this approach is that initial experimental data are not required for these predictions; reversed-phase, ion-exchange, or size-exclusion separation can be explored by computer simulation with little investment of time or effort. Unfortu-

[13] K. D. Nugent, W. G. Burton, T. K. Slattery, B. F. Johnson, and L. R. Snyder, *J. Chromatogr.* **443**, 381 (1988).

[14] A. N. Hodder, M. I. Aguilar, and M. T. W. Hearn, *J. Chromatogr.* **476**, 391 (1989).

[15] C. T. Mant and R. S. Hodges, *in* "High-Performance Liquid Chromatography of Peptides and Proteins" (C. T. Mant and R. S. Hodges, eds.), p. 705. CRC Press, Boca Raton, FL, 1991.

nately, accurate predictions of retention and resolution based on Eq. (4) are not possible, owing in part to the secondary and tertiary structure of typical peptides or proteins.[16] As a result, the use of Eq. (4) for quantitative predictions of chromatographic separation appears to be of limited value.

Changes in Retention Order Due to Changes in Gradient Steepness

By optimizing gradient conditions, it is possible to take advantage of potential changes in relative retention as a function of gradient steepness. Peptide and protein samples often exhibit such changes in retention when gradient steepness is varied. Gradient steepness can be defined by the parameter b[11]:

$$b = V_m \, \Delta\phi S/(t_G F) \tag{5}$$

Usually gradient steepness is varied by changing t_G or $\Delta\phi$, but changes in column volume (V_m) or flow rate (F) provide equivalent effects on relative retention.[17] Note that gradient steepness [Eq. (5)] is also affected by the solute parameter S; only when values of S are different for two adjacent bands will a change in relative retention result from a change in gradient steepness. Values of S are determined approximately by solute molecular weight:

$$S \approx 0.48(M)^{0.44} \tag{6}$$

Two peptides or proteins that are (1) adjacent in a chromatogram and (2) have molecular weights that differ by more than about 20% will exhibit significant changes in relative retention and separation as gradient steepness (b) is varied. The reason is that the S values of these two compounds will differ significantly. This then means [Eq. (1)] that relative retention will vary with mobile-phase composition ϕ during elution; ϕ in turn varies with gradient steepness.[11]

A difference in S values for two adjacent bands means that an intermediate, optimum gradient steepness will often exist for such a sample. Computer simulation allows this optimum gradient time or steepness to be determined without carrying out trial-and-error simulations. This can be achieved with the use of a computer-generated *resolution map* as shown in Fig. 2 for the sample of Fig. 1.

The map in Fig. 2a shows a plot of minimum sample resolution (for the most overlapped or "critical" pair of bands) as a function of gradient time. These critical band pairs are shown in the map in Fig. 2b (dark bands), where parts of the chromatogram are displayed for different gradient times.

[16] R. A. Houghton and S. T. DeGraw, *J. Chromatogr.* **386**, 223 (1987).
[17] J. L. Glajch, M. A. Quarry, J. F. Vasta, and L. R. Snyder, *Anal. Chem.* **58**, 280 (1986).

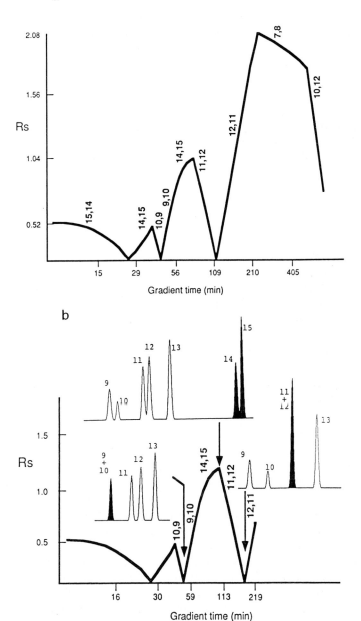

Fig. 2. Resolution map for the separation of a tryptic digest of recombinant human growth hormone (Fig. 1). (a) Minimum predicted sample resolution; (b) the dark bands are the critical band pairs (see text for details). (Reprinted from *J. Chromatogr.* **594,** R. C. Chloupek, W. S. Hancock, and R. L. Snyder, p. 65, 1992, with kind permission of Elsevier Science–NL, Sara Burgerhartstraat 25, 1055 KV Amsterdam, The Netherlands.)

158 LIQUID CHROMATOGRAPHY [7]

The map in Fig. 2a indicates that the best overall resolution ($R_s = 2.1$) can be obtained for a gradient time of about 4 hr. Since this is rather long, a second-best gradient time of 90 min was selected ($R_s = 1.3$). The separation of Fig. 1 uses this optimized gradient steepness for the first segment of the gradient.

Another example of changes in solute retention order as gradient steepness is varied is shown in Fig. 3a for a myoglobin digest and flow rates of 0.5 and 1.5 ml/min.[17] Corresponding portions (*) of each chromatogram in Fig. 3a are expanded and shown in Fig. 3b. From Eq. (5), the steeper gradient corresponds to the slower flow rate (0.5 ml/min). Bands 5 and 5a are seen to be resolved at 0.5 ml/min but coalesce at 1.5 ml/min; bands 6 and 6a exhibit the reversed behavior; bands 6b and 7b change their separation order as the flow rate is varied. Computer simulation indicates that all six of these bands can be resolved with a flow rate of 3 ml/min.

Uses of Computer Simulation

The reversed-phase separation of peptide or protein samples typically involves gradient elution. A successful separation then depends on the right choice of various gradient conditions: initial and final % B, gradient time, and gradient shape. Although it is common practice to use trial-and-error experiments as a means for optimizing these gradient conditions, this approach has several practical limitations. Such separations may require one or more hours per run, and the chromatograms are frequently difficult to evaluate because of their complexity. While many workers expect a continuous improvement in separation as gradient steepness is reduced, often this is not observed: Some bands may coalesce as the gradient becomes flatter, while very flat gradients often mean long run times and lower sample recoveries. As we will see, computer simulation addresses these and other problems associated with the reversed-phase separation of peptide and protein samples, by allowing the user to examine a wide range of gradient conditions so as to achieve optimized separations with minimal effort. The evaluation and interpretation of individual simulations are also facilitated by the computer through the use of chromatograms and tables.

30S Ribosomal Proteins

Figure 4a shows the separation of a sample of 30S ribosomal proteins (240-min gradient). This chromatogram can be divided into three groups (A–C); the separation of each group as a function of gradient time is also shown in Fig. 4b. Several inversions of separation order are noted between the 60- and 240-min runs: bands 11/20, 3/8, 3/10, 5/7, and 5/9. It is also

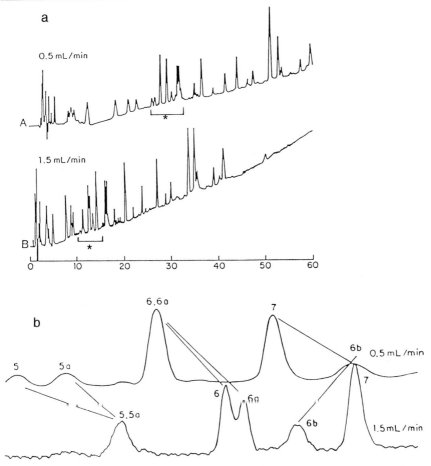

FIG. 3. Separation of a myoglobin tryptic digest as a function of gradient steepness (flow rate varies). Conditions: 10–70% (v/v) acetonitrile/(0.1% trifluoroacetic acid in water) in 60 min. (a) Experimental resolution at flow rates of 0.5 and 1.5 ml/min; (b) an expanded view of the bracketed portions in (a). (Reprinted with permission from *Anal. Chem.*, **58,** J. L. Glajch, M. A. Quarry, J. F. Vasta, and L. R. Snyder, p. 280, Copyright 1986, American Chemical Society.)

seen that groups A and C are better separated by the flatter 240-min gradient, while group B is better separated by the steeper 60-min gradient. This suggests that the best overall separation will result from a segmented gradient (flat/steep/flat).

On the basis of the initial 60- and 240-min experimental runs for this sample, computer simulation was used to explore the effect of gradient

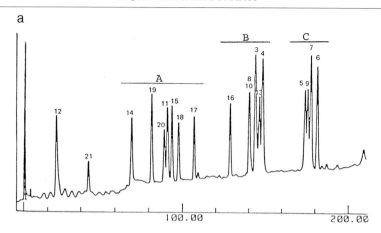

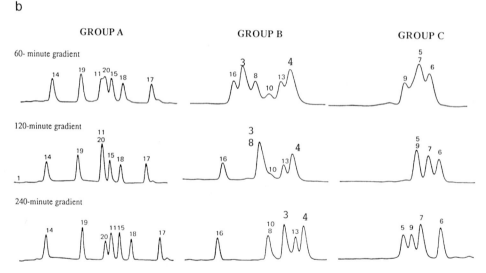

FIG. 4. Separation of 30S ribosomal proteins as a function of gradient steepness. (a) 26–46% B in 240 min; (b) same, but gradient time varies as shown. See text for details. Conditions: 26–46% (v/v) acetonitrile/(0.1% trifluoroacetic acid in water) at 0.7 ml/min. (Reprinted from Ref. 9, p. 403, by courtesy of Marcel Dekker, Inc.)

conditions on separation. A computer-generated resolution map was first determined for this 30S sample, and it was found that no single linear gradient was capable of resolving all 19 bands (i.e., providing $R_s > 0.6$ for all band pairs). Trial-and-error simulations did result in the predicted separation ($R_s > 0.8$) of all 19 proteins in this sample, using a 4-segment gradient (Fig. 5a); the confirmatory experimental run (Fig. 5b) is in close agreement with the prediction of Fig. 5a. The separation shown in Fig. 5b marked the first time that all of the 30S ribosomal proteins were resolved from each other by means of a single reversed-phase run. Resolution maps as illustrated in Fig. 2 can also be helpful in the systematic development of optimized multisegment gradients; see Refs. 18 and 19 for details.

The similar separation of the thirty-four 50S ribosomal proteins using computer simulation has also been described.[19] In this case it could be shown that two proteins were inseparable (for any gradient) with the particular column used for these studies. The remaining 33 proteins were resolved using a 4-segment gradient in a time of just over 5 hr. Predicted and observed retention times for the optimized separation of the 30S and 50S ribosomal proteins agreed within 1% (average). Comparisons of experimental and simulated chromatograms for various peptide, protein, and other samples show similar agreement.[12,20–23]

Standardized Separation Procedures: Corrections for Different Equipment

When a gradient elution procedure is developed for routine use, it is common to encounter differences in the separation among different laboratories. These changes in separation arise mainly from differences in the holdup volume ("dwell volume") of different HPLC systems. Computer simulation can be used to explore rapidly the effects of equipment dwell volume on the final chromatogram[24]; in most cases, it is possible to select conditions that ensure a satisfactory separation regardless of the choice of equipment.

Complex Samples: Separation of Storage Proteins
    from Wheat Sample

The preceding peptide digest and ribosomal protein samples involve mixtures with a defined number of known components. Many protein samples are initially not so well characterized and may contain a large number (>50) of components present in significant concentrations. In these cases

[18] B. F. D. Ghrist and L. R. Snyder, *J. Chromatogr.* **459,** 25 (1988).
[19] B. F. D. Ghrist and L. R. Snyder, *J. Chromatogr.* **459,** 43 (1988).
[20] L. R. Snyder, *in* "High-Performance Liquid Chromatography of Peptides and Proteins" (C. T. Mant and R. S. Hodges, eds.), p. 725. CRC Press, Boca Raton, FL, 1991.
[21] J. W. Dolan, L. R. Snyder, and M. A. Quarry, *Chromatographia* **24,** 261 (1987).
[22] J. W. Dolan, D. C. Lommen, and L. R. Snyder, *J. Chromatogr.* **485,** 91 (1989).
[23] B. F. D. Ghrist, B. S. Cooperman, and L. R. Snyder, *J. Chromatogr.* **459,** 1 (1988).
[24] L. R. Snyder and J. W. Dolan, *LC-GC Mag.* **8,** 524 (1990).

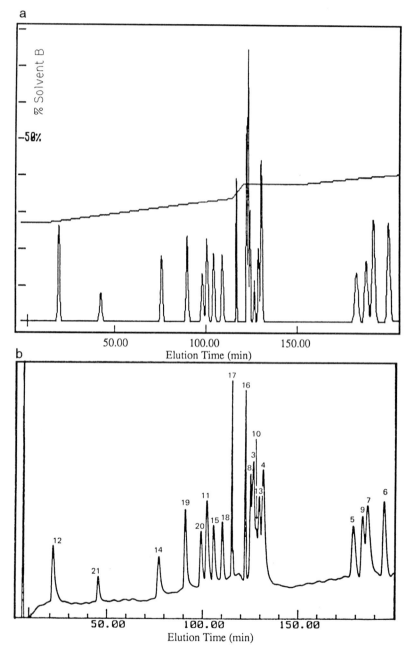

FIG. 5. Separation of 19 ribosomal proteins by means of optimized gradient conditions. (a) Predicted chromatogram (DryLab G/*plus*); (b) experimental chromatogram. Conditions: 25×0.46 cm C_3 column; 27/34/38/38/41% B in 0/98/104/133/193 min, 0.7 ml/min; A is water and B is acetonitrile, each with 0.1% added trifluoroacetic acid. (Reprinted from Ref. 9, p. 403, by courtesy of Marcel Dekker, Inc.)

the complete separation of the sample may require some combination of separation techniques, e.g., multidimensional separation. However, such procedures are often time consuming and experimentally demanding. Alternatively, computer simulation allows the rapid development of an optimized gradient elution method that may provide adequate resolution for a given sample, especially when only some of the sample components are of interest. Preliminary studies for a protein sample isolated from wheat will be used to illustrate some of the advantages and limitations of this approach.

Marchylo and co-workers[25–27] have used reversed-phase HPLC (RP-HPLC) for the analysis of cereal storage proteins. Protein fingerprints generated by RP-HPLC can be used to identify wheat varieties, and qualitative and quantitative analysis of these proteins has provided insight into their relationship with wheat quality. High-resolution separations are required to provide fingerprints that can be used to differentiate among wheat varieties, many of which have similar genetic backgrounds, and to provide separation of individual quality-related proteins. From a practical standpoint, it is desirable to achieve high-resolution separations in as short a time as possible in order to maximize throughput for routine applications. Prior to the present study, high-resolution and/or routine separations were developed by the traditional trial-and-error experimental approach.

Experimental chromatograms for the reversed-phase gradient elution separation of the storage proteins from a particular wheat sample were supplied by B. A. Marchylo (Canadian Grain Commission, Winnepeg, Manitoba) and are shown in Fig. 6 (gradient times of 51, 102, 153, and 204 min). The complexity of this sample is immediately apparent; more than 100 individual bands can be observed in each of the separations of Fig. 6. These chromatograms can be divided into the distinct groups A–F of Fig. 6.

It is instructive to examine the number of recognizable bands in each group (A–F) as a function of gradient time or steepness. If we count only bands with areas > 0.5% of the total, the data of Table I result. The number of major bands separated in each run varies from 51 to 57, but note that the number of resolved bands does not increase continuously as gradient time is increased. Group D is best separated by the 51-min gradient, while the 204-min gradient gives the largest number of bands for groups A and F. If a single gradient could be developed to provide an optimum separation of each group, it can be calculated that at least 63 major protein bands larger than 0.5% would be resolved for quantitation.

[25] B. A. Marchylo, K. E. Kruger, and D. W. Hatcher, *J. Cereal Sci.* **9,** 113 (1989).
[26] B. A. Marchylo, D. W. Hatcher, and J. E. Kruger, *Cereal Chem.* **65,** 28 (1988).
[27] B. A. Marchylo, D. W. Hatcher, J. E. Kruger, and J. J. Kirkland, *Cereal Chem.* **69,** 371 (1992).

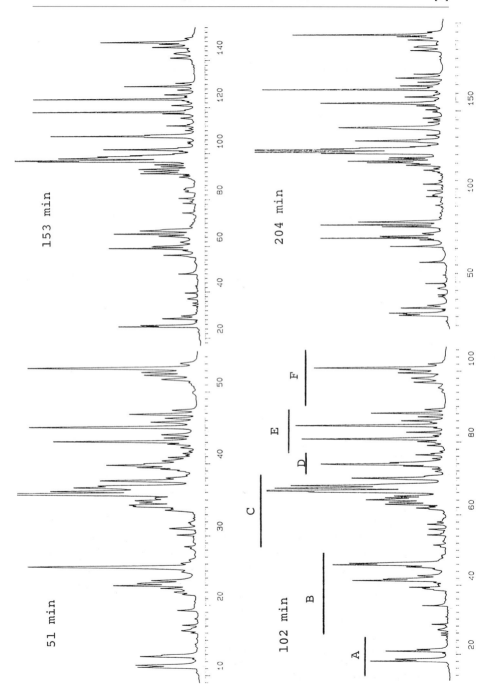

TABLE I
NUMBER OF DISCERNIBLE MAJOR BANDS IN WHEAT
PROTEIN SAMPLE[a]

Group	Number of bands in each group (different runs)			
	51 min	102 min	153 min	204 min
A	4	4	4	5
B	12	14	11	10
C	14	17	19	21
D	7	5	4	4
E	10	10	10	10
F	4	5	5	6
Total[b]:	51	57	53	56

[a] Sample shown in Fig. 6. Only bands with areas > 0.5% are counted, and the number of bands is classified by group and gradient time.
[b] Total protein bands (separated or unseparated) is 63.

Computer Simulation: Preliminary Validation

Computer simulation for the sample in Fig. 6 was begun by entering retention data and experimental conditions for the 51- and 153-min runs into the DryLab program. For samples as complex as this, it is important to verify the accuracy of the input data and the assignment of bands between the two runs (peak tracking); otherwise errors in subsequent predictions are likely. The latest version of this software provides automatic peak tracking, but as discussed below, peak assignments should not be taken for granted. Only bands with relative areas > 1% were used, in order to simplify computer simulation; there are 40 such bands in the sample.

The reliability of computer simulation for a given sample can be checked in two different ways, each of which requires one or more additional experimental runs; in the present study, the 102- and 204-min runs of Fig. 6 were used for this purpose. Using the 51- and 153-min runs as input, it was possible to predict retention times for the 102- and 204-min runs. The

FIG. 6. Separation of wheat protein sample as a function of gradient steepness. Conditions: 15 × 0.46 cm Zorbax Rx-300-C8 column (Rockland Technologies), 23–48% acetonitrile–water gradients (0.1% TFA in each solvent), 1 ml/min, 50°. A–F, Groups of wheat protein bands. See Ref. 26 for experimental details.

overall accuracy of the predicted retention times was equal to ±0.3–0.5%, which is acceptable. No individual retention time values were obviously out of line, implying that reliable computer simulations will be possible for the present system.

A second check on these computer simulations is provided by visual comparisons of predicted and experimental chromatograms. Such comparisons involve both retention times and bandwidths; they therefore require an estimate of the (isocratic) column plate number N for this sample and conditions. This plate number estimate can be obtained by the trial-and-error use of different values of N in computer simulation, with comparison of the resulting simulated chromatograms vs experimental runs. In the present example, $N = 1900$ was obtained in this fashion. This value can be compared with values predicted from experimental conditions as described in Ref. 11: $N = 1500–2300$ depending on gradient time. This implies "ideal" chromatography[13] and a column that exhibits state-of-the-art performance.

Figure 7a–d presents computer simulations for group C at different gradient times. A comparison of these simulations with the chromatograms of Fig. 6 shows reasonable agreement. The resolution of the smaller bands in Fig. 7 appears generally better than in Fig. 6, owing to the presence of minor bands that overlap these bands and are not included in these simulations. This is a general limitation on the use of computer simulation for samples as complex as this.

An additional advantage of comparisons as in Fig. 7a–d is that it allows for the discovery of small, "hidden" bands that are overlapped by different larger bands in the two input runs. Thus imagine that compounds A, B (small area), and C are present in a sample, but A and B overlap in one run and B and C overlap in a second run; it would therefore be concluded that only bands A and C are present in the sample. A comparison of simulated and experimental chromatograms for a third run will usually exhibit a discrepancy that in turn leads to the discovery of the missing band B. Comparisons as in Fig. 7 were carried out for the other groups of this sample, from which it was concluded that no additional major bands are present in the sample. Thus the present HPLC conditions are able to separate all of the primary proteins in this sample (although not necessarily during a single run).

Separation of Groups A–C

Consider next the separation of group C of Fig. 6 (and Fig. 7), which contains the largest number of major bands. Because of the bunching of bands in this cluster, segmented gradients will be of little value; maximum resolution can be achieved by a single linear gradient. The resolution map

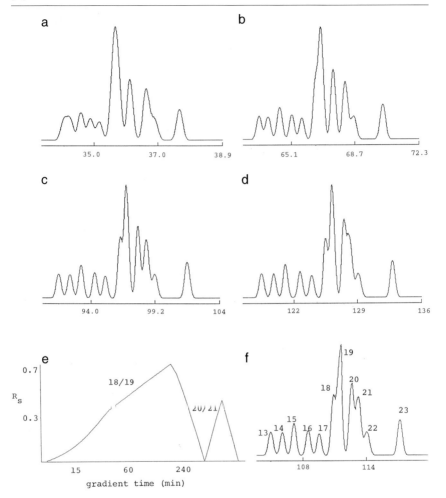

Fig. 7. Computer simulations for group C of wheat protein sample. Conditions as in Fig. 6 except where noted otherwise. Gradient times: (a) 51 min, (b) 102 min, (c) 153 min, (d) 204 min, (f) 168 min. (e) Resolution map for group C.

for group C (Fig. 7e) indicates that the best resolution ($R_s = 0.7$) will result with a gradient time of 180 min (23–48% B) or 0.14% B/min. The predicted separation under these conditions is shown in Fig. 7e.

Resolution maps for groups A and A–C are shown, respectively, in Fig. 8a and b. The maximum possible resolution for either group A or A + B + C is $R_s = 0.6$, owing to the difficult separation of bands 1–3. Where two or three adjacent bands are critical regardless of gradient conditions, the

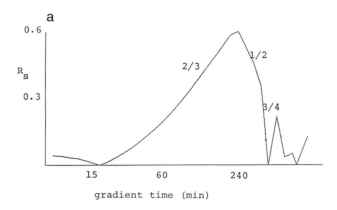

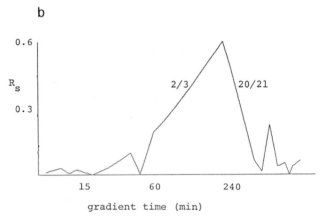

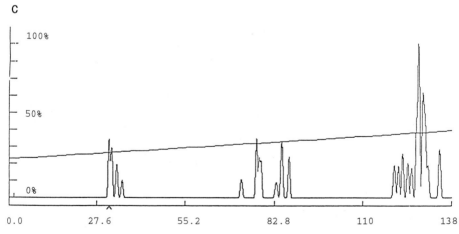

FIG. 8. Computer simulations for groups A–C of wheat protein sample. Conditions as in Fig. 6 except where noted otherwise. (a) resolution map for group A; (b) resolution map for groups A + B + C as sample; (c) optimized separation ($R_s = 0.6$) for groups A–C in a gradient time of 207 min (0.12% B/min).

use of a multisegment gradient cannot result in an overall increase in sample resolution. The optimized separation of this sample using a linear gradient of 0.12% B/min is shown in Fig. 8c. A multisegment gradient will in some cases be advantageous as a means of increasing resolution (e.g., Fig. 5), in other cases as a means of shortening run time (Fig. 1).

The design of a complete gradient for the separation of the total wheat protein sample is still under study. Preliminary computer simulations suggest that the separation of the total sample ($R_s > 0.6$) cannot be achieved with any one gradient and the present column. The primary problem is the need for a rather flat gradient for group C as opposed to a steep gradient for group D (see Table I and Fig. 6). Owing to the close proximity of these two groups within the chromatogram, the elution of group C results in significant migration of group D through the column (under unfavorable gradient conditions). A subsequent increase in gradient steepness for elution of group D is then unable to reverse the unfavorable partial separation effected in the prior gradient segment. See the further discussion in Ref. 18.

When it can be shown that no one gradient is able to achieve the desired separation, the best alternative is to increase resolution by increasing column length. While this will increase run time proportionately, the desired separation can then generally be attained. Alternatively, a decision must be made on which bands in the chromatogram are of primary interest and therefore require some minimum resolution.

Potential Problems in Use of Computer Simulation

The use of computer simulation as an alternative to experimental studies has two practical requirements: (1) reliable predictions of separation and (2) fast and convenient entry of input data into the computer. That is, simulated separations must be as useful as "real" chromatograms, and the total work required for computer simulation should be significantly less than that needed by a purely experimental approach. At the present time, computer simulation can be justified for some separations but not for others. In this section some possible problems or difficulties in the use of computer simulation for samples of biological interest are reviewed, especially as regards complex samples such as that of Fig. 6; see also the discussion in Ref. 23. We also examine some future developments that are likely to address these concerns and increase the relative value of computer simulation.

Inaccurate Input Data

Errors in the input data (two experimental runs) used for computer simulation can result in much larger errors in subsequent simulations.

Therefore every measure should be taken to ensure the accuracy of the retention times and the conditions of separation. Accurate retention time data require (1) constant or reproducible conditions (see below) and (2) accurate measurements by the data system or operator. The main error in the measurement of retention time is likely to occur for the case of overlapping bands. If two bands are unresolved, but their retention times differ, a single retention time measurement will be recorded for the two bands— resulting in a small error for one or both bands. Similar errors can occur for the smaller of two overlapped bands, even when a maximum can be seen for each band; band 18 of Fig. 7f is an example. Unless the data system provides for band deconvolution, the reported retention time for band 18 in this example will be larger than the "true" value (corresponding to the elution of a sample of pure protein).

There are two possible means for correcting retention times for band overlap. First, future enhancements in data systems would allow for better deconvolution of overlapped bands and the determination of more accurate retention times. Second, the use of additional input runs allows the selection of any two of these (three or four) runs for the determination of values of k_w and S. In this case, once the various bands have been matched between runs (see Peak Tracking, below), two runs (differing gradient times) are chosen for each sample component, such that reliable retention time values (minimal overlap) can be obtained in each run.

Accurate values for certain run conditions are also required in computer simulation, specifically, the equipment holdup volume (dwell volume) V_D and column dead volume V_m. A previous discussion of simulation errors due to error in V_D or V_m[23] suggests that these errors are not likely to be significant, except for simulations in which isocratic elution (e.g., preelution) is involved. However, as sample molecular weight increases, we have observed an increased dependence of simulation accuracy on the accuracy of values of V_m and especially V_D. Values of the dwell volume V_D that result in the most accurate simulations can be obtained from a third experimental run; the best value of V_D then provides the most accurate predictions of retention for this third run.[24] This approach was used to determine V_D for the HPLC system used in Fig. 6 and is now an automatic feature of the latest DryLab software. Accurate values of V_m are readily available from the measurement of column dead time t_0, the time required for a nonretained peak to move through the column; $V_m = t_0/F$, where F is flow rate.

Variable (Imprecise) Experimental Conditions

Modern HPLC equipment allows a precise control of variables such as flow rate and gradient conditions. However, a lack of column equilibration at the beginning of separation or changes in column performance with time can lead to small (but important) run-to-run changes in retention. Prior to the beginning of a gradient elution run, it is necessary to equilibrate the column with the starting mobile phase, because incomplete column equilibration can result in variable retention for early bands in the chromatogram. Procedures for column equilibration that may be adequate for general applications of HPLC may not suffice for the needs of computer simulation, especially in view of the need for even more accurate retention times for early bands. Passage of 15–20 column volumes of starting mobile phase through the column will usually be adequate for column equilibration, but this should be confirmed by carrying out duplicate runs and comparing retention time reproducibility for early bands. Fast equilibration of the column is favored by starting the gradient with a mobile phase that contains some organic solvent.

Changes in column performance are a well-documented problem for computer simulation. In one study involving the 30S ribosomal protein sample of Fig. 4,[23] it was found for 11 different computer simulations (11 different pairs of experimental runs as input) that the accuracy of predicted retention times varied from ±0.4 to ±5%, depending on whether (1) a new column was used, and/or (2) data for computer simulation and verification were collected within 2 days. The most accurate simulations resulted with columns that had been used previously (not a new column) and for data that were collected within a 2-day period. Each of these effects is probably the result of a loss of bonded phase during column use, which occurs most rapidly for some new columns.

For the low-pH conditions commonly used for peptide and protein separations by reversed phase, it is well documented that alkylsilane groups are subject to hydrolysis and loss, especially for the case of C_3 and C_4 packings commonly used for these separations.[28] New, "sterically protected" reversed-phase packings have been developed (StableBond; Rockland Technologies, Newport, DE) that are much more stable[28] and that have been demonstrated to furnish protein retention time values that are highly reproducible during extended column use. Columns of this type can largely solve previous problems due to column instability.

Failures of Retention Model

The retention model used in the DryLab computer simulation software is based on certain assumptions: (1) applicability of Eq. (1), (2) negligible dispersion of the gradient within the equipment, (3) an absence of solvent

---

[28] J. J. Kirkland, J. L. Glajch, and R. D. Farlee, *Anal. Chem.* **61,** 2 (1988).

demixing during the gradient, and (4) no change in the conformation of large biomolecules during separation. Each of these assumptions has been discussed in detail.[23] Equation (1) is an empirical relationship that is known to be at best approximate for many samples and experimental conditions. While its relative accuracy has been documented in numerous studies, including separations of peptides and proteins, plots of log k' vs ϕ that should be linear according to Eq. (1) typically exhibit a slight concave curvature. It has been shown that systems of this kind are more accurately represented by a variation of Eq. (1)[29]:

$$\log k' = \log k_w - S\phi + A\phi^2 \tag{7}$$

where A is a constant that depends on the sample and experimental conditions. Reliable computer simulations can nevertheless be obtained even for systems in which Eq. (1) is less reliable, by limiting simulation to gradients that are intermediate in steepness between the two experimental runs used as input (interpolation as opposed to extrapolation). Alternatively, the use of three experimental runs as input (vs the present two runs) would allow the determination of values of k_w, S, and A of Eq. (7), in turn leading to more accurate predictions of separation for gradient conditions that involve interpolation (but not extrapolation). This improvement is currently under investigation in our laboratory.

The gradient mixer and other system components in the flow path between the mixer and column lead to some dispersion or "rounding" of the gradient. This can in turn result in error in predicted retention times, especially for the case of segmented gradients.[30] HPLC systems, especially those designed for "microbore" operation, exhibit relatively little gradient dispersion and are unlikely to cause errors in computer simulation. Solvent demixing, another example of model failure, was shown to have a generally negligible effect on predictions of retention time and separation.[24]

Changes in conformation of a peptide or protein molecule can occur during reversed-phase separation,[13] and this could result in inaccurate simulations. Changes in conformation have been implicated, however, in a number of other problems that have been encountered in protein separations. "Good" chromatography can be obtained for most protein samples by a suitable choice of experimental conditions; e.g., one study[13] suggests prior denaturation of the sample with urea or guanidine, separation at 50–60° using acetonitrile–water gradients with 0.1% trifluoroacetic acid, use of a 5- to 15-cm, 30-nm pore column that is stable at pH 2. Conditions similar

[29] P. J. Schoenmakers, H. A. H. Billiet, and L. De Galan, *J. Chromatogr.* **185,** 179 (1979).
[30] D. D. Lisi, J. D. Stuart, and L. R. Snyder, *J. Chromatogr.* **555,** 1 (1991).

to these should also minimize any simulation errors due to changes in conformation during separation.

It should be emphasized that these various "failures" of the retention model often lead to errors in retention time that are generally similar for adjacent bands in the chromatogram. This then results in a shift of retention times for all bands to higher or lower values, but no change in relative retention or resolution. Because the prediction of resolution is of primary importance in computer simulation, these retention time errors are of less practical importance. It should also be noted that the use of retention data from three runs to determine a value of V_D (see above) tends to correct for errors caused by "model failure," mobile-phase demixing, and other artifacts.

Peak Tracking

Computer simulation (based on two initial experimental separations) requires that bands containing the same compound be matched between the two runs. For the example in Fig. 3b, this means that the retention times must be paired as follows:

0.5-ml/min run	1.5-ml/min run	Peptide
Band 1	Band 1	5
Band 2	Band 1	5a
Band 3	Band 2	6
Band 3	Band 3	6a
Band 4	Band 5	7
Band 5	Band 4	6b

When all the bands in the chromatogram are well resolved and band areas are not quite similar for two or more components, peak tracking is not a problem. For more complex chromatograms as in Fig. 6, involving a very large number of bands having a wide range in size, peak tracking is usually difficult. This presently constitutes the main limitation to the application of computer simulation for such samples.

Peak tracking is straightforward, but tedious, when standards exist for all the components of a sample. The identity of each band in the chromatogram can be established by comparison with the retention time of a standard. This approach is rarely available for the case of peptide and protein samples. Comparisons of relative retention and peak size can often be used to match bands between two runs, although this becomes impractical for samples such as that of Fig. 6, especially if band areas are less reproducible.

The use of a diode array detector allows ultraviolet (UV) spectra to be obtained for each band in the chromatogram, in turn providing a further basis for peak tracking. Unfortunately, the UV spectra of many peptides and proteins do not differ significantly, which limits this approach to peak tracking. The use of a mass spectrometer as detector (LC-MS) can provide the ultimate in peak tracking, because each band can be identified unequivocally.

Present DryLab software provides automatic peak tracking based on comparisons of relative retention and band area. For "simple" samples such as the peptide digest of Fig. 1, all of the bands for the two input runs will typically be matched—providing that band areas for the two runs are reproducible within about 5%. When the software fails to match all of the bands in each run, residual unmatched bands are displayed for a decision by the user. In most cases of this kind, peak matching will not require much effort on the part of the user. Any error in initial peak matching can be detected by comparing experimental and predicted retention times for a third run.

More challenging samples, such as that of Fig. 6, are complicated by frequent band overlaps and the presence of both large and very small bands in the same chromatogram. Under these conditions, band areas are less reliable, and the recognition of band overlaps is more difficult. The net result is a considerable increase in the effort required for peak tracking. In the future it may be possible to carry out peak tracking for such samples by the use of three or four experimental runs under conditions that allow accurate measurements of band area. Preliminary peak tracking would be done as at present (relative band retention and area), followed by the use of computer simulation to confirm initial band matches in experimental runs not used as input. Discrepancies uncovered in this way would then be resolved automatically by the computer.

Conclusion

Computer simulation is presently a useful tool that can complement method development for the separation of biological macromolecules by reversed-phase or ion-exchange gradient elution. The selection of near-optimum gradient conditions for a given sample will often require a large number of trial-and-error runs, which becomes impractical because of the time and effort involved. With only two or three experimental runs, however, the use of computer simulation allows a complete investigation of separation as a function of gradient conditions in a short time; i.e., usually no more than half a day. This in turn can lead to better resolution of the sample, a decrease in run time, and a more complete understanding of the separation.

Greater care must be taken in carrying out experimental separations that will be used for computer simulation. With appropriate precautions, however, simulated separations are found to be quite reliable; simulated retention times usually agree with experimental values within $\pm 1\%$ or better, and sample resolution R_s is generally predicted to better than $\pm 10\%$.

Acknowledgment

The assistance of Dr. Brian Marchylo in providing the experimental data for the wheat storage protein example and for critical comments on the final manuscript is gratefully acknowledged.

[8] Detection in Liquid Chromatography

By Michael Szulc, Rohin Mhatre, Jeff Mazzeo, and Ira S. Krull

Principles of Biopolymer Detection

Sensitive and selective detection of biological macromolecules is the goal of almost every biotechnology laboratory, and chromatography is an important tool in biopolymer characterizations. However, no one detector is capable of completely characterizing an eluting biopolymer from a chromatographic column. The detector is a tool that provides the researcher with different information based on the requirements of the sample. While numerous texts have appeared dealing with detectors for high-performance liquid chromatography (HPLC) or biopolymer separations, most of these have not specifically emphasized biopolymer detection. This chapter discusses the detection and characterization of components in a biological sample.

Each detector provides a different type of information, in terms of spectroscopic, electrochemical, and light-scattering properties, molecular weight, etc. The real problem in the unambiguous identification of a given biopolymer lies in the fact that within each class of biopolymers (proteins, nucleic acids, and carbohydrates), properties are similar to one another from a detection point of view. The lack of specificity in spectroscopic and electrochemical detectors is particularly problematic in biotechnology, where minor variants of proteins are frequently the analytes of interest. Therefore, although single-wavelength ultraviolet (UV) absorbance detection is among the most universally applied detection techniques, it often does not provide the selectivity necessary to characterize a biopolymer. If more selectivity is needed, dual-wavelength or linear diode array spectra can be employed. Fluorescence (FL) and electrochemical (EC) detection can also be employed for improved selectivity of appropriate samples. Derivatization can be utilized for improved sensitivity and selectivity. Molecular weights can be determined with mass spectrometric and light-scattering techniques. Mass spectrometry (MS) can yield further structural information about the biopolymer from fragmentation studies (see Section III in this volume), while low-angle laser light-scattering (LALLS) techniques can give molecular size and may detect aggregation formation. In this chapter, we emphasize UV, FL, EC, and LALLS detectors.

Spectrophotometric Methods

Spectrophotometric techniques are the most widely applied methods for the detection of biopolymers. However, because of concomitant impurities that may elute with the analyte in a biological sample, the techniques often lack the selectivity for accurate quantitation of an analyte, positive determination of the purity of an analyte, and specific determination of the structure of an analyte. There are spectrophotometric techniques that can be employed for improved sensitivity and selectivity of the analyte of interest, beyond the native UV absorbance of the biopolymer. These include dual-wavelength absorbance ratioing, derivative spectra, fluorescence detection, and derivatization.

Determination of Native Species

Single-Wavelength Detection. The simplest detection method for sensitive determination of biopolymers is UV absorbance at a single wavelength. The actual absorbance maximum for the peptide bond is around 185 nm, but most solvents and buffers are not transparent below 200 nm. Therefore, peptides and proteins are most often detected via UV absorbance at wavelengths between 210 and 220 nm. By lowering the detection wavelength from 220 to 210 nm, the sensitivity for most peptides is increased by a factor of two to four. Trifluoroacetic acid (TFA), the most often used ion-pairing agent in reversed-phase chromatography, absorbs below 210 nm and interferes with detection. It is possible to improve sensitivity of detection below 210 nm by using alternate ion-pairing and ion-suppression agents to TFA. Dilute HCl was used as an ion-pairing reagent, improving sensitivity by lowering the detection wavelength to 195 nm.[1]

---

[1] P. M. Young and T. E. Wheat, *J. Chromatogr.* **512,** 273 (1990).

Detection in the low-UV region near 210 nm is relatively unselective, so that a host of other compounds that may be present in a biological sample may coelute and interfere with quantitation or purity determinations. The amino acids tryptophan, tyrosine, and phenylalanine are also sensitive at 280 nm, so that this wavelength may be used for detection of peptides and proteins containing these aromatic amino acids. Detection at 280 nm is often performed in series with detection at 210 nm with peptide mapping, in order to distinguish aromatic amino acid-containing peptides while maintaining sensitive detection of the entire sample. While aromatic amino acid residues are often monitored at lower wavelengths (254 nm) because of available inexpensive fixed-wavelength instrumentation, wavelengths above 280 nm are less reliable because of changing molar absorptivities in alkaline media, owing to the phenol group on tyrosine.

The native fluorescence of tryptophan and tyrosine can also be utilized for sensitive peptide and protein determinations. By exciting at 220 nm, and monitoring the emission at 330 nm, tryptophan-containing peptides can be detected with six times more sensitivity than is obtainable by UV spectroscopy.[2] The native fluorescence of these proteins can also be used to monitor changing conformation of the protein under different conditions. For example, as the protein changes conformation, and the relative exposure of tryptophan residues changes, the fluorescence response changes accordingly.

Proteins containing prosthetic groups, such as metalloproteins, often may be selectively detected at wavelengths in the far-UV or visible region. The wavelength is dependent on the oxidation state of the metal. Cytochrome-c oxidase may be detected near 410 nm, while myoglobin is sensitive at 405 nm, cytochromes at 550–560 nm, and hemoglobin at 415 nm.[3] Amino acids and small peptides not containing tryptophan or tyrosine have also been detected as their copper(II) complexes at 254 nm, using ligand-exchange chromatography.[4]

For other biopolymer determinations, oligonucleotides are UV active over the range of 180–300 nm, with maxima at 190–200 and 250–280 nm. Molar absorptivities at 250–280 nm are on the order of 10^4, so that inexpensive fixed-wavelength detectors (254 nm) are usually employed. Oligosaccharides must be detected below 200 nm, so that UV detection of the native oligosaccharide is seldom performed. Oligosaccharides are most often de-

[2] T. D. Schlabach and C. T. Wehr, in "Proceedings of the First International Symposium on the HPLC of Proteins and Peptides" (M. T. W. Hearn, F. E. Regnier, and C. T. Wehr, eds.), pp. 221–232. Academic Press, New York, 1983.
[3] S. C. Powell, E. R. Friedlander, and Z. K. Shihabi, J. Chromatogr. 317, 87 (1984).
[4] A. Foucault and R. Rosett, J. Chromatogr. 317, 41 (1984).

tected via electrochemical or mass spectrometric detection, and spectrophotometrically via derivatization.

Spectrophotometric detection at a single wavelength is adequate for quantitation and determination of known analytes well characterized by previously validated methods of analysis. For unknown samples, single-wavelength detection communicates nothing about the identity of the compound or its purity. Other, more selective methods of detection must be employed to demonstrate purity and to aid in structural characterization.

Photodiode Array and Other Multiwavelength Detection Techniques. Photodiode array (PDA) detectors can assist in the characterization of an analyte via absorbance ratioing, spectral comparison, and derivative spectra. Dual-wavelength absorbance ratioing can be used to show purity of the eluting peak. This technique relies on the fact that the ratio of the absorbances for an analyte at two different wavelengths is actually the ratio of the molar absorptivities of the analyte, so that the ratio output signal is independent of sample concentration and characteristic of a pure analyte. The dual-wavelength absorbance ratio of a chromatographic peak must be constant and identical to that of a pure, known standard of the analyte injected under identical conditions. The plot of absorbance ratios from a chromatogram should show a square wave signal over the elution of the entire peak to demonstrate 100% purity, so that even for coeluting peaks, where overlap is so severe that no shoulder is evident, the ratio data will be constant only when the pure analyte is eluting. If the two components have exactly the same retention time, the signal will be constant at a level intermediate to the two analytes, so that this ratio can be compared to pure standards for purity assessment.

As an example of absorbance ratioing, the absorbance ratios at 254/280 nm for standard reference nucleosides and bases have been determined.[5] These ratios of unknowns in a chromatogram were compared to standard nucleosides to distinguish between closely eluting species, and to determine purity of the eluting nucleoside.

Absorbance ratioing has similarly been used to distinguish peptides containing tyrosine and tryptophan.[6] Tryptophan has a higher absorptivity at 280 nm, but tyrosine is capable of shifting its wavelength maximum to 293 nm at basic pH. Peptides not containing tyrosine are expected to show identical 293/280-nm absorbance ratios with and without postcolumn addition of base, while these ratios are expected to change depending on

[5] A. M. Krstulovic, P. R. Brown, and D. M. Rosie, *Anal. Chem.* **49,** 2237 (1977).
[6] A. F. Fell, J. B. Castledine, B. Sellberg, R. Modin, and R. Weinberger, *J. Chromatogr.* **535,** 33 (1990).

the relative amounts of tyrosine. In this study, for six tyrosine-containing dipeptides, absorbance ratios for the unshifted spectra were similar, while the absorbance ratios for the pH-shifted difference spectra allowed better peak identification.

In amino acid, peptide, and protein analysis, absorbance ratioing works best with lower molecular weight species. A problem with using dual-wavelength absorbance ratios for purity determination of proteins is that the ratio may change with the elution of conformers. We determined that for a pure eluting monomer protein in hydrophobic interaction chromatography, conformational changes of the protein as a function of protein concentration injected, temperature, and gradient conditions resulted in changing dual-wavelength absorbance ratios, owing to the changing relative exposures of tryptophan and tyrosine residues.[7] Other techniques, such as spectral comparisons and low-angle laser light scattering, were necessary to describe the purity of the protein.

Photodiode array detection may also be used to acquire on-line UV spectra of an eluting analyte. These spectra may then be compared to standards to assist in structural characterization of a protein, peptide, or polynucleotide. However, because UV spectra within classes of biopolymers are so similar, it is often difficult to distinguish between analytes with only minor structural differences. Photodiode array detection has been used to distinguish between peptides yielded from a tryptic digest of recombinant DNA-derived human growth hormone (rhGH).[8] A library of standard spectra was compiled for each of the peptides in the tryptic map, so that, for each peptide, the library contained the UV spectrum, retention time, and area and height values for the peptide at 200 nm. An algorithm was developed to match unknown samples to these standards, and the method was applied to normal and oxidized rhGH samples under a variety of chromatographic modes. The technique showed good correlation between unknowns and standards and was able to distinguish the oxidized products. The major limitation of the method was that the algorithm could not match precisely coeluting peptides. Figure 1 shows the detected differences for the spectral comparisons of two peptides from a peptide map.

Many times, differences in UV spectra alone are difficult to distinguish for the eluting peptides, because tryptophan and tyrosine are the only amino acids contributing to absorption above 220 nm. In these cases, first and second derivative spectra are helpful in uncovering poorly separated peaks. By determining peak-to-trough values between 283 and 287 nm, and

---

[7] I. S. Krull, H. H. Stuting, and S. C. Krzyko, *J. Chromatogr.* **442,** 29 (1988).
[8] H. P. Sievert, S.-L. Wu, R. Chloupek, and W. S. Hancock, *J. Chromatogr.* **499,** 221 (1990).

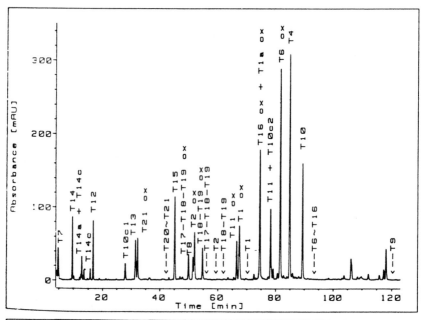

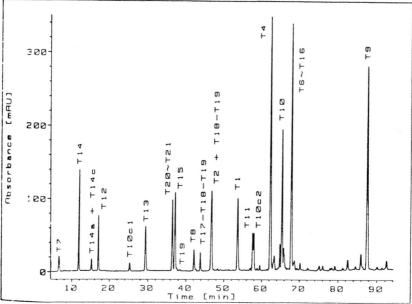

Fig. 1. Chromatograms of peptides differing by a single amino acid substitution detected by on-line spectral detection using PDA. (Top) Tryptic map for oxidized r-hGH analyzed with gradient II (phosphate). (Bottom) Tryptic map of r-hGH separated with gradient I (TFA). (Reprinted from *J. Chromatogr.* **499,** H. P. Sievert, S.-L. Wu, R. Chloupek, and W. S. Hancock, p. 221, Copyright 1990, with kind permission of Elsevier Science–NL, Sara Burgerhartstraat 25, 1055 KV Amsterdam, The Netherlands.)

between 290 and 295 nm, the environment of tyrosine and tryptophan can be probed.[9] Thus, derivative spectra provide additional confirmatory information that can be masked in similar UV spectra.

Derivatization

Chemical derivatization of an analyte improves the sensitivity and detectability of an analyte by introduction of a chromophore or fluorophore. Derivatization also increases the selectivity of an analyte if it can be shown that the derivatizing reagent reacts in the biological matrix solely with the analyte of interest.

Derivatizations can be performed both before and after a chromatographic separation. Postcolumn techniques allow improved detection of analytes after separation in their native state, while precolumn derivatizations change both detection and chromatographic properties. Postcolumn derivatizations require a rapid reaction time with a reagent that does not share the detection properties of the derivative, so that reagent blank interference is minimal. Because the reaction is being performed on the analyte chromatographically resolved from its matrix, it is not necessary for the reaction to go to completion, or for only a single derivative to be formed; however, the percent derivatization should be reproducible and should not change with concentration. These reactions are often easy to automate for on-line detection of the derivatives. Disadvantages of the technique are possible reagent instability, generally longer chromatographic run times, more complex hardware requirements, and lower sensitivities than with precolumn derivatization reagents.

For precolumn derivatizations, rapid reaction conditions are not necessary. However, percent derivatization of the analyte should be high, 100% if possible, because derivatization may be matrix dependent, and should yield only a single derivative. Precolumn derivatization reagents also often share the detection properties of the derivative, so that it is necessary to remove excess reagent before separation, or to resolve the excess reagent from the derivative. Both pre- and postcolumn methodologies have been developed for the derivatization of biomolecules.

Derivatization of Peptides and Proteins. A variety of fluorescence derivatization techniques exist for improving sensitivity and selectivity of peptide fragments. Because peptides contain several functional groups, derivatization generally yields a mixture of products, so that reactions are usually carried out postcolumn, after the separation of the native peptides. Precolumn derivatizations are limited to small peptides. In addition, it is

---

[9] I. Sukuma, N. Takai, T. Dohi, Y. Fukui, and A. Ohkubo, *J. Chromatogr.* **506,** 223 (1990).

desirable for the derivative to fluoresce at wavelengths high enough so that the native fluorescence of tryptophan or tyrosine does not interfere. Many of the derivatization reagents used for amino acids are not useful for peptide derivatizations. *o*-Phthaldehyde (OPA) does not react significantly with the N-terminal amino group of the peptide, except for dipeptides and other small peptides, and dansyl chloride does not react except at high concentrations of peptides.[10] Fluorescamine has been shown to derivatize the amino group of peptides for sensitive detection in the low picomolar range for peptides less than 20 amino acids in length.[11] 1,2-Diamino-4,5-dimethoxybenzene was used for the precolumn fluorescent derivatization of Leu-enkephalin,[12] and naphthalene-2,3-dicarboxaldehyde (NDA) in the presence of cyanide has been used for the precolumn derivatization of both Leu- and Met-enkephalin.[13]

In addition to these derivatization reagents, which react with N-terminal and lysine ε-amino groups, others have been developed for the selective detection of specific peptide amino acid residues. Arginine-containing peptides have been selectively derivatized using benzoin[14] and ninhydrin.[15] Cysteine residues have been determined with 4-(aminosulfonyl)-7-fluoro-2,1,3-benzoxadiazole, a fluorescent reagent for thiols,[16] as well as dansylaziridine, maleimide, bromobimane, and fluorescein isothiocyanate.[17,18] Prolyl dipeptides have been selectively derivatized precolumn with 4-chloro-7-nitrobenzofurazan after reaction of the N-terminal group with OPA.[19] The biuret and Lowry reactions have been compared for the selective detection of chromatographically resolved proteins using ion-exchange, size-exclusion, and reversed-phase chromatography.[20] Thiamin has been used for the postcolumn derivatization of peptide fragments.[21] Each peptide bond was chlorinated with hypochlorite, then reacted with thiamin, so that the fluorescence intensity was directly proportional to the number of peptide bonds. Figure 2 compares the detection of the peptides using UV ab-

[10] R. Newcomb, *LC-GC* **10**, 34 (1992).
[11] D. S. Brown and D. R. Jenke, *J. Chromatogr.* **410**, 157 (1987).
[12] M. Kai, J. Ishida, and Y. Ohkura, *J. Chromatogr.* **430**, 271 (1988).
[13] P. DeMontigny, C. M. Riley, L. A. Sternson, and J. F. Stobaugh, *J. Pharm. Biomed. Anal.* **8**, 419 (1990).
[14] D. Liu and D. J. MacAdoo, *J. Liq. Chrom.* **13**, 2049 (1990).
[15] V. K. Boppana and G. R. Rhodes, *J. Chromatogr.* **507**, 779 (1990).
[16] T. Toyo'oka and K. Imai, *Anal. Chem.* **57**, 1931 (1985).
[17] E. C. Klasen, *Anal. Biochem.* **121**, 230 (1982).
[18] A. F. Wilderspin and N. M. Green, *Anal. Biochem.* **132**, 449 (1983).
[19] M. Codini, C. A. Palmerini, C. Fini, C. Lucarelli, and A. Floridi, *J. Chromatogr.* **536**, 337 (1991).
[20] T. D. Schlabach, *Anal. Biochem.* **139**, 309 (1984).
[21] T. Yokoyama and T. Kinoshita, *J. Chromatogr.* **518**, 141 (1990).

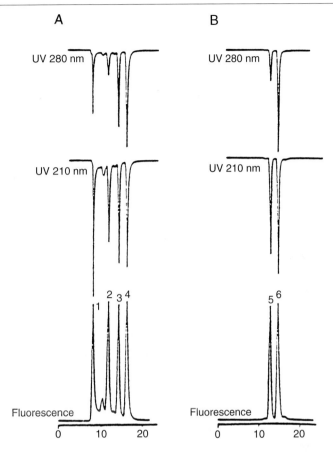

FIG. 2. Chromatograms of a mix of standard proteins (2 μg each), obtained by dual UV detection at 210 or 280 nm and by thiamin derivatization. (A) Thyroglobulin, BSA, myoglobin and lysozyme; (B) ovalbumin and cytochrome *c*. Column, TSKgel-G3000SW (30 cm × 7.5 mm i.d.); mobile phase, 0.1 *M* phosphate buffer (pH 7.5) containing 0.1 *M* sodium sulfate; flow rate, 0.8 ml/min. Peaks: 1, thyroglobulin; 2, bovine serum albumin (BSA); 3, myoglobin; 4, lysozyme; 5, ovalbumin; 6, cytochrome *c*. (Reprinted from *J. Chromatogr.*, 518, T. Yokoyama and T. Kinoshita, p. 141, 1990, with kind permission of Elsevier Science–NL, Sara Burgerhartstraat 25, 1055 KV Amsterdam, The Netherlands.)

sorbance at 280 and 210 nm with FL detection after derivatization with thiamin reagent.

Finally, fluorescent dyes that bind to proteins may also be used for selective detection. Indocyanine green dye was used for detection of proteins after gel-filtration chromatography, using a semiconductor laser for excitation.[22,23] Because the wavelengths available for semiconducter lasers were limited to 750–1300 nm, the ability of the fluorescent dye to couple with proteins, and its high molar absorptivity at 780 nm, made it ideal for sensitive detection, with detection limits one to two orders of magnitude better than that of conventional UV–Vis or FL techniques.

Derivatization of Nucleic Acids and Their Components. Oligonucleotides and nucleosides are most often detected in their native form, and derivatization is not often employed. Guanine residues in nucleosides and nucleotides can be selectively derivatized by phenylglyoxal.[24] Another specific reagent is chloroacetaldehyde, which is specific to adenine and cytosine and their derivatives. Molybdenum blue has also been used for postcolumn complexation between orthophosphate groups and molybdenum, followed by detection at 880 nm.[25] While detection limits with this complexation were higher than those obtained by UV absorbance at 250 nm, the method was selective for phosphate-containing components of a biological sample.

Another highly selective reagent, 1,2-bis(4-methoxyphenyl)ethylenediamine, was used to react with ribonucleosides and ribonucleotides in acidic media to yield fluorescent derivatives. The nucleoside pseudouridine in urine and serum was derivatized with excellent selectivity, significantly reducing the need for cleanup of the biological fluid samples.[26]

Derivatization of Oligosaccharides. Because of the reactivity of multiple functionalities on oligosaccharides, derivatizations are usually performed postcolumn. Derivatizations are also most often performed on reducing carbohydrates. Nonreducing oligosaccharides are analyzed after acid hydrolysis to the component monosaccharides. By using an acid hydrolysis reactor postcolumn with the derivatzation reagent, reducing sugars can easily be distinguished from nonreducing sugars, by simply switching the hydrolysis reactor on or off. Common derivatization reagents for oligosaccharides include p-hydroxybenzoic acid hydrazide,[27] 1-phenyl-3-methyl-5-pyrazolone,[28] 2-aminopyridine[29] and thymol[30] for improved UV detection, and 2-cyanoacetamide,[31] ethylenediamine, ethanolamine, and lutidine for im-

[22] K. Sauda, T. Imasaka, and N. Ishibashi, *Anal. Chem.* **58**, 2649 (1986).

[23] R. J. Williams, M. Lipowska, G. Patonay, and L. Stekowski, *Anal. Chem.* **65**, 601 (1993).

[24] M. Kai, Y. Ohkura, S. Yonekura, and M. Iwasaki, *Anal. Chim. Acta* **207**, 243 (1988).

[25] W. Hu, H. Haraguchi, and T. Takeuchi, *J. Chromatogr.* **557**, 441 (1991).

[26] Y. Umegae, T. Nohta, and Y. Ohkura, *J. Chromatogr.* **515**, 495 (1990).

[27] P. Vratny, U. A. T. Brinkman, and R. W. Frei, *Anal. Chem.* **57**, 224 (1985).

[28] K. Kakehi, M. Ueda, S. Suzuki, and S. Honda, *J. Chromatogr.* **630**, 141 (1993).

[29] N. O. Maness and A. J. Mort, *Anal. Biochem.* **178**, 248 (1989).

[30] M. Kramer and H. Engelhardt, *HRC&CC* **15**, 24 (1992).

[31] S. Honda, T. Konishi, S. Suzuki, M. Takahashi, K. Kakehi, and S. Ganno, *Anal. Biochem.* **134**, 483 (1983).

proved fluorescence detection. These methods have been reviewed by Honda.[32]

Table I provides a summary of the selectivities available using spectrophotometric methods for various biopolymers.

Electrochemical Detection

The use of electrochemical detection after liquid chromatographic separation (LCEC) has now been widely popularized and practiced. The technique has been successfully applied to amino acids, peptides, and proteins,[33] generating subnanogram detection limits and high degrees of selectivity. It has also been applied to nucleic acids and sugars/glycopeptides. We first discuss the principles and instrumentation of LCEC, followed by selected applications relevant to the biotechnology field.

In all forms of LCEC, current generated at a working electrode is actually being measured; thus, EC detection in LC is a form of voltammetry. Usually, the working electrode is held at a fixed potential, which, when placed in a flowing stream of mobile phase, will generate a background current due to any oxidation/reduction of the mobile phase. When an electroactive analyte passes the working electrode, it will be oxidized (or reduced) by the working electrode, increasing the background current. The selectivity of EC detection is tuned by choosing the appropriate potential of the working electrode, such that only the analyte(s) of interest is detected. In some cases, it is advantageous to pulse the potential of the working electrode, especially when oxidized or reduced analytes foul the electrode surface, with the current only measured during a specified voltage, a technique known as pulsed amperometric detection (PAD).[34]

The most commonly employed cell design for LCEC is the thin-layer amperometric cell. Positioning the auxiliary electrode directly across from the working electrode helps to minimize current × resistance (iR) drop between the two. As a direct result, a wider linear dynamic range is achieved, because higher concentrations of injected analyte will not lead to significant changes in the working electrode potential. In this type of cell, only about 1% of the analyte in the eluent undergoes an electrochemical reaction. Coulometric electrochemical detectors use a porous graphite working electrode, such that all of the mobile phase and analyte come into contact with the electrode, and 100% of the analyte is electrochemically converted. Intuitively, it may seem that coulometric detectors are more sensitive than

---

[32] S. Honda, *Anal. Biochem.* **140,** 1 (1984).

[33] L. Dou, J. Mazzeo, and I. S. Krull, *BioChromatography* **5,** 704 (1990).

[34] L. E. Welch, W. R. LaCourse, D. A. Mead, Jr., and D. C. Johnson, *Anal. Chem.* **61,** 555 (1989).

TABLE I

SELECTIVITIES FOR BIOMOLECULES USING SPECTROPHOTOMETRIC DETECTION

Analyte and method of detection	Selectivity	Refs.
Peptides and proteins		
UV below 220 nm	Peptide bond	—
UV 280 nm	Trp, Tyr residues	—
UV–Vis > 280 nm	Metalloprotein prosthetic groups	—
FL emission 330–350 nm	Native FL of Trp, Tyr residues	—
PDA dual-wavelength absorbance ratios	Characteristic ratios for pure compounds	5, 6
PDA UV spectra	Identification based on differences in spectra	8
PDA derivative spectra	Further characterization of peptides based on differences in derivative spectra and peak to trough ratios	9
OPA, NDA, fluorescamine derivatization	Reaction with free amino groups	10–12
Benzoin, ninhydrin derivatization	Selective reaction with Arg residue	14, 15
4-(Aminosulfonyl)-7-fluoro-2,1,3-benzoxadiazole dansylaziridine, maleimide, bromobimane, fluorescein isothiocyanate	Selective reaction with Cys residue	16–18
4-Chloro-7-benzofurazan	Selective reaction with Pro residue	19
Thiamin reagent	Reaction at each peptide bond	21
Biuret reagent	Copper complexation with peptide linkage and Tyr residues	20
Lowry reagent	Copper–protein complex reduces phosphomolybdic–phosphotungstate reagent	20
FL dyes	Covalent and noncovalent binding	22, 23
Oligonucleotides		
260 nm	Native species	—
PDA dual-wavelength absorbance ratios	Characteristic ratios for pure compounds	5
Chloroacetaldehyde	Selective reaction with adenine and cytosine residues	26
Phenylglyoxal	Selective reaction with guanine residue	26
Molybdenum blue	Complexation with orthophosphate groups	27
Oligosaccharides		
Native species	Not applicable to spectrophotometric methods	—
2-Cyanoacetamide	Carbohydrate derivatization for UV and FL detection	32
p-Aminobenzoic acid hydrazide	Carbohydrate derivatization for UV detection	29
Thymol	Carbohydrate derivatization for Vis (500 nm) detection	32
2-Aminopyridine	Carbohydrate derivatization for UV detection	31
1-Phenyl-3-methyl-5-pyrazolone		30

amperometric detectors, because more of the analyte is converted. However, noise also scales with the increased signal. Because it is the signal-to-noise ratio that determines the sensitivity, it is generally acknowledged that in most cases amperometric detectors are more sensitive. Noise is further minimized by ensuring that the background current due to the mobile phase is as low as possible. Thus, high-purity solvents and salts are necessary for maximum sensitivity.

An approach that offers added selectivity is the use of two working electrodes, either in parallel or series. In the parallel case, the electrodes are directly across from each other in the flowing stream, held at slightly different potentials (0.1–2.0 V). This technique is similar to purity confirmation by dual-wavelength absorbance ratioing. The ratio of the current produced at the two electrodes can be compared to that of a pure standard for confirmatory work, because this ratio should be different for most compounds. In the series approach, one electrode is placed downstream of the other. Here, the reversibility or quasireversibility of the electrochemical reaction is exploited. An example would be when the upstream electrode is set at a potential to affect reduction of an analyte, with the downstream electrode set at a more positive potential for reoxidation. Because reductive LCEC suffers from poor signal to noise (S/N) due to reduction of dissolved oxygen, the downstream electrode would be monitored, generating better sensitivity than if only a one-electrode, reductive mode LCEC approach was employed.

Liquid Chromatography–Electrochemical Detection of Peptides

LCEC analysis of peptides using conventional carbon electrodes is limited to those peptides that contain tyrosine, tryptophan, methionine, cysteine, or cystine.[33] Much work has been done on selective determinations of peptides containing such amino acids, such as endorphins, bradykinin, opioids, hormones, and others.

One application of LCEC in peptide analysis is the determination of disulfide bond locations in proteins.[35] It has been reported by many that disulfides can be converted to free thiols by electrochemical reduction at mercury electrodes, so that a series dual-electrode approach for detecting both disulfides and free thiols can be used.[36] A gold upstream electrode is used to reduce disulfides to free thiols, with the thiols detected oxidatively downstream at a gold–mercury amalgam electrode.

Peptides not containing these electroactive amino acids do not show

---

[35] C. Lazure, J. Rochemont, N. G. Seiden, and M. Chretien, *J. Chromatogr.* **326,** 339 (1985).
[36] L. A. Allison and R. E. Shoup, *Anal. Chem.* **55,** 8 (1983).

electrochemical activity with typical working electrode materials like glassy carbon. Thus, the use of chemically modified electrodes (CMEs) has been widely studied. These electrodes are modified with groups that will allow oxidation or reduction of otherwise unconvertible species. Two of the most common chemically modified electrodes for amino acids are the nickel oxide and copper electrodes.[37-40] More general EC detection strategies for peptides are now appearing. The copper electrode yields low-nanogram detection limits with model peptides.[41] However, this method is not totally general, as the kinetics of the complexation between the copper oxide layer and peptide amino acids are often too slow for sensitive detection. In an interesting use of the classic biuret reaction, postcolumn copper(II) addition in HPLC of peptides was used to generate peptide–copper(II) complexes, which could be detected oxidatively and then reductively at series glassy carbon electrodes.[42]

Liquid Chromatography–Electrochemical Detection of Proteins

In electrochemistry, it is important that the analyte be able to diffuse to the electrode surface for a reaction to occur. Once at the surface, it is also important that those functional groups that may undergo electron transfer be able to orient properly onto the electrode surface. The general use of LCEC for protein analysis is limited by these two requirements, because proteins possess low diffusion coefficients and because electroactive groups may be buried inside the folded structure. Several chemically modified electrodes have been reported for the detection of certain specific proteins, with high sensitivity. These electrodes, rely on specific functionalities in the protein, such as heme groups in cytochrome c[43] or in hemoglobin and myoglobin.[44] A polyaniline electrode has also been reported for detection of ceruloplasmin, a copper-containing protein.[45]

We have described a more general approach for peptide and protein analysis in LCEC, using postcolumn photolysis, whereby degradation or conformational change in the protein structure may take place, leading to more electroactive photoproducts.[41,46] This method has been interfaced to

[37] S. Dong and Y. Yang, *Electroanalysis* **1,** 99 (1989).
[38] J. B. Kafil and C. O. Huber, *Anal. Chim. Acta* **175,** 275 (1985).
[39] W. T. Kok, U. A. T. Brinkmann, and R. W. Frei, *J. Chromatogr.* **256,** 17 (1983).
[40] W. T. Kok, U. A. T. Brinkmann, and R. W. Frei, *J. Pharm. Biomed. Anal.* **1,** 369 (1983).
[41] L. Dou and I. S. Krull, *Anal. Chem.* **62,** 2599 (1990).
[42] K. Stulik and V. Pacakova, *J. Chromatogr.* **436,** 334 (1988).
[43] J. W. Schlager and R. P. Baldwin, *J. Chromatogr.* **390,** 379 (1987).
[44] J. Ye and R. P. Baldwin, *Anal. Chem.* **60,** 2263 (1988).
[45] J. A. Cox and T. J. Gray, *Anal. Chem.* **61,** 2462 (1989).
[46] L. Dou, A. Holmberg, and I. S. Krull, *Anal. Biochem.* **197,** 377 (1991).

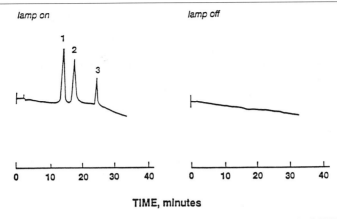

FIG. 3. Photoderivatization-EC detection of proteins. Equipment: 1.4-ml KOT reactor; mercury lamp; glassy carbon working electrode, +0.90 V vs Ag|AgCl. Peaks: (1) ribonuclease A, (2) lysozyme, (3) chymotrypsinogen A. (Reprinted from L. Dou and I. S. Krull, *Electroanalysis*, **4**, 381, 1992, with permission of VCH Publishers.)

size-exclusion, reversed-phase, ion-exchange, and hydrophobic interaction chromatographic separations of proteins, with detection limits in the parts per million range, 6 to 20 times lower than for UV detection. Figure 3 shows a typical separation. The mechanism of detection in this system is believed to result from complex events that happen when a protein is irradiated, such as reduction of disulfide bonds, denaturation, and modification of side-chain amino acids.

Liquid Chromatography–Electrochemical Detection of Nucleic Acids

LCEC has been applied to the quantitative analysis of nucleic acids.[47] This method involves hydrolyzing the nucleic acid and then measuring the released adenine and guanine using a glassy carbon electrode at +1.1 V vs Ag|AgCl.

LCEC has also been used to determine nucleotides and restriction fragments using a silver electrode.[48] Reversed-phase, ion-exchange, and size-exclusion separations have all been used. The operative mechanism is oxidation of the purine bases. It was found that the two naturally occurring purines, guanidine and adenine, could be readily distinguished owing to different EC oxidation potentials, allowing detection of one purine when the other purine is present in large excess. This proved useful in the analysis

[47] J. B. Kafil, H. Y. Cheng, and T. A. Last, *Anal. Chem.* **58**, 285 (1988).
[48] D. P. Malliaros, M. J. DeBenedetto, P. M. Guy, T. P. Tougas, and E. G. E. Jahngen, *Anal. Biochem.* **169**, 121 (1988).

of restriction fragments, where the leader sequence, poly(A), can interfere with the analysis of guanidine-rich fragments.

Like proteins, large restriction fragments possess low diffusion coefficients, which eventually lead to poor sensitivity for larger oligonucleotides. Hence, using EC detection, the only way to quantitate large oligonucleotides sensitively is to break them down to smaller molecules.

Liquid Chromatography–Electrochemical Detection of Sugars and Glycopeptides

Pulsed amperometric detection (PAD), in conjunction with high-pH anion-exchange chromatography, has become the method of choice for separating and sensitively detecting sugars and glycopeptides.[49–51] The direct electrochemistry of carbohydrates on noble metal electrodes suffers from electrode fouling due to absorption of oxidized species on the electrode surface. PAD overcomes this problem by using a triple-step potential waveform. In the first step, the gold electrode is held at a potential suitable for oxidation of the analyte. During this step, data are collected. The second step raises the potential to some higher value, where adsorbed oxidation products are oxidized further into mobile phase-soluble products, thereby cleaning the electrode. Finally, the electrode is brought down to a reduction potential to generate a new gold oxide surface for subsequent use in the sampling cycle. Using this technique, closely related oligosaccharides can be separated and detected in the 10- to 100-pmol range. Furthermore, glycopeptides and glycoproteins may also be analyzed, and glycopeptide maps generated. Oligosaccharide compositional analysis can also be done using this powerful combination, making it one of the most important tools in biotechnology today.

A summary of the different selectivities for biomolecules using EC detection is listed in Table II.

Detection of Biopolymers by Light Scattering

Light scattering (LS) is a well-defined, absolute, and theoretically rigorous technique that has been extensively used for the detection of synthetic polymers and biopolymers in a batch or dynamic mode. The physical parameters of the biopolymer that can be obtained are (1) molecular weight, (2)

---

[49] M. R. Hardy and R. R. Townsend, *Proc. Natl. Acad. Sci. U.S.A.* **85,** 3289 (1988).
[50] J. D. Olechno, S. R. Carter, W. T. Edwards, D. G. Gillen, R. R. Townsend, Y. C. Lee, and M. R. Hardy, in T. E. Hugli (eds.), "Techniques in Protein Chemistry." Academic Press, San Diego, 1989.
[51] A. S. Feste and I. Khan, *J. Chromatogr.* **630,** 129 (1993).

TABLE II
SELECTIVITIES WITH ELECTROCHEMICAL DETECTION

Analyte and method of detection	Selectivity	Refs.
Peptides and proteins		
Glassy carbon electrode	Native electroactivity of Trp, Tyr, Met, and Cys residues	34–36
Nickel oxide CME[a]		40
Copper CME	Copper ion–peptide complex on electrode surface	41
Pyridine–gold CME	Specific for heme prosthetic groups	44, 45
Methylene blue CME	Specific for heme prosthetic groups	45
Polyaniline CME	Specific for copper prosthetic groups	46
Postcolumn photolysis	Photolysis reactor generates more electroactive photoproducts	41, 46
Oligonucleotides		
Ag|AgCl electrode	Specific for adenine and guanine residues	47, 48
Oligosaccharides		
Au electrode with PAD	Oxidation of carbohydrate higher potential replenishes electrode surface	49–51

[a] CME, Chemically modified electrode.

particle size, (3) radius of gyration, (4) degree of aggregation, and (5) conformational change. The determination of absolute molecular weight is an important factor in the characterization of large polymers.

There are only a handful of techniques for determining the absolute molecular weight of macromolecules: membrane osmometry, sedimentation analysis, mass spectrometry, and LS. Each of these methods has its advantages and disadvantages. Light scattering is, however, most widely used for polymer characterization. Most of the earlier applications of LS have dealt with characterization of synthetic polymers. An increasing number of applications can now be found on the characterization of biopolymers by LS.

Theory of Light Scattering

The interaction of light with matter results in the scattering of light, along with other phenomena such as reflection and refraction. Light is a combination of oscillating electric and magnetic fields. When light interacts with matter, it induces a temporary dipole in the molecule due to the polarizable electron cloud of the molecule. The induced dipole now oscillates at the frequency of incident light and reemits the light in different directions as it returns to the ground state. If the frequency of emitted light is the same as the incident light, the process is termed *elastic scattering*.

However, if the emitted frequency is different, then the process is termed *inelastic scattering*. We limit our discussion to elastic LS, particularly Rayleigh scattering and quasielastic LS (quasielastic light scattering), because these are the most widely used and published techniques.

In Rayleigh scattering, the intensity of the light scattered by the biopolymer is proportional to the excess Rayleigh factor, R'_θ.

$$R'_\theta = R_\theta \text{solution} - R_\theta \text{solvent} \tag{1}$$

When using LALLS, the calculation of molecular weight can be simplified by using Eq. (2):

$$Kc/R'_\theta = 1/\text{MW} + 2A_2 c \tag{2}$$

where K, the polymer optical constant, is a function of the differential refractive index (dn/dc) of the sample. The dn/dc can be easily calculated using a differential refractometer. To determine the molecular weight of a biopolymer, an appropriate concentration of the sample, usually 0.5–2 mg/ml, is injected onto an HPLC column. The effluent from the column then flows through an in-line filter into an LS cell. A concentration-sensitive detector is placed in series with the LALLS. The LALLS signal is used to determine R'_θ and the concentration is determined from the UV detector. A_2, the second virial coefficient, is determined by plotting Kc/R'_θ vs c. The slope is A_2 and the intercept is 1/MW. The advantage of using LS is that the molecular weight can be determined without any form of calibration. For more comprehensive discussions of the theory, applications, and other aspects of LS, please see Ref. 52.

Instrumentation

An example of the optical system in a LALLS photometer is shown in Fig. 4. A 633-nm He/Ne laser is used for the scattering process. The laser is focused onto the sample cell held between two quartz windows. The scattered light is collected at a certain angle and focused through a set of optics on a photomultiplier. The scattering angle is defined using an annulus located after the sample cell. The instrument is also equipped with a microscope that can be used to observe scattered light, as well as the incident light, and also for aligning the cell in the optical path.

All modern light-scattering photometers can be interfaced with computers and operated using software from their respective manufacturers. For determination of molecular weight, a concentration-sensitive detector [UV

[52] P. Kratochvil, "Classical Light Scattering from Polymer Solutions." Elsevier, Amsterdam, The Netherlands, 1987.

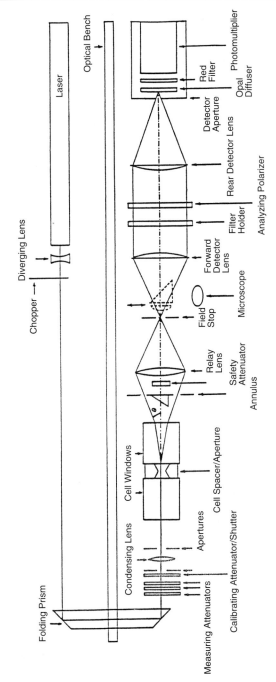

Fig. 4. Optical diagram of a low-angle laser light-scattering photometer.

or refractive index (RI)] is used in series with a light-scattering photometer. The concentration and scattering signals are later processed by a computer, usually to generate the weight-average molecular weight.

Many reports have been published on the applications of LS detectors for the determination of biopolymers off-line, as well as those separated by size-exclusion chromatography (SEC) or HPLC. Low-angle laser light-scattering detectors have been used for the determination of molecular weight and particle size of various types of biopolymers, including proteins, nucleic acids, and polysaccharides. However, the most extensive applications have been in the analysis of proteins and their aggregates. Some of the more recent applications are summarized as follows.

Proteins and Polypeptides

Takagi has reviewed the applications of LALLS detection in the field of biochemistry and concluded that "the application of GPC-LALLS technique in biochemistry has made molar mass measurement of biopolymers easier and has opened up approaches leading to new concepts, such as structure–function relationship of the membrane proteins."[53]

With the rapid growth of recombinant protein technology, the analysis and characterization of proteins are of utmost importance. Some of the possible contaminants in recombinant proteins are aggregates of proteins, DNA fragments, and other cell materials from the fermentors. Because LS is a molecular weight-sensitive detector, it is sensitive to aggregates of proteins that may go almost undetected by UV or fluorescence. Mass spectrometry is limited to about 100K molecular weight species. Also, it is not quite clear if MS reflects the actual compositions of the aggregates in the protein solutions.

We have published extensively on interfacing HPLC with LALLS for determining the molecular weight of proteins and their aggregates. One of the practical applications reported was the characterization of recombinant hGH and its aggregate.[54] Recombinant hGH had a small percentage of aggregate that could not be well resolved from the monomer by conventional SEC. In spite of the poor resolution, LALLS was able to confirm that the aggregate was a dimer. Species that are not completely resolved can be characterized because the molecular weight is calculated at specific, predetermined intervals along the chromatograms. The computer then generates a printout and a plot of retention volume versus molecular weight. If a certain aggregate is partially resolved from its monomer, then the

---

[53] T. Takagi, *J. Chromatogr.* **506**, 409 (1990).
[54] H. H. Stuting and I. S. Krull, *J. Chromatogr.* **539**, 91 (1991).

molecular weight of the high aggregate can be estimated from the signal at the beginning of the chromatogram, which would be predominantly due to the aggregate. The next example illustrates this point better. Figure 5 shows the SEC chromatogram of bovine serum albumin (BSA). The LALLS signal clearly indicates the presence of two additional species besides the monomer. Although the species are not completely resolved, the molecular weights determined at the peak apex indicated the presence of a trimer, dimer, and monomer of BSA.

Most LS photometers can be easily interfaced with an HPLC system. Fast gradient HPLC with LALLS has been interfaced in separating proteins and characterizing them on-line. LALLS was able to provide information on protein aggregation that was virtually undetected by UV. Figure 6 shows the typical LALLS and UV traces obtained for a mixture of proteins separated by reversed-phase HPLC. The LALLS peaks look asymmetrical and indicate the presence of unresolved aggregates. The UV peaks are quite symmetrical and one would not suspect the presence of additional species coeluting with the main peak. Because LALLS is a molecular weight-sensitive detector, higher molecular weight species that are partially resolved can be easily detected.

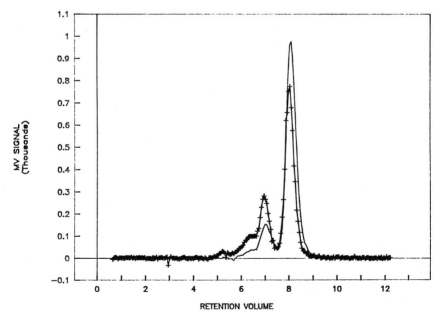

FIG. 5. SEC/LALLS/UV chromatogram of BSA. Column: G3000SW. Mobile phase: 0.05 M Na_2SO_4 at 0.5 ml/min. LALLS: G_0 = 300 mV, 0.2 mmFS, 6–7 annuli. UV: 280 nm, 0.2 AUFS. Concentration: 3 mg/ml injected through a 100-μl loop.

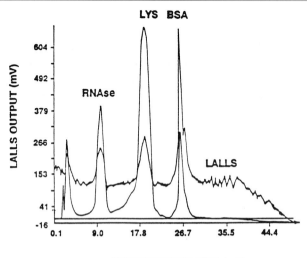

RETENTION VOLUME (mL)

FIG. 6. Superimposed RP-LALLS (top) and RP-UV (bottom) chromatograms for a 20-μl injection of ribonuclease A (9.95 mg/ml), lysozyme (10.10 mg/ml), and bovine serum albumin (3.25 mg/ml). Column: C$_8$ nonporous. Solvent A: 0.15% TFA, pH 3.0. Solvent B: 0.15% TFA in 95% acetonitrile, pH 3.0. Gradient: 20–55% B over 20 min. (Reprinted from *J. Chromatogr.,* 502, R. Mhatre, I. S. Krull, and H. H. Stuting, p. 21, 1990, with kind permission of Elsevier Science–NL, Sara Burgerhartstraat 25, 1055 KV Amsterdam, The Netherlands.[55a])

Low-angle laser light scattering has been interfaced with fast gradient ion-exchange chromatography.[55] The molecular weights for three of the four proteins studied were in excellent agreement with literature values. However, in the case of β-lactoglobulin A, the molecular weight increased as steeper gradients were used. In the absence of a molecular weight-sensitive detector, the peaks could be misidentified as those from the monomer. The ability to study the aggregation at various stages of chromatography is a major advantage of LS. Because LS does not in any way alter the species during detection, the percentage of aggregation is truly representative of what is present in the solution. This is important to determine recombinant proteins that are dissolved in various buffers prior to their being injected into patients. Proteins can aggregate very easily with minor changes in pH and ionic strengths of a solution, and LS is one of the ideal ways to study such changes in protein solutions.

A similar aggregation study of β-lactoglobulin was conducted by interfacing hydrophobic interaction chromatography (HIC) with LALLS/

[55] R. Mhatre and I. S. Krull, *J. Chromatogr.* **591**, 139 (1992).
[55a] R. Mhatre, I. S. Krull, and H. H. Stuting, *J. Chromatogr.* **502**, 21 (1990).

UV.[56] Once again, LALLS was useful in determining exactly the aggregate state of the protein in solution, during chromatography, and in the lyophilized state. Under HIC conditions, at pH 4.5 and 4°, β-lactoglobulin formed a mixture of aggregates, these being a dodecamer, tetramer, and an octamer. SEC/LS for the same protein at room temperature showed mainly a dimer. When the protein was dissolved in a nonionic surfactant, the molecular weight of a monomer was obtained. The nonionic surfactant usually solubilizes the protein and eliminates protein–protein interactions. Most proteins therefore do not aggregate in the presence of the surfactant, and the molecular weights of the proteins should therefore reflect the molecular weight of the solid protein sample.

The study of aggregation/dissociation phenomena of proteins as a function of temperature has been reported.[57] The molecular weights of ovalbumin were found to increase when the sample was heated. The degree of aggregation increased with exposure to higher temperatures. The aggregates eluted as a wide peak in the retention time span of about 60 min. This suggested the presence of a broad molecular weight distribution. The light-scattering peak was slightly shifted to the left (lower retention volume), indicating that a large population of high molecular weight aggregates was present in the broad peak. Also, when a sample of ovalbumin that had been heated to a temperature of 80° was allowed to stand at room temperature, the UV peaks for the 0-, 6-, and 20-hr incubation looked identical. However, the light-scattering response for the 22-hr incubation increased significantly. The molecular weights suggested that the aggregates formed on heating had aggregated further during incubation.

One other practical application for recombinant proteins is the determination of percent glycosylation, and LALLS can be used for determining the degree of glycosylation.[58] Takagi et al. have used LALLS, UV, and RI to determine the percent glycosylation of recombinant human stem factor. By determining the molecular weight of the glycosylated protein and the molecular weight of the protein moiety alone, the authors were able to determine the amount of glycosylation present. The glycosylated protein had a molecular weight of 53,000 and the protein had a molecular weight of 34,000, indicating that the protein was 34% glycosylated. Maezawa and Takagi have discussed the molecular weight determination of glycoproteins by SEC/LALLS. The molecular weights showed excellent agreement with the molecular weights obtained by amino acid and carbohydrate composi-

[56] H. H. Stuting, I. S. Krull, R. Mhatre, S. C. Krzysko, and H. G. Barth, LC-GC Mag. 7(5), 402 (1989).
[57] A. Kato and T. Takagi, J. Agric. Food Chem. 35, 633 (1987).
[58] T. Arakawa, K. E. Langley, K. Kameyama, and T. Takagi, Anal. Biochem. 203, 53 (1992).

tional analysis. More recently, Takagi *et al.* have used SEC/LALLS for determining the molecular weight of recombinant human hepatitis B virus surface antigen particles.[59] The two particles differed in that one consisted of a polypeptide with an additional peptide of nine amino acid residues. Accurate molecular weights were obtained for both particles, indicating that the method was appropriate for the characterization of such antigen particles, which will be widely used as vaccines.

Although LALLS is now being used increasingly for characterization of proteins and other biopolymers, the sample requirements for analysis often restrict the routine use of this technique. The sensitivity of light-scattering instruments is often poor with low molecular weight proteins. Also, the calculation of dn/dc, which is done off-line in a differential refractometer, is tedious and requires additional sample. Typically, determining the molecular mass of a 25-kDa protein requires approximately 5–7 mg of sample. With increasing molecular weights, the sample requirements are accordingly reduced. The calculation of dn/dc can be performed on-line to simplify the procedure and also to minimize sample requirements.[60] A three-detector system can be used to determine dn/dc and then the molecular weight. We have interfaced LALLS, UV, and RI detectors in series. The RI signal is used to determine dn and dc is determined from the UV signal. The two signals can then be ratioed across the peak to determine dn/dc. More recently, we have determined dn/dc on-line for gradient HPLC.[61] The RI baseline was stabilized by using isorefractive (same refractive index) solvents. In a gradient ion-exchange mode, dextrose was added to one of the low ionic strength buffers and the refractive index of the two buffers was matched.

In spite of its many advantages, light scattering is not being used extensively. The analysis of biomolecules in aqueous systems requires expertise in sample and buffer clarification to avoid scattering from dust or impurities in the water. Often, the filters used in the light-scattering instruments can adsorb large biomolecules and cause problems in recovery of expensive biomolecules. Also, binding of salts or other buffer additives to the sample molecules must be taken into consideration. Ideally, a salt concentration less than 0.5 M is recommended for analysis of proteins. It is also crucial to keep the optics and cells in the instrument clean. The use of complex buffers, commonly used in formulations, may require additional corrections to the equations of light scattering.

[59] Y. Sato, N. Ishikawa, and T. Takagi, *J. Chromatogr.* **507,** 25 (1990).
[60] S. Maezawa and T. Takagi, *J. Chromatogr.* **280,** 124 (1983).
[61] R. Mhatre and I. S. Krull, *Anal. Chem.* **65,** 283 (1993).

Nucleic Acids and Polysaccharides

Light scattering can also be used for measuring the molar mass of other classes of biopolymers, such as polysaccharides and DNA fragments. It is ideal for measuring the molecular weight of large DNA fragments because the molecular weights of certain large fragments are in the range of several million, and the sample requirements are minimal. For DNA fragments of molecular weights $>10^7$, a sample concentration of 10 μg/ml is sufficient.

The number of reports on using LS for characterizing DNA is limited. Light scattering has been used for determining the molecular weight of DNA fragments[62] and also to study the size and shape in various solutions.[63] Several polysaccharides have been routinely characterized by LS for the determination of the molecular weight distribution.[64–66] The molecular weight distribution can be used to determine the polydispersity (M_w/M_n) and the heterogeneity in a sample.

Future Developments

What does the future hold? Although not discussed at length in this chapter, perhaps the greatest advances in the detection and characterization of biopolymers will evolve from advances in mass spectrometry. The current ability to identify, with little ambiguity, virtually any high molecular weight biopolymer via the electrospray interface between HPLC and MS is going to continue and improve, in terms of ease of use, lower sample requirements, higher molecular weight ranges, and sequencing abilities for longer chain peptides. Improvements will also come about with regard to the mapping of the glycoportions of glycoproteins, or in mapping carbohydrates and polysaccharides (see Section III in this volume).

What newer detection principles might evolve? Postcolumn reaction chemistry may never be specific for an individual protein or nucleic acid, although such chemistry can be specific for proteins or nucleic acids (e.g., phosphorus content). The emphasis for the future does not appear to reside in reaction detection for HPLC of biopolymers, but rather in more sophisticated instrumental methods of detection. The continued use of circular dichroism or optical rotary dispersion for detection of biopolymer confor-

[62] T. Nicolai, L. van Dijk, J. van Dijk, and J. Smit, *J. Chromatogr.* **389,** 286 (1987).
[63] K. S. Schmitz and J. M. Schurr, *Biopolymers* **12,** 1543 (1973).
[64] B. R. Viyayendran and T. Bone, *Carbohydr. Polymers* **4,** 299 (1984).
[65] D. Lecacheux, R. Panaras, G. Brigand, and G. Martin, *Carbohydr. Polymers* **5,** 423 (1985).
[66] D. Lecacheux, Y. Mustiere, and R. Panaras, *Carbohydr. Polymers* **6,** 477 (1986).

mations on-line in HPLC will continue to grow, but it can only provide conformation information.

Electrochemistry seems to hold much appeal, and yet it has not realized its full potentials. It holds exquisite sensitivity and detectability for suitably EC active small molecules, but not for large biomolecules. Pulsed amperometric detection methods do not really identify the biopolymer and provide little more information than UV methods. Chemically modified electrodes can be much more selective and sensitive than glassy carbon or Au|Hg-type electrodes, but still do not provide absolute, qualitative information about the nature of the biopolymers. Multiple array detectors, perhaps with chemically modified electrodes or different noble metal electrodes (Ni, CuO), may provide more information of a qualitative nature, such as cyclic voltammograms, but that type of information is also not very specific for an individual protein or nucleic acid. EC may always be relegated to providing ambiguous structural information, but it will still be useful as a qualitative indicator of the presence of a protein and have the ability to quantitate with high sensitivity.

The other current detection methods, UV–Vis, FL, LS will always be useful for qualitative information, identification of the chromatographic performance of a peak (retention times, capacity factors), and for absolute quantitation, but they may never provide 100% specific information about the structure of a biopolymer. Further advances using these detection techniques will be realized by incorporating the strengths of each detection technique in sequential detection schemes. UV detection prior to MS is already common. Other combinations may also show utility, especially for specific applications. Thus, as new techniques appear to mature, they will likely augment rather than replace existing techniques.

Acknowledgments

Much of our own work in areas of biochromatography detection, especially for biopolymers, was supported by several outside sources. In the areas of low angle laser light scattering detection of proteins and nucleic acids, instrumentation, materials, service, and financial support were provided by ThermoSeparation Products (TSP), Inc. (Riviera Beach, FL). In the areas of electrochemical detection for proteins/peptides, that work was supported via instrumentation, materials, service, and graduate student support by Bioanalytical Systems, Inc. (West Lafayette, IN). In the areas of photometric (UV, FL, LDA) detection of biopolymers, instrumentation donations were provided by Hewlett Packard Corporation (Waldbronn, Germany), EM Science, Inc. (Gibbstown, NJ), and TSP Analytical. Funding was provided, over several years, by an NIH Biomedical Research Support Grant to Northeastern University, No. RR07143, Department of Health and Human Services (DHHS). Additional funding was provided, over many years, by a grant from Pfizer, Inc., Pfizer Central Research, Analytical Research Department (Groton, CT). Additional funding from The United States Pharmacopeial Convention, in the form of a USP Fellowship to M.E.S., is also greatly appreciated.

[9] Diode Array Detection

By GERARD P. ROZING

Introduction

Ultraviolet–visible (UV–Vis) spectrophotometric detection methods used in separation techniques provide quantitative and qualitative information about the species separated. Quantitative determination mandates the use of standards, which in separations of biological macromolecules may be unavailable during research phases and therefore is of importance in quality assessment/quality control (QA/QC) of biopharmaceuticals. Alternatively, qualitative data of separated, high molecular weight biological molecules such as UV–Vis spectra, can give information about the identity, homogeneity, and structure of the molecule. The spectra are provided by a spectrophotometric technique such as photodiode array detection, in which UV–Vis spectra are acquired on-line with the recording of the chromatogram. As the chromatographic techniques used in bioseparations may not have sufficient resolving power to separate all substances in the sample or to remove the complex matrix from the sample, spectral data are important in the analysis of complex biological samples. UV–Vis spectra will be helpful in such cases in assessing purity and identity of the particular band in the chromatogram.

The usefulness of UV–Vis spectra for qualitative assessments of high molecular weight species has been questioned. UV–Vis spectra in general barely show fine structure. In the case of peptides and proteins this lack of information is enhanced by the low molecular extinction of bands above 250 nm (originating from the amino acids phenylalanine, tyrosine, and tryptophan). In high molecular weight biological molecules, the molecular extinction contributions of these amino acids are "averaged," and spectra may become even more alike. However, progress in instrument design for detection by diode array technology has provided the utmost accuracy and precision in spectral elucidation. Spectral differences thus may be small, but can be used in combination with sophisticated algorithms that enhance these small differences. Furthermore, UV–Vis spectra of biological macromolecules will depend on the solvent. This provides an opportunity to study the dynamics of structural and chemical interchanges of the molecule, and therefore of its activity, as a function of its environment. The column used in the separation may provide an attractive vehicle for this purpose.

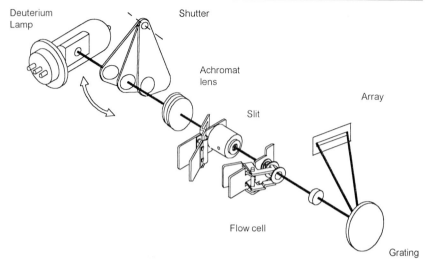

FIG. 1. Diagram of a commercial diode array detector for HPLC. (Courtesy of Hewlett-Packard GmbH, Waldbronn, Germany.)

Instrumental Design

A schematic representation of a commercial photodiode array (PDA) detector is given in Fig. 1. Polychromatic light from a deuterium and/or tungsten lamp is focused by a lens system on a flow cell. The light beam transversing the flow cell is dispersed on a diffraction grating and falls on an array of light-sensitive diodes. Typically these diodes are 10–50 μm wide and 1–2 mm high. Approximately 35–1000 diodes can be arranged linearly on the chip. The physical dimensions of the array therefore will be a few centimeters in length and a few millimeters in height.

The wavelength range of the diode array spectrophotometers is governed by the light source and is limited at the upper range by the decaying response of the diodes above 1100 nm. The typical wavelength range of a diode array detector is 190–600 nm. Spectral resolution is determined by the number of diodes "segmenting" the spectrum and by the optical bandwidth of the spectrograph. In commercial instruments the bandwidth varies between 2 and 20 nm.

The instrument shown in Fig. 1 has a "reversed optics" arrangement of flow cell and grating; reversed, because in conventional scanning spectrophotometers the grating is placed in front of the flow cell. A consequence of this arrangement is the absence of moving parts in the spectrograph (but for the shutter), which allows high precision in spectral measurements. In addition, such instruments can be calibrated simply (in Fig. 1 via the narrow

bands of a holmium oxide filter present in the shutter). All optical components are carefully positioned and permanently fixed in the spectrograph
by the manufacturer. The lamp and flow cells are easily exchangeable and
self-adjusting, so that the highest accuracy of wavelength settings is attained.

Data Acquisition

The intensity of the light on the diodes is measured by charging an
associated capacitor. All capacitors are then sequentially read. As the capacity is small, fast readouts of 10–25 msec are mandatory. This means that
the basic information to generate the spectrum becomes available with a
rate that obviates corrections if the concentration of the species changes
during the spectrum measurement.

In principle, multiple, parallel signals are obtained. At least one signal
is used in most detectors for on-line display and output. Simple arithmetic-
like signal ratio calculation is possible and can be outputted as well. Likewise, the signals obtained can be transferred to a computer for further processing.

Besides signals, a photodiode array detector delivers continuous snapshots of absorbance versus wavelength (spectrum) in the eluate in the flow
cell at a high rate (up to 40 Hz). The detector can be set up to acquire
spectra continuously or triggered by a peak detector that uses one of the
on-line signals. The spectra can be displayed on-line, stored in the memory
of the detector for output after the run, and/or transferred to the PC for
further evaluation. Combination of the snapshots provides a raw (continu-
ous) A-λ-t data matrix. An example is given in Fig. 2. The A-λ-t matrix is
used for elaborate off-line data presentation, comparison of spectra, peak
purity determination, and quantitation on the attached computer. Most
commercial photodiode array detectors provide this capability (see Table
I). In the next sections, some of these manipulations are explained and
illustrated with examples of relevance for separation of biological molecules.

Diode array detectors were introduced several years ago for HPLC
and have become available for capillary electrophoresis. Table I gives an
overview of the commercial instrumentation available. A number of reference books have been published that describe the technique and applications (see, e.g., the book by Huber and George.[1])

[1] L. Huber and S. A. George, in "Diode Array Detection in HPLC," Chromatographic
Science Series, Vol. 62. Marcel Dekker, New York, 1993.

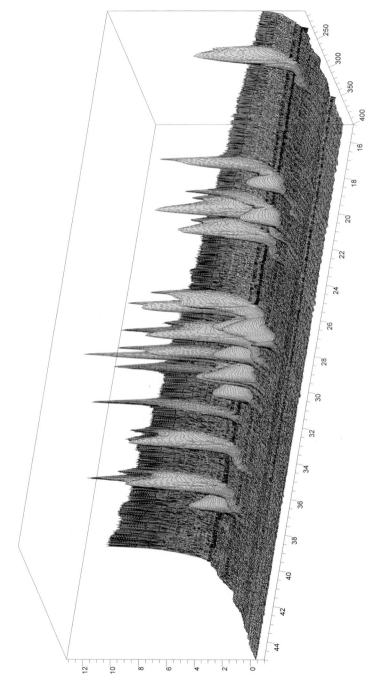

FIG. 2. Three-dimensional data matrix for a pesticide standard. Instrument: Hewlett-Packard 1040A diode array detector. (Courtesy of Hewlett-Packard GmbH, Waldbronn, Germany.)

Optimization of Data Acquisition

As was stated, detector hardware and data acquisition control settings must be selected carefully to fit the requirements of the user. One needs to decide up front whether utmost chromatographic integrity, highest sensitivity, or best spectral resolution is designed. Some important aspects are dealt with in the next sections.

Detector Cell Volume and Optical Path Length. Commercial photodiode array detectors are available with a selection of flow cells allowing optimal adaptation to demands imposed by the separation column. In all separation techniques described in this volume, using spectrophotometric detection one faces the dilemma to combine longest optical path length of the detection cell with minimal detection volume of the detection cell in order to obtain the best sensitivity. Low-volume flow cells and typical cylindrical cuvettes, 6–10 mm long and 1 mm in diameter (8–12 μl), will maintain the integrity of the separation obtained on normal-bore LC columns (3- to 8-mm i.d.). Narrower columns, however, mandate smaller cell volumes. But narrowing the cuvette diameter and maintaining optical path length will reduce the amount of light traversing the cell and therefore increase noise. Shortening the path length will maintain chromatographic integrity but decrease response. Alternatively, one may need a very short path length in the case of preparative separations in order to remain in the linear range of a detector. Besides these considerations, pressure rating of the flow cells is of importance for eventual coupling of the LC column effluent to a mass spectrometer. In capillary electrophoresis, where only the capillary diameter is available as path length, path length elongation (three to five times the capillary diameter) is achieved through the formation of a detection bubble (Table I), which improves sensitivity.

An example of the effect when too large a flow cell is selected is given in Fig. 3.[2] As this example is a gradient elution, and both flow cells have the same optical path, one expects the peak heights to be equal. The smaller flow cell, though, generates narrower peaks, in particular exemplified by the deeper valleys (better resolution) between the peaks.

Optical Bandwidth Selection; Diode Bunching. The optical bandwidth of diode array detectors is governed by the width of the slit aperture and the number of adjacent diodes one selects to record the sample signal. These parameters can be set independently. Most commercial detectors have an exchangeable aperture that allows wider optical bandwidths and/ or allows selection the detection wavelength range.

Broader bandwidth will increase the amount light passing through the flow cell and/or allow the measurement of more light. Consequently noise

[2] M. Herold, D. N. Heiger, and R. Grimm, *Am. Lab.* August (1993).

TABLE I

COMMERCIAL PHOTO DIODE ARRAY DETECTION SYSTEMS

Characteristic	Manufacturer					
	Applied Biosystems	Beckman Instruments	Gynkotek	Hewlett-Packard	Hewlett-Packard	Hitachi
Model	1000S	168	UVD320	G1306A	G1601A (for CE-System)	L-4500
Lamp	Deuterium	Deuterium	Deuterium	Deuterium	Deuterium	Deuterium
Wavelength range	190–555 nm in two ranges	190–600 nm	200–355 nm	190–600 nm	190–600 nm	190–800 nm
Bandwidth	Minimum 5 nm, adjustable	Minimum 4 nm, adjustable	2 nm	2, 4, and 6 nm	Minimum 4 nm, adjustable	Minimum 2 nm, adjustable
Flow cell (volume, path length, pressure)	6 μl, 8 mm, 40 bar	11 μl, 10 mm, 4 μl, 5 mm, 2 μl, 4 μl, 2 mm<; all 1000 psi	12 μl, 10 mm, 50 atm	13 μl, 10 mm, 1.7 μl, 6 mm, 50 atm; 8 μl, 6 mm, 120 atm	Elongated path cell (bubble cell)	8 μl, 10 mm, 1 μl, 0.5 mm, 10 atm, 8 μl, 10 mm, 150 atm
Absorbance range	0.001–2.0 AUFS	0.001–2.0 AU	0.0005–2.0 AU	NA	0–0.5 AU	−0.5–2.0 AU
Linearity	NA[a]	<1% at 1.5 AU	NA	<1% at 1.5 AU	NA	NA
Noise	<±2 × 10^{-5} AU	10^{-4} AU	±1.0 × 10^{-5} AU	±2.0 × 10^{-5} AU	4 × 10^{-5} AU	<5 × 10^{-5} AU
Drift	<2 × 10^{-4} AU/hr	2 × 10^{-3} AU/hr	5 × 10^{-4} AU/hr	<2 × 10^{-3} AU/hr	NA	3 × 10^{-3} AU/hr
Time constant	0.01–5.0 sec	0–5.0 sec	0.5–4 sec	0.1–20 sec	0.1–20 sec	0.1–3.2 sec
Output/control	Integrator/recorder, contact closure, RS232	Integrator/recorder, PC for control/data analysis, System Gold (Windows) HPLC and CE	Integrator (2), start/stop, RS232, Centronics	Integrator, external contacts, RS232, IEEE interface for PC control/data analysis (Windows)	Integrator, external contacts, RS232, IEEE interface for PC control/data analysis (Windows)	PC interface
Other	Autozero, marker, display, pump control; double-beam optics	Autozero, marker, display, HPLC and CE	Autozero, display	Autozero	Autozero	

Characteristic	LDC	Waters	Perkin-Elmer	Perkin-Elmer	Shimadzu	Unicam	Varian
Model	Spectromonitor 5000	996 PDA	LC-235C	LC 480	SPD-M10A	240, 250	9065
Lamp	Deuterium	Deuterium	Deuterium	Deuterium	Deuterium	Deuterium	Deuterium
Wavelength range	190–360 nm	190–800 nm	195–365 nm	190–430 nm	195–600 nm	190–800 nm	190–367 nm
Bandwidth	Minimum 5 nm, adjustable	1, 3 nm	5 nm	≤4 nm	NA	1, 3, 4 (20) nm	Minimum 4 nm adjustable
Flow cell (volume, path length, pressure)	15 μl, 8 mm, 70 atm	8 μl, 10 mm, 1000–3000 psi	8 μl, 10 mm, 2200 psi	12 μl, 10 mm, 50 atm	8 μl, 10 mm, 50 atm; 3 μl, 6 mm, 400 atm	8 μl, 10 mm	4.7 μl, 6 mm, 140 atm
Absorbance range	0.0005–2.0 AUFS	0.01–2.0 AUFS	—	0.0005–2.0 AUFS	0.001–3 AUFS	−0.1–1.6 AUFS	NA
Linearity	<1% at 1.9 AU, <5% at 2.6 AU	NA	NA	NA	NA	NA	NA
Noise	2×10^{-5} AU	5×10^{-5} AU	$\pm 2.5 \times 10^{-5}$ AU	10^{-5} AU	$\pm 2.5 \times 10^{-5}$ AU	5×10^{-5} AU	6×10^{-5} AU
Drift	10^{-4} AU/hr	10^{-4} AU/hr	$<10^{-4}$ AU/hr	5×10^{-4} AU/hr	$<10^{-3}$ AU/hr	10^{-3} AU/hr	$<10^{-3}$ AU/hr
Time constant	0.05–10 sec	0–5 sec	—	0.5–4 sec	0.24–2.4 sec	—	0.05–2 sec
Output/control	Integrator/recorder, 4 channels; signals, spectra, arithmetic	Integrator/recorder, PC control system Millenium (Windows)	Integrator (2), start/stop, RS232	Integrator, RS232 and IEEE interface PC control/data analysis (Windows)	Interface to IBM PC for control/data analysis	Integrator/recorder	Integrator/recorder start/stop, not ready, contacts, PC control/data analysis (Windows)
Other	Autozero, marker, display; double-beam optics	Autozero, marker, display	Autozero, marker, display; double-beam optics	Autozero, marker, display	Autozero	Display, PC control with model 250	Autozero, marker

[a] NA, Not available.

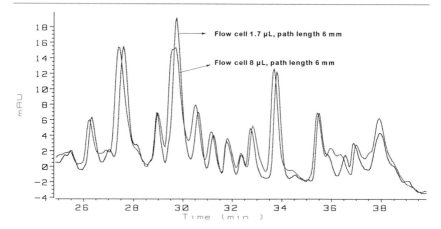

Fig. 3. Comparison of resolution in a peptide map performed on a 1-mm bore column using an 8-μl and a 1.7-μl flow cell. Column: SPHERI 5, 250 × 1.0 mm. Mobile phase: gradient 0.05% trifluoroacetic acid (TFA) in water versus 0.045% TFA in acetonitrile, 1–30% in 85 min; flow, 0.05 ml/min; detection, 214 nm. Sample: Complete tryptic digest of aspartate aminotransferase from *E. coli*, 2 μl, 50 pmol. (Courtesy of Hewlett-Packard GmbH, Waldbronn, Germany.)

will be reduced. The effects of each approach are given in Fig. 4a and b, respectively. One must realize that the optical bandwidth cannot be widened without penalty. In the optimum, the sample wavelength equals the maximum absorption wavelength of the solute and optical bandwidth matches the bandwidth of the absorption band. By widening the bandwidth of the instrument beyond the spectral bandwidth the average absorption coefficient and therefore the signal-to-noise ratio will decrease.[3]

The magnitude of slit aperture and bandwidth will affect the spectral fine structure. This is illustrated in Fig. 5, where the slit was varied from 2 to 4 nm and the optical bandwidth—the sample diode range—was varied from 2 to 8 nm.

As can be seen, best spectral resolution is obtained when the aperture and the number of diodes in spectral recording are at a minimum. On the other hand, when only minute amounts of solute are available, setting the bandwidth for spectral reading at a higher value will reduce spectral noise and will provide a spectrum with a better "spectrum-to-noise" ratio.

Reference Wavelength/Bandwidth Selection. The reference wavelength in single-beam diode array detectors is selected as close as possible to the sample wavelength, with the broadest bandwidth setting that the detector

---

[3] S. A. George and H. Elgass, *Am. Lab.* July, 24 (1984); A. Maute, Waldbronn, Germany, personal communication, 1984.

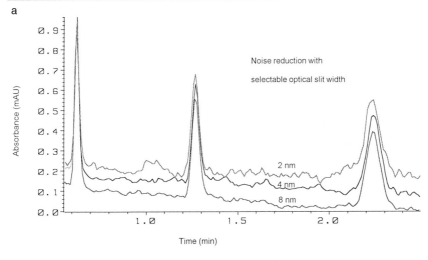

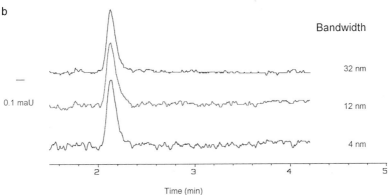

Fig. 4. Influence of optical slit width and diode bunching on detector noise. (a) Noise with varying aperture; (b) noise with varying diode range. (Courtesy of Hewlett-Packard GmbH, Waldbronn, Germany.)

allows. The function of the reference signal is to correlate as closely as possible with all fluctuations occurring at the sample wavelength without subtraction of significant signal absorption. The effect of reference wavelength and bandwidth selection is illustrated in Fig. 6.

Time Averaging. As the data in diode array are available in digital form after front-end processing, on-line signal processing allows noise reduction by elaborate routines in contrast to conventional analog signal filtering. Bunching the data in the time domain reduces the noise by the inverse of the square root of the number of data points. Figure 7 gives an example.

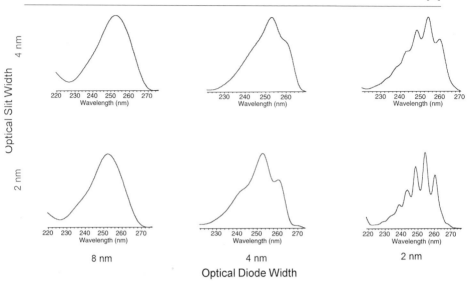

FIG. 5. Improved spectral resolution by minimization of the spectral bandwidth. (Courtesy of Hewlett-Packard GmbH, Waldbronn, Germany.)

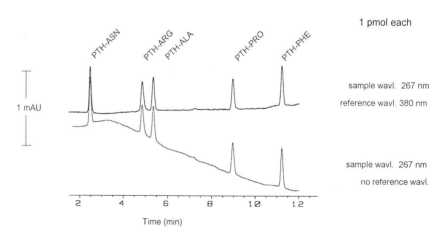

FIG. 6. Influence of reference wavelength selection on noise and baseline drift. *Top:* Influence of reference wavelength selection. *Bottom:* Influence of reference wavelength selection on baseline drift. Gradient conditions: 0.02 *M* phosphate buffer/acetonitrile, 12–45% B in 12 min. (Courtesy of Hewlett-Packard GmbH, Waldbronn, Germany.)

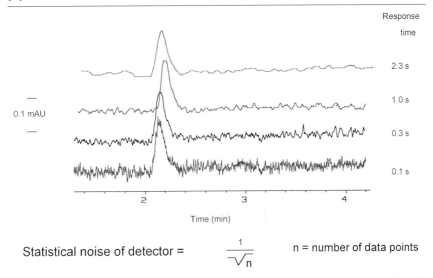

$$\text{Statistical noise of detector} = \frac{1}{\sqrt{n}} \qquad n = \text{number of data points}$$

FIG. 7. Influence of time averaging on signal noise. (Courtesy of Hewlett-Packard GmbH, Waldbronn, Germany.)

The filter rate is expressed as response time. Obviously, selection of a too-high value of the response time will distort the exact reading of a fast peak. Selection of a low value of the response time conserves signal integrity but will lead to a high data rate that makes on line and off-line data reduction slow. The optimal value of the response time is selected from the peak width at the base of the first eluting peaks. As a rule of thumb, response time is set at about 1/20 the width of the peak at base.

Following the same argument, the noise in the spectrum will be reduced. It has been explained that the raw spectrum is recorded with a high rate, e.g., 10 msec. Setting the response time to, e.g., 1 sec effectively means that all the spectra recorded in this time frame are averaged to one output spectrum. Time bunching of spectra thus improves the "spectrum-to-noise" ratio and is set by the response time selected.

Likewise it has been found that adaption of the column effluent temperature to the temperature of the cuvette is of importance to reduce noise. In most detectors a heat exchanger is available for this purpose. Obviously such a heat exchanger contributes dead volume to the system, which adversely affects chromatographic integrity. In practice, though, heat exchangers are only needed when a large difference ($>20°$) in temperature between the column and detector cell exists and a high flow rate is used (≥ 0.5 ml/min).

TABLE II
OPTIMAL SETTINGS FOR DIODE ARRAY DETECTOR

Parameter	Setting
Best sensitivity of detection	Widest slit aperture available
	Sample wavelength and bandwidth match with maximum and bandwidth of spectral band
	Reference wavelength close to sample wavelength, wide bandwidth
	Response time matching peak width
	Flow cell pairing longest optical path with widest diameter
	Heat exchanger in front of the flow cell
Best spectral resolution	Smallest slit aperture available
	Highest diode resolution available
	Fastest response time
Best chromatographic integrity	Minimal cell volume
	Fastest response time

Optimal Setting of Data Acquisition Parameters. In Table II, optimal data acquisition settings of photodiode array detectors for various purposes have been summarized.

Data Analysis Techniques and Applications

After acquisition and preprocessing, the data are available in digital form and can be further processed and/or stored on the detector or transferred to a computer for this purpose. Some commercial systems output on-line signal or spectral arithmetic, e.g., signal ratios or a peak purity parameter. In practice, however, much computing overhead must be designed into such detectors. Data evaluation on a computer provides much greater flexibility for data handling, for preparation of presentation graphics and documentation, and for sophisticated calculations.

The three-dimensional data matrix lends itself to versatile data handling and can provide detailed information about the chemical structure of the species detected. This is explained in the next sections and is illustrated by examples relevant to the use of photodiode array detection in HPLC and capillary electrophoresis of biological (macro)molecules.

Signal Ratio

The ratio of the absorbances measured at two different wavelengths is a constant, characteristic for a given solute [Eq. (1)]:

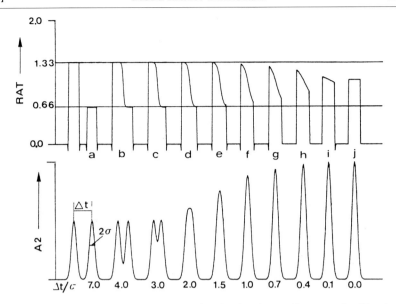

F$_{\text{IG}}$. 8. Influence of peak overlap on the signal ratio of two adjacent peaks. (Reprinted with permission from Ref. 4. Copyright 1984 American Chemical Society.)

$$\text{Ratio} = \frac{\text{absorbance}_1}{\text{absorbance}_2} = \frac{a_1 bc(t)}{a_2 bc(t)} = \frac{a_1}{a_2} = \text{constant} \qquad (1)$$

In Eq. (1), a is the molar absorptivity at wavelength 1 and 2, respectively, b is the path length of the cuvette, and $c(t)$ is the concentration elution profile. A plot of this ratio over time will generate a block-shaped profile with an amplitude that is characteristic of the substance when the solute is pure. A significant distortion of the ideal rectangular form is an indication of an inhomogeneity in the elution profile and therefore of the presence of a closely eluting impurity. Drouen $et\ al.$ have systematically investigated the usage of wavelength absorption ratios for solute recognition and peak purity assessment in liquid chromatography.[4] The effect of reduction of peak separation on the signal ratio was given (see Fig. 8) and usefulness for qualitative assessment of peak purity demonstrated.

Although in principle a simple and straightforward technique, there are practical limitations on the usage of signal ratio. Baseline shifts and solvent effects as in gradient elution may induce distortion of the rectangular profile. Moreover, to minimize the influence of signal noise, in most cases a relatively high threshold is required (height of impurity $>5\%$ of main peak).

[4] A. C. J. H. Drouen, H. A. H. Billet, and L. De Galan, $Anal.\ Chem.$ **56,** 971 (1984).

Therefore, the usage of signal ratio for peak purity verification has been replaced by more sophisticated algorithms.

Even so, some authors have shown that the signal or absorbance ratio can be used for identification of aromatic amino acids in peptides separated by HPLC. Nyberg et al.,[5] for example, have used the absorbance ratio of the aromatic amino acids in short (neuro)peptides for identification. They have measured the absorbance ratios at 255/270 nm for tyrosine and tryptophan and at 255/265 nm for phenylalanine in 0.1% trifluoroacetic acid in water (Table III). Yang et al.[6] have taken a similar approach for identification and quantitation of aromatic amino acids present in tryptic peptides derived from human apolipoprotein A. They also used the absorbance ratio 255/270 nm. Their data are given in Table III. Similar data from spectra of the amino acids tyrosine, tryptophan, and phenylalanine measured by the author have been added to Table III.

Consistent data were reported for the absorbance ratios of tyrosine (Y) and tryptophan (W). There is good agreement between the absorbance ratio 255/270 nm for tyrosine and tryptophan reported by Nyberg and Yang for those peptides containing only Y or W and from our own work, namely 0.33–0.37 and 0.56–0.61, respectively. For phenylalanine (F), Nyberg used the ratio 255/265 because the relatively low absorbance at 255 nm would be divided by the noise absorbance values at 270 nm and would give rise to poor precision of the ratio. However, the absorbance ratio 255/265 deviates from the ratio obtained in our own work. In addition, the ratio 255/270 reported by Yang for a peptide containing only F deviates from the value reported by Nyberg (0.98 vs >1, respectively). This is indicative of the limitations of using absorbance ratios as an absolute means of identification.

Nyberg et al. have used the ratios to help discriminate neuropeptides formed by enzymatic conversion and degradation of cerebrospinal fluid peptides such as substance P, the enkephalins, a C-terminally extended analog of dynorphin, and delta sleep-inducing peptide (DSIP). The ratios found in the peptides formed and separated or in the direct comparison of DSIP and its tyrosine analog, in which the terminal tryptophan is replaced by tyrosine, unambiguously identified the presence of the particular aromatic amino acid or the presence of both phenylalanine and tyrosine.

Table III also illustrates the results when two or more aromatic amino acids are present in a peptide, e.g., tyrosine and phenylalanine. In this case an average absorbance spectrum from these amino acids will provide an intermediate value for the ratio. The data from both Nyberg et al. and

[5] F. Nyberg, C. Pernov, U. Moberg, and R. B. Erikson, J. Chromatogr. **359**, 541 (1986).
[6] C.-Y. Yang, H. J. Pownall, and A. M. Gotto, Jr., Anal. Biochem. **145**, 67 (1985).

TABLE III
ABSORBANCE RATIOS OF SYNTHETIC PEPTIDES AND AROMATIC AMINO ACIDS

Substance	Absorbance ratio		Corresponding ratios[a]
	255 nm/270 nm	255 nm/265 nm	
Y[b]	0.34 ± 0.01[d]		0.33 ± 0.01
YG[b]	0.34 ± 0.01		
YGG[b]	0.33 ± 0.01		
F[b]	>1	1.80 ± 0.13	1.40 ± 0.01
W[b]	0.61 ± 0.01		0.56 ± 0.001
YGG**F**[b]	0.44 ± 0.01		
GG**FL**[b]	>1	1.62 ± 0.10	
G**FL**[b]	>1	1.62 ± 0.03	
YGGFL[b]	0.42 ± 0.03		
DEPPQSPWDR[c]	0.56		
LAEYHAK[c]	0.35		
DLATVYVDVLK[c]	0.37		
THLAPYSDELR[c]	0.36		
QGLLPVLESFK[c]	0.98		
LLDNWDSVTSTFSK[c]	0.57		
LLDNWDSVTSTF[c]	0.59		
EQLGPVTQEFWDNLQK[c]	0.58		
WQEEMELYR[c]	0.51		
WQEEMELYREK[c]	0.52		
WQEEMELY[c]	0.53		
VQPYLDDFQK[c]	0.41		
DYVSQFQGSALGK[c]	0.44		
VSFLSALEEYK[c]	0.41		

[a] From our data.

[b] Instrumental data from Ref. 5; HPLC system LKB 2150 with LKB 2140 rapid scanning spectral detector interfaced with an IBM personal computer. Chromatography: column, 100×4.6 mm, 3-μm Spherisorb ODS2; mobile phase, 0.1% aqueous trifluoroacetic acid adjusted to pH 2.5 with NaOH; flow rate 1.0 ml/min; gradient to 50% acetonitrile in 20 min. Letters in boldface indicate aromatic amino acids.

[c] Instrumental data from Ref. 6; HPLC system Hewlett-Packard 1090 with built-in diode array detector and HP85 microcomputer. Chromatography: column, 250×4.6 mm, 5-μm Vydac C_{18}; mobile phase, 0.1% aqueous trifluoroacetic acid (TFA) and 0.08% TFA in 95% acetonitrile/water; gradient 0–40% in 40 min at 1.5 ml/min.

[d] Values given as means ± standard deviation.

Yang *et al.* are comparable. For a peptide containing one tyrosine (Y) and one phenylalanine residue (F), for the absorbance ratio 255/270 nm, values of 0.41–0.44 are reported. This allows unambiguous identification of such a peptide. Likewise, a peptide containing one tyrosine and one tryptophan residue can be distinguished from a peptide containing only tyrosine or tryptophan (absorbance ratio 255/270 nm, 0.51–0.53 versus 0.33–0.37 or

0.56–0.61) if the precision of the ratio measurement is high. In contrast, the value of the ratio of a peptide containing one tryptophan (W) and one phenylalanine does not significantly differ from that of a peptide containing only W (0.57–0.59 versus 0.56–0.61).

van Iersel et al.[7] used the absorbance ratio 280/205 nm of 25 known proteins to determine the absorbance value of a 0.1% protein solution at 280 nm. The method can be of potential use for the determination of unknown proteins. The specific problem with the technique thus far has been the limited precision of results caused by the preparation of the dilutions required. These authors considered that liquid chromatography was a much more reproducible "dilution method" and provided following empirical relationship to obtain the A_{280} value:

$$A_{280}^{0.1\%} = 34.14 \, A_{280}/A_{205} - 0.02 \tag{2}$$

Obviously in this application it is mandatory that the eluting peak be pure protein. If the ratio obtained shows relatively high fluctuation, closely eluting impurities may be present.

Peak Purity Assessment

As mentioned previously, signal absorbance ratios have found application in peak purity assessment but are limited for chromatographic reasons. Therefore, more sensitive and selective tools to evaluate the purity of an eluting band over the profile have been developed. These methods use the available spectral information in the eluting band.

One can easily conceive that if a chromatographic band were composed of two closely eluting peaks, the UV–Vis spectra obtained during elution would not be constant provided that the individual substances have different spectra. UV–Vis spectra of proteins and peptides differ in most cases only slightly. Therefore, sophisticated algorithms are required to detect inhomogeneities in the eluting chromatographic band. In essence these algorithms quantitate differences between spectra. Sievert and Drouen have discussed the algorithms in detail.[8] Determination of the correlation between spectra serves as an example.

$$r = \frac{\sum\limits_{i=1} [(a_i)(b_i)]}{\sqrt{\sum\limits_{i=1} (a_i)^2 \sum\limits_{i=1} (b_i)^2}} 1000 \tag{3}$$

[7] J. van Iersel, J. Frank Jzn, and J. Duine, *Anal. Biochem.* **151,** 196 (1985).
[8] H.-J. P. Sievert and A. C. J. H. Drouen, *in* "Diode Array Detection in HPLC" (L. Huber and S. A. George, eds.), Chromatographic Science Series, Vol. 62. Marcel Dekker, New York, 1993.

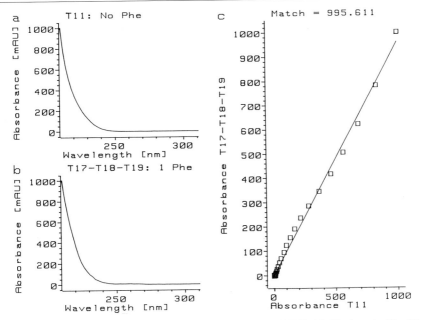

FIG. 9. Spectral match between tryptic peptides T11 and T17–18–19 given in Fig. 12. (Conditions as in Fig. 14.)[9] (Printed with permission of H.-J. P. Sievert.)

where a_i and b_i are the absorbance values in each spectrum at wavelength i, respectively. The correlation coefficient times 1000 is called the *match factor*. As an example the spectra of two peptides are compared in Fig. 9. These peptides stem from a tryptic digest of recombinant human growth hormone (rhGH) and differ by the presence of a phenylalanine residue.[9] On visual inspection, no difference between the spectra can be observed. However, the plot of the absorbance values shows systematic deviations from a straight line and the match factor of 995.6 is significantly different from 1000, which would be expected for identical spectra.

Based on this concept, one can follow the correlation of spectra over an eluting band vs an average or peak apex spectrum to allow determination of the spectral homogeneity of the eluting band. Obviously, this requires that the impurity present have a spectrum different from that of the overlapping component and not be 100% coeluting. Moreover, below a certain relative concentration of the impurity, it will not be distinguishable from

[9] H.-J. P. Sievert, S.-L. Wu, R. Chloupek, and W. S. Hancock, Automated Evaluation of Tryptic Digest from Recombinant Human Growth Hormone Using Ultraviolet Spectra and Numeric Peak Information, presented as paper #892 at the 39th Pittsburg Conference and Exhibition, New Orleans, Feb. 22–26, 1988.

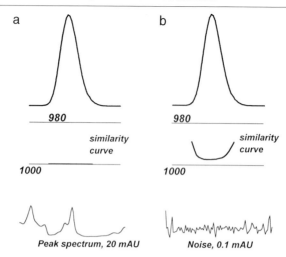

Fig. 10. Similarity curves for a pure peak (b) with and (a) without noise, plotted in relation to the ideal similarity factor (1000) and a user-defined threshold (980). (Courtesy of Hewlett-Packard GmbH, Waldbronn, Germany.)

spectral noise. Along the same line as the comparison of two spectra [Eq. (4)] a similarity factor can be defined according to

$$r = \frac{\sum\limits_{i=1}^{n=1} [(a_i - a_{av})(b_i - b_{av})]}{\sqrt{\sum\limits_{i=1}^{n=1} (a_i - a_{av}) \sum\limits_{i=1}^{n=1} (b_i - b_{av})^2}} \quad \text{and} \quad \text{Similarity} = 1000 r^2 \quad (4)$$

In Eq. (4), b represents the mean spectrum over a chromatographic peak (or another spectrum specified by the user) and a represents an individual spectrum recorded during peak elution. The absorbance values are mean centered by subtraction of a_{av} and b_{av}, which are the average of all absorbance values of spectrum a or b.[10] Depending on the spectral noise, a similarity >995 indicates that the spectrum in a peak is very similar to the mean spectrum and therefore one can consider the peak spectrally homogeneous and pure. In practice, all spectra in a band are compared with the average spectrum. The similarity factor obtained is plotted over the time during elution. An ideal profile of a pure peak is a flat line at a match factor of 1000, as demonstrated in Fig. 10a. However, at the beginning and end of a band, where the ratio of the band spectrum to spectral background noise decreases, the similarity plot deviates as shown in Fig. 10b.

[10] J. C. Reid and E. C. Wong, *Appl. Spectrosc.* **20**, 220 (1960).

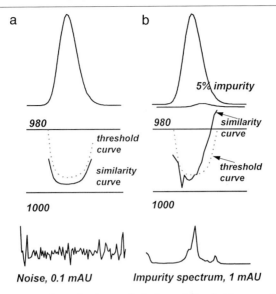

FIG. 11. Effect of impurity and noise on similarity and threshold curves. (a) Peak without impurity but with noise; (b) peak with impurity and noise. (Courtesy of Hewlett-Packard GmbH, Waldbronn, Germany.)

The spectral noise is used to calculate a threshold curve. When impurities are absent, the similarity curve will be below the threshold curve. However, when an impurity is closely eluting, e.g., at the rear of the peak, the similarity curve will exceed the threshold curve (see Fig. 11a and b). In fact, the spectral noise provides a practical limit to detection of impurities of 1% relative and can be used to calculate a threshold curve.

Identification of Peptides and Peptide Substructures with UV/Visible Spectra

The main emphasis in this section is on the usage of diode array spectrophotometric detectors for identification of peptides and peptide substructures and proteins by their UV–Vis spectra. Usage of spectra for identification of peptides needs additional comments.

The successful identification of a substance by visual comparison of the UV–Vis spectra with a reference spectrum and of peak purity verification techniques depends entirely on the presence of significant differences in spectral properties of the compounds involved. In view of the similarity between spectra of proteins and peptides it can be questioned whether the technique is useful for their identification. Use of the least-squares fit correlation coefficient of the spectrum of the unknown and the standard as a metric for this purpose[11,12] has been suggested. This principle has been adapted for diode array-based HPLC detectors by Demorest[12] and Hill[13] and incorporated into the software of several commercial workstations supporting diode array detectors in different ways.[15–18]

The unambiguous identification of peptides is a particular problem in the technique of peptide mapping, where a large protein is broken down into smaller peptide fragments by chemical or enzymatic digestion. The peptides so obtained form a typical pattern after separation on an HPLC column (see Fig. 12). In the early phase of recombinant protein research, the peptides obtained are of importance in the elucidation of the primary structure of the original protein. Diode array detection is invaluable in the detection of peptide fragments containing aromatic amino acids. Accurate comparison of the peptide map of a newly manufactured protein therapeutic drug with the map of a reference protein is mandatory prior to release of the manufacturing batch.[19]

This requirement necessitates the unambiguous identification of the peptides in the map of a manufactured protein. Chromatographic comparison alone of a manufactured sample versus a reference sample requires careful control of chromatographic conditions and skillful operation. Even with these precautions, only the identification of the largest peaks in the map is unequivocal. Peptide identification is further impeded in practice because the elution of the peptides off the HPLC column strongly depends on brand and type of column and, even when these do not change, on the stationary phase batch. It is common in such cases that the separation needs to be reoptimized and the identification of peptides of the reference protein repeated.

With the advent of diode array detectors, it was expected that the additional UV–Vis spectral information could eliminate the ambiguities.

[11] K. Tanabe and S. Saeki, *Anal. Chem.* **47,** 118 (1975).
[12] D. M. Demorest, J. C. Fetzer, I. S. Lurie, S. M. Carr, and K. B. Chatson, *LC-GC Mag. Liq. Gas Chromatog.* **5**(2), 128 (1987).
[13] D. H. Hill, T. R. Kelly, and K. J. Langner, *Anal. Chem.* **59,** 350 (1987).
[14] Deleted in proof.
[15] J. Schaefer, *Int. Chromatogr. Lab.* **1,** 6 (1990).
[16] "LC 235 Diode Array Detection System," Publication Number L1014. Perkin Elmer, Norwalk, CT, 1987.
[17] "Applied Biosystems 1000S: A Photo-Diode Array Detector That Is Sensitive Enough to Use Every Day." Applied Biosystems, Inc., Foster City, CA, 1988.
[18] R. L. Garnick, N. J. Solli, and P. A. Pappa, *Anal. Chem.* **60,** 2546 (1988).
[19] H.-J. P. Sievert, S.-L. Wu, R. Chloupek, and W. S. Hancock, *J. Chromatogr.* **499,** 221 (1990).

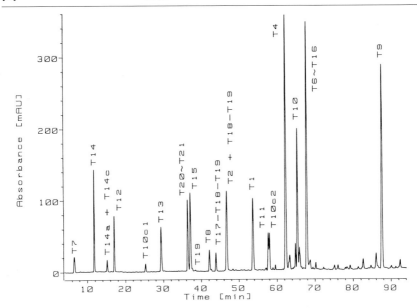

Fig. 12. Peptide map of recombinant human growth hormone (rhGH). Column: Nucleosil C_{18}, 150×4.6 mm, 5 μm. Mobile phase A, 0.1% trifluoroacetic acid (TFA) in water; mobile phase B, 0.08% TFA in acetonitrile; gradient, 0–60% in 120 min; flow rate, 1 ml/min. Detection wavelength: 220 nm. HPLC system is a Hewlett-Packard 1090M with a built-in diode array detector and HP 79994 ChemStation. The peptide map of rhGH was generated by digestion with trypsin (tryptic map); peptides have been identified by FAB-MS. Actual amino acid composition is given in Figs. 13 and 14 for peptides T12, T13, and T14. (Reprinted from *J. Chromatogr.*, **499**, H.-J. P. Sievert, S.-L. Wu, R. Chloupek, and W. S. Hancock, p. 221, Copyright 1990 with kind permission of Elsevier Science–NL, Sara Burgerhartstraat 25, 1055 KV Amsterdam, The Netherlands.)

However, the problems of low fine structure and spectral averaging mandate the use of elaborate algorithms to extract this information from the spectra. Sievert *et al.* demonstrated an elegant approach to solving this problem.[19] They developed a procedure that allows unequivocal assignment of the peptide in the unknown map from a library and have proposed an overall similarity score for the map based on that identification.

The basic principle of their method is elucidated in Figs. 13 and 14. The examples are taken from the peptide map of recombinant human growth hormone, rhGH (Fig. 12). Point-to-point comparison of two UV–Vis spectra gives a linear relationship when the spectra indeed are identical with the slope depending on the concentration ratio of the solutes and an intercept depending on the difference in background (see previous section). In Fig.

[20] Deleted in proof.

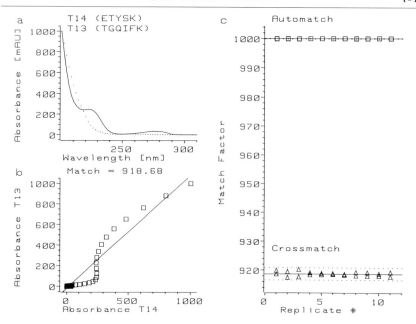

FIG. 13. Spectral match between the tryptic peptide T14 and T13 as an example of moderate similarity. (a) UV/visible spectra of the two peptides (T13 shown as dotted line). (b) Point-to-point spectrum comparison of both peptides; the solid line is the least-squares linear fit through these points. The match factor is given at the top of (b). (c) Correlation of all spectra recorded for each peptide with its average spectrum over its peak (automatch) or with the average spectrum for the other peptide peak (cross-match). The mean match factors are given as solid lines; the dotted lines represent the $\pm 3\sigma$ limits. (Personal communication, H.-J. Sievert.)

13, an example is given for two peptides, T14 and T13, from the map of rhGH. Their spectra are clearly different. These peptides differ in aromatic amino acid composition, with a phenylalanine in T13 and tyrosine in T14. The match factor is 919. With the naked eye, the spectral differences can be observed, and the correlation is clearly nonlinear.

However, when peptide T13 is compared with T12, which does not contain aromatic amino acids, spectral differences cannot be detected unambiguously. The regression seems linear and the match factor has increased to 997.3 (Fig. 14). This example demonstrates that two chromatographically well-spaced peptides, that are definitively different species, may have very similar match factors.

This differentiation is achieved by the so-called automatch and cross-match as indicated in Figs. 13c and 14c. In these examples the factors are obtained by correlating the average spectrum of the peptide T13 (Fig. 14) with all spectra recorded over its peak (automatch) and with all spectra

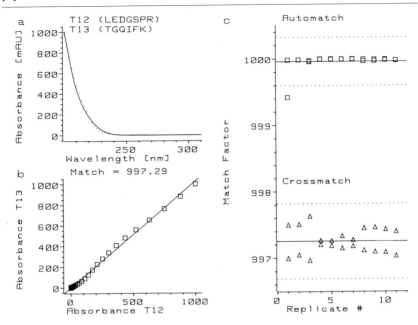

FIG. 14. Spectral match between tryptic peptides T12 and T13 as an example of strong similarity. See Fig. 12 for details. (Reprinted from *J. Chromatogr.*, **499**, H.-J. P. Sievert, S.-L. Wu, R. Chloupek, and W. S. Hancock, p. 221. Copyright 1990 with kind permission of Elsevier Science–NL, Sara Burgerhartstraat 25, 1055 KV Amsterdam, The Netherlands.)

recorded for the peptide T12 (cross-match). It can be seen that, where the visual comparison of spectra and the regression do not allow an unambiguous differentiation, the automatch and cross-match factors differ significantly (Student's *t* value is 57, which means a confidence level of >99.99% that the spectra are indeed different).

The basis for the comparison is provided by a library of standard spectra for the various fragments in the map of the reference protein. This library is generated from baseline-corrected spectra of the fragments (interpolated from two baseline spectra adjacent of the peak) and validated by repetitive injections of the reference map. The final library of the rhGH tryptic peptides separated with the aqueous TFA–acetonitrile mobile phase contained 19 entries. The coefficient of variation (CV) of the match factors of these peptides with the respective library entry in repetitive runs was less than 0.13%. The CV was highest for low peaks, therefore match criteria were adapted to reflect the relative abundance of each peptide. If now all peaks found in the map are compared with each library entry within a time window of ±0.5 min, only the correct matches are found. If retention times are used for this comparison, typically five to eight matches are found

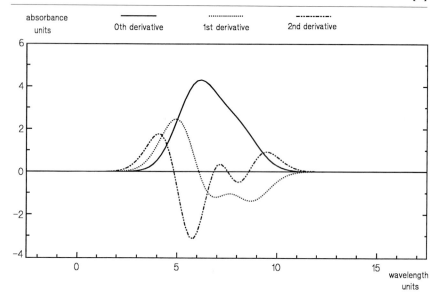

FIG. 15. Enhancement of spectral fine structure by differential calculation.

within the same window. Thus, spectral information in the library provides clear advantages in reduction of the possibilities for incorrect peak matching.

Derivative Spectroscopy

Electronic spectra of dissolved compounds frequently do not show a pattern of well-resolved sharp absorption bands. Instead, in most cases overlapping bands are observed even under conditions of the highest spectral resolution, which leads to a pattern of broad peaks, valleys, and shoulders. The excitation of electrons is accompanied by changes in vibrational and rotational states so that, instead of one narrow line, a broad peak with fine structure should be obtained. Owing to interactions of the solute with solvent molecules, these absorptions are blurred and a smooth band is observed.[21] UV–Vis spectra of amino acids, peptides, and proteins do not have a well-expressed fine structure. Derivatization of UV–Vis spectra can be used to reveal the fine structure. In particular, the second derivative is helpful as it converts peaks and shoulders into minima, valleys into maxima, and inflexion points into zero. This is illustrated in Fig. 15, in which a UV–Vis spectrum is simulated consisting of two Gaussian absorption bands.

[21] B. Wettlaufer, *Adv. Protein Chem.* **23,** 21 (1968).

The units at the abscissa and ordinate are arbitrary normalized wavelength and absorption values. The two bands have a bandwidth of one, are separated by one bandwidth, and differ by a factor of two in molar absorptivity. In the resulting spectrum hardly any differention at the high-wavelength side of the band can be observed. The second derivative spectrum clearly isolates the fine structure that is present, with a minimum where the second band has its maximum (at wavelength units = 8), a peak for the valley between the two bands, and a minimum for the higher extinction band.

Further differentiation would, in principle, reveal even more fine structure but precision rapidly deteriorates. In addition, real spectra will have noise superimposed, and this noise will increase by repetitive differentiation. Therefore, third and higher order derivatives are not meaningful and the second derivative is an optimal choice for revealing spectral fine structure.

UV–Vis derivative spectroscopy has been used for the study of conformational effects of solvents on proteins.[22–24] It is possible to transfer a solution of the protein of interest into a cuvette and record a spectrum. With conventional spectrophotometers this takes seconds to minutes and is laborious. It can be anticipated that fast, structural changes occur unnoticed and important information may be lost. Diode array spectrophotometers used as HPLC detectors join the convenience of fast, on-line UV–Vis spectral acquisition with the advantage of the coupled chromatographic separation at the cost of a lower spectral resolution than in conventional spectrophotometers. Karger and co-workers were the first to apply this potential in conformational studies of proteins in combination with HPLC.[24,25] In later publications, Hearn et al.[26] and Ackland et al.[27] have expanded on this work.

Subtle changes in the UV–Vis spectrum may occur as proteins undergo conformational changes. In particular, the aromatic amino acid residues, which have appreciable absorption above 240 nm, may become more or less exposed to the solvent and therefore to the incident light. Generally the aromatic amino acid residues, which are hydrophobic, will be found in the interior of a water-soluble protein. Conformational changes may bring

[22] J. W. Donovan, *Methods Enzymol.* **27**, 497 (1973).
[23] D. Freifelder, "Physical Biochemistry—Application to Biochemistry and Molecular Biology." W. H. Freeman, San Francisco, CA, 1982.
[24] S. L. Wu, K. Benedek, and B. L. Karger, *J. Chromatogr.* **359**, 3 (1986).
[25] S. L. Wu, A. Figueroa, and B. L. Karger, *J. Chromatogr.* **371**, 3 (1986).
[26] M. T. W. Hearn, M. I. Aguilar, T. Nguyen, and M. Fridman, *J. Chromatogr.* **435**, 271 (1988).
[27] C. E. Ackland, W. G. Berndt, J. E. Frezza, B. E. Landgraf, K. W. Pritchard, and T. L. Giradelli, *J. Chromatogr.* **540**, 187 (1991).

these residues to the surface of the protein and therefore lead to changes in the spectrum, e.g., wavelength of (sub)maxima, absorbance.

Karger and co-workers anticipated that the absorbance ratio at specific wavelengths would be sensitive enough to manifest such changes. They selected an absorbance ratio, 292/254 nm, to indicate changes in tryptophan exposure. Increase in the ratio would mean that tryptophan residues had moved from the interior to the surface of the protein. Similarly, an increase in the 272/292-nm ratio accompanied by decrease of the 292/254-nm absorbance ratio suggested greater tyrosine exposure to the mobile phase. Finally, an increase of the 280/254-nm ratio with a concomitant increase of the 292/254-nm ratio suggested either tyrosine or tryptophan exposure. These authors regarded the absorption of phenylalanine as too low to be of diagnostic value.

The problem in the interpretation of the spectral data is the interference of the tryptophan main absorption band with the tyrosine side band in the 278- to 280-nm range. Second derivative spectroscopy provides a possibility for better differentiation between these two residues. As explained, maxima in the zero-order spectrum will turn into minima in the second derivative spectrum; valleys will turn into maxima in the second derivative. Two amplitudes can be defined in a second derivative spectrum in which tyrosine and tryptophan are present (see Fig. 16). The amplitude a originates from the maximum at 280 nm and the valley at 285 nm in the zero-order spectrum and the amplitude b from the tryptophan side maximum at approximately 290 nm and the next valley after this maximum. Therefore one can anticipate that the ratio $\gamma = a/b$ will be discrimative for a conformation change involving tryptophan or tyrosine, i.e., the larger the γ, the higher the exposure of tyrosine to the aqueous solvent. The actual wavelengths selected for the reported amplitude ratio vary between groups. Karger used the amplitudes 282.5 nm/288.5 nm for a and 292.5 nm/296.5 nm for b. Hearn used 284 nm/289 nm for a and 292 nm/295 nm for b. Ackland used 283 nm/287 nm for a and 290.5 nm/295 nm for b. It is suggested to use the minimum at 280–282 nm and maximum at 285–287 nm for a and the minimum 288–290 nm and maximum at 294–296 nm for b. These differences may be caused by selection of different substrates and conditions for registration of the reference UV–Vis spectra as well as by wavelength accuracy differences between instruments. Within one experimental setup, the ratio γ, however, has good reproducibility. Karger reported the error in γ in replicate experiments to be ± 0.15.

Besides the ratio γ, Karger used the experimentally established value of a factor Z. The Z value is related to the contact area of a protein to the stationary phase surface. It is well known that a protein in the folded state is less retained than in an unfolded state. This is because the number of

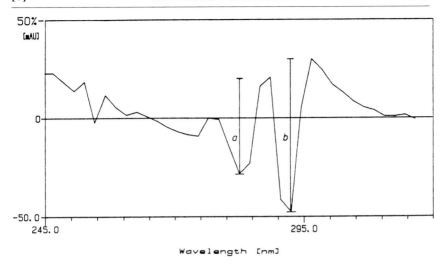

FIG. 16. On-line second derivative spectrum of lysozyme eluted from a hydrophobic interaction column (ether type). Conditions: mobile phase A, 2 M ammonium sulfate, 0.5 M ammonium acetate, pH 6; mobile phase B, 0.5 M ammonium acetate, pH 6; flow rate, 1.0 ml/min; gradient, 0–100% B in 20 min. Instrumentation: Hewlett-Packard 1040A diode array detector with HP85B personal computer and DPU multichannel integrator. The peak-to-peak distance between the maximum at 288.5 and the minimum at 282.5 (a) and between the maximum at 296.5 nm and the minimum at 292.5 nm (b) defines γ ($= a/b$). (Reprinted from *J. Chromatogr.*, **359**, S.-L. Wu, K. Benedek, and B. L. Karger, p. 3, Copyright 1986 with kind permission of Elsevier Science–NL, Sara Burgerhartstraat 25, 1055 KV Amsterdam, The Netherlands.)

points or areas of contact of the protein with the surface is larger in an unfolded state. Consequently, a change in conditions resulting in a conformational change may result in an increase of the contact area. This leads to higher Z values. The Z values of a solute thus are coupled to retention, in a logarithmic or semilogarithmic relationship depending on the separation mode, and can be measured experimentally. Therefore experimental measurement of Z and γ provides meaningful information about conformational changes of the protein separated. In Table IV the most relevant data that were obtained by Karger are summarized.

These results illustrate the potential of the technique. The absorbance ratio, the second derivative ratio, and Z value obtained for lysozyme in the temperature range 10–40° show no changes. This result suggests that lysozyme is quite stable under the conditions of hydrophobic interaction chromatography and undergoes no conformational changes.

In contrast it is known that β-lactoglobulin is more easily denatured (thermally), which is in accordance with the data in Table IV. From 25 to 40° a significant decrease in the 292/254-nm absorbance ratio and concomitant increase of the 274/292-nm ratio indicate exposure of tyrosine to the outside

TABLE IV

CHROMATOGRAPHIC AND SPECTROSCOPIC CHARACTERISTICS OF PROTEINS IN HYDROPHOBIC
INTERACTION CHROMATOGRAPHY AS FUNCTION OF COLUMN TEMPERATURE[a]

Protein	Temperature (°C)	Absorbance ratio			γ^b	Z^c
		292 nm/254 nm	280 nm/254 nm	274 nm/292 nm		
Lysozyme	10	1.42	2.10	1.41	0.61	4.50
	25	1.41	2.10	1.42	0.63	4.89
	40	1.42	2.10	1.42	0.63	4.56
β-Lactoglobulin	10	1.45	2.50	1.64	1.24	2.11
	25	1.42	2.48	1.66	1.11	2.16
	40	1.3	2.30	1.71	1.64	4.36
α-Lactalbumin	0.5	1.41	1.43	1.78	0.82	2.1
	5	1.42	1.42	1.79	0.83	2.2
	10	1.43	1.43	1.80	0.85	2.3
	18	1.40	1.44	1.77	0.89	2.6
	25	1.41	1.43	1.77	0.87	5.3
	32	1.45	1.41	1.87	0.70	5.6
	50	1.40	1.46	1.84	0.85	8.5

[a] Experimental conditions as given in Fig. 16. (Reprinted from *J. Chromatogr.,* **359,** S.-L. Wu, K. Benedek, and B. L. Karger, p. 3, Copyright 1986 with kind permission of Elsevier Science–NL, Sara Burgerhartstraat 25, 1055 KV Amsterdam, The Netherlands.)
[b] Second derivative spectrum characteristic as defined in Fig. 16.
[c] Slope of the plot log k' vs log %B.

of the protein. This is corroborated by the significant increase in the value of γ and further substantiated by the doubling of the Z value. It should be emphasized that β-lactoglobulin contains two tryptophan and four tyrosine residues, thus it is easier to detect changes in the tyrosine environment.

Calcium-depleted α-lactalbumin is known to be a very labile protein. In the low temperature range of 0.5 to 18° no change in spectroscopic values, and a small change in Z value, are observed. From 18 to 25° no change in the spectroscopic features is seen but the Z value doubles. This may indicate that there is no change in the microenvironment of the aromatic amino acid residues when the protein starts to unfold. With further increase in temperature, there is a slight but significant change in γ, indicating exposure of tryptophan residues, without an increase in Z. Finally, at 50°, the molecule may have unfolded completely, accounting for the further increase in Z while the spectroscopic values have reached their final values for α-lactalbumin, in which tyrosine and tryptophan are present in a one-to-one ratio.

In a subsequent paper,[28] Karger and co-workers have expanded this work in detail for the labile, calcium-depleted α-lactalbumin. They have

[28] S.-L. Wu, A. Figueroa, and B. L. Karger, *J. Chromatogr.* **371,** 3 (1986).

investigated the influence of the lipophilicity of the stationary phase, the contact time on the stationary phase, and the addition of Ca^{2+} and Mg^{2+} ions to the mobile phase on the folding behavior of this protein. By measuring the second derivative ratio over the α-lactalbumin peak, it was demonstrated that the peak actually consisted of two species depending on the experimental conditions.

Compositional Analysis of Oligonucleotides

The precise spectral data from a diode array detector has been used by Sievert in an attempt to analyze the composition of short oligonucleotides.[29] If one assumes or knows that an oligonucleotide eluting from an HPLC column is chemically homogeneous, the UV–Vis spectra obtained from this peak will be composed of the spectra of the individual nucleotides. Therefore, it should be possible to derive the nucleotide composition by multicomponent analysis (MCA).

The following example from Sievert illustrates the feasibility of this idea. HPLC separations of oligonucleotides were performed on a Beckmann Ultrapore C3 column with a gradient of 0.1 M triethylammonium acetate versus acetonitrile. Spectra were acquired by the diode array detector installed in a Hewlett-Packard 1090M system. In a first step, Sievert verified that the precision of the spectra obtained over the peak suffices for this purpose and even allows differentiation between oligomers of adenosine. Poly(A)$_n$ (n = 2–5) were separated and on each peak five spectra extracted around the maximum. These spectra were then compared against their own average (automatch) and against the average of each other peak (cross-match) as described in the section on peptide identification. A discriminator value D was defined as

$$D = \frac{\text{Mean}_{\text{automatch}} - \text{Mean}_{\text{cross-match}}}{\text{Variance} \cdot t_{n-2,0.01}} \tag{5}$$

The bar chart in Fig. 17 shows that the D values for cross-correlation are significantly higher than those for the autocorrelation for poly(A), differing only in the number of A's. This indicates that the dissimilarity of spectra of a dimer and a trimer of adenosine is significant and suffices to differentiate between them even though the naked eye can hardly differentiate the spectra (Fig. 18, left).

However, in practice oligonucleotides are a mixture of G, C, T, and A nucleosides. Spectra of poly(G), poly(C), poly(T), and poly(A) are distinctly

---

[29] H.-P. Sievert, Hewlett-Packard Little Falls Division, Little Falls, DE, personal communication.

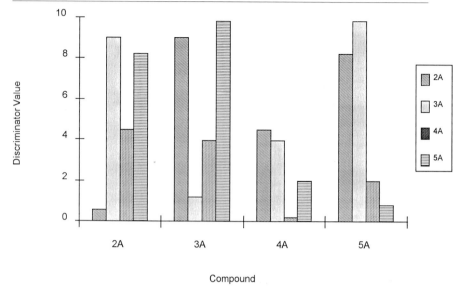

FIG. 17. Comparison of the automatch and cross-match of poly(A) ($n = 2$–5).

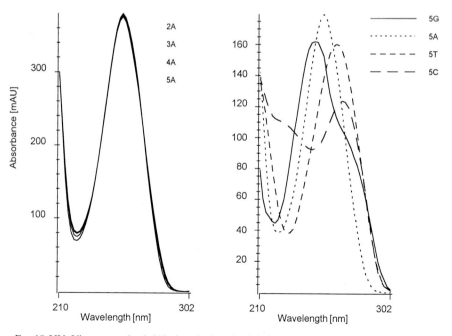

FIG. 18. UV–Vis spectra of poly(A)$_n$ ($n = 2$–5) and poly(A)$_5$, poly(C)$_5$, poly(T)$_5$, and poly(G)$_5$.

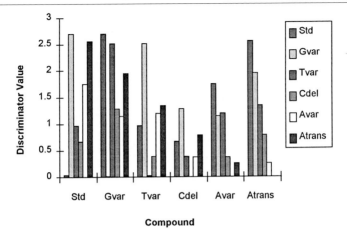

FIG. 19. Automatch and crossmatch of mixed oligonucleotides.

different, as is shown in Fig. 18, right. Here the spectra obtained for the four pentamers are given besides the spectra of the four oligomers of adenosine used in the previous example. Thus it can be expected that with the attained precision of spectral acquisition, short oligomers with differing numbers of the individual nucleotides will have significantly different spectra. This is illustrated in Fig. 19 where six different oligonucleotides are compared in the same way as the poly(A)s in Fig. 17. In the bar chart, one automatch and five cross-matches are given.

Thus it can be expected that with the attained precision of spectral acquisition, short oligomers with differing numbers of individual nucleotides, will have significantly different spectra. With a homogeneous chromatographic peak of a oligonucleotide, the composition analysis problem

TABLE V
COMPARISON OF NUCLEOTIDE COMPOSITION OBTAINED FROM MULTICOMPONENT ANALYSIS

	A		C		T		G	
Value	Theory	Found	Theory	Found	Theory	Found	Theory	Found
Std	5.0	5.0	6.0	6.0	6.0	6.0	3.0	3.0
G_{var}	5.0	4.8	5.0	4.8	6.0	6.1	4.0	3.9
C_{del}	5.0	5.1	5.0	5.0	6.0	5.9	3.0	3.0
T_{var}	5.0	5.2	5.0	5.1	7.0	6.9	3.0	3.0
A_{var}	6.0	5.9	5.0	5.1	6.0	5.8	3.0	3.0
A_{trans}	6.0	6.2	5.0	4.9	6.0	6.0	3.0	2.8

can be reduced to a multicomponent analysis problem as in normal UV–Vis spectroscopy. The particular problem, however, is that one needs a calibration up front in order to deconvolute the composition of an unknown. This was achieved by selecting one known oligonucleotide as a standard and adjusting the concentration of the individual components—nucleotides—to obtain the theoretical composition for this standard. The spectra for the unknown nucleotides are then normalized to the standard nucleotide. The result of the MCA is given in Table V.

As can be seen, the accordance between the experimentally found nucleotide composition of the "unknowns" and the values known from the literature is excellent. This is a first illustration of the potential of photodiode array detection for the purpose of structure clarification in DNA/RNA research.

Conclusion

The examples in this chapter of photodiode array, spectrophotometric detection for HPLC and CE are a brief illustration of the versatility of this detection system for the solution of biochemical problems. Although the full exploitation of the capabilities of such detection is demanding, the data evaluation may provide strong complementary information about the problem under investigation. Research chemists in the life sciences may become attracted by this tool as an alternative to other spectrometric techniques because the data are generated in conjunction with liquid chromatography. There is ample literature about the usage of photodiode array detection relevant to the biosciences (see, e.g., the book by Huber and George).[1]

Section II

Electrophoresis

A. Slab Gel Electrophoresis: High Resolution
Articles 10 through 12

B. Capillary Electrophoresis
Articles 13 through 19

[10] Isoelectric Focusing in Immobilized pH Gradients

By PIER GIORGIO RIGHETTI, CECILIA GELFI, and MARCELLA CHIARI

Conventional isoelectric focusing (IEF) in carrier ampholyte (CA) buffers has already been covered in this series both at the analytical[1] and preparative[2] level. Several manuals have also appeared covering CA-IEF in depth.[3–5] Therefore, we deal here only with its most advanced version, the immobilized pH gradient (IPG), which has turned out to be an entirely new technique. Immobilized pH gradients are unique in that they are based on the insolubilization of the entire set of buffers and titrants responsible for generating and maintaining the pH gradient along the separation axis. Contrary to conventional IEF, in which a multitude (hundreds, perhaps thousands) of soluble, amphoteric buffers is expected to create and sustain the pH gradient during the electrophoretic run, IPGs are based on only a few, well-defined nonamphoteric buffers and titrants, able to perform in a highly reproducible manner. Quite a few advantages of the IPG technique are immediately apparent: (1) increased resolution (by at least one order of magnitude); (2) unlimited stability of the pH gradient; (3) flexibility in the choice of pH interval; (4) increased loading ability; (5) high reproducibility; (6) minimal distortion by salts in the sample; (7) full control of pH, buffering capacity, and ionic strength; and (8) easy separation of sample from buffering ions in preparative runs.

However, there are a number of problems with the conventional IEF technique, as listed in Table I. Some of them are quite severe: e.g., low ionic strength often induces near-isoelectric precipitation and smears of proteins, occurring even in analytical runs, at low protein loads. The problem of uneven conductivity is magnified in poor ampholyte mixtures, e.g., the Poly Sep 47 (Polysciences, Warrington, PA) (a mixture of 47 amphoteric and nonamphoteric buffers, claimed to be superior to CAs): owing to their poor composition, large conductivity gaps form along the migration path, against which proteins of different p*I* values are stacked; the results are

[1] D. E. Garfin, *Methods Enzymol.* **182,** 459 (1990).
[2] B. J. Radola, *Methods Enzymol.* **104,** 256 (1984).
[3] P. G. Righetti, "Isoelectric Focusing: Theory, Methodology and Applications." Elsevier, Amsterdam, The Netherlands, 1983.
[4] P. G. Righetti, *in* "Protein Structure: A Practical Approach" (T. E. Creighton, ed.), pp. 23–63. IRL Press, Oxford, 1989.
[5] P. G. Righetti, E. Gianazza, C. Gelfi, and M. Chiari, *in* "Gel Electrophoresis of Proteins" (B. D. Hames and D. Rickwood, eds.), pp. 149–216. IRL Press, Oxford, 1990.

TABLE I
PROBLEMS WITH CARRIER AMPHOLYTE FOCUSING

1. Medium of very low and unknown ionic strength
2. Uneven buffering capacity
3. Uneven conductivity
4. Unknown chemical environment
5. Not amenable to pH gradient engineering
6. Cathodic drift (pH gradient instability)

often unacceptable. The cathodic drift is also a major unsolved problem of CA-IEF, resulting in extensive loss of proteins at the gel cathodic extremity on prolonged runs. For all these reasons, in 1982 we launched the technique of IPGs, as a result of an intensive collaborative work with B. Bjellqvist's group in Stockholm and A. Görg's group in Munich.[6]

In this chapter, after the basic chemistry of the Immobiline chemicals is given, the preparation of narrow and extended IPG intervals is described. This is followed by a section on IPG methodology, covering all major aspects of the analytical technique. The chapter closes with a section on advanced versions of preparative IPGs.

Chemicals Used for Immobiline Matrix

Immobilized pH gradients are based on the principle that the pH gradient, which exists prior to the IEF run itself, is copolymerized, and thus insolubilized, within the fibers of the polyacrylamide matrix. This is achieved by using, as buffers, a set of seven nonamphoteric weak acids and bases, having the following general chemical composition: $CH_2 = CH - CO - NH - R$, where R denotes either three different weak carboxyls, with pK values of 3.6, 4.4, and 4.6, or four tertiary amino groups, with pK values of 6.2, 7.0, 8.5 and 9.3 (available under the trade name Immobiline from Pharmacia-LKB AB [Uppsala, Sweden]). A more extensive set, comprising 10 chemicals (a pK 3.1 buffer and two strong titrants, a pK 1 acid and a $pK > 12$ quaternary base) is available as pI select from Fluka AG (Buchs, Switzerland). A more extensive list, for a total of 17 compounds, is given in Tables II–IV, which also offer the relevant physicochemical data (structural formula, mass).[7] During gel polymerization, these buffering species are efficiently incorporated into the gel (84–86% conversion efficiency at 50° for 1 hr). Immobiline-based pH gradients can be cast

[6] B. Bjellqvist, K. Ek, P. G. Righetti, E. Gianazza, A. Görg, W. Postel, and R. Westermeier, *J. Biochem. Biophys. Methods* **6**, 317 (1982).
[7] M. Chiari and P. G. Righetti, *Electrophoresis* **13**, 187 (1992).

TABLE II

ACIDIC ACRYLAMIDO BUFFERS

pK^a	Formula	Name	M_r	Source
1.0	CH₃ | $CH_2=CH-CO-NH-C-CH_3$ | CH_2-SO_3H	2-Acrylamido-2-methylpropanesulfonic acid	207	b
3.1	$CH_2=CH-CO-NH-CH-COOH$ | OH	2-Acrylamidoglycolic acid	145	c
3.6	$CH_2=CH-CO-NH-CH_2-COOH$	N-Acryloylglycine	129	d
4.4	$CH_2=CH-CO-NH-(CH_2)_2-COOH$	3-Acrylamidopropanoic acid	143	d
4.6	$CH_2=CH-CO-NH-(CH_2)_3-COOH$	4-Acrylamidobutyric acid	157	d

[a] The pK values for the three Immobilines and for 2-acrylamidoglycolic acid are given at 25°; for AMPS (pK 1.0) the temperature of pK measurement is not reported.
[b] Polysciences, Inc. (Warrington, PA).
[c] P. G. Righetti, M. Chiari, P. K. Sinha, and E. Santaniello, *J. Biochem. Biophys. Methods* **16**, 185 (1986).
[d] Pharmacia-LKB Biotechnology (Uppsala, Sweden).

in the same way as conventional polyacrylamide gradient gels, by using a density gradient to stabilize the Immobiline concentration gradient, with the aid of a standard two-vessel gradient mixer. As shown in the structural formulas, these buffers are no longer amphoteric, as in conventional IEF, but are bifunctional: At one end of the molecule is located the buffering (or titrant) group, and at the other the acrylic double bond, which will disappear during the grafting process. Owing to the different temperature coefficients (dpK/dT) of acidic and alkaline Immobilines, for reproducible runs and pH gradient calculations all experimental parameters have been fixed at 10°.

Temperature is not the only variable that affects Immobiline pK values (and therefore the actual pH gradient generated): Additives in the gel that change the water structure (chaotropic agents, e.g., urea) or lower its dielectric constant, and the ionic strength of the solution, alter the pK values on the gel. The largest changes are due to the presence of urea: Acidic Immobilines increase their pK, in 8 *M* urea, by as much as 0.9 pH units, while the basic Immobilines are increased by only 0.45 pH unit.[8] Detergents in the gel (up to 2%) do not alter the Immobiline pK, suggesting that acidic and basic groups attached to the gel are not incorporated into

[8] E. Gianazza, F. Artoni, and P. G. Righetti, *Electrophoresis* **4**, 321 (1983).

TABLE III
BASIC ACRYLAMIDO BUFFERS

pK^a	Formula	Name	M_r	Source
6.2	$CH_2{=}CH{-}CO{-}NH{-}(CH_2)_2{-}N$ (morpholine)	2-Morpholinoethylacrylamide	184	b
7.0	$CH_2{=}CH{-}CO{-}NH{-}(CH_2)_3{-}N$ (morpholine)	3-Morpholinopropylacrylamide	198	b
8.5	$CH_2{=}CH{-}CO{-}NH{-}(CH_2)_2{-}N(CH_3)_2$	N,N-Dimethylaminoethylacrylamide	142	b
9.3	$CH_2{=}CH{-}CO{-}NH{-}(CH_2)_3{-}N(CH_3)_2$	N,N-Dimethylaminopropylacrylamide	156	b
10.3	$CH_2{=}CH{-}CO{-}NH{-}(CH_2)_3{-}N(C_2H_5)_2$	N,N-Diethylaminopropylacrylamide	184	c
>12[d]	$CH_2{=}CH{-}CO{-}NH{-}(CH_2)_2{-}\overset{+}{N}(C_2H_5)_3$	N,N,N-Triethylaminoethylacrylamide	198	e
>12	$CH_2{=}CH{-}CO{-}NH{-}(CH_2)_2{-}\overset{+}{N}(CH_3)_3$	N,N,N-Trimethylaminoethylacrylamide	153	e

[a] All pK values (except for pK 10.3) measured at 25°. The value of pK 10.3 was measured at 10°.
[b] Pharmacia-LKB Biotechnology (Uppsala, Sweden).
[c] P. K. Sinha and P. G. Righetti, *J. Biochem. Biophys. Methods* **15**, 199 (1987).
[d] QAE, Quaternary aminoethylacrylamide.
[e] IBF (Villeneuve La Garenne, France).

TABLE IV
NEW BASIC ACRYLAMIDO BUFFERS

pK^a	Formula	Name	M_r	Ref.
6.6	$CH_2=CH-CO-NH-(CH_2)_2-N$ (thiomorpholine, S)	2-Thiomorpholinoethylacrylamide	200	b
6.85	$CH_2=CH-CO-$ (piperazine, N—CH_3)	1-Acryloyl-4-methylpiperazine	154	c
7.0	$CH_2=CH-CO-NH-(CH_2)_2-C$ (imidazole ring, N, NH)	2-(4-Imidazolyl)ethylamine-2-acrylamide	165	d
7.4	$CH_2=CH-CO-NH-(CH_2)_3-N$ (thiomorpholine, S)	3-Thiomorpholinopropylacrylamide	214	b
8.05	$CH_2=CH-CO-NH-(CH_2)_3-N(CH_2CH_2OH)_2$	N,N-Bis(2-hydroxyethyl)-N'-acryloyl-1,3-diaminopropane	200	e

[a] All pK values measured at 25°.
[b] M. Chiari, P. G. Righetti, P. Ferraboschi, T. Jain, and R. Shorr, *Electrophoresis* 617 (1990).
[c] M. Chiari, C. Ettori, A. Manzocchi, and P. G. Righetti, *J. Chromatogr.* **548**, 381 (1991).
[d] M. Chiari, M. Giacomini, C. Micheletti, and P. G. Righetti, *J. Chromatogr.* **558**, 285 (1991).
[e] M. Chiari, L. Pagani, P. G. Righetti, T. Jain, R. Shorr, and T. Rabilloud, *J. Biochem. Biophys. Methods* **21**, 165 (1990).

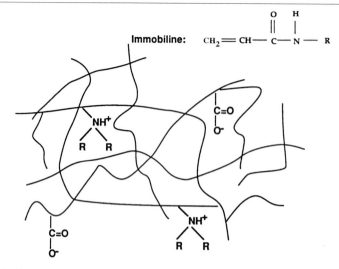

Fig. 1. Diagrammatic representation of part of a polyacrylamide gel with buffering and titrant ions covalently bound to the backbone. The cross-over points of the chains can be regarded as cross-links. *Upper right:* The basic structural formula of the Immobiline chemicals (acrylamido acids and bases). (Courtesy of LKB Produkter AB.)

the surfactant micelle. For generating extended pH gradients, we use two additional chemicals, strong titrants having pK values well outside the desired pH range: one is QAE [(quaternary aminoethyl)acrylamide (pK > 12)] and the other is AMPS (2-acrylamido-2-methylpropanesulfonic acid, $pK \approx 1$). We have also produced two additional acrylamido buffers: one with pK 10.3, for casting alkaline (pH 10–11) IPG gradients,[9] and one with pK 3.1, for generating acidic (pH 2.5–3.5) IPG intervals.[10]

Figure 1 depicts a hypothetical structure of an IPG matrix. Because they are copolymerized within the matrix, the Immobiline buffers no longer migrate in the electric field: This means that the pH gradient is stable indefinitely, although this condition must be established before the onset of polymerization; stable gradients can be destroyed only if the polyacrylamide gel is destroyed. At conventional matrix concentration (4% T) and at the standard Immobiline levels (10 mM buffering ion + 5 mM titrant at pH equal to pK) there is about 1 Immobiline group every 37 acrylamide

[9] C. Gelfi, M. L. Bossi, B. Bjellqvist, and P. G. Righetti, *J. Biochem. Biophys. Methods* **15**, 41 (1987).
[10] P. G. Righetti, M. Chiari, A. Santaniello, and P. K. Sinha, *J. Biochem. Biophys. Methods* **16**, 185 (1988).

residues, which means that, statistically, 2 Immobiline species are about 74 carbon atoms apart in a single polyacrylamide chain.

Narrow and Ultranarrow pH Gradients

We define the gradients from 0.1 to 1 pH unit as ultranarrow gradients. Within these limits only two acrylamido buffers are needed: a buffering Immobiline, either a base or an acid, with its pK within the desired pH interval, and a nonbuffering Immobiline, i.e., an acid or a base, respectively, with its pK at least 2 pH units removed from either the minimum or maximum of the desired pH range. The titrant will thus provide equivalents of acid or base to titrate the buffering group but will not itself buffer in the desired pH interval. Recipes have been tabulated[11] listing 58 gradients, each 1 pH unit wide and separated by increments of 0.1 pH unit, starting with the pH 3.8–4.8 interval and ending with the pH 9.5–10.5 range. These recipes provide the Immobiline volumes (for 15 ml of mixture) needed in the acidic (mixing) chamber to obtain pH$_{min}$ and the corresponding volumes for the basic (reservoir) chamber of the gradient mixer needed to generate pH$_{max}$ of the desired pH interval. If a narrower pH gradient is needed, it can be derived from any of the 58 pH intervals tabulated by a simple linear interpolation of intermediate Immobiline molarities.

Extended pH Gradients

Recipes for 2 pH unit-wide gradients up to 6 pH unit spans have been calculated and optimized by computer simulation[12] and are listed in Ref. 11. All the formulations are normalized to give the same average value of buffering power (β) of 3 mEq liter^{-1} pH^{-1}, a β value ample to produce highly stable pH gradients. It will be noted that, in practice, for pH intervals covering >4 pH units, the best solution is to mix a total of 10 Immobiline species, 8 of them being buffering ions and 2 being the strong acidic and basic titrants.

While all the formulations up to 1985 were calculated for producing rigorously linear pH gradients, for two-dimensional maps of complex samples (cell lysates, biopsies) there would in fact be a need for nonlinear gradients, flattened in the pH 4–6 region (where more than 50% of the proteins are isoelectric) and steeper in the alkaline region of the pH scale.

[11] P. G. Righetti, "Immobilized pH Gradients: Theory and Methodology." Elsevier, Amsterdam, The Netherlands, 1990.
[12] E. Gianazza, S. Astrua-Testori, and P. G. Righetti, *Electrophoresis* **6,** 113 (1985).

TABLE V

ISOELECTRIC FOCUSING IN IMMOBILIZED pH GRADIENTS

1. Assemble the gel mold; mark the polarity of the gradient on the back of the supporting plate
2. Mix the required amounts of Immobilines; fill to one-half of the final volume with distilled water
3. Check the pH of the solution and adjust as required
4. Add acrylamide, glycerol (0.2–0.3 ml/ml to the "dense" solution), and TEMED and bring to the final volume with distilled water
5. For ranges removed from neutrality, titrate to about pH 7.5, with either Tris or formic acid
6. Transfer the denser solution to the mixing chamber, the lighter solution to the reservoir of a gradient mixer; center the mixer on a magnetic stirrer; check for the absence of air bubbles in the connecting duct
7. Add ammonium persulfate to the solutions
8. Pour the gradient into the mold
9. Allow the gel to polymerize for 1 hr at 50°
10. Disassemble the mold, weigh the gel
11. Wash the gel for 1 hr in 0.5 liter of distilled water on a shaking platform (two water changes)
12. Dry the gel to its original weight (in front of a blower)
13. Transfer the gel to the electrophoresis chamber (temperature, 10°); apply the electrode strips
14. Load the samples
15. Apply current
16. Stain the gel

Such nonlinear gradients have also been tabulated.[13] All these recipes, extending in general from pH_{min} 4 to pH_{max}[10], do not contain the hydrolytic products of water in the formulations; however, outside these limits H^+ and OH^- are accounted for, as their contribution to conductivity and β power becomes appreciable. In a more recent release of the computer program (which extends also to oligoprotic buffers) H^+ and OH^- are automatically present in all equations.[14–16]

Immobilized pH Gradient Methodology

The overall procedure is outlined in the flow sheet of Table V. We treat here particularly: (1) gel polymerization, (2) reswelling of dried Immobiline gels, (3) storage of the Immobiline chemicals, and (4) mixed-bed, CA-IPG

[13] E. Gianazza, P. Giacon, B. Sahlin, and P. G. Righetti, *Electrophoresis* **6**, 53 (1985).
[14] C. Tonani and P. G. Righetti, *Electrophoresis* **12**, 1011 (1991).
[15] P. G. Righetti and C. Tonani, *Electrophoresis* **12**, 1021 (1991).
[16] E. Giaffreda, C. Tonani, and P. G. Righetti, *J. Chromatogr.* **630**, 313 (1993).

gels. It should be noted that the basic equipment is the same as for conventional CA-IEF gels, consisting of a thermostat, a horizontal focusing chamber (best with movable electrodes), and a power supply (the latter, however, should be able to deliver up to 5000 V).

Casting Immobiline Gel

When preparing an IPG experiment, two pieces of information are required: the total liquid volume needed to fill up the gel cassette, and the desired pH interval. Once the first is known, this volume is divided into two halves: One half is titrated to one extreme of the pH interval, the other to the opposite extreme. As the analytical cassette usually has a thickness of 0.5 mm and, for the standard 12 × 25 cm size, contains 15 ml of liquid to be gelled, in principle two solutions, each of 7.5 ml, should be prepared. For better precision, however, we recommend preparing a double volume, enough for casting two gel slabs. These two volume aliquots are termed "acidic dense" and "basic light" solutions: this is because the polymerization cassette is filled with the aid of a two-vessel gradient mixer and thus the liquid elements that fill the vertically standing cassette must be stabilized against remixing by a density gradient (by convention, the acidic pH extreme is chosen as the dense solution). For the sequence of steps needed, refer to Table V and to Fig. 2 for the gel cassette assembly.

Preparation of Gel Mold and Gel Polymerization

For molding sample application slots in the gel, apply suitably sized pieces of Dymo (also known as "Tesa") tape to the glass plate with the U frame; a 5 × 3 mm slot can be used for sample volumes between 5 and 20 µl (Fig. 2A). To prevent the gel from sticking to the glass plates with U frame and slot former, coat them with Repel-Silane. Use a drop of water on the Gel Bond PAG (polyacrylamide gel) film to determine the hydrophilic side. Apply a few drops of water to the plain glass plate and carefully lay the sheet of Gel Bond PAG film on top with the hydrophobic side down (Fig. 2B). Roll the film flat to remove air bubbles and to ensure good contact with the glass plate. Clamp the glass plates together with the Gel Bond PAG film and slot former on the inside, by means of clamps placed all along the U frame (Fig. 2C).

Figure 3 presents the final assembly for the cassette and gradient mixer. In modern gel molds (with the cover plate bearing three V-shaped indentations) inserting the capillary tubing conveying the solution from the mixer is greatly facilitated. As for the gradient mixer, one chamber contains a magnetic stirrer, while in the reservoir is inserted a plastic cylinder having

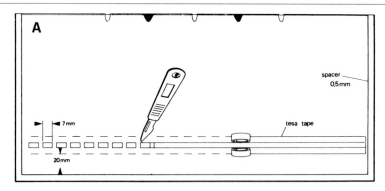

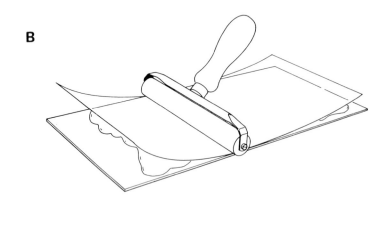

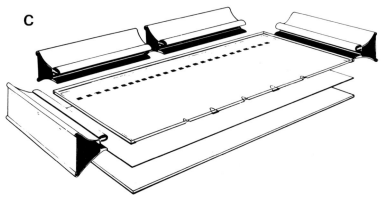

Fig. 2. Assembly of the gel cassette. (A) Preparation of the slot former: onto the cover plate (bearing the rubber gasket U frame) is glued a strip of Tesa tape out of which rectangular tabs are cut with a scalpel. (B) Application of the Gel Bond PAG film to the supporting glass plate. (C) Assembling the gel cassette. (Courtesy of LKB Produkter AB.)

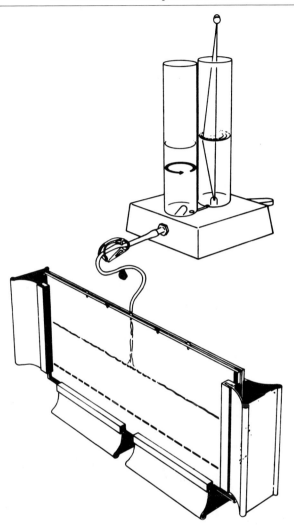

FIG. 3. Setup for casting an IPG gel. A linear pH gradient is generated by mixing equal volumes of a dense and light solution, titrated to the extremes of the desired pH interval. Note the "compensating" rod in the reservoir, used as a stirrer after addition of catalysts and for hydrostatically equilibrating the two solutions. With modern cover plates, bearing three V-shaped indentations, insertion of the capillary conveying the solution from the mixer is greatly facilitated. (Courtesy of LKB Produkter AB.)

the same volume, held by a trapezoidal rod. The latter, in reality, is a "compensating cone" needed to raise the liquid level to such an extent that the two solutions (in the mixing chamber and in the reservoir) will be hydrostatically equilibrated; it can also be utilized for stirring.

A linear pH gradient is generated by mixing equal volumes of the two starting solutions in a gradient mixer. Prepare the acidic, dense solution and the basic, light solution for the pH gradient according to published recipes.[11] Prepared gel solutions must not be stored. Gels, however, with a pH less than 8 can be stored in a humidity chamber for up to 1 week after polymerization. Transfer 7.5 ml of the basic, light solution to the reservoir chamber and fill the connecting channel in between. Then transfer 7.5 ml of the acidic, dense solution to the mixing chamber. Place the prepared mold upright on a leveled surface, 5 cm below the gradient mixer outlet. Open the clamp of the outlet tubing, fill the tubing halfway with the dense solution, and close the clamp again. Switch on the stirrer, and set to a speed of ca. 500. Add the catalysts to each chamber as specified.[11] Insert the free end of the tubing between the glass plates of the mold at the central V indentation (Fig. 3). Open the clamp on the outlet tubing and the valve between the dense and light solutions so that the gradient solution starts to flow down into the mold by gravity. The mold will be filled within 5 min. When the gradient mixer is empty, carefully remove the tubing from the mold; after leaving the cassette to rest for 5 min, place it on a leveled surface in an oven at 50° for 1 hr. When polymerization is complete, remove the clamps and carefully take the mold apart. Weigh the gel and then place it in 0.3 liter of distilled water (three rises of 20 min each) to wash out any remaining ammonium persulfate, N,N,N',N'-tetramethylethylenediamine (TEMED) and amounts of unreacted monomers and Immobilines. After washing the gel, carefully blot any excess water from the surface with a moist paper tissue and remove the extra water absorbed by the gel during the washing step by using a nonheating fan placed at ca. 50 cm from the gel. Check the weight of the gel after 5 min and from this, estimate the total excess liquid reducing time. This weight-restoring process is essential, as a gel containing too much water will "sweat" during the electrofocusing run and droplets of water will form on the surface.

Note: it is preferable to use soft gels, i.e., with a low % T (4% T or even 3% T). These "soft" gels can be easily dried without cracking and allow better entry of larger proteins. In addition, the local ionic strength along the polymer coil is increased, and this permits sharper protein bands due to increased solubility at the pI. It is a must, for any gel formulation removed from neutrality (pH 6.5–7.5), to titrate the two solutions to neutral pH, so as to ensure reproducible polymerization conditions and avoid hydrolysis of the four alkaline Immobilines. If the pH interval used is acidic, add Tris; if it is basic, add formic or acetic acid.

Reswelling of Dry Immobiline Gels

Precast, dried Immobiline gels, encompassing a few acidic ranges, are now available from Pharmacia-LKB AB: they all contain 4% T and span the following pH ranges: pH 4–7, pH 4.2–4.8 (e.g., for α_1-antitrypsin analysis), pH 4.5–5.4 (e.g., for Gc [group-specific component or vitamin D binding protein] screening), pH 5.0–6.0 (e.g., for transferrin analysis), and pH 5.6–6.6 (e.g., for phosphoglucomutase screening). Precast, dried IPG gels in the alkaline region have not been introduced as yet, possibly because at high pH values the hydrolysis of both the gel matrix and the Immobiline chemicals bound to it is much more pronounced. However, gel strips covering the pH 3–10 interval, to be used as first dimension of two-dimensional maps, have been produced.

On reswelling, even under isoionic conditions, acidic ranges regain water four to five times faster than alkaline ones. Thus, it is preferable to reswell dried Immobiline gels in a cassette similar to the one for casting the IPG gel. A special reswelling chamber is available from Pharmacia-LKB AB: the dried gel is inserted in the cassette, which is clamped and allowed to stand on the short side; via a small hole in the lower right side and a cannula, the reswelling solution is gently injected in the chamber, until filling is complete. As the system is volume controlled, it can be left to reswell overnight, if needed. Gel drying and reswelling is the preferred procedure when an IPG gel containing additives is needed: in this case it is always best to cast an "empty" gel, wash it, dry it, and then reconstitute it in the presence of the desired additive (e.g., urea, alkyl ureas, detergents, carrier ampholytes, and mixtures thereof).

Storage of Immobiline Chemicals

There are two major problems with the Immobiline chemicals, especially with the alkaline ones: (1) hydrolysis and (2) spontaneous autopolymerization. The first problem is quite a nuisance because, on hydrolysis, only acrylic acid is incorporated into the IPG matrix, with a strong acidification of the calculated pH gradient. Hydrolysis is an autocatalyzed process for the basic Immobilines, as it is pH dependent[17]: For the pK 8.5 and 9.3 species, such a cleavage reaction on the amido bond can occur even in the

---

[17] P. G. Pietta, E. Pocaterra, A. Fiorino, E. Gianazza, and P. G. Righetti, *Electrophoresis* **6**, 162 (1985).

frozen state, at a rate of about 10%/year.[18] Autopolymerization,[19,20] is also quite deleterious for the IPG technique. Again, this reaction occurs particularly with alkaline Immobilines and is purely autocatalytic, as it is greatly accelerated by deprotonated amino groups. Oligomers and n-mers are formed, which stay in solution and can even be incorporated into the IPG gel, as in principle such oligomers still contain a double bond at one extremity. These products of autopolymerization, when added to proteins in solution, are able to bridge proteins via two unlike binding surfaces; a lattice is formed and the proteins (especially larger ones, such as ferritin, α_2-macroglobulin, and thyroglobulin) are precipitated out of solution.

Owing to all these problems, alkaline Immobilines are today supplied dissolved in propanol (containing a maximum of 60 ppm water), which prevents both hydrolysis and autopolymerization for a virtually unlimited amount of time (less than 1% degradation per year; simply store them at +4°; the acidic Immobilines, being much more stable, continue to be available as water solutions, with added inhibitor, 5 ppm hydroquinone monomethyl ether).[21]

Mixed-Bed, Carrier Ampholyte-Immobilized pH Gradient

In CA-IPG gels the primary, immobilized pH gradient is mixed with a secondary, soluble carrier ampholyte-driven pH gradient. When working with membrane proteins[22] and with microvillar hydrolases, partly embedded in biological membranes,[23] it has been found that addition of CAs to the sample and IPG gel would enhance protein solubility, possibly by forming mixed micelles with the detergent used for membrane solubilization or by directly complexing with the protein itself. In the absence of CAs, these same proteins would essentially fail to enter the gel and mainly precipitate or give elongated smears around the application site (in general cathodic sample loading). On a relative hydrophobicity scale, the basic Immobilines (pK values of 6.2, 7.0, 8.5, 9.3, 10.3, and >12) are decidedly more hydrophobic than their acidic counterparts (pK values of 3.1, 3.6, 4.4, and 4.6): On

[18] S. Astrua-Testori, J. J. Pernelle, J. P. Wahrmann, and P. G. Righetti, *Electrophoresis* **7**, 527 (1986).
[19] P. G. Righetti, C. Gelfi, M. L. Bossi, and E. Boschetti, *Electrophoresis* **8**, 62 (1987).
[20] T. Rabilloud, C. Gelfi, M. L. Bossi, and P. G. Righetti, *Electrophoresis* **8**, 305 (1987).
[21] B. M. Gåveby, P. Pettersson, J. Andrasko, L. Ineva-Flygare, U. Johannesson, A. Görg, W. Postel, A. Domscheit, P. L. Mauri, P. Pietta, E. Gianazza, and P. G. Righetti, *J. Biochem. Biophys. Methods* **16**, 141 (1988).
[22] M. Rimpilainen and P. G. Righetti, *Electrophoresis* **6**, 419 (1985).
[23] P. K. Sinha and P. G. Righetti, *J. Biochem. Biophys. Methods* **12**, 289 (1986).

incorporation in the gel matrix, the phenomenon becomes cooperative and could lead to formation of hydrophobic patches on the surface of such a hydrophilic gel as polyacrylamide. Thus, CAs could act as shielding molecules, coating on the one side the polyacrylamide matrix studded with Immobilines (especially the basic ones) and, on the other side, the protein itself. This strongly quenches potential protein–IPG matrix interactions, effectively detaches the protein from the surrounding polymer coils, and allows good focusing into sharp bands.

In answer to the basic question concerning when and how much CA to add, we suggest the following: (1) if your sample focuses well as such, ignore the mixed-bed technique (which presumably will be mostly needed with hydrophobic proteins and in alkaline pH ranges); (2) add only the minimum amount of CAs (in general around 0.3 to 0.5%) needed for avoiding sample precipitation in the pocket and for producing sharply focused bands.

Effect of Salts on Proteins in Immobilized pH Gradients

We have seen in the previous sections some of the problems afflicting the IPG technique. We have also discussed the adverse phenomena of hydrolysis and autopolymerization of alkaline Immobilines, and how this is circumvented by dissolving these chemicals in propanol. There is still one important problem plaguing focusing methods. Since its inception, the IPG technique was recognized to be quite tolerant of salt levels present in the sample. This was publicized as one of the greatest advantages of the IPG technique as opposed to CA-IEF, known to be quite sensitive even to low salt levels in the sample. Thus biological samples (containing high salt and dilute proteins) could be run in IPGs without prior dialysis or concentration. This statement is only partially true: It is true when referring to the IPG matrix, which in principle can stand any amount of salt; but it is not true when referring to the protein sample. Salts formed from strong acids and bases (e.g., NaCl, Na_2SO_4, and Na_2HPO_4), present in a protein sample applied to an IPG gel, induce protein modification (e.g., oxidation of iron moiety in hemoglobin) even at low levels (5 mM) and irreversible denaturation (precipitation) at higher levels (>50 mM). This effect is due to production of strongly alkaline cationic and strongly acidic anionic boundaries formed by the splitting of the ion constituents of the salt, as the protein zone is not and cannot be buffered by the surrounding gel until it physically migrates into the IPG matrix. Substitution of "strong" salts in the sample zone with salts formed by weak acids and bases, e.g., Tris–acetate, Tris–glycinate, and Good's buffers, essentially abolishes both phe-

TABLE VI

GUIDELINES FOR USE OF IMMOBILIZED pH GRADIENTS

1. Avoid high salt levels in the sample (>40 mM)
2. Avoid salts formed by strong acids and bases (e.g., NaCl, Na_2SO_4, Na_2HPO_4)
3. In the presence of high salt levels, add high levels of CA (e.g., 10% CA to 100 mM salt)
4. If salt is needed for sample solubility, use salts formed from weak acids and bases [e.g., Tris–acetate, Tris–glycinate, any of the Good's buffers (e.g., ACES, ADA, MOPS)] titrated around the pK of their amino groups
5. In the presence of high salt levels, run the sample at low voltage for several hours (e.g., 500 V for 4 hr) so as to prevent formation of strongly acidic and alkaline boundaries

nomena, oxidation and irreversible denaturation.[24] Suppression of strong salt effects is also achieved by adding to the sample zone carrier ampholytes in amounts proportional to the salt present (e.g., by maintaining a salt : CA molar ratio of ca. 1 : 1). Low-voltage runs for extended initial periods (e.g., 4 hr at 500 V) are also beneficial. Table VI summarizes all of these recommendations: With these precautions, the IPG technique is today a trouble-free operation.

Illustrating the Resolving Power

In CA-IEF, a resolving power of 0.01 pH unit (in ΔpI, i.e., in pI difference between a protein and the nearest resolved contaminant) is routinely achieved,[3] whereas in IPGs a ΔpI of 0.001 between closely related species can be attained.[11] This is a remarkable resolving power, hardly equaled by any other separation technique. Figure 4 illustrates the progress in separation capability, with a sample of cord blood lysate from an individual heterozygous for fetal hemoglobin Sardinia (an Ile → Thr mutation in residue 75 of the γ chains). The two hemoglobins are not separated by a 1-pH unit CA-IEF (Fig. 4, top), but are fully resolved in an IPG gel spanning 0.25 pH units. The latter, however, could not resolve a more subtle mutation. Normal fetal hemoglobin is a mixture of two components, called A_γ and G_γ, having an alanine or glycine residue at position 136, respectively. The two tetramers, normal components during fetal life, are found in approximately a 20 : 80 ratio. If the pH gradient is further decreased to 0.1 pH unit (over a standard 10-cm migration length), even these two tetramers can be separated (Fig. 4, bottom), with a resolution close to the practical limit of ΔpI of 0.001, even though the γ polypeptide chains differ by a single methyl group![25]

[24] P. G. Righetti, M. Chiari, and C. Gelfi, *Electrophoresis* **9**, 65 (1988).
[25] G. Cossu and P. G. Righetti, *J. Chromatogr.* **300**, 165 (1987).

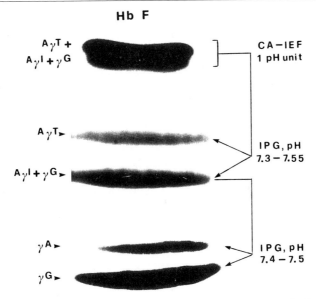

FIG. 4. Focusing of umbilical cord lysates from an individual heterozygous for fetal hemoglobin (HbF) Sardinia (for simplicity, only the HbF bands are shown, and not the two other major components of cord blood, i.e., HbA and HbF$_{ac}$). *Top:* Focusing performed in a 1-pH unit span in CA-IEF. Note that broadening of the HbF occurs, but not the splitting into well-defined zones. *Middle:* Same sample as above, but focused over an IPG range spanning 0.25 units. *Bottom:* Same as above, but in an IPG gel spanning 0.1 pH unit. The resolved A$_\gamma$/G$_\gamma$ bands are in a 20:80 ratio, as theoretically predicted from gene expression. Their identity was established by eluting the two zones and fingerprinting the γ chains. (Reprinted from *J. Chromatogr.* **300,** G. Cossu and P. G. Righetti, p. 165, Copyright 1987 with kind permission of Elsevier Science–NL, Sara Burgerhartstraat 25, 1055 KV Amsterdam, The Netherlands.)

Preparative Aspects

Just as IPGs exhibit a resolving power one order of magnitude greater than that of CA-IEF, they offer a much improved protein load ability as well. The latest preparative IPG variant is based on a unique purification concept: the use of zwitterionic membranes as isoelectric traps for each isoform of a family of proteins.[26–29] Figure 5 gives an enlarged view of a peculiar electrophoresis chamber, dubbed a *multicompartment electrolyzer.*

[26] P. G. Righetti, E. Wenisch, and M. Faupel, *J. Chromatogr.* **475,** 293 (1989).

[27] P. G. Righetti, E. Wenisch, A. Jungbauer, H. Katinger, and M. Faupel, *J. Chromatogr.* **500,** 681 (1990).

[28] P. G. Righetti, M. Faupel, and E. Wenisch, *in* "Advances in Electrophoresis" (A. Chrambach, M. J. Dunn, and B. J. Radola, eds.), Vol. 5, pp. 159–200. VCH, Weinheim, Germany, 1992.

[29] C. Ettori, P. G. Righetti, C. Chiesa, F. Frigerio, G. Galli, and G. Grandi, *J. Biotechnol.* **25,** 307 (1992).

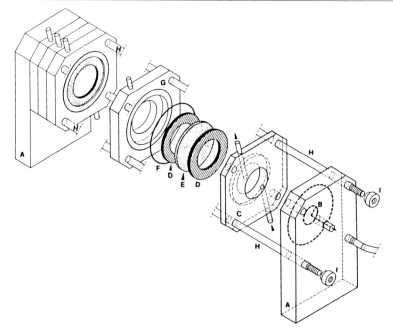

FIG. 5. Enlarged view of the multicompartment electrolyzer. A, Rectangular supporting legs; B, Pt electrode; C, terminal flow chamber; D, rubber rings for supporting the membrane; E, isoelectric Immobiline membrane cast onto the glass fiber filter; F, O ring; G, one of the sample flow chambers; H, four threaded metal rods for assembling the apparatus; I, nuts for fastening the metal bolts. (Reprinted from *J. Chromatogr.* **300,** P. G. Righetti *et al.* p. 681, Copyright 1990 with kind permission of Elsevier Science–NL, Sara Burgerhartstraat 25, 1055 KV Amsterdam, The Netherlands.)

It consists of a series (up to eight) recycling chambers, separated by isoelectric membranes. Once the pI of a given protein species has been determined in an analytical IPG run, such protein can be trapped in any chamber having membranes satisfying the condition: pI_a ≤ pI_p ≤ pI_c, where the subscripts a and c denote the anodic and cathodic membranes, respectively, and pI_p indicates the protein isoelectric point. The experimental setup is further illustrated in Fig. 6: The electrolyzer, a multichannel peristaltic pump, a number (up to eight) of sample and electrolyte reservoirs, and relative stirrers are housed in a safety box (the power supply, not shown, is connected to the end chambers of the electrolyzer, which thus act as anodic and cathodic reservoirs). The sample to be purified is thus subjected to two perpendicular flows: a hydraulic flow, recycling the analyte from the reservoirs into the electric field, and a flow of electricity, driving each protein species to the correct chamber delimited by two isoelectric membranes. At

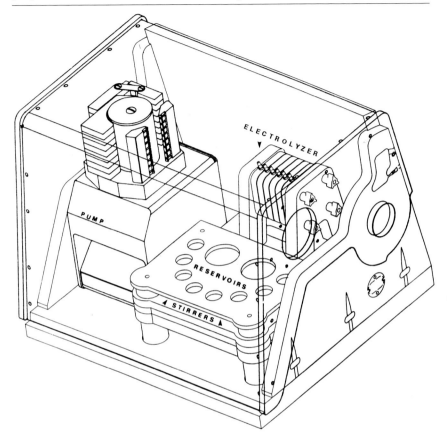

Fig. 6. Commercially available experimental setup for recycling isoelectric focusing with zwitterionic membranes. A safety box houses a multichannel peristaltic pump, a platform for sample and electrodic reservoirs with individual stirrers, and the multicompartment electrolyzer. Note that stirring in each vessel is needed in order to avoid electrodecantation. The vessels and the connecting tubings are not shown. (Courtesy of Hoefer Scientific Instruments.)

steady state, each isoform ideally collects into a reservoir, free of adjacent proteinaceous contaminants and salt free as well (note that only in such a technique is a protein both isoelectric and isoionic!). A number of advantages are immediately apparent.

1. Large sample loads and large liquid volumes can be applied, because they are stored in the reservoirs outside the electric field.
2. The recycling technique provides an uncoupling between the electric and liquid flows, thereby allowing Joule heat to be dissipated in the reservoirs, rather than in the electrolyzer.

3. The use of isoelectric membranes allows protein recovery in a liquid vein, thereby avoiding adsorption onto solid surfaces (as typical of chromatography).
4. In turn, the use of sanitized isoelectric membranes prevents contamination of sample macroions from unreacted acrylamide monomers (neurotoxic).
5. It goes without saying that the use of insoluble buffers and titrants avoids the contamination of sample by soluble carrier ampholytes, a typical drawback of CA-IEF.

Figure 6 shows the experimental setup, as now commercially available from Hoefer Scientific Instruments (San Francisco, CA): A multichannel peristaltic pump, a platform for reservoirs and stirrers, and the electrophoretic chamber are housed in a safety box (note that the tubings for recycling the sample from the reservoirs to the electric field and back are not shown). The unit is compact and can be easily operated in a cold room (which will be adequate for dissipating the Joule heat produced). The power supply is

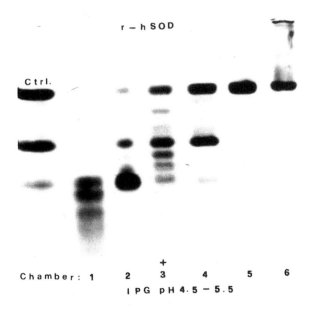

Fig. 7. Analytical IPG gel of a preparative rhSOD run in the multicompartment electrolyzer. IPG gel: 5% T, 4% C, pH range 4.5–5.5, run for 60,000 V · hr at 10°. Staining with Coomassie Brilliant Blue in Cu^{2+}. Ctrl., Control; unfractionated rhSOD. Tracks 1–6: content of chambers 1–6 in the electrolyzer. The cathode is uppermost. Note that chamber 5 contains a single homogeneous SOD band. (From E. Wenisch and P. G. Righetti, unpublished data, 1994.)

usually kept outside the safety box (not shown in Fig. 6) and is connected to the electrolyzer via a port at the base of the box. An example of the unique purification capability of this instrument is given in Fig. 7: A highly purified preparation of recombinant human superoxide dismutase (SOD) was found to consist of at least three isoforms, with the following isoelectric points: 4.80, 4.92, and 5.07. When monitoring the purification progress, it is seen that a pure pI 5.07 form collects in chamber 5, with the lower pI species focusing in the chambers delimited by the proper set of two isoelectric membranes.

Acknowledgments

Supported by grants from Radius in Biotechnology (ESA, Paris) to P.G.R.

[11] Pulsed-Field Gel Electrophoresis

By JANET C. WRESTLER, BARBARA D. LIPES, BRUCE W. BIRREN, and ERIC LAI

Introduction

Pulsed-field gel electrophoresis (PFGE) of agarose gels enables the reproducible separation of large DNA fragments.[1] In concept, PFGE is an extension of conventional electrophoresis, in which two alternating (or pulsed) electric fields are used instead of the traditional single static field. Separation occurs when these fields are oriented at an obtuse angle to one another. In a pulsed-field gel, the end of each molecule migrates in a new direction with each change of the electric fields. The DNA molecules thus migrate through the agarose matrix in a zigzag motion. The tardiness of the larger molecules in turning corners (e.g., in PFGE) or in running forward and backward [e.g., in field-inversin gel electrophoresis (FIGE), see below] separates them from the smaller size fragments. The effectiveness of PFGE, however, is not limited to the separation of very large DNA molecules. PFGE can improve the resolution of DNA molecules of only a few hundred bases and permits separation up to 12,000 kilobase pairs (kb).[2] Figure 1 shows the effectiveness of PFGE in separating yeast chromosomes from 200 to 2000 kb.

[1] D. C. Schwartz and C. R. Cantor, *Cell* **37**, 67 (1984).
[2] M. Bellis, M. Pages, and G. Roizes, *Nucleic Acids Res.* **15**, 6749 (1987).

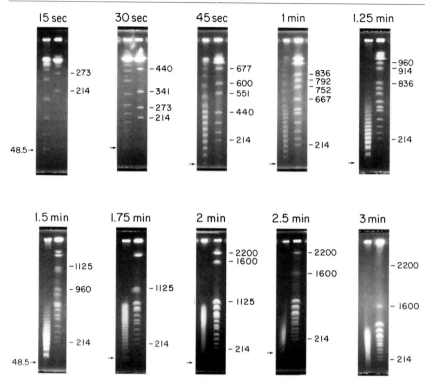

Fig. 1. Separation of 50- to 1000-kb DNAs with different switch intervals. λ ladders and chromosomes of the yeast *Saccharomyces cerevisiae* (strain YNN295) were separated in gels, using identical conditions except for the different switch intervals. All gels were 1% SeaKem LE agarose in 0.5× TBE run at 14° in 6-V/cm fields with a reorientation angle of 120°. With each increase in switch interval the new size limit for resolution (based on the separation of the markers) is indicated.

A number of models and theories have been proposed to explain some of the more complex behavior of DNA molecules in PFGE.[3] However, biologists rarely need to consult these physical models or equations for practical PFGE applications. This chapter outlines optimum PFG electrophoretic conditions for the separation of DNA fragments from 1 to 6000 kb.

---

[3] E. Lai and B. W. Birren (eds.), "Current Communications in Cell and Molecular Biology," Vol. 1: Electrophoresis of Large DNA Molecules: Theory and Applications. Cold Spring Harbor Laboratory Press, Cold Spring Harbor, NY 1990.

Unique Features

PFGE of high molecular weight DNA involves additional factors that are not encountered in conventional agarose gels. We outline some of these factors and the modifications required for running PFGE.

Sample Preparation, Gel Loading, and DNA Recovery

High molecular weight DNA samples for PFGE must be carefully prepared to minimize degradation resulting from mechanical breakage or nuclease activity. This is achieved by using a method developed by Schwartz and Cantor of embedding intact cells in an agarose matrix prior to lysis.[1] This provides support for the DNA while allowing digestion of membranes and subsequent diffusion of non-DNA particles. The DNA can then be separated in gels or subjected to restriction digestion. The DNA plugs can be loaded directly into the wells or placed in front of a comb before pouring the gel. Molecules as large as 12 megabases (Mb) have been successfully prepared by these methods,[2] which have been extensively reviewed.[4-6] DNA recovery from PFGE can be achieved by using low melting point agarose and treatment with agarase after electrophoresis.

Electrophoretic Parameters

The presence of at least two electric fields introduces many electrophoretic variables that are unique to PFGE. Chief among these is the switch time (or switch interval), which is the duration of each of the alternating electric fields. In general, the larger the DNA molecules, the longer the switch times required for separation. This is because larger molecules take longer to reorient before they begin migrating with each field switch. Thus, large molecules that spend a great portion of each switch interval reorienting will not migrate sufficiently to be resolved. The switch interval is chosen according to the size range of fragments to be resolved. This effect is clearly demonstrated in Fig. 1, in which lengthening the switch time separates larger fragments.

One problem associated with PFGE is band inversion.[7] This refers to the migration of certain fragments at a faster rate than some smaller fragments, resulting in an inverted fragment size order in specific regions of the gel.

---

[4] C. L. Smith, P. E. Warburton, A. Gaal, and C. R. Cantor, *in* "Genetic Engineering" (J. K. Setlow and A. Hollaender, eds.), Vol. 8, p. 45. Plenum, New York, 1986.
[5] G. F. Carle and M. V. Olson, *Methods Enzymol.* **155,** 468 (1987).
[6] B. Birren and E. Lai, "Pulsed Field Gel Electrophoresis: A Practical Guide," 1st Ed. Academic Press, Orlando, FL, 1993.
[7] G. F. Carle, M. Frank, and M. V. Olson, *Science* **232,** 65 (1986).

This is particularly apparent when the reorientation angle between the fields is 180° (i.e., field inversion; see as follows). A technique for minimizing band inversion and improving the linearity of resolution in a gel is switch time ramping.[7] This refers to a progressive increase in the switch interval during the duration of a gel run. DNA fragment migration will reflect the average of the mobilities over all the different switch intervals used. In this way, resolution will be improved by employing favorable separation conditions for each portion of the size range of DNA molecules being separated for at least part of the run.

Other variables that must be adjusted in accordance with DNA size are the voltage gradient and the reorientation angle.[8] The voltage gradient is the electrical potential applied to the gel, measured in volts (V) per centimeter (cm). Gradients from 6 to 10 V/cm are suitable for molecules up to 1 Mb. If large molecules such as *Schizosaccharomyces pombe* yeast chromosomes (3.5, 4.7, and 5.7 Mb) are to be separated, gradients no higher than 2 V/cm can be used. The reorientation angle, which is the angle between the direction of the pulsed fields, may also be varied to affect resolution and the rate of separation. Angles between 105 and 165° give comparable resolution of molecules smaller than 1 Mb. The most common angle is 120°, which is suitable for all applications. Smaller reorientation angles are desired for separation of very large DNA fragments (>1000 kb) because this increases mobility.

Pulsed-Field Gel Electrophoresis Systems

A number of pulsed-field systems have been described,[6] which differ mainly in the electrode geometry and method of reorientation of the electric fields. Despite these differences, the theory and the mechanism of PFGE separation are similar among all systems. In this chapter, we discuss only asymmetric voltage field-inversion gel electrophoresis (AVFIGE)[9,10] and contour-clamped homogeneous electric field (CHEF)[11] electrophoresis systems because of their superior resolution and widespread usage. Systems such as the transverse alternating-field electrophoresis (TAFE)[12] gel box

---

[8] S. M. Clark, E. Lai, B. W. Birren, and L. Hood, *Science* **241**, 1203 (1988).

[9] M. Y. Graham, T. Otani, I. Boime, M. V. Olson, G. F. Carle, and D. D. Chaplin, *Nucleic Acids Res.* **15**, 4437 (1987).

[10] C. Turmel, E. Brassard, R. Forsyth, K. Hood, G. W. Slater, and J. Noolandi, *in* "Current Communications in Cell and Molecular Biology," Vol. 1: Electrophoresis of Large DNA Molecules: Theory and Applications (E. Lai and B. W. Birren, eds.), p. 101. Cold Spring Harbor Laboratory Press, Cold Spring Harbor, NY, 1990.

[11] G. Chu, D. Vollrath, and R. W. Davis, *Science* **234**, 1582 (1986).

[12] K. Gardiner, W. Laas, and W. Patterson, *Somatic Cell. Mol. Genet.* **12**, 185 (1986).

or the rotating electrophoresis systems[13,14] are too limited in their functions and thus are not recommended. For example, neither system is well suited for separating DNA from 1 to 100 kb.

Asymmetric voltage field-inversion gel electrophoresis involves periodically inverting a uniform electric field (180° angle of reorientation) using different voltage gradients in the forward and backward fields. For some switch time and voltage conditions, this is also referred to as zero-integrated field electrophoresis (ZIFE).[10] Schematic diagrams for constructing an AVFIGE apparatus have been published in detail[6] and commercial units are available (CHEF-MAPPER and Q-Life Autobase; Bio-Rad, Hercules, CA). We do not discuss earlier FIGE systems that use the same forward and reverse voltages (also referred to as standard FIGE) because of the severe problem of band inversion. AVFIGE provides the best possible resolution of fragments in the 1- to 50-kb range[15] and adequate resolution of molecules in the 50- to 1000-kb range.[10] However, it is too slow for separation of fragments larger than 1 Mb.

CHEF is the most versatile and widely used PFGE system. This versatility stems from its ability to separate fragments from 1 to 6000 kb without band inversions. Also, CHEF can speed up separation of large fragments by using smaller reorientation angles.[8] A great many electrophoretic data have been generated with the CHEF systems,[16] which guide selection of optimal running conditions. The other reason for the popularity of CHEF apparatuses is that excellent systems are available commercially (Bio-Rad).

Experimental Considerations

This chapter is designed to lead someone familiar with conventional electrophoresis through the steps needed to run a PFG. Because various size ranges of DNA require different voltage gradients for optimal separation, the experimental section is divided to cover three size ranges of DNA: 1- to 100-kb, 100- to 1000-kb, and 1- to 6-Mb DNA. For each size range, a preferred method and an alternate procedure are provided. Along with the procedure, recommended materials and necessary data are given. These protocols provide optimal conditions for running gels in each of the specific size ranges.

[13] P. Serwer, *Electrophoresis* **8**, 301 (1987).
[14] E. M. Southern, R. Anand, W. R. A. Brown, and D. S. Fletcher, *Nucleic Acids Res.* **15**, 5925 (1987).
[15] B. W. Birren, E. Lai, L. E. Hood, and M. Simon, *Anal. Biochem.* **177**, 282 (1989).
[16] B. W. Birren, E. Lai, S. M. Clark, L. Hood, and M. Simon, *Nucleic Acids Res.* **16**, 7563 (1988).

Selecting a Size Range

One major difference between conventional electrophoresis and PFG electrophoresis is that a DNA size range must be chosen prior to running a PFG. In conventional electrophoresis the same window of resolution (e.g., 0.2 to 20 kb) is nearly always obtained. With PFG electrophoresis there are many parameters that can be altered so as to determine the width of the window of resolution within the continuum of DNA fragment sizes. This window can be fine tuned to obtain high resolution, for example, to separate a 400-kb fragment from a 450-kb fragment. Alternately, the window can be set for lower resolution of a wider size range, as in the separation of a 100-kb fragment from a 700-kb fragment. In general, the narrower the size range of molecules being separated, the higher the resolution obtained. Extremely high resolution requires long runs; usually resolution is compromised to obtain a convenient run time. Every PFG run thus begins with the selection of the desired size range, and this choice determines what fragments will be separated as well as the resolution obtained.

Using Multiple Size Markers

The conditions presented in each section are known not to produce band inversion within the range they are designed to separate. However, it is important to use two differently sized markers on every gel to ensure that band inversion has not occurred in a region of interest.

Chemicals

To replicate the data in this chapter we recommend using low electroendosmosis (EEO) agarose such as SeaKem LE from FMC BioProducts (Rockland, ME). TAE buffer (1×) is composed of 40 mM Tris–acetate and 1 mM EDTA. The composition of TBE buffer (0.5×) is 44.5 mM Tris–borate and 12.5 mM EDTA.

Resolution of DNA Fragments from 1 to 100 kb

DNA larger than 20 kb cannot be resolved easily by conventional electrophoresis. Initially, standard FIGE (equal voltage for forward and backward switch times) was the method of choice for separating small DNA because of the simplicity of the equipment. However, severe band inversion problems limit the usefulness of standard FIGE. Asymmetric voltage FIGE (AVFIGE) is the preferred method for separating small (1–100 kb) DNA because of superior resolution. Owing to the widespread availability of CHEF boxes in laboratories now, we have also included information on separating small DNA using CHEF.

Materials

Agarose gel, 1% (w/v)
TBE buffer, 0.5×
Markers:
   1. Under 50 kb:
        8.2- to 48.5-kb range (Bio-Rad; BRL, Gaithersburg, MD)
        4.9–49 kb [5 kb ladder; New England Biolabs (NEB), Bev-
        erly, MA]
   2. Above 50 kb: 15–291 and 24–291 kb (Mid-Range I, II; NEB)

Procedures

Preferred Method: AVFIGE

1. Set the voltage gradient at 9 V/cm in the forward direction, and at
   6 V/cm in the backward direction.
2. Select a size range.
3. Choose switch times from Fig. 2.
   a. Using the smallest sized DNA fragment of interest, choose the
      initial switch time. (*Note:* Forward switch time equals backward
      switch time).
   b. Using the largest sized DNA fragment of interest, choose the final
      switch time for the forward and backward direction.

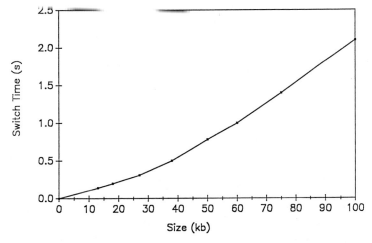

Fɪɢ. 2. Switch time selection for 1- to 100-kb DNA by AVFIGE. Maximum size of fragments
that can be resolved at the different switch times is shown. All data represent migration in
1% agarose gels run at 15° in 0.5× TBE with a 180° reorientation angle, a 9-V/cm forward
voltage gradient, and a 6-V/cm reverse voltage gradient.

TABLE I
MOBILITY OF 1- TO 100-kb DNA SEPARATED BY AVFIGE[a]

Size (kb)	Switch time					
	0.2 sec	0.5 sec	0.8 sec	1 sec	1.5 sec	2 sec
1	0.58	0.58	0.58	0.58	0.58	0.57
5	0.42	0.40	0.45	0.48	0.47	0.48
10	0.24	0.31	0.37	0.42	0.40	0.41
15	0.19	0.28	0.34	0.40	0.38	0.37
20	0.15	0.24	0.33	0.39	0.37	0.36
25		0.20	0.31	0.37	0.36	0.36
30		0.18	0.28	0.34	0.35	0.36
35		0.19	0.24	0.29	0.33	0.36
40			0.20	0.25	0.31	0.35
45			0.16	0.20	0.28	0.34
50			0.16	0.17	0.25	0.32
60				0.13	0.21	0.27
75					0.15	0.23
100						0.20

[a] The mobility (cm/hr) is indicated for DNA of a specific size range separated by a variety of switch times. All data represent migration in 1% agarose gels run at 15° in 0.5× TBE with a 180° reorientation angle and a 9-V/cm forward gradient and 6-V/cm reverse gradient.

   c. Ramp linearly between the initial and final switch times.
4. Determine run time from Table I.
   a. Find the mobility of the smallest fragment at the initial switch time. Interpolate if the size or switch time is not explicitly listed in Table I.
   b. Find the mobility of the smallest fragment at the final switch time.
   c. Average the mobilities.
   d. Determine the run time by dividing the desired distance of migration by the average mobility. The desired distance is three-quarters the length of the gel measured from the well.

Example: To Separate DNA Samples in the Range of 10–60 kb

1. Set the voltage gradient at 9 V/cm in the forward direction. In a typical horizontal gel box (30 cm between electrodes) this would be 270 V. Set the voltage gradient at 6 V/cm in the backward direction. In a 30-cm gel box this would be 180 V.
2. Choose switch times from Fig. 2.
   a. The initial switch time (for forward and backward fields) is 0.2 sec, based on 10 kb being the smallest fragment of interest.

b. The final switch time (for forward and backward fields) is 1 sec, based on 60 kb being the largest fragment of interest.

c. Set the ramping function of the switcher to linear ramp.

3. Determine run time from Table I.

  a. From Table I, a 10-kb fragment, under the influence of a 0.2-sec switch time, has a mobility of 0.24 cm/hr.

  b. From Table I, a 10-kb fragment, under the influence of a 1-sec switch time, has a mobility of 0.42 cm/hr.

  c. The average of these mobilities is 0.33 cm/hr.

  d. If the gel is 12.5 cm long, then the 10-kb band should migrate three-quarters the distance from the wells to the end of the gel, or approximately 9 cm. To determine the run time divide 9 cm by 0.33 cm/h. The run time should be 27.7 hr.

Alternative Method: CHEF

1. Set the voltage gradient at 6 V/cm.
2. Select a size range.
3. Choose switch times from Fig. 3.

  a. Using the smallest sized DNA fragment of interest, choose the initial switch time.

  b. Using the largest sized DNA fragment of interest, choose the final switch time.

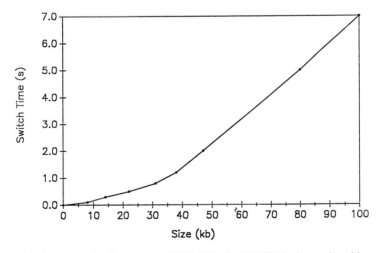

FIG. 3. Switch time selection for 1- to 100-kb DNA by CHEF. Maximum size of fragments that can be resolved at the different switch times is shown. All data represent migration in 1% agarose gels run at 15° in 0.5× TBE with a 120° reorientation angle and a 6-V/cm voltage gradient.

TABLE II
MOBILITY OF 1- TO 1000-kb DNA SEPARATED BY CHEF[a]

Size (kb)	Switch time								
	0.1–2 sec	3 sec	5 sec	15 sec	30 sec	45 sec	60 sec	75 sec	90 sec
1	1.06	1.06	1.06						
5	0.51	0.51	0.51						
10	0.38	0.38	0.38						
20	0.30	0.30	0.30						
30	0.26	0.26	0.26						
40	0.23	0.23	0.23						
50	0.20	0.22	0.25	0.32	0.30	0.30	0.30	0.31	0.31
75		0.21	0.29	0.29	0.29	0.29	0.30	0.30	
100				0.27	0.27	0.28	0.28	0.30	0.30
150				0.22	0.25	0.26	0.27	0.29	0.29
200				0.17	0.23	0.24	0.26	0.28	0.28
300					0.17	0.21	0.23	0.26	0.26
400					0.11	0.17	0.20	0.24	0.24
500						0.12	0.17	0.21	0.22
600						0.07	0.13	0.18	0.20
700							0.10	0.15	0.18
800								0.12	0.15
900								0.08	0.13
1000									0.10

[a] The mobility (cm/hr) is indicated for DNA of a specific size range subjected to a variety of switch times. All data represent migration in 1% agarose gels run at 15° in 0.5× TBE with a 120° reorientation angle and a 6-V/cm gradient.

    c. Ramp linearly from the initial to the final switch time.
4. Determine run time from Table II (see previous example).
    a. Find the mobility of the smallest fragment at the initial switch time. Interpolate if the size or switch time is not explicitly listed in Table II.
    b. Find the mobility of the smallest fragment at the final switch time.
    c. Average the mobilities.
    d. Determine the run time by dividing the desired distance of migration by the average mobility. The desired distance is three-quarters the length of the gel measured from the well.

Resolution of DNA Fragments from 100 to 1000 kb

Separation of DNA fragments from 100 to 1000 kb can best be achieved using a contour-clamped homogeneous electric field (CHEF) electrophoresis system. Such a system can separate a large number of DNA samples in straight lines with shorter run times than other systems. Standard field inversion offers good resolution in this range but band inversion can occur with this technique. Asymmetric voltage FIGE can minimize band inversion, but requires much longer run times than CHEF gels. Protocols for separation of DNA from 100 to 1000 kb by CHEF electrophoresis and by asymmetric voltage FIGE are provided.

Materials

> Agarose gel, 1% (w/v)
> TBE buffer, 0.5×
> Markers: Wild-type λ ladders (48.5 to more than 1000 kb; Bio-Rad, BRL, FMC, NEB) or *Saccharomyces cerevisiae* chromosomes (200–2000 kb; Bio-Rad, BRL, FMC, NEB)

Procedures

Preferred Method: CHEF

1. The voltage gradient to be used is 6 V/cm, and the reorientation angle is 120°.
2. Determine the size range of DNA fragments to be resolved.
3. Determine switch times, using Fig. 4.

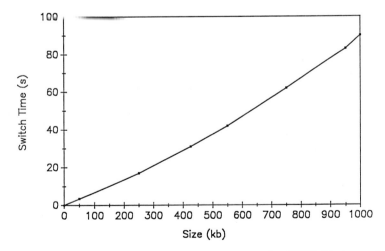

Fig. 4. Switch time selection for 100- to 1000-kb DNA by CHEF. Maximum size of fragments that can be resolved at the different switch times is shown. All data represent migration in 1% agarose gels run at 15° in 0.5× TBE with a 120° reorientation angle and a 6-V/cm voltage gradient.

a. Determine the initial switch time from Fig. 4, using the smallest fragment of interest.

b. Determine the final switch time from Fig. 4, using the largest fragment of interest.

c. Ramp linearly from the initial value to the final value.

4. Calculate the run time using Table II (see example above).

a. Determine the initial velocity of the smallest fragment (i.e., its mobility at the initial switch time) from Table II. Interpolate if the size or switch time is not explicitly listed in Table II.

b. Determine the final velocity of the smallest fragment (i.e., its mobility at the final switch time) from Table II.

c. Average these two mobility values.

d. Divide the desired migration distance (three-quarters the length of the gel measured from the well) of the smallest fragment by the average mobility value to obtain the necessary run time.

Alternative Procedure: AVFIGE

1. Use a forward voltage gradient of 2.8 V/cm and a reverse gradient of 0.8 V/cm.

2. Determine the size range of DNA fragments to be resolved.

3. Select switch times using Fig. 5.

a. Determine the initial forward switch time from Fig. 5, using the

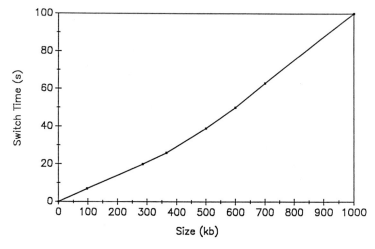

FIG. 5. Switch time selection for 100- to 1000-kb DNA by AVFIGE. Maximum size of fragments that can be resolved at the different switch times is shown. All data represent migration in 1% agarose gels run at 15° in 0.5× TBE with a 180° reorientation angle, a 2.8-V/cm forward gradient, and a 0.84-V/cm reverse voltage gradient.

TABLE III
Mobility of 100- to 1000-kb DNA Separated by AVFIGE[a]

Size (kb)	Switch time					
	10 sec	25 sec	40 sec	50 sec	70 sec	100 sec
50	0.06	0.13	0.12	0.12	0.12	0.12
100	0.04	0.11	0.11	0.11	0.11	0.11
150		0.09	0.10	0.10	0.11	0.11
200		0.08	0.09	0.09	0.10	0.10
300		0.04	0.07	0.08	0.08	0.10
400			0.05	0.06	0.07	0.09
500			0.03	0.04	0.05	0.08
600				0.03	0.04	0.07
700				0.01	0.03	0.06
800					0.01	0.06
900						0.05
1000						0.04

[a] The mobility (cm/hr) is indicated for DNA of a specific size range subjected to a variety of switch times. All data represent migration in 1% agarose gels run at 15° in 0.5× TBE with a 180° reorientation angle and a 2.8-V/cm forward gradient and 0.84-V/cm reverse gradient.

smallest fragment of this range. The initial reverse switch time is 1.4 times the forward switch time.

   b. Determine the final forward switch time from Fig. 5, using the largest fragment of this range. The final reverse switch time is 1.4 times the forward switch time.
   c. Ramp linearly from the initial switch time to the final switch time.
4. Calculate the run time using Table III (see example above).
   a. Determine the initial velocity of the smallest fragment (mobility at the initial forward switch time) from Table III. Interpolate if the size or switch time is not explicitly listed in Table III.
   b. Determine the final velocity of the smallest fragment (mobility at the final forward switch time) from Table III.
   c. Average these two mobility values.
   d. Divide the desired migration distance (three-quarters the length of the gel measured from well) of the smallest fragment by the average mobility value to obtain the necessary run time.

Resolution of DNA Fragments from 1 to 6 Mb

Separation of megabase DNA involves additional problems compared to the previously described separation of smaller DNA. Most notably, the voltage gradient must be decreased and the switch times increased. Consequently, to keep run times within reasonable time frames other parameters must be altered.

The data presented in this section were collected from gels made of SeaKem LE. To decrease gel run times lower EEO agarose, such as Bio-Rad CGA or FMC SeaKem GOLD PFG-agarose DNA, can be used. To ensure proper use of such special agaroses check with the manufacturer for decreases in run time. To reduce further the time required to separate megabase-sized DNA, the concentration of agarose can be decreased. For 1- to 3-Mb DNA, 1% gels provide acceptable separation in a reasonable amount of time while retaining the mechanical strength of the gel. However, the ease of handling a 1% gel can be sacrificed in order to separate larger DNA more quickly by using 0.6–0.7% gels.

Megabase DNA moves extremely slowly. Decreasing the reorientation angle to 106° lessens the sideways migration of DNA and thus shortens the run times. For gel boxes lacking the ability to change the reorientation angle, 1- to 6-Mb DNA can be separated using a 120° angle. The gel will run approximately 25% longer with this modification.

DNA in the megabase size range is best separated using CHEF setups. Owing to the very long run times required, AVFIGE is not recommended for separation of DNA over 1000 kb. We present two different protocols that employ different voltage gradients depending on the maximum size of the DNA to be separated.

Materials

>   Agarose, 1% (w/v)
>   TAE buffer, 1×
>   Markers: *S. cerevisiae* chromosomes (0.2–1.9 Mb; Bio-Rad, FMC, BRL, NEB)
>   *Candida albicans* chromosomes (1–3 Mb; Clontech, Palo Alto, CA)
>   *Hansenula wingei* chromosomes (1–3.3 Mb; Bio-Rad, BRL)

Procedure

CHEF Separation of 1- to 3-Mb DNA

1. Use a reorientation angle of 106° and a voltage gradient of 3 V/cm.
2. Select a size range.
3. Select switch times from Fig. 6.
   a. Choose the initial switch time based on the smallest sized DNA fragment of interest.

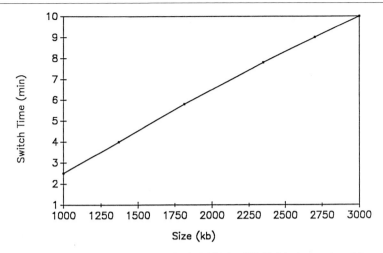

FIG. 6. Switch time selection for 1- to 3-Mb DNA by CHEF. Maximum size of fragments that can be resolved at the different switch times is shown. All data represent migration in 0.8% agarose gels run at 15° in 1× TAE with a 106° reorientation angle and a 3-V/cm voltage gradient.

    b. Choose the final switch time based on the largest sized DNA fragment of interest.
    c. Ramp linearly between the initial and the final switch time.
4. Determine the run time from Table IV (see example above).

TABLE IV
MOBILITY OF 1- TO 3-Mb DNA SEPARATED BY CHEF[a]

	Switch times				
Size (kb)	2.5 min	4 min	6 min	8 min	10 min
1000	0.06	0.11	0.14	0.14	0.15
1250		0.08	0.12	0.13	0.14
1500		0.04	0.11	0.12	0.13
1750			0.10	0.11	0.13
2000				0.10	0.12
2250				0.08	0.11
2500					0.11
2750					0.09
3000					0.09

    [a] The mobility (cm/hr) is indicated for DNA of a specific size range subjected to a variety of switch times. All data represent migration in 0.8% agarose gels run at 15° in 1× TAE with a 106° reorientation angle and a 3-V/cm gradient.

a. Find the mobility of the smallest fragment at the initial switch time. Interpolate if the size or switch time is not explicitly listed in Table IV.
b. Find the mobility of the smallest fragment at the final switch time.
c. Average the mobilities.
d. Determine the run time by dividing the desired distance of migration by the average mobility. The desired distance is three-quarters the length of the gel measured from the well.

Materials

Agarose, 0.8% (w/v)
TAE buffer, 1×
Markers: *S. pombe* chromosomes (3.5–5.7 Mb; Bio-Rad); *Hansenula wingei* chromosomes (1–3.3 Mb; Bio-Rad, BRL)

Procedure

CHEF Separation of 3- to 6-Mb DNA

1. Use a reorientation angle of 106° and a voltage gradient of 2 V/cm.
2. Select a size range.
3. Select switch times from Fig. 7.

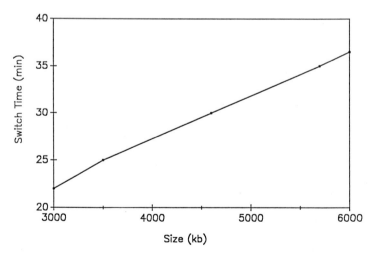

Fig. 7. Switch time selection for 3- to 6-kb DNA by CHEF. Maximum size of fragments that can be resolved at the different switch times is shown. All data represent migration in 0.8% agarose gels run at 15° in 1× TBE with a 106° reorientation angle and a 2-V/cm voltage gradient.

TABLE V
MOBILITY OF 3- TO 6-Mb DNA SEPARATED BY CHEF[a]

Size (kb)	Switch time			
	20 min	25 min	30 min	35 min
2500	0.06	0.06	0.06	0.07
3000	0.04	0.05	0.06	0.06
3500		0.05	0.05	0.05
4000			0.04	0.05
5000			0.03	0.03
6000			0.01	0.02

[a] The mobility (cm/hr) is indicated for DNA of a specific size range subjected to a variety of switch times. All data represent migration in 0.8% agarose gels run at 15° in 1× TAE with a 106° reorientation angle and a 2-V/cm gradient.

    a. Using the smallest sized DNA fragment of interest, choose the initial switch time.

    b. Using the largest sized DNA fragment of interest, choose the final switch time.

    c. Ramp linearly between the initial and final switch times.

4. Determine the run time from Table V (see example above).

    a. Find the mobility of the smallest fragment at the initial switch time. Interpolate if the size or switch time is not explicitly listed in the table.

    b. Find the mobility of the smallest fragment at the final switch time.

    c. Average the mobilities.

    d. Determine the run time by dividing the desired distance of migration by the average mobility. The desired distance is three-quarters the length of the gel measured from the well.

Future Directions

In this chapter we have summarized the current optimum electrophoretic conditions for the separation of DNA fragments from 1 to 6000 kb. However, we believe that there are many improvements to be made. For example, we still do not understand the basis for the apparent limit in sizes that can be effectively separated. Nor do we understand fully why higher voltages cannot be used to separate large DNA molecules. Most publications have concentrated on PFG separations using only two electric fields.

In theory, PFG separations involving multiple fields should be even more advantageous.

The conditions described in this chapter were determined experimentally from large numbers of gel runs. A more logical approach for the future would be the development of a comprehensive model that can correctly simulate all the observed behavior of large DNA molecules under PFGE. A number of computer models consistent with some aspects of PFG separation have been published. However, no current model can explain all the aspects of the behavior of DNA under various PFG conditions. For example, none of the models explain the dependence of separation on the reorientation angle, the voltage gradient limits for the resolution of large DNA, or the effect of using multiple electric fields.

Direct microscopic observation of individual DNA molecules in the gel has provided a better picture of how DNA molecules move in electrophoresis. In the future, real-time optimization of DNA electrophoresis might provide major improvements over existing separation procedures.

[12] Migration of DNA through Gels

By GARY W. SLATER, PASCAL MAYER, and GUY DROUIN

Introduction

Gel electrophoresis has become a major laboratory tool for separating biological macromolecules. For example, large (megabase) double-stranded DNA (dsDNA) molecules can readily be separated on agarose gels using pulsed-field gel electrophoresis (PFGE), while subkilobase single-stranded DNA (ssDNA) molecules can be sequenced on polyacrylamide gels (see [14], [16], and [17] in this volume, and [17] in Vol. 271 of this series[1]). These two techniques are essential to map and sequence the human genome. Although great technological advances have been made, the process itself, i.e., the electrophoretic migration of large flexible polyelectrolytes in pseu-

[1] N. Matsubara and S. Terabe, *Methods Enzymol.* **270,** Chap. 14, 1996 (this volume); T. Wehr, M. Zhu, and R. Rodriguez, *Methods Enzymol.* **270,** Chap. 16, 1996 (this volume); L. Křivánková, P. Gebauer, and P. Boček, *Methods Enzymol.* **270,** Chap. 17, 1996 (this volume); W. S. Hancock, A. Apfell, J. Chakel, C. Sounders, T. M'Timkulu, E. Pungor, Jr., and A. W. Guzetta, *Methods Enzymol.* **271,** 403 (1996).

dorandom networks, is surprisingly ill understood. The fact that some new technologies were suggested by theoretical studies [e.g., zero-integrated field electrophoresis (ZIFE)[1a]] indicates that much can be gained from a fundamental study of gel electrophoresis.

Agarose electrophoresis has been studied extensively since the development of PGFE. Analytical models,[2,3] detailed computer simulations,[4,5] mobility studies,[6,7] and video microscopy[8–10] have led to a good semiquantitative description of the migration of dsDNA in these very open gels (see Ref. 11 for a review).

However, polyacrylamide gel electrophoresis of ssDNA is less well characterized. Video microscopy is not yet available, while mobility measurements are somewhat unreliable, as we show below. The theories developed for agarose electrophoresis seem to apply but they show that the situation requires a somewhat different approach.[12,13]

In this chapter, we first review the different models of DNA migration through gels, and we discuss their limitations and the conditions under which they should apply. We then describe how experimental data should be obtained and analyzed in order to (1) identify the migration mechanism that is taking place, (2) estimate the parameters of the relevant model, and (3) optimize separation. As an example, we apply this approach to the separation (sequencing) of ssDNA in polyacrylamide gels. We conclude by briefly reviewing the challenges still open in the study of DNA sequencing and we briefly introduce some new ideas being explored. This should provide a useful practical guide for experimentalists who want to analyze their results in the framework of a particular theory and to use their results efficiently to enhance DNA separation.

[1a] C. Turmel, E. Brassard, R. Forsyth, K. Hood, G. W. Slater, and J. Noolandi, in "Current Communications in Molecular Biology: Electrophoresis of Large DNA Molecules" (B. Birren and E. Lai, eds.), p. 101. Cold Spring Harbor Laboratory, Cold Spring Harbor, NY, 1990.

[2] O. J. Lumpkin, P. Déjardin, and B. H. Zimm, Biopolymers 24, 1573 (1985).

[3] G. W. Slater and J. Noolandi, Biopolymers 25, 431 (1986).

[4] J. M. Deutsch and T. L. Madden, J. Chem. Phys. 90, 2476 (1989).

[5] T. A. J. Duke, J. Chem. Phys. 93, 9049 (1990).

[6] S. P. Edmonsson and D. M. Gray, Biopolymers 23, 2725 (1984).

[7] G. W. Slater, J. Rousseau, J. Noolandi, C. Turmel, and M. Lalande, Biopolymers 27, 509 (1988).

[8] D. C. Schwartz and M. Koval, Nature (London) 338, 520 (1989).

[9] S. Gurrieri, E. Rizzarelli, D. Beach, and C. Bustamante, Biochemistry 29, 3396 (1990).

[10] S. B. Smith, P. K. Aldridge, and J. B. Callis, Science 243, 203 (1989).

[11] S. D. Levene and B. H. Zimm, Q. Rev. Biophys. 25, 171 (1992).

[12] P. Mayer, G. W. Slater, and G. Drouin, Appl. Theor. Electrophoresis 3, 147 (1992).

[13] G. W. Slater and G. Drouin, Electrophoresis 13, 574 (1992).

Gel Electrophoresis Theory and DNA Separation

Nucleic acids usually migrate with size-independent mobilities in free solution[14] and the "sieving" properties of gels are necessary to separate DNAs according to their size. In the following discussion of the different migration mechanisms found in DNA gel electrophoresis, we often refer to Fig. 1, which shows a schematic diagram of mobility vs molecular size, as well as a schematic illustration of "typical" molecular conformations in the different regimes.

Choosing Variables

The natural unit of length is the mean gel pore size a. Unfortunately, the concept of a mean pore size has been frequently abused: It should be kept in mind that the mean pore size depends on the method used to measure it. The natural unit of molecular size is the size M_a of a globular (unperturbed) DNA molecule that fits exactly into a pore of size a, i.e., a molecule whose effective radius is given by

$$R(M_a) = a \tag{1}$$

The natural unit of electric field intensity is E_a, for which the drop in potential energy of a molecule of size M_a across a pore of size a is equal to the thermal energy:

$$(1/2)M_a q E_a = k_B T \tag{2}$$

where q is the net electric charge per base, k_B is the Boltzmann constant, and T is the temperature in degrees Kelvin.

The physics of gel electrophoresis is better described using the following dimensionless variables[2,3]:

$$N = M/M_a; \qquad \varepsilon = E/E_a \tag{3}$$

Therefore, N measures the number of pores (or fraction thereof) occupied by a molecule of size M. The field ε measures the relative strengths of the applied field and of the Brownian motion for a molecule of size $N = 1$.

Measuring M_a and E_a should be the initial step for any nonempirical optimization of the DNA separation process. These parameters are necessary in order to identify the mode of migration of the DNA molecules through the gel structure and to obtain the optimal separation conditions from the relevant transport equations.

---

[14] B. M. Olivera, P. Baine, and N. Davidson, *Biopolymers* **2**, 245 (1964).

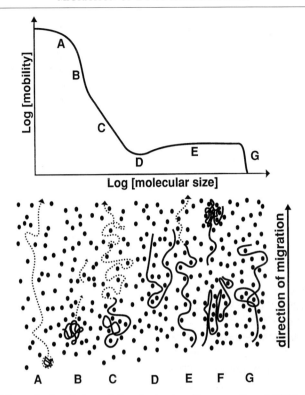

FIG. 1. Schematic description of the electrophoretic properties of DNA in gels. *Top:* Schematic log–log plot of DNA mobility vs molecular size. *Bottom:* Schematic drawings showing molecular conformations characterizing the various migration regimes. These regimes may not all be present for a given choice of experimental conditions. (A) Ogston sieving: The random coil DNA is sieved by the gel. (B) Entropic trapping: The random coil DNA jumps between the larger pores. (C) Near-equilibrium reptation: The random coil DNA migrates head-first through the gel. (D) Reptation trapping: The DNA molecule becomes trapped in U-shaped conformations and is slowed by the tug-of-war between the two arms. (E) Oriented reptation: The DNA is aligned in the direction of the electric field and reptates head-first. (F) Geometration: For high field intensities, migration may be characterized by hernias and "bunching" instabilities. (G) Complete trapping: Very large DNA molecules do not migrate through gels, possibly because they form knots around gel fibers.

Ogston Regime: N < 1

For "small" molecules, the gel is an open structure that retards migration (Fig. 1A). The so-called Ogston model is based on four assumptions: (1) the molecule has a rigid spherical shape; (2) the field intensity is low enough that the molecule can easily backtrack if it is engaged into a dead end; (3) the mobility is proportional to the fraction of the total gel volume that is actually available to a sphere of radius $R(N)$. The problem is then to calculate this fraction for a given gel architecture; (4) Ogston's theory simply assumes that the gel is a random array of fibers.[15] This gel structure naturally leads to a Poisson distribution and the mobility is given by

$$\mu = \mu_0 e^{(-\pi/4)\{[R(N)+r]/a_{\text{Ogs}}\}^2} \tag{4}$$

where $a_{\text{Ogs}} \propto 1/C^{1/2}$ is the Ogston mean pore size, C is the gel concentration, r is the radius of the gel fibers, and μ_0 is the mobility in free solution. Equation (4), which can be used to obtain μ_0, $a_{\text{Ogs}}(C)$, and r, has been frequently criticized; however, it is often used in situations where some of its basic assumptions are violated.

The most obvious violation comes from the fact that gels are not simple random arrays of fibers. Indeed, a_{Ogs} does not seem to vary as $C^{-1/2}$ as expected for a truly random gel, even when $\log[\mu]$ is found to decrease linearly with gel concentration C. Published values[7,16] indeed give $a_{\text{Ogs}} \propto C^{-y}$, with $y = 0.5$ to 1.0. Moreover, agarose gel fibers orient in the presence of an electric field, which affects the migration of DNA molecules.[17]

Another limitation comes from the possible DNA–gel interactions, because the latter may invalidate the assumption that the mobility is limited by volume exclusion only.

A third weakness is that a DNA molecule is neither rigid nor spherical. For example, a random-walk conformation is only spherical or globular "on the average": in reality, instantaneous conformations are anisotropic and may be very elongated. It is thus quite remarkable that the Ogston model works so well.

Two effects limit the validity of this model to low fields ($N\varepsilon < 1$). First, high fields may deform the molecule (and the gel). Second, even for rigid spheres, the Ogston model does not apply in the high field limit because it assumes that the objects can backtrack easily. For instance, strong trapping in dead ends occurs when $N\varepsilon > 1$, and the model fails.[18]

The Ogston model is a useful concept when its assumptions are valid. However, many aspects remain ill understood. For instance, the diffusion constant (band broadening) has not been studied either theoretically or experimentally. High field intensities, realistic gel architectures, and the flexibility of the DNA molecules also need to be considered by theoreticians.

[15] A. G. Ogston, *Trans. Faraday Soc.* **54,** 1754 (1958).
[16] N. C. Stellwagen and D. N. Holmes, *Electrophoresis* **11,** 649 (1990).
[17] N. C. Stellwagen, *in* "Pulsed-Field Gel Electrophoresis" (M. Burmeister and L. Ulanovsky, eds.), p. 285. Humana Press, Totowa, NJ, 1992.
[18] G. A. Griess and P. Serwer, *Biopolymers* **29,** 1863 (1990).

The Ogston regime is found both for agarose and polyacrylamide gel electrophoresis of DNA, typically for molecules smaller than ~2–3 kb and ~50–300 bases, respectively. In principle, DNA separations can be optimized, in this regime, by increasing the field intensity in order to reduce the time duration and hence the diffusional broadening of the bands (eventually, the initial bandwidth becomes the limiting factor). However, note again that non-Ogston molecular deformation and trapping may occur at large field intensities, and that this may limit the resolution.

Entropic Trapping: $N \approx 1$, $\varepsilon \ll 1$

Computer simulations[19] have suggested that the motion of a flexible polymer may be severely hindered in random structures. This "entropic" trapping occurs because molecules spend most of their time in the larger pores and must therefore fight strong entropic forces to cross the narrow passages joining these large and rare voids (Fig. 1B). This effect has been reported by two groups[12,20] and the mobility was found to scale like

$$\mu \propto 1/N^{\nu}; \qquad \nu \sim 1.5\text{--}2.5 \qquad (5)$$

It should be noted that the mean pore size a_{Ogs} of the Ogston theory cannot be used here because the molecules may deform and select preferentially the larger pores.

This regime is potentially useful because the mobility exhibits a strong dependence on molecular size. Unfortunately, band broadening is also much increased and, in practice, this regime actually defines the lowest electric field intensity that can be used for sequencing.

Reptation: $N > 1$, $\varepsilon < 1$

Both preceding models predict negligible mobilities when $N \gg 1$. The fact that this is not the case suggests a new mechanism, usually called *biased reptation*,[2,3,11] which assumes that the DNA moves by creeping like a snake. This seems reasonable because the molecule is so big that it must be divided among many gel pores, which collectively define a "tube" into which the motion occurs.

The biased reptation model is the only model that is simple enough to allow for efficient computer simulations and analytical results even in the most complicated situations. Its predictions are remarkably good (see Ref. 21), and it provides a simple framework within which to interpret experi-

[19] M. Muthukumar, *J. Non-crystalline Solids* **131**(3), 654 (1991).
[20] E. Arvanitidou and D. Hoagland, *Phys. Rev. Lett.* **67**, 1464 (1991).
[21] J. L. Viovy and A. D. Défontaines, *in* "Pulsed-Field Gel Electrophoresis" (M. Burmeister and L. Ulanovsky, eds.), p. 403. Humana Press, Totowa, NJ, 1992.

mental data. Its basic assumptions are (1) $N > 1$ and (2) $\varepsilon < 1$, as indicated previously. Unfortunately, this model is frequently criticized in cases in which these assumptions are violated.

In this model, the instantaneous mobility is related to the projection of the end-to-end vector of the molecule on the field axis and is thus strongly affected by molecular deformations.[2,3,21] The ends of the DNA molecule are entirely responsible for the selection of the new pores that define the tunnel in which the molecule migrates. Because these ends are charged, they are oriented by the electric field, and the tube they define aligns in the field direction; in other words, the macromolecular conformation is no longer random. In fact, the molecular conformations and the electrophoretic transport process affect each other in a self-consistent way; this makes the problem difficult to treat, and the reptation model is indeed the only approach that allows for an analytical model. The parameter $N\varepsilon^2$ plays an important role in the theory, e.g., it serves as a parameter for the series expansions [see Eqs. (6) and (7)].[2,3] At least three reptation regimes have been identified.

Near-Equilibrium Reptation: $N > 1$, $N\varepsilon^2 < 1$, $\varepsilon < 1$

When the electric forces are weak, i.e., when $N\varepsilon^2 < 1$, the model predicts[22,23]

$$\mu(N) = \mu_0 \left(\frac{1}{3N} + \frac{\varepsilon^2}{27} + \frac{2N\varepsilon^4}{1215} + \cdots \right) \tag{6}$$

Vanishing electric fields ($\varepsilon \rightarrow 0$) do not modify the (random-walk) conformation of a reptating molecule (hence the name "near-equilibrium reptation"; see Fig. 1C), and there is an approximate inverse relationship between mobility and size.[6,7] Note that the small (leading) correction factor ($\varepsilon^2/27$), which is due to molecular orientation, contains a 27 and not a 9, as suggested by earlier theories.[2] In principle, the transition between the Ogston and reptation regimes must be found for $N \approx 1$. Equations (4) and (6) do indeed give $\mu/\mu_0 \approx 1/3$ at this point [assuming $r \ll a_{\mathrm{Ogs}} = R(N = 1)$], which offers a consistent definition for the transition regime[24] (note that entropic trapping may affect this transition). Finally, we stress the fact that the mean pore size a_{rep0}, which can be obtained from the $1/N$ term in Eq. (6), is not equal to a_{Ogs} because Ogston sieving and reptation do not "measure" the same pore size.

[22] P. Déjardin, *Phys. Rev.* **A40**, 4752 (1989).
[23] G. W. Slater, *J. Phys. II (Paris)* **2**, 1149 (1992).
[24] G. W. Slater and J. Noolandi, *Biopolymers* **28**, 1781 (1989).

This regime is generally found when "long" DNA molecules are successfully separated under constant electric field conditions. It can be shown that the lower the field intensity the longer the reptating molecules that can be separated on a gel of a given length. This is now well known for dsDNA separation in agarose gels, but should also apply for DNA-sequencing conditions, as we show in the section, Optimizing Separation of DNA-Sequencing Ladders. Note that the rate of increase of interband spacing, $dV(N)/dN \approx -\mu_0 E/3N^{-2}$, is a linear function of the field intensity in this regime. This means that the distance between adjacent bands will remain constant if the product of the field intensity and the duration of the separation (i.e., the total distance migrated) are held constant. However, the number of well-separated bands contained on a gel of a given length will decrease with increasing field intensity.

Oriented Reptation: $N > 1$, $N\varepsilon^2 \gg 1$, $\varepsilon < 1$

When the electric forces are large enough, molecular orientation dominates the thermal forces (see Fig. 1E) and we obtain[22,23]

$$\mu = \mu_0 \left(\frac{\varepsilon^2}{9} - \frac{1}{3N} - \frac{2}{N^2\varepsilon^2} + \cdots \right) \qquad (7)$$

The theory thus predicts that the mobility becomes essentially molecular size independent for $N \gg 3/\varepsilon^2$, a well-known experimental limitation of gel electrophoresis. Three remarks must be made: (1) this is a series expansion and saturation is in fact predicted when $\varepsilon \to \infty$ (with $\mu \approx \mu_0$); (2) the coefficient of the (leading) ε^2 term is now 1/9; and (3) the sign of the $1/3N$ term is now negative. Care must therefore be taken when the mobility is studied as a function of field intensity: The predicted dependence is not a simple ε^2 if the conditions are such that data points corresponding to both Eqs. (6) and (7) are used (to our knowledge, this point has been completely overlooked until now). Note also that $\varepsilon < 1$ is required to avoid saturation and to remain in the reptation regime. Moreover, when a mean pore size a_ε is estimated from the experimental value of ε^2, this value should not be expected to agree with either a_{Ogs} or a_{rep0} because ε^2 is proportional to a^6 (assuming that $M_a \propto a^2$).

This regime corresponds to the well-known mobility plateau found for molecules larger than ~20–50 kb in the case of dsDNA agarose gel electrophoresis, and for molecules larger than ~2–5 kb in the case of ssDNA polyacrylamide gel electrophoresis.

Reptation with Self-Trapping: $N\varepsilon^2 \approx 14$, $\varepsilon < 1$

Equations (6) and (7) predict that the mobility should be minimum for an intermediate molecular size. This surprising prediction has been confirmed in agarose[25] and polyacrylamide[26] gels and is probably one of the greatest successes of the biased reptation model. Computer simulations indicate that the minimum should be found for a molecular size $N_{min} = 14/\varepsilon^2$ and that it is due to self-trapping in metastable U-shaped conformations[25] (Fig. 1D).

This transition regime, found between the near-equilibrium reptation and the oriented reptation regimes described previously, effectively defines the explicit limit of resolution of gel electrophoresis. The mobility minimum can easily be seen on standard sequencing gels as a dark smear in the range of ~1.5–3 kb. This phenomenon also limits the value of the highest electric field intensity one can use for sequencing, especially in capillary gel electrophoresis where the high fields can make this regime appear for molecules as small as ~300–500 bases.

Reptation and Diffusion: ε < 1

The reptation model has been used to predict the behavior of the diffusion constant during gel electrophoresis[27]; the analytical and simulation results provide the following scaling laws:

$$D(N, E) \propto E^0 N^{-2}, \qquad N < \varepsilon^{-2/3} \tag{8a}$$
$$D(N, E) \propto E^1 N^{-1/2}, \qquad \varepsilon^{-2/3} < N < \varepsilon^{-2} \tag{8b}$$
$$D(N, E) \propto E^2 N^0, \qquad N > \varepsilon^{-2} \tag{8c}$$
$$D(N, E) = \text{maximum}, \qquad N = 28/\varepsilon^2 = 2N_{min} \tag{8d}$$

The results of the first theory of electrophoretic diffusion clearly show that, except for extremely small molecular sizes [Eq. (8a)], the diffusion constant is not related to the mobility by an Einstein relation. Therefore, the latter assumption should not be used to study band broadening. The theory predicts much increased diffusion in the presence of a field and a fairly weak molecular size dependence. An interesting local maximum in diffusion is also predicted for $N = 2N_{min}$; this maximum can clearly be seen in the experimental results presented in Ref. 25.

Band broadening, which is generally (but not always) of little importance for dsDNA agarose gel electrophoresis, is a major source of problems for DNA sequencing. The results of the biased reptation model (BRM) clearly show that increasing the electric field intensity does not always produce sharper bands. This is counterintuitive and different from what happens during the separation of smaller molecules migrating in the Ogston regime.

[25] J. Noolandi, J. Rousseau, G. W. Slater, C. Turmel, and M. Lalande, *Phys. Rev. Lett.* **58,** 2428 (1987).

[26] É. Brassard, C. Turmel, and J. Noolandi, *Electrophoresis* **13,** 529 (1992).

[27] G. W. Slater, *Electrophoresis* **14,** 1 (1993).

According to the biased reptation model, the easiest way to increase the size of the largest molecule that can be sequenced on a gel of a given length might be to reduce the field intensity and increase the time duration. However, the field intensity must remain strong enough to overcome entropic trapping, and hence there is in fact an optimal low field intensity.

High Fields: $\varepsilon > 1$

The most remarkable failure of the reptation model is found for field-inversion gel electrophoresis (FIGE),[28] for which it predicts no effect (unless the fields are of different amplitudes, in which case it is again satisfactory[1a,21]). This is not unexpected because the electric forces are then large enough to deform the molecule at the level of each pore, which violates the assumptions of the model. Computer simulations[29] and video microscopy[8-10] have shown that the molecule tends to form hernias (loops of DNA trying to reptate on their own between gel fibers) and often collapses because of the instability of the reptation tube (Fig. 1F): This new mode has been called *geometration*.[4,29] As suggested in Ref. 21, the reptation tube concept is still valid, but the tube is clearly not linear anymore (it leaks!). Computer simulations give results that agree with FIGE.[5,29-31]

It is interesting to remark that the reptation model is adequate for many PFGE systems.[21,32,33] This may indicate that hernias can be accounted for by using renormalized reptation parameters.[21] In fact, the reptation model captures the essential feature of the migration (motion in tubes) as long as the pulse durations do not match the time scales associated with the formation and growth of hernias and bunching instabilities.

This regime corresponds to fields of a few volts per centimeter in agarose gels and is more apparent for large (>50 kbp) dsDNA molecules. In the case of denaturing polyacrylamide gel electrophoresis, high fields probably correspond to $E > 70$ V/cm and are thus quite relevant for capillary electrophoresis.

Trapping

Finally, very large DNA molecules seem to have major migration problems in gels.[34,35] For example, it is well known that intact human chromo-

[28] G. F. Carle, M. Frank, and M. V. Olson, *Science* **232**, 65 (1986).
[29] J. M. Deutsch, *J. Chem. Phys.* **90**, 7436 (1989).
[30] B. H. Zimm, *J. Chem. Phys.* **94**, 2187 (1991).
[31] T. A. J. Duke, *Phys. Rev. Lett.* **62**, 2877 (1989).
[32] J. L. Viovy, *Electrophoresis* **10**, 429 (1989).
[33] G. W. Slater and J. Noolandi, *Electrophoresis* **10**, 413 (1989).
[34] C. Turmel, É. Brassard, G. W. Slater, and J. Noolandi, *Nucleic Acids Res.* **18**, 569 (1990).
[35] J. L. Viovy, F. Miomandre, M. C. Miquel, F. Caron, and F. Sor, *Electrophoresis* **13**, 1 (1992).

somes do not migrate into gels. High-frequency modulations of the electric field have been shown to help[34]; reducing the field intensity also improves the situation. Although the physics of this effect is not yet understood, it is believed that DNA knots (see Fig. 1G) might be involved.[35] Because these effects are found for $N > 1000$–10,000, it is difficult to study them using computer simulations, although a Monte Carlo algorithm may allow such studies.[36]

Pulsed-Field Gel Electrophoresis

A detailed analysis of PFGE is beyond the scope of this chapter. However, we would like to stress the following points. A fundamental difference exists between the crossed field electrophoresis methods [e.g., contour-clamped homogeneous electric field (CHEF) electrophoresis and orthogonal field alternating gel electrophoresis (OFAGE)] and the one-dimensional pulsed-field methods (FIGE and ZIFE). In the first case, enhanced separation is obtained when we use geometric constraints to affect the migration path of the molecules. As long as the molecules are large enough to reptate (i.e., $N > 1$), crossed fields will have a notable effect on their mobility, even in concentrated gels (small M_a and large E_a) and low field intensities (small ε) where tube leakage and bunching instabilities are probably less important. The BRM then accounts for most of the observed mobilities. The situation is quite different for one-dimensional pulsed-field methods that rely on the asymmetry of the molecular conformations one obtains when using high field intensities and low concentration gels (large M_a and small E_a, i.e., small N and large ε). When the electric field changes in intensity and/or direction, this asymmetry changes with strongly molecular weight-dependent characteristic times.[21] Note that although these effects can also amplify crossed field separations, they are not an essential part of their action.

Field inversion seems to be unproductive for DNA-sequencing conditions. Part of the reason is that E_a is large in polyacrylamide gels (also, N is rather small). Such one-dimensional pulsed-field methods might require low concentration gels and high electric field intensities, which would favor asymmetric conformations. Such conditions are typical of capillary electrophoresis.

Experimental Methods

The basic methods involved in the preparation of agarose and polyacrylamide gels are well known, and detailed protocols can be found in molecular

[36] T. A. J. Duke and J. L. Viovy, *Phys. Rev. Lett.* **68,** 542 (1992).

biology experimental textbooks (e.g., Ref. 37). Our goal in this section is to point out some specific details that may affect the reliability of the data for quantitative analysis and reduce the resolution of high-performance separations. We also present a protocol to label a single strand of a 100-bp DNA ladder. Using such a ladder ensures that all the fragments being analyzed have an identical base composition. This is important because ssDNA fragments of different base compositions can have different mobilities even in denaturing polyacrylamide gels.

Preparation of Agarose Gels

1. The apparatus must be sealed and the electrodes (plates or wires) should be centered horizontally and vertically with respect to the gel. The buffer must be circulated with a pump through an external reservoir and through a heat exchanger immersed in a thermostat in order to reduce "buffer aging" and to keep the temperature constant.
2. Gelation must be carefully controlled to obtain uniform gels. We suggest the following method: mix the agarose powder with the appropriate buffer, weigh, and add 2 to 5 ml of water. Boil on a Bunsen burner until the solution is optically clear and evaporation has brought the solution back to its initial weight. Allow the agarose solution and the gel mold (with its comb) to equilibrate at 50° for 1 hr. Finally, gelation should take place in a closed box at a controlled temperature of 20–30°. The gel should stay immersed in its buffer for 24 hr before it is used. A plenum is not required.
3. Depending on the design of the system, the electric field in the gel is generally different from the ratio of the applied voltage to the distance between electrodes, and there is an electrode potential drop, e.g., 2.7 V for platinum electrodes, independent of the applied voltage. The effective electric field in the gel and its homogeneity along the gel must be measured with a voltmeter.
4. One should always use a constant voltage in order to compare with theory.
5. Because ethidium bromide affects the mobility of DNA, it should be added after electrophoresis is completed.
6. The properties of photosensitive films and optical scanners may affect the shape of the bands and thus the measurement of diffusion constants. The procedure must therefore be carefully calibrated.

---

[37] J. Sambrook, E. F. Fritsch, and T. Maniatis, "Molecular Cloning: A Laboratory Manual." Cold Spring Harbor Laboratory Press, Cold Spring Harbor, NY, 1989.

Preparation of Single-Stranded DNA Ladder

The ends of the double-stranded 100-bp DNA ladder sold by Pharmacia LKB Biotechnologies (Piscataway, NJ) have different 5′ extensions. One end has a 5′ TCGG 3′ extension, whereas the other has a 5′ CCGA 3′ extension. Therefore, one can label a specific strand. We usually perform the following single-strand labeling.

1. Mix 10 pmol of DNA ladder and 10 pmol of [α-^{32}P]deoxythymidine triphosphate (NEG-005H; New England Nuclear, Boston, MA) in a final volume of 10 μl, which is 40 mM Tris-HCl (pH 7.5), 20 mM MgCl$_2$, and 50 mM NaCl.
2. Add 0.25 μl (2.5 units) of DNA polymerase I (Promega, Madison, WI).
3. Incubate at room temperature (20°) for 15 min.
4. Add an equal volume of loading buffer made of 95% (v/v) formamide, 1% (w/v) xylene cyanol, 1% (w/v) bromphenol blue, and 10 mM EDTA.
5. Boil for 5 min before loading.

Preparation of Polyacrylamide Gels

1. Thermostatic gel plates and 0.1- to 0.4-mm thick gels should be used for good heat exchange and uniform gel temperatures.
2. As the polymerization is chemically initiated, little control over the homogeneity of the gel is possible. We suggest equilibrating the plates and the acrylamide solution at room temperature prior to molding, and calibrating the amount of ammonium persulfate and N,N,N',N'-tetramethylethylenediamine (TEMED) for the polymerization to take about 20 min.
3. The gel must be prerun at the voltage and temperature used for the actual run, and until the current becomes constant. The required prerun duration is inversely proportional to the applied voltage. For a 6% T, 5% C, 0.5× Tris–borate–EDTA buffer (TBE), 8 M urea gel that is 0.2 mm thick and 53 cm long, it takes 8000 hr/V to reach a constant current.[12] The current variations follow the changes in the ionic strength of the gel buffer. Because the conformations of the DNA molecules are affected by the ionic strength, any change in the value of the current during the run will make the instantaneous velocities time dependent.
4. Again, one should carry out the electrophoresis at constant voltage in order to compare with theory.
5. Add water to keep constant the buffer level and ionic strength in the tanks.

6. There is a linear relationship between the real field intensity and the ratio between the applied voltage and the gel length (not the distance between the electrodes, as in agarose gels, because here the deep buffer tanks are much more conductive than the thin gel).

7. Automated sequencers use fluorescence detection where the bands are integrated over a certain scan width. If they are irregular, tilted, or "smiling," measured positions may be inaccurate and bandwidths overestimated.

8. Radioactive labeling with ^{35}S is to be preferred over ^{32}P labeling because it gives less deformed images of the actual DNA bands. The properties of X-ray films and optical scanners make quantitative bandwidth measurements questionable. The procedure must therefore be carefully calibrated.

Measuring Velocities

In agarose gels, the velocity can be taken as the ratio between the migrated distance and the migration time.[38] This relation may not be valid in denaturing polyacrylamide gels, in which velocity gradients at the beginning and at the end of the gels must be taken into account.[12]

With manual sequencers, make multiple loadings (done at regular intervals) and measure the distances x_i for different migration times t_i. The resulting $x(t)$ vs t diagram shows that the velocity is a function of x, but that it is uniform in the center of the gel, where the field gradients are absent. These gradients are present in the first few centimeters of gel near the electrodes. Because the field gradients only affect the velocity over a few centimeters, it is easy to obtain reliable velocity data by ensuring that all loadings have migrated passed the initial gradient. Given a set of (t_i, x_i) pairs, differential velocities $V(N, E)$ can be given by

$$V(M, E) = \mu(M, E)E = (x_j - x_i)/(t_j - t_i) \qquad (9)$$

If one measures $n \geq 3$ time–position pairs for each value of molecular size M and field intensity E, $n(n - 1)/2$, such differential velocities are obtained. The mean value and the standard deviation provide reliable estimates of the velocity and its scattering. Alternatively, one can fit the set of (x, t) data points to a straight line (where appropriate) with software that provides the uncertainties on the fitting parameters.

With automated sequencers (and capillaries) one should measure passage times at three or more different detector positions in order to be certain that reliable velocities are obtained. Note that measuring final migra-

[38] N. C. Stellwagen, *Biochemistry* **22**, 6180 (1983).

tion times for different gel lengths would not necessarily solve the gradient problems.

Measuring Diffusion Constant

When the electrophoretic bands are approximately Gaussian, they can be analyzed using the mathematical form

$$I(x, t) = (I/[\sqrt{\pi}\, w(t)]) e^{-[(x - t\mu E)/w(t)]^2} \tag{10}$$

where x is the position of the detector, I is the total area under the curve, and $w(t)$ is the dispersion

$$w^2(t) = w_0^2 + 2Dt \tag{11}$$

with D the diffusion constant and w_0 the initial width of the band. Because w_0 is equal to the product of the loading width and the ratio $\mu(N, E)/\mu_0$, it is expected to be a function of molecular size and field intensity.

Dispersion $w(t)$ is often measured using the bandwidth at half-height. Because the latter involves an extra $\log[w(t)]$ term, it might be more appropriate to measure the peak area I and the maximum peak height I_p, which give directly

$$(1/\pi)(I/I_p)^2 = w^2(t) = w_0^2 + 2Dt \tag{12}$$

The diffusion constant is then obtained by measuring I and I_p for different migration times, and by taking the derivative of $w^2(t)$ vs time, i.e., by using the differential method described earlier. Note that the differential method automatically subtracts the widths due to the loading process and to the initial field gradient.

Data Analysis

The data obtained with the methods previously described must be analyzed differently in the various migration regimes already mentioned (the traditional semilog plot is not enough).

Ogston Sieving Regime

The analysis relies on the Ferguson plot giving log[mobility at zero field] vs gel concentration C. This should lead to straight lines with slopes related to the molecular size M and to the diameter of the gel fibers, r [Eq. (4)]. The free solution mobility μ_0 is obtained by extrapolating the lines to zero gel concentration. In polyacrylamide gels, these values should also be extrapolated to zero cross-linker concentration.[39]

[39] D. L. Holmes and N. C. Stellwagen, *Electrophoresis* **12**, 612 (1991).

There is no simple relationship between $\log[\mu]$ and molecular size M because the relevant size is in fact the "effective" radius $R(M)$. A possible choice for $R(M)$ is the radius of gyration:

$$R_g^2 = \frac{1}{3} pL \left[1 - \frac{3p}{L} + \frac{6p^2}{L^2} - \frac{6p^3}{L^3} (1 - e^{-L/p}) \right] \tag{13}$$

where p is the persistence length (half the length of chain necessary to assure that the orientations of its ends become uncorrelated) and L ($= 3.4$ Å $\times M$ for dsDNA) is the contour length[40] of the chain. In principle, Eq. (4) could be used to fit the mobility data in order to estimate parameters p and L; however, the always-present experimental errors make this approach dubious. Identifying the Ogston sieving regime should thus rely on the Ferguson plot.

Entropic Trapping Regime

This regime is best displayed on log–log plots of mobility μ vs molecular size M. It should lead to straight lines with slopes $\nu < -1$ [Eq. (5)]. The near-equilibrium reptation regime gives a slope of -1, while the oriented reptation and Ogston regimes both lead to curved lines with local slopes > -1 (see Fig. 1).

Reptation Regimes

The reptation model suggests fitting the experimental data with the following mathematical form:

$$\mu(M, E) = (A/M) + B(E) \tag{14}$$

where $A = \mu_0 \times M_a/3$. The model actually predicts that $B(E) \propto E^2$ in a limited range of field intensities.

Log–log plots and μ vs $1/M$ plots only give indirect indications of reptation. The best way to verify the mathematical form suggested by Eq. (14) is to plot $\mu E \times M$ vs M: On this "reptation plot," the reptation regime is characterized by a straight line with a positive slope $E \times B(E)$ that measures the degree of orientation of the molecules. The extrapolation at zero length gives $AE = \mu_0 E \times M_a/3$. The Ogston regime then appears for small molecular sizes as a curve with a different positive slope independent of the field intensity. The entropic trapping regime appears for intermediate molecular sizes as a curve with a negative slope. Note that if reptation self-trapping leads to band inversion, the expected straight line will show a kink, i.e., a change in slope.

[40] H. Benoit and P. Doty, *J. Phys. Chem.* **51**, 924 (1969).

Plotting $B(E)$ versus E on a log–log plot allows us to follow the increase in molecular orientation as field intensity is increased. The reptation model predicts that $B(E) \propto E^2$ for low field intensities such that $\varepsilon < 1$ (saturation is expected for $\varepsilon > 1$). The value of $B(E)$ at $E = 1$ (in the field units used, e.g., V/cm) gives $\mu_0/(27 \times E_a^2)$ if band inversion and the plateau mobility have not been reached ($N\varepsilon^2 < 14$), from which we can obtain E_a. In principle, the field E at which the saturation appears must satisfy $\varepsilon = E/E_a \approx 1$, which offers a second (but much less precise) and independent way to estimate E_a.

Diffusion Constant

Assuming that diffusion is the leading cause of band broadening, Eq. (8) provides a theoretical framework within which to analyze experimental data. For example, a log–log plot of the diffusion constant D vs molecular size M should lead to straight lines with slope -2, $-1/2$, or 0. Similarly, a log–log plot of D vs field intensity E should give straight lines with slope 0, 1, or 2.

Optimizing Separation of DNA-Sequencing Ladders

The theoretical approach to data analysis presented in this chapter suggests that the optimal separation conditions will be found for an electric field intensity lying somewhere between the field intensities where entropic trapping and molecular orientation are first observed. Hence, we suggest the following systematic method to determine these two field intensities and the optimal conditions for sequencing.

1. Measure the velocity V of the DNA fragments of the 100-base ladder, using the differential method described in this chapter.
2. Draw a reptation plot (e.g., $V \times M/E$ vs M) and fit the data to a straight line. Extrapolate the line to $M = 0$.
3. Repeat steps 1 and 2 until you have a collection of plots with slopes $B(E)$ [see Eq. (14) for the definition] ranging from 0 (horizontal lines, typical of reptation without orientation) to a constant (maximum) value (which indicates the onset of the high field regime). If the maximum slope is nearly 0, assume $E_a = +\infty$ and skip steps 5 and 6.
4. For each reptation plot, measure the extrapolated value of $V \times M/E$ at $M = 0$. Calculate the mean of these values and divide the result by $\mu_0/3$ to obtain the experimental value of the parameter M_a.
5. Draw a log–log plot of $B(E)/\mu_0$ vs E. Determine the range of field intensities for which the slope of this log–log plot is about 2. Draw

a straight line through these points and extrapolate it to $E = 1$ V/cm to calculate $E_a = [\mu_0/27B(1 \text{ V/cm})]^{1/2}$.

6. Calculate E_{max}, the highest electric field intensity that should be used for the separation of fragments of size smaller than M_{max}: $E_{max} \approx E_a(M_a/M_{max})^{1/2}$. Note that this value is defined to within a factor of about 2.

7. Next, find E_{min}, the lowest electric field intensity that can be used. Measure the velocity of the DNA fragments of the 100-base ladder, using the differential method described in this chapter, and draw a log–log plot of V vs M. Repeat this process with lower field intensities until the data exhibit a slope less than -1. This indicates the onset of the entropic trapping regime where band broadening becomes a major problem.

8. Make a normal sequencing ladder with a known sequence.

9. The optimal sequencing field intensity is to be found between the values of E_{max} and E_{min} obtained in the previous steps. To find this optimal field, separate the sequencing ladder, using field intensities chosen systematically between these extrema, until the field intensity that allows the maximum number of bases for this sequence to be read is found. A strategy that leads to a good survey of this range of field intensities should be used. The durations $t(M, E)$ of the separations should allow the $M = 200$-base molecule to migrate to the bottom of the gels, with $t(M, E)$ given approximately by

$$t(M, E) = \frac{\text{gel length}}{\mu_0 \times E \times \left[\dfrac{M_a}{3M} + \dfrac{1}{27}\left(\dfrac{E}{E_a}\right)^2\right]} \tag{13}$$

If E_a was not measured in step 5, increase the field intensity by increments of 5 V/cm starting from E_{min}. If E_{min} could not be found, start at the lowest electric field intensity used in step 7.

Practical Example: Polyacrylamide Gels

Gel Edge Effects

The inhomogeneities of the system are easily studied when new sample loadings are made periodically (Fig. 2A). Molecules go faster at the beginning of the gel, and their steady state trajectories (i.e., in the middle of the gel) do not extrapolate to the same starting position.[12]

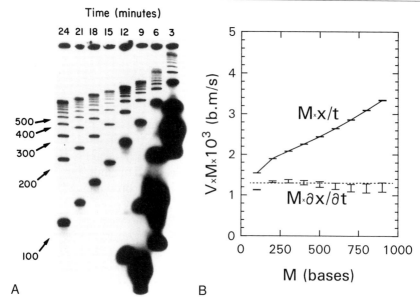

FIG. 2. (A) Autoradiogram of a 100-base ssDNA ladder electrophoresed on a denaturing polyacrylamide gel. The gel was made of 6% T:5% C polyacrylamide, 8 M urea, 0.5× TBE, and was 53 cm long and 0.2 mm thick. The prerun lasted for 2 hr at 50° and 4000 V, and the run itself was performed at the same voltage. The samples were loaded every 3 min as indicated. (B) Velocity V times molecular size M vs M for 100- to 1000-base ssDNA molecules. The gel was as described in (A), but prerun for 16 hr at 1000 V and run at 1000 V; the loadings were made at 40-min time intervals, The upper line shows the migrated distance $x(t)$ divided by the electrophoresis duration t, whereas the lower line shows the derivative of migrated distance against time, $dx(t)/dt$. [Adapted, with permission, from *Appl. Theor. Electrophoresis*, **3**, 147 (1992).]

Differential Velocities

The gel edge effects may easily hide the true migration mechanism. Figure 2B shows that x/t and $dx(t)/dt$ give dramatically different "velocities." In this example, the differential method shows that molecular orientation is negligible while the classic method wrongly suggests strong molecular orientation. Different setups (gel length, manual or automated sequencers, slab or capillary systems) may lead to different results for the same experimental conditions (DNA length, gel concentration, buffer, and temperature) when such gel edge effects exist.

Ogston Regime

Figure 3 shows a Ferguson plot obtained using very small molecular sizes and the differential method described previously. The extrapolated

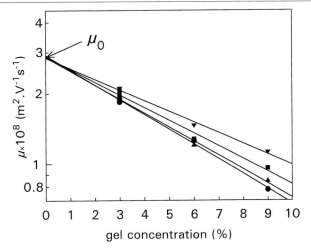

FIG. 3. Ferguson plot of mobility μ for single-stranded oligonucleotides (▼, 17 bases; ■, 25 bases; ▲, 32 bases; ●, 38 bases). The gels were as described in Fig. 2, with various acrylamide concentrations. They were prerun for 4 hr at 2000 V and run at 2000 V for up to 2 hr, with loadings made every 20 min. Mobilities were obtained from the derivative of the migrated distance vs time. The extrapolated value of μ_0 is 2.9×10^{-8} cm^2 V^{-1} sec^{-1}.

value of μ_0 gives 2.9×10^{-8} m^2 sec^{-1} V^{-1}, which is comparable to the value obtained for double-stranded DNA.[39,41]

Entropic Trapping Regime

Figure 4 shows a slope of -1.6 for a field intensity of 9.4 V/cm. In practice, this regime is not easy to observe for three reasons. First, because large electric forces kill entropic trapping, only low field intensities will show this effect. Second, the bands are found to broaden extensively (P. Mayer, unpublished), which reduces the number of bands that can be seen on a gel. Finally, because the molecules must migrate passed the zone where edge effects exist, long electrophoresis durations are necessary and again few bands can be observed.

Reptation Regime

Figure 5 shows $V \times M$ vs M "reptation plots." The mathematical form given by Eq. (14) is verified for a broad range of field intensities. Extrapolation at zero molecular size gives a mean value of $\frac{1}{3}\mu_0 M_a \approx 7 \times 10^{-7}$ bases m^2 V^{-1} sec^{-1}, independent of E (data not shown). If we use the value of

[41] N. C. Stellwagen, *Biopolymers* **24**, 2243 (1985).

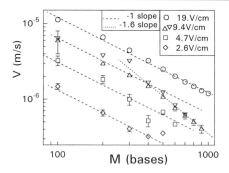

FIG. 4. Log–log plot of the velocity V of 100- to 1000-base ssDNA molecules vs molecular size M for different field intensities. The gels were as described in Fig. 2 but were 0.4 mm thick (except for ∇, the 9.4-V/cm experiment, in which the gel was 0.2 mm thick), and prerun at the experimental voltage V for 150/V · hr. Loadings were made in each case every 750/V · min. The longest fragment migrated at least 5 cm along the gel. Velocity data were obtained from the derivative of migrated distance vs time. [Adapted, with permission, from *Appl. Theor. Electrophoresis*, **3**, 147 (1992).]

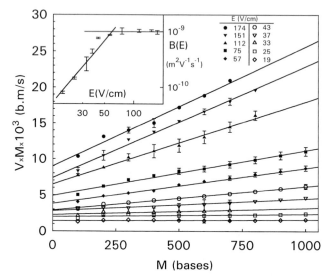

FIG. 5. Velocity $V = \mu E$ times molecular size M vs M for 100- to 1000-base ssDNA molecules and different field intensities E. The gels were as described in Fig. 2, except for the 174-V/cm experiment, in which the gel was 0.4 mm thick. Other conditions were as described in Fig. 4. The inset is a log–log plot of $B(E)$, i.e., the slope of the straight lines here divided by the field intensity, vs field intensity E. The lines have slopes 2 and 0. [Adapted, with permission, from *Appl. Theor. Electrophoresis*, **3**, 147 (1992).]

μ_0 obtained from the previous Ferguson plot, the latter result indicates that there is an average of $M_a = 72$ ssDNA bases per gel pore. The slopes of these straight lines lead to the $B(E)$ vs E plot shown in the inset (Fig. 5): $B(E)$ scales as E^2 for low field intensities, in agreement with the prediction of the reptation model. Using the previous value of μ_0 and the value of $B(E)$ extrapolated at $E = 1$ V/cm, we find that E_a is approximately 70 V/cm. Hence, the saturation of $B(E)$ at higher fields begins roughly when the scaled field ε approaches 1 ($E \approx 50$–70 V/cm in Fig. 5, which corresponds to $\varepsilon = E/E_a = 0.7$–1.0), i.e., when nonreptation modes appear (the tube starts leaking).

Sequence-Dependent Separations

In nondenaturing polyacrylamide gels, or in small-pore agarose gels, the electrophoretic mobility may depend on the secondary structure (ssDNA and RNA) or the local stiffness (bent DNA) of the chain. Parameters M_a and E_a may also depend slightly on the sequence of the DNA molecules. Note, however, that most of the different regimes seen above should be recovered. For example, the persistence length of dsDNA is known (about 50 nm or 150 bases under normal electrophoretic conditions), and its large value implies that in tight gels (e.g., 6% polyacrylamide) hernias and bunching instabilities should not be expected. The relevant migration regimes are thus the Ogston regime for molecules of size smaller than M_a and the reptation regime(s) for longer molecules.

Agarose versus Polyacrylamide Gel Electrophoresis

Our results indicate that DNA is much less oriented under normal sequencing conditions than during normal agarose gel electrophoresis. For instance, the scaled electric field intensity ε is smaller than unity up to about 70 V/cm in a 6% polyacrylamide gel, while it is on the order of unity for fields as low as 1–3 V/cm in agarose gels. Moreover, the number of reptation segments is $N = M/M_a = M/72$ in our example (typically 1–20), while it is 20–10,000 in the case of agarose gel electrophoresis (and PFGE) of DNA restriction fragments. These differences may explain why pulsed fields are less efficient when applied to ssDNA sequencing.[13,42]

Agarose gels are used to separate 10–20 dsDNA molecules whose molecular sizes differ by many kilobases, while polyacrylamide gels are used to separate hundreds of ssDNA molecules differing by a single base. While band broadening is not normally a problem in the former case, it clearly

---

[42] E. Lai, N. A. Davi, and L. E. Hood, *Electrophoresis* **10**, 65 (1986).

becomes the limiting factor in the latter even when there is little molecular orientation.[13]

Gel Characteristics and DNA Separation

For electrophoresis, the most important characteristic of a gel (providing it is suitable for DNA separations) is the pore size distribution. Different varieties (preparation, cross-linker content, or even chemical substitutes) of agarose or polyacrylamide will in general exhibit different values of M_a and E_a because these parameters are related to different moments of the pore size distribution. However, two different gels characterized by the same values of M_a and E_a should lead to the same separation, except eventually for the entropic trapping regime, where the details of the distribution of pore sizes are expected to be more important. Gels characterized by large values of both M_a and E_a would greatly enhance resolution. Unfortunately, comparison between different gels (and buffer compositions) is always made in qualitative terms rather than by measuring the values of M_a and E_a.

Conclusion

Although the resolution power of agarose gel electrophoresis has improved enormously in the last decade, some problems still remain, e.g., the complete trapping of >10-Mbp molecules.[34,35] It is clear that theory and analytical studies of the process will play an important role in solving such problems. For example, theoretical studies have previously suggested approaches such as ZIFE,[1a] while megabase trapping might be tractable by new Monte Carlo algorithms.[36]

Most theories for the electrophoresis of ssDNA in denaturing polyacrylamide gels have been borrowed from the twin science of agarose electrophoresis of dsDNA. However, these theories are often abused. Moreover, we have shown that strong gradient effects exist in these systems and that this interferes with our ability to obtain reliable data.

The first step toward the optimization of sequencing would in fact be a good empirical study of the electrophoretic process. In particular, we now need reliable velocity and diffusion data as well as a knowledge of the flexibility (persistence length) of ssDNA in denaturing gels. We stress the fact that there is no reason to believe that diffusion and mobility are related by an Einstein relation because electrophoretic migration is a highly nonequilibrium process.[27] For these reasons, it is clear that the current sequencing techniques are far from being optimized.[13]

New ideas are being explored. For example, field gradients can affect band spacing and width, e.g., by leading to "self-focusing" bands.[43] It was also suggested that neutral molecular ends could kill the reptation molecular orientation (much like pulsed fields do in agarose gels) and increase the number of bases read per gel.[23]

Gels are still poorly characterized. A better knowledge of gel structure may lead to better predictions of the migration properties of DNA and better control over the parameters that limit resolution. Although new chemical gels are being proposed, solutions of linear neutral polymers seem to be more promising as retarding media because changing the concentration of the solution, the length, and/or the flexibility of the polymers directly affect the mean entanglement spacing of the medium in a predictable way.[44]

It has been shown that high fields and high gel concentrations can be used simultaneously in order to reduce diffusion and molecular orientation,[45,46] and micron-size nonoverlapping bands were observed in minutes (under a microscope).[46]

Finally, a new process, called *trapping electrophoresis*,[47] is being tested. A large neutral object (the streptavidin protein), whose radius matches the mean pore size of the gel, is attached to the end of each DNA molecule. Because this duo is often trapped in passages too narrow for the steptavidin to pass, it must move backward (against the molecular size-dependent electric forces) for detrapping to occur, and the velocity decreases very rapidly with size M.

In conclusion, the next great challenge is to understand diffusion, because the technology is now at the point where the slightest band-broadening process limits resolution.[48,49] New technological breakthroughs are likely to arise from methods that can control band broadening. Theoretical and experimental studies of electrophoretic migration and diffusion are thus needed. It is hoped that this chapter serves as a useful practical guide to obtaining and analyzing the relevant data.

Acknowledgments

This work was supported by a Strategic Grant from the National Science and Engineering Research Council (NSERC) of Canada to G.D. and G.W.S. The authors thank Drs. Paul Grossman, David Hoagland, and Jean Louis Viovy for useful discussions.

[43] G. W. Slater and J. Noolandi, *Electrophoresis* **9**, 643 (1988).
[44] P. D. Grossman and D. S. Soane, *J. Chromatogr.* **559**, 257 (1991).
[45] G. W. Slater, P. Mayer, and G. Drouin, *Electrophoresis* **14**, 961 (1993).
[46] M. J. Heller and R. H. Tullis, *Electrophoresis* **13**, 512 (1992).
[47] L. Ulanovsky, G. Drouin, and W. Gilbert, *Nature (London)* **343**, 190 (1990).
[48] J. A. Luckey, T. B. Norris, and L. M. Smith, *J. Phys. Chem.* **97**, 3067 (1993).
[49] G. W. Slater, P. Mayer, and G. Drouin, *Analusis* **21**, M25 (1993).

[13] Capillary Electrophoretic Separation in Open and Coated Tubes with Special Reference to Proteins

By Stellan Hjertén

I. Introduction

There are several methods designed for free zone electrophoresis, i.e., zone electrophoresis in a carrier-free medium. This chapter describes in detail only those in which the electrophoresis chamber is a capillary with an inside diameter of less than 0.1 mm.

The applied sample zone and the electrophoretically migrating zones most often have a density differing from that of the surrounding buffer. The density difference gives rise to sedimentation (or flotation) of the solutes. This resolution-suppressing movement is called *convection*. Solutions of proteins often have a density higher than those of low molecular weight compounds and, therefore, exhibit greater convection (electrophoresis experiments have been performed in space, because convection is eliminated at zero gravity). The most common way to suppress convection is to perform the electrophoresis in a medium containing narrow channels, because sedimentation decreases as the diameter of the channel decreases. Such media include gels of agarose, cellulose acetate strips, and powders of cellulose or plastics.[1,2] When the size of the channels approaches that of the solutes (proteins) a molecular sieving effect is obtained, for example, with gels of starch and polyacrylamide. In many cases, however, the presence of an anticonvection medium is disturbing due to adsorption of solutes. Therefore, zone electrophoresis in the absence of a stabilizing agent is a preferred technique and is treated herein, particularly from a practical point of view.

On the basis of the previous discussions, in 1967, the author suggested carrier-free zone electrophoresis in narrow-bore tubes,[3] although no practical experiments were described, mainly because highly sensitive ultraviolet (UV) detectors were not available. The experiments were performed in 1- to 3-mm wide quartz tubes (the relatively long optical path length facilitated detection of the solutes). Stabilization against convection was achieved

---

[1] C. J. O. R. and P. Morris, "Separation Methods in Biochemistry," 2nd Ed. Pitman Publishing, London, 1976.

[2] S. Hjertén, *in* "Topics in Bioelectrochemistry and Bioenergetics" (G. Milazzo, ed.), Vol. 2, p. 89. John Wiley & Sons, New York, 1978.

[3] S. Hjertén, *Chromatogr. Rev.* **9,** 122 (1967).

by rotating the electrophoresis capillary slowly around its long axis. The problems encountered in the studies of electrophoresis in these capillaries[3] were the same as those encountered when the diameter of the capillary was reduced to 0.025–0.2 mm.[4–6] For instance, the following topics were treated both theoretically and experimentally and should be of interest to users of capillary electrophoresis: suppression of electroendosmosis and adsorption by coating the inside of the capillary with a nonionic, hydrophilic polymer; thermal zone broadening; indirect detection; tracing of the detecting light beam traversing the capillary; suppression of thermal and electroendosmotic zone broadening by a hydrodynamic counterflow; and asymmetry of a peak caused by a difference in the electrical field strength between the zone and the surrounding buffer.

II. Some Definitions, Fundamental Equations, and Determination of Mobilities

A voltage V (volts, V) applied over a tube generates a current I (amperes, A) determined by Ohm's Law:

$$V = RI \tag{1}$$

where R (ohms, Ω), the resistance of the medium in the tube, can be expressed in terms of κ, the conductivity of the electrophoresis medium (Ω^{-1} cm^{-1}); L, the length of the tube (cm); and q, the cross-sectional area of the tube (cm^2)·

$$R = L/\kappa q \tag{2}$$

The field strength, E, defined by the expression

$$E = V/L \quad (\text{V cm}^{-1}) \tag{3}$$

can, by a combination of Eqs. (1)–(3), be written in the alternative form

$$E = I/q\kappa \tag{4}$$

The mobility μ of a solute is defined by the equation

$$\mu = v/E \tag{5}$$

i.e., the mobility is the migration velocity at unit field strength. Smoluchowski has derived the following relationship for free electrophoresis of macromolecules:

[4] R. Virtanen, *Acta Polytech. Scand. Chem.* **123**, 1 (1974).
[5] F. E. P. Mikkers, F. M. Everaerts, and T. P. E. M. Verheggen, *J. Chromatogr.* **169**, 11 (1979).
[6] J. W. Jorgenson and K. D. Lukacs, *J. Chromatogr.* **218**, 209 (1981).

$$\mu = \varepsilon\zeta/4\pi\eta \qquad\qquad (6)$$

where ε is the dielectric constant and ζ is the zeta potential of the solute, i.e., the potential at the slipping plane, a plane inside which the water molecules form a layer assumed to adhere to the surface of the solute and therefore move with the same speed as the solute. This is a hypothetical plane, because in practice there is a smooth transition from freely moving water molecules to those firmly attached to the solute.

It should be emphasized that Eq. (6) does not contain any parameter related to the size of the solute. Accordingly, all solutes with the same ζ potential should migrate at the same speed, irrespective of the size. This has been verified experimentally. For instance, monomers, dimers, trimers, tetramers, etc., of albumin give a single peak on electrophoresis in free solution (but several peaks in size-sieving gels). Other examples are that collodion particles coated with albumin have the same mobility as have free albumin molecules and that no proteins can be separated in a sodium dodecyl sulfate (SDS)-containing buffer, except those proteins that have an unusual structure, i.e., are exceptionally rich in carbohydrates or are strongly basic or acidic and therefore bind SDS differently than do "normal" proteins (1.4 g of SDS per gram protein). One should also observe that the ζ potential in Eq. (6) refers to the potential at the slipping plane of the protein molecule. Therefore, intact virus particles have the same mobility as have virus particles depleted of nucleic acid, which is located within the particle. Nucleic acids are free-draining polymers, i.e., all parts of the molecule are exposed to the electrophoresis buffer. Consequently, all nucleic acid molecules have the same ζ potential, irrespective of length, and cannot be separated in free solution according to size.[3,7] The same is true for many carbohydrates. Therefore this chapter deals only with capillary free zone electrophoresis of proteins.

The mobility of a solute can be calculated from Eq. (5) following an experimental determination of v, the migration velocity (l/t, where l is the electrophoretic migration distance from injection to detector and t is the migration time), and E, the field strength. E can be calculated simply from Eq. (3). However, as emphasized below, it is safer to utilize Eq. (4). The conductivity κ and the current I can be determined easily with a conductometer and an ammeter, and q, the cross-sectional area, by weighing the capillary empty and filled with a heavy liquid of known density. It should be emphasized that the mobility obtained is the mobility at the temperature at which the conductivity is measured—not that at the temperature of the

[7] H. A. Abramson and L. S. Moyer, "Electrophoresis of Proteins." Reinhold Publishing, New York, 1942.

buffer in the capillary.[3] The conversion of the mobility from one temperature t_1 to another temperature t_2 can be made by means of the relation

$$\mu_{t_1}/\mu_{t_2} = \eta_{t_2}/\eta_{t_1} \tag{7}$$

as is evident from Eq. (6). The viscosities η_{t_1} and η_{t_2} can, for normal buffers,
be put equal to those of water.[8] It is emphasized that the mobility, and not
the migration time (although often employed), should be used to characterize a solute, because the mobility under a particular set of conditions is an
inherent property of the solute, just as is the molecular weight.

The following approach for the calculation of mobilities is an attractive
alternative, based on the use of the expression

$$\mu = lq\kappa/tI \tag{8}$$

which is obtained by a combination of Eqs. (5) and (4) and $v = l/t$. Using
Eq. (2), Ohm's law, [Eq. (1)] can be transformed to

$$V = (L/q\kappa)I \tag{9}$$

The capillary is filled with the buffer used in the electrophoresis experiment
and I is read for different values of V. From the slope $L/q\kappa$ of the straight
line obtained the product $q\kappa$ can be calculated (because L is known) and
is inserted into Eq. (8) to calculate the mobility μ of the solute. In this
preexperiment one should not use such high voltages that the temperature
rises in the capillary and causes an increase in κ and thereby a deviation
from the straight line.

III. Different Techniques for Injection of Sample

A. Electrophoretic Injection

The sample solution, into which an electrode is immersed, is brought
into contact with one end of the capillary. When a voltage is applied over
the capillary the sample migrates into it. The length, ΔX_0, of the starting
zone can be calculated from the expression

$$\Delta X_0 = vt = \mu Et \tag{10}$$

where t is the duration of the application (seconds). Equation (10) is valid
under the assumption that the sample is equilibrated with the electrophoresis buffer (see VII,B) and has a conductivity virtually equal to that of
the buffer.

---

[8] R. C. Weast (ed.), "Handbook of Chemistry and Physics," 68th Ed., Table F-39. CRC Press,
Boca Raton, FL, 1987–1988.

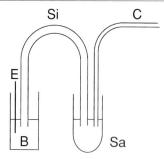

Fɪɢ. 1. Electrophoretic injection of the sample. The sample, Sa, is separated from the electrode via a siphon, Si, filled with sample or a gel. E, Electrode; B, buffer; C, electrophoresis capillary. The same arrangement can with advantage be used for electrophoresis in low-conductivity buffers (the siphon and the sample vial are then filled with buffer; see Section VI).

In this common injection technique the electrode often is in direct contact with the sample, which is not to be recommended, because the electrode reactions can both denature the sample and cause changes in the pH and composition of the buffer. To eliminate these disadvantages, and thereby increase the reproducibility in peak area and migration time, the electrode and sample should be separated by a siphon, as illustrated in Fig. 1. The siphon can be a piece of fused silica tubing filled with sample or with a gel of agarose prepared in the buffer used for the solubilization of the sample (the latter alternative is preferable when the sample volume is limited). A somewhat different approach, wherein the sample is in the form of a small droplet, is described in Ref. 9.

Equation (10) shows that the width of the starting zone of any particular component in the sample is directly proportional to the mobility μ of the component, which means that it may be difficult to detect slowly migrating components when they are present at very low concentrations. It should be emphasized, however, that the peak area is nevertheless proportional to the concentration of the component in the original sample. For instance, assume that the sample contains two components, A and B, at the same concentrations and with the same absorption coefficients and that the mobility of B is one-third that of A (see also Section V,C). The width of the starting zone of B is then also one-third that of A, and so are the amounts of B and A injected. Neglecting the diffusion-induced zone broadening the heights of the two peaks will be the same, as will be the widths, because component B migrates threefold more slowly than does component A, i.e.,

[9] S. Hjertén, in "HPLC of Proteins, Peptides and Polynucleotides" (M. T. W. Hearn, ed.), p. 737. VCH Publishers, New York, 1991.

the peak areas will be the same. It should be emphasized that with pressure injection of the sample the peak areas should be divided by the migration time to become proportional to the concentration of the components in the applied sample.[10]

B. Pressure Injection

When a pressure P expressed in dynes per centimeter squared (1 dyn/cm^2 = 10^{-6} atm = 15×10^{-6} psi; 1 mm H_2O/cm^2 = $0.1g \approx 100$ dyn/cm^2, where g is the gravitational acceleration) is applied over a capillary with diameter ϕ (centimeters) and length L (centimeters) for t sec, the width of the starting zone, ΔX_0 (centimeters), is determined by

$$\Delta X_0 = P\phi^2 t / 32\eta L \qquad (11)$$

where η is the viscosity in poise (0.01 for water[8] at 20°).

C. Thermal Application

This method can be employed if the capillary is thermostatted by a flowing coolant.[11] On lowering the temperature of the coolant the buffer in the capillary contracts and the sample is sucked in.

$$\Delta X_0 = \alpha l \Delta T \qquad (12)$$

where α is a constant equal to 1.8×10^{-4} for aqueous solutions, l is the length of the cooled section of the capillary, and ΔT is the temperature difference. Equation (12) is valid when one end of the capillary is closed. When both ends are open, the width of the starting zone is about one-half of that calculated by Eq. (12).

IV. Coating to Suppress Adsorption and Electroendosmosis

A. Need for Coating and Effect of Electroendosmosis in Open and Closed Tubes

In most high-performance capillary electrophoresis (HPCE) experiments the capillaries are made of fused silica. They are commercially available from several companies, for instance, MicroQuartz (Munich, Germany), Polymicro Technologies (Phoenix, AZ), and Scientific Glass Engineering Pty, Ltd. (Victoria, Australia).

[10] S. Hjertén, K. Elenbring, F. Kilár, J.-L. Liao, A. Chen, C. Siebert, and M.-D. Zhu, *J. Chromatogr.* **403,** 471 (1987).
[11] S. Hjertén, *Electrophoresis* **11,** 665 (1990).

From chromatographic experiments on silica beads it is well known that the surface silanol groups interact strongly with the solutes, especially proteins, even after end capping. This interaction is not only of electrostatic origin due to the negatively charged silanol groups, as is evident from the fact that many proteins adsorbed to a silica support cannot be completely released at either low or high pH or at high salt concentrations. From these considerations one can draw the conclusion that it is seldom possible to avoid adsorption of all proteins to the fused silica tubing used in capillary electrophoresis by adjusting the pH and the ionic strength of the buffer. Furthermore, these latter parameters should be selected freely to attain maximum resolution. For these reasons, relatively few successful CE analyses of proteins in naked fused silica tubing have been reported. In most of these experiments the silica tubing had been washed with 0.1 M sodium hydroxide for 1–2 min followed by water and fresh buffer prior to each run to release adsorbed proteins. However, it is possible to suppress adsorption of a protein by coating the inner wall of the fused silica tubing with a hydrophilic layer with a charge opposite to that of the protein. A more generally applicable coating, useful for both acidic and basic proteins, is one consisting of a cross-linked hydrophilic nonionic polymer. The methyl cellulose coating described as follows has the advantage of permitting washing of the capillary between runs at extreme pH values and in the presence of SDS, which is of importance in order to release any adsorbed protein (a modified procedure has recently been described[11a]).

Because this coating consists of a highly viscous polymer layer, the electroendosmosis is also suppressed efficiently because the electroendosmotic flow, which is generated at the negatively charged surface of the silica tubing, is inversely proportional to the viscosity in the double layer at the tubing wall.[3] All discussions throughout this chapter and all experiments described herein refer to capillaries coated with nonionic polymers in order to eliminate or strongly suppress adsorption and electroendosmosis. Other methods for surface modifications have also been developed, including (1) addition of nonionic polymers to the buffer to provide a neutral dynamic coating or of cationic compounds or polymers to decrease the negative charge of the capillary wall or make it positive, and (2) electrostatic or hydrophobic attachment of polymers or detergent micelles to the capillary wall. Some of these methods were introduced not only to eliminate adsorption but also to adjust the electroendosmotic flow to a desired value. It is far beyond the scope of this chapter to discuss all of these methods and therefore only that based on a neutral polymer coating is treated, especially because we have experience only with this kind of coating and we know

---

[11a] J.-L. Liao, J. Abramson, and S. Hjerten, *J. Cap. Elec.* **2,** 191 (1995).

that it works satisfactorily.[12] However, other coatings have their own merits. Some of them are designed particularly for certain types of analyses. For different types of coatings, see Refs. 12–28.

Coated fused silica tubing is commercially available from several companies, for instance, Applied Biosystems (Foster City, CA), Beckman Instruments (Fullerton, CA), Bio-Rad Laboratories (Hercules, CA), and Scientific Resources, Inc. (Eatentown, NJ). From the information sheets of these companies one can get some idea of the types of experiments for which their capillaries are intended.

Because all discussions herein refer to electroendosmosis-free capillaries, the conclusions drawn are the same for open tubes and for tubes closed with a gel or a membrane. These capillaries can with advantage be closed to avoid zone broadening caused by hydrodynamic flow due to a difference in buffer level between the electrode vessels. However, in a closed tube exhibiting electroendosmosis the zones will become deformed parabolically because a hydrodynamic counterflow is generated, which is equivalent to a loss in resolution.[3]

B. Coating Procedure

The coating to be described permits washing of the capillary between runs at pH 12 in the presence of SDS and at pH 1. This harsh treatment of the coating serves to recondition it by desorbing any adsorbed proteins. It should be stressed that this coating procedure or any other (for instance, that in Ref. 15, which gives a somewhat less pH-stable and probably a little less hydrophilic coating) does not always work satisfactorily with all batches

[12] S. Hjertén and K. Kubo, *Electrophoresis* **14**, 390 (1993).

[13] S. Hjertén, *Arkiv Kemi* **13**, 151 (1958).

[14] J. Jorgenson and K. D. Lukacs, *Science* **222**, 266 (1983).

[15] S. Hjertén, *J. Chromatogr.* **347**, 191 (1985).

[16] R. M. McCormick, *Anal. Chem.* **60**, 2322 (1988).

[17] G. Bruin, J.-P. Chang, R. Kuhlman, K. Zegers, J. Kraak, and H. Poppe, *J. Chromatogr.* **471**, 429 (1989).

[18] J. K. Towers and F. E. Regnier, *J. Chromatogr.* **516**, 69 (1990).

[19] S. A. Swedberg, *Anal. Biochem.* **185**, 51 (1990).

[20] K. A. Cobb, V. Dolnik, and M. Novotny, *Anal. Chem.* **62**, 2478 (1990).

[21] W. Nashabeh and Z. E. Rassi, *J. Chromatogr.* **559**, 367 (1991).

[22] M. Huang, W. P. Vorkink, and M. L. Lee, *J. Microcol. Sep.* **4**, 233 (1992).

[23] J. K. Towns, J. Bao, and F. Regnier, *J. Chromatogr.* **599**, 227 (1992).

[24] M. A. Strege and A. L. Lagu, *J. Chromatogr.* **630**, 337 (1993).

[25] X.-W. Yao and F. Regnier, *J. Chromatogr.* **632**, 185 (1993).

[26] X.-W. Yao, D. Wu, and F. Regnier, *J. Chromatogr.* **663**, 21 (1993).

[27] J. T. Smith and Z. E. Rassi, *Electrophoresis* **14**, 396 (1993).

[28] D. Schmalzing, C. A. Piggee, F. Foret, E. Carrilho, and B. L. Karger, *J. Chromatogr.* **A652**, 149 (1993).

of fused silica tubing, and sometimes not even with pieces from the same batch.

1. Synthesis of Allyl Methyl Cellulose. The synthesis is based on the reaction at high pH between an OH group in the methyl cellulose and an epoxide group in allyl glycidyl ether.[12] Sodium borohydride (2 g) is dissolved in 100 ml of a freshly prepared 2.5 M solution of sodium hydroxide. With stirring, this solution is added to 100 ml of a 2% (w/v) aqueous solution of methyl cellulose with the designation "viscosity 7000 cps" (Dow Chemical Company, Midland, MI). Some other types of methyl cellulose tested did not give a satisfactory coating. Allyl glycidyl ether (20 ml) is not added until the viscous polymer solution becomes homogeneous, which requires stirring for about 10 min. With continuous stirring, the reaction between the epoxy group in the allyl glycidyl ether and the OH groups in the methyl cellulose should be allowed to proceed for about 20 hr. For removal of salts and low molecular weight compounds, the solution is dialyzed repeatedly against distilled water until the pH has dropped to that of distilled water, which is below pH 7 due to uptake of carbon dioxide. The dialysis time is about 2 days. The allyl methyl cellulose solution is finally freeze-dried and the dry material seems to be stable indefinitely (probably for years). The coating procedure described in Sections IV,B,2 and IV,B,3 is similar to that presented in Ref. 15.

2. Activation of Inner Wall of Electrophoresis Tube with Methacryl Groups. Following washing with water the fused silica tubing is treated with 0.1 M sodium hydroxide for 5 min, flushed with water, 0.1 M hydrochloric acid, and finally with water. The tubing is washed with acetone for 5 min. A 1:1 mixture of γ-methacryloxypropyltrimethoxysilane and acetone is then sucked into the tubing and allowed to react for 15 hr before flushing the tubing with acetone, and finally with water.

3. Covalent Binding of Allyl Methyl Cellulose to Methacryl-Activated Inner Wall of Fused Silica Tubing. The allyl groups in the allyl methyl cellulose are allowed to react with the immobilized methacryl groups in the presence of a catalyst system, as described in detail in the following paragraphs.

The allyl methyl cellulose, synthesized as described above, is dissolved in 0.5 ml of 0.05 M potassium phosphate, pH 7.4, to a concentration of 0.7% (w/v). A 7.5-μl volume of a fresh 10% (w/v) aqueous solution of ammonium persulfate is added. The solution is deaerated, mixed with 7.5 μl of a 5% (v/v) aqueous solution of N,N,N',N'-tetramethylethylenediamine, and pressed into the methacryl-activated fused silica tubing with the aid of a syringe, the piston of which should have a small diameter to generate a relatively high pressure for a small force on the piston. Following polymerization for 17 hr the tubing is connected to a high-performance liquid chromatography (HPLC) pump to displace the highly viscous polymer phase with 0.005 M sodium hydroxide containing 5% (w/v) SDS. After flushing with water, the tubing is filled with water and placed in an ultrasonic bath for 15–30 min to remove loosely, noncovalently attached allyl methyl cellulose. The coated tubing is stored in water. The concentration of allyl methyl cellulose should be lower than 0.7%. It is difficult to displace it from the tubing following polymerization.

After each run the coated tubing is washed sequentially with distilled water, 0.01–0.05 M sodium hydroxide supplemented with 5% (w/v) SDS (30 sec), distilled water, 0.1–1 M hydrochloric acid, and water.

V. Factors Affecting Reproducibility of Peak Areas and Migration Times

A. Differences in Buffer Levels in Anode and Cathode Vessels during Run

A level difference of h cm of buffer gives rise to a flow in the capillary. The linear average velocity of the flow, $\bar{\nu}$ (centimeters per second), is determined by

$$\bar{\nu} = \rho g h R^2/8\eta L \approx 1000 h R^2/8\eta L \tag{13}$$

where ρ is the density of the buffer.

This flow velocity can be in the same direction or opposite to that of the electrophoretic velocity, i.e., decrease or increase the true migration time (for $h = 0$). The true and apparent migration times are defined by the following expressions:

$$t_{\text{true}} = l/\nu_{\text{true}} \tag{14}$$
$$t_{\text{app}} = l/(\nu_{\text{true}} \pm \bar{\nu}) \tag{15}$$

Accordingly,

$$t_{\text{true}} = t_{\text{app}}[1 \pm (\bar{\nu}/\nu_{\text{true}})] \tag{16}$$

The percent error in the determination of migration time can be calculated from the formula

$$\frac{t_{\text{true}} - t_{\text{app}}}{t_{\text{true}}} \cdot 100 \approx \frac{t_{\text{true}} - t_{\text{app}}}{t_{\text{app}}} \cdot 100 = \frac{\bar{\nu}}{\nu_{\text{true}}} \cdot 100 \approx \frac{125 h R^2}{\eta L \nu_{\text{true}}} \cdot 100 \, (\%) \tag{17}$$

according to Eq. (13).

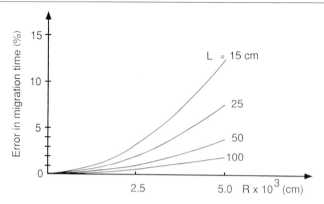

Fig. 2. The error in migration time, caused by a level difference in the electrode vessels of 1 mm, as a function of the radius R of the capillary. L, Length of the capillary.

A plot of this error against the radius of the capillary is presented in Fig. 2 for four different lengths (L) of the capillary when the difference h between the buffer levels is 1 mm—a value that is difficult to come below in practice. In an automated CE apparatus the height difference can be relatively constant in consecutive runs. The reproducibility in apparent migration time can thus be satisfactory (often <0.5% rsd), although the error in the true values of migration times, and thereby mobilities, can be considerable, as Fig. 2 indicates. To come below an error of 1% in the determination of the true values of the migration times one should, according to Fig. 2, choose a diameter of 25 μm for 15- to 25-cm long capillaries. A capillary length of 50–100 cm permits a diameter of 50 μm.

The influence of differences in buffer levels on plate heights is treated in Ref. 29.

B. Temperature

Combination of Eqs. (3), (5), and (6) gives

$$\nu = \frac{\varepsilon\zeta}{4\pi\eta} \cdot \frac{V}{L} \tag{18}$$

For high day-to-day reproducibility in migration velocity ν it is, accordingly, important that not only the voltage V over the capillary but also the temperature of the buffer in the capillary be constant from run to run, because the viscosity η of water changes by 2.7% per degree Celsius. However, if constant current is used instead of constant voltage the migration velocity

[29] E. Grushka, J. Chromatogr. **559,** 81 (1991).

is almost independent of variations in temperature.[3] This statement can easily be proved by combining Eqs. (4)–(6):

$$\nu = \frac{\varepsilon \zeta}{4\pi\eta} \cdot \frac{I}{q\kappa} \tag{19}$$

Because η decreases (increases) by 2.7% per degree Celsius and κ increases (decreases) at the same rate the product is independent of temperature and so is ν, according to Eq. (19), if I is kept constant (the product $\varepsilon\zeta$ is also nearly independent of temperature[3]). Consequently, more reproducible migration velocities and migration times are obtained despite variations in temperature in the capillary if the current is kept constant during the run. This conclusion should be kept in mind because the end sections of the capillary are seldom cooled efficiently.

The width of the starting zone on electrophoretic application is determined by Eq. (10), which, with the aid of Eq. (19), can be written as $\Delta X_0 = \varepsilon\zeta/4\pi\eta(I/q\kappa)t$. Using the same argument as in the discussion about how to obtain reproducible migration velocities one can conclude that the sample should be applied at constant current to avoid thermal variations in the width of the starting zone and thereby in peak area. It is important that the current used for the electrophoretic application be so low that there is no significant increase in temperature inside the capillary. Otherwise, the thermal expansion of the buffer in the capillary will press part of the sample backward out of the capillary, with attendant loss of reproducibility in peak area.[30,31]

On pressure application of the sample, the width of the starting zone is inversely proportional to the viscosity of the buffer in the capillary [Eq. (11)]. Therefore, the capillary should be thermostatted, preferably with a streaming liquid (air cooling is less efficient at high field strengths). For two runs performed at 21 and 23° the widths of the starting zones will differ by $2 \times 2.7\% = 5.4\%$. When the subsequent electrophoresis experiments are done at 21 and 23° and at constant current the migration velocities will be the same (see above) and therefore the peak areas will also differ by 5.4%. The situation is different when the analysis is performed at constant voltage. The zones then pass the detector with a 5.4% higher mobility and velocity at 23° than at 21° [Eqs. (6) and (8)], which means a 5.4% narrower peak, compensating for the 5.4% wider starting zone. In other words: The peak areas will be the same in experiments performed at different tempera-

[30] R. M. McCormick, *Anal. Chem.* **60,** 2322 (1988).
[31] H. Wätzig, *Chromatographia* **33,** 445 (1992).

tures when the sample is applied by pressure injection and the run is performed at constant voltage (provided that the temperature is the same at the sample application and during the subsequent run and that the ohmic resistance does not change by air bubbles, etc., during the run, as discussed in Section V,C).

C. Local Variations in Electroendosmosis and in Resistance in the Capillary

A requirement for uniform electroendosmotic flow in a capillary is that the ζ potential, and consequently the net surface charge density, have the same value (zero for an ideal electrophoresis) at any point at the capillary wall. In a region where this requirement is not fulfilled, the electroendosmotic flow will differ from that outside this region, resulting in a hydrodynamic flow, because the net flow through all cross-sections of the capillary must be the same. This compensating hydrodynamic flow will distort any zone and decrease the resolution.[12] To eliminate or suppress these local variations in electroendosmosis it is necessary to wash the capillary between runs to remove adsorbed solutes. An efficient desorption requires washing at pH 0–1 (0.1–1 M HCl) and at pH 11–13 (0.001–0.1 M NaOH) in the presence of detergents, preferably 5% (w/v) SDS. The electroendosmosis-free polymer-coated capillaries described herein withstand such harsh conditions. The washing with 0.001–0.1 M NaOH should not be longer than 30 sec. It should be emphasized that only a few types of coatings withstand this treatment.

Air bubbles may form in the capillary at high field strengths, even if the buffer is deaerated, particularly in noncooled sections of the capillary where the Joule heat is not dissipated efficiently. The generation of bubbles is often attended by an increase in the ohmic resistance in the capillary, thereby decreasing the current and consequently increasing the migration time when the voltage is kept constant during a run. However, if the current is constant, the field strength E ($I/q\kappa$) and the migration velocity v (μE) will be constant everywhere in the capillary except where the bubbles are located. If a bubble is large and positioned so that a zone migrates past it, the zone will become distorted.

From the previous discussion, it should be evident that constant current is preferable to constant voltage, because variations in temperature (see Section V,B) and local fluctuations in the resistance in the capillary caused, for instance, by bubble formation will only slightly influence the migration velocity and time (and peak areas) (the current regulator will not function if the change in the resistance exceeds a critical value, which happens, for instance, when the cross-section of a bubble approaches that of the capillary).

At very high field strengths part of the current can pass outside the capillary (electrical discharge). Because a constant-current regulator keeps the total current constant, a current leakage will decrease the current in the capillary and thereby increase the migration time. In such cases it is an advantage to use constant voltage.

It should be emphasized that a decrease (an increase) in migration time gives rise to a decrease (an increase) in peak area, because the width of a peak is proportional to the time it takes for the solute to pass the detector. To eliminate this cause of irreproducible peak areas the observed peak area should be normalized by division by the migration time.[10]

VI. Choice of Buffer

The resolution of the components in an unknown sample ought to be studied in buffers of different pH, because the mobilities of most solutes are dependent on this parameter. When the sample consists of proteins the pH of the electrophoresis medium should not be close to the pI value of any of the proteins in order to avoid long migration times and the risk of protein precipitation.

The concentration of the buffer should not be higher than that required to maintain the pH and the electric conductivity virtually the same in the migrating solute zone and the surrounding buffer. This lower limit can be determined by diluting the buffer until the peaks in the electropherograms begin to become asymmetrical. However, at high protein concentrations the peaks may become asymmetrical even at normal buffer concentrations. The reason is that the difference in mobilities between the protein and the buffer ion of like charge is large.[1,32,33] For instance, in a Tris-HCl buffer the mobility of chloride ions is much greater than those of proteins. Therefore, Tris should not be titrated to the desired pH with HCl, but with acetic or propionic acid, because acetate and propionate ions have lower mobilities than do chloride ions.

The advantage of using a buffer of low conductivity is that it gives a lower current (I) at a given voltage (V) than does a buffer of high conductivity. The Joule heat (VI) is thus lower, which means less thermal zone broadening. We are developing such buffers.[34] One example is lysine. When

[32] J. de Wael and E. Wegelin, *Rec. Trav. Chim. Pays-Bas* **71**, 1035 (1952).

[33] S. Hjertén, *in* "Topics in Bioelectrochemistry and Bioenergetics" (Milazzo, G., ed.), Vol. 2, pp. 89–128. John Wiley & Sons, London, 1978.

[34] S. Hjertén, L. Valtcheva, K. Elenbring, and J.-L. Liao, *Electrophoresis* **16**, 584 (1995).

this amino acid in its free form—not as a salt—is dissolved in deionized, deaerated (carbon dioxide-free) water the pH is equal to the pI (9.7) of the amino acid (at pI the net charge is zero and the conductivity, therefore, very low). Because the pK values of the amino groups are about 9.1 and 10.4 the buffer capacity is satisfactory. We often use lysine at a concentration of 0.005 M.

Another type of a low-conductivity buffer is 0.38% (w/v) polyoxyethylene bis(3-amino-2-hydroxypropyl)–0.031% (w/v) polyoxyethylene bis (acetic acid). This solution gives pH 8.6. Because the molecular weights of these compounds (available from Sigma Chemical Company, St. Louis, MO) are relatively high (above 3000) their mobilities are low, which is equivalent to low conductivity. An experiment is presented in Fig. 3. Note that the recorded resolution at a field strength as high as 2000 V/cm is the same as that at 330 V/cm.

The buffer 0.01 M 2,6-diaminopimelic acid–0.005 M 2-amino-2-methyl-1,3-propanediol has a pH of 8.6. It has a relatively low conductivity and a good buffer capacity, because the first buffer constituent has one pK of about 8.8 and the other constituent has a pK of about 8.4.

Several examples of fast high-resolution HPCE experiments in these and other buffers designed for high field strengths can be found in Ref. 34, which also discusses theory. All of these buffers can be used with field strengths up to at least 2000 V/cm without any observable loss in resolution in 50-μm capillaries. The application of such high field strengths requires efficient insulation to avoid current leakage and electric shock.

Among the conventional buffers, phosphate (pK 2.2, 7.2, 12.3) and borate (pK 9.2) are of particular interest, because they cover a broad pH range and have low UV absorption at wavelengths down to 195 nm. They are, therefore, useful when solutes are to be detected by absorption measurements in the UV region, although they have a relatively high conductivity. The polyoxyethylene buffers mentioned previously also have low UV absorption. For all buffers, particularly low-conductivity buffers, it is essential that the volume of the buffer in the electrode vessel not be so small that the electrolysis changes the composition and the pH of the buffer significantly. If this happens, the peak areas and the migration times will be irreproducible. For the same reason, the electrodes should be placed as far as possible from the ends of the capillary. For optimum reproducibility the electrode vessels should be filled with fresh buffer prior to every run. We often employ the electrode arrangement outlined in Fig. 1 for electrophoresis in low-conductivity buffers. The siphon and the sample vial are then filled with buffer.

Dust and other particles should be removed from the buffers by filtration through a 0.2-μm pore size filter. It is also important that the buffers be

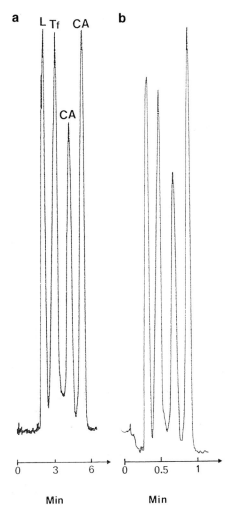

FIG. 3. Free zone electrophoresis of model proteins in a low-conductivity buffer permitting very high field strengths (2000 V/cm). Sample: A 1.7-mg/ml concentration of each of the proteins β-lactoglobulin (L), human transferrin (Tf), and carbonic anhydrase (CA) dissolved in the buffer, diluted 1:10. Buffer: 0.38% (w/v) polyoxyethylene bis(3-amino-2-hydroxypropyl)–0.031% (w/v) polyoxyethylene bis(acetic acid), pH 8.6. The sample was applied with the aid of capillary forces.[9] The width of the starting zone obtained was around 3 mm, which was strongly reduced in the initial phase of the electrophoresis, because the conductivity of the sample solution was much lower than that of the buffer.[9] Dimensions of the coated tube: 150 (135) × 0.05 (i.d.) × 0.14 (o.d.) mm. Detection wavelength: 210 nm. (a) 5000 V (330 V/cm), 0.42 μA; chart speed, 0.5 cm/min. (b) 30,000 V (2000 V/cm), 2.8 μA; chart speed, 3 cm/min. The Joule heat is very small owing to the low conductivity of the buffer (around 6×10^{-5} Ω^{-1} cm^{-1}). Therefore, there is no observable difference in resolution between (a) and (b). An agarose gel plug was inserted into the cathodic end of the capillary to prevent hydrodynamic flow from one electrode vessel to the other.[35] (The experiment was performed in the author's laboratory by K. Elenbring.)

degassed by connection to a water pump for a few minutes. Excessively long aspiration should be avoided, because it increases the concentration of the buffer with attendant decrease in the mobilities of the solute ions.

VII. Concentration and Desalting of Proteins

In displacement electrophoresis (isotachophoresis) and isoelectric focusing, the solute concentration at steady state is so high that the solutes can be monitored easily by UV detection, even when the diameter of the capillary is as small as 25 μm. In zone electrophoresis the situation is different, because the zones broaden rather than sharpen during a run. Therefore, one often must concentrate the sample prior to an analysis, either in or off tube. We have developed several methods for this purpose,[35-37] although only two are described here. These methods are also suitable for desalting macromolecules. The off-tube method described in Section VII,A permits concentration of 2–15 μl of a protein solution down to about 0.5 μl. If further enrichment is required this volume can be transferred to a capillary for on-tube concentration to a zone of 0.2–0.5 mm, i.e., a few nanoliters (see Section VII,B).

A. Off-Tube Concentration

In an off-tube version the sample is sucked up into a gel syringe[35] (Fig. 4). The pores of the gel are so narrow that only small ions (e.g., sodium chloride), but not peptides or proteins, can penetrate the gel. If the gel is prepared in the electrophoresis buffer, diluted 10-fold, the sample can be desalted quickly by diffusion and at the same time transferred to the diluted buffer (2–5 min). Owing to the Donnan effect the sample will also become concentrated about 10- to 25-fold, which takes an additional 20–25 min. The concentrated sample will become concentrated still further, because the low salt concentration in the sample affords zone sharpening at the start (about 10-fold).[9] The final degree of concentration is, accordingly, 100- to 250-fold.

The gel syringe is prepared as follows (Fig. 4). Part of the tip of a Pasteur pipette is burned off and closed 4 cm from the end. A piece of fused silica tubing (outside diameter, 0.35 mm) is inserted through a stopper and positioned at the axis of the pipette. The modified pipette is filled with a deaerated acrylamide solution consisting of 18.4 (19.0) g of acrylamide, 1.6 (1.0) g of N,N'-methylenebisacrylamide, 0.01 g of potassium persulfate,

[35] S. Hjertén, L. Valtcheva, and Y.-M. Li, *J. Cap. Elec.* **1,** 83 (1994).
[36] S. Hjertén, J.-L. Liao, and R. Zhang, *J. Chromatogr.* **676,** 409 (1994).
[37] J.-L. Liao, R. Zhang, and S. Hjertén, *J. Chromatogr.* **676,** 421 (1994).

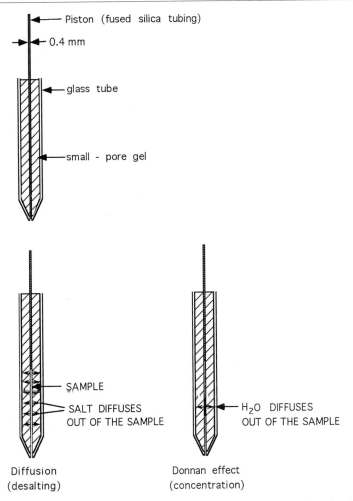

FIG. 4. Off-tube desalting and concentration. The sample is sucked into the gel syringe by drawing the piston upward. Peptides and proteins will become both desalted and concentrated, because the gel is impermeable to large solute molecules.

and 100 ml of the electrophoresis buffer, diluted 1 : 10 with water (the figures without and within parentheses refer to gels for desalting and desalting in combination with concentration, respectively). Polymerization takes place at 70° for 60 min (observe that N,N,N',N'-tetramethylethylenediamine is not used in the polymerization). Most of the remaining part of the tip of the Pasteur pipette is cut off and the sample is sucked into the "syringe" by drawing the piston (the fused silica tubing) upward. Following concentration and desalting, the sample is pressed out by means of the "piston" into a plastic vial (with lid) for application into the electrophoresis capillary. The above-cited dimensions of the "syringe," including those of the "piston," are often used in our laboratory, but should be adjusted according to the volume of the sample.

The desalting and concentration technique described is flexible, because the shape of the gel can be varied easily. For instance, we have prepared 3- to 10-mm high gels in the sample vials used in automated CE apparatus. The gels have a concave surface (created by means of a plastic mold inserted in the monomer solution) to accommodate the sample. The vials are filled with buffer diluted 1 : 5 or 1 : 10 with water to extract low molecular weight compounds from the gel and to accomplish a sharpening of the starting zone. Following a second extraction and withdrawal of the buffer the gel is exposed to air to render it "semidry." With these gel-filled vials desalting, concentration, zone sharpening of the starting zone, and the electrophoretic analysis can be performed automatically in one step with a commercial CE apparatus.

Samples to be analyzed with high resolution in isoelectric focusing (IEF) should have a very low ionic strength. The off-tube method described can be employed with advantage for desalting of the sample. An on-line desalting method, particularly designed for IEF experiments, is described in Ref. 38.

B. In-Tube Concentration

The method is illustrated in Fig. 5.[36,37] The entire coated electrophoresis tube is filled with the sample to be concentrated, and for highly dilute solutions the left electrode vessel as well. When the sample is a powder, it is dissolved in the leading buffer. A small volume of a highly concentrated solution of the leading buffer is added to dissolved samples (the volume added should be that giving approximately the required final concentration of the leading buffer). A third alternative is to equilibrate the sample with the leading buffer by use of the "syringe" in Fig. 4 with the polyacrylamide gel prepared in the leading buffer. The methods described for adjustment of the concentration of the leading buffer are suitable when the sample has a volume of only 5–100 μl. For larger volumes conventional dialysis methods are applicable. However, for most samples the in-tube concentration method described as follows works without any pretreatment of the sample.

The left electrode vessel is filled with the leading buffer, L', and the right one with the terminator, T' (Fig. 5a). To create a hydrodynamic buffer

[38] J.-L. Liao and R. Zhang, J. Chromatogr. A **684**, 143 (1994).

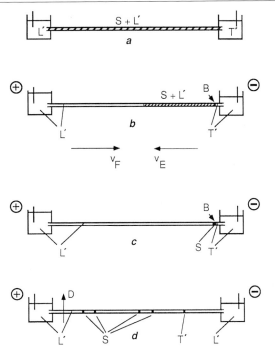

FIG. 5. The principle for in-tube concentration by displacement electrophoresis in combination with a hydrodynamic counterflow. L', Leading buffer; T', terminating buffer. The experimental conditions are shown at the start (a), during the concentration (b), at completed concentration (c), and during the zone electrophoretic analysis (d).

flow in the capillary, the buffer level in the left electrode vessel should be higher than that in the right electrode vessel. When a voltage is applied over the capillary, the proteins in the sample (dissolved in the leading buffer) become concentrated in the layer between the leading and terminating ions (Fig. 5b). This layer migrates electrophoretically toward the anode, but is stationary with respect to the capillary wall when the hydrodynamic counterflow is adjusted properly. Following the concentration (Fig. 5c) the terminator in the right electrode vessel is replaced by the leading buffer, the level of which should be the same as that of the leading buffer in the left electrode vessel (Fig. 5d). When a voltage is applied the displacement electrophoresis step is transformed to an analytical zone electrophoresis (Fig. 5d). Part of the population of rapidly migrating proteins may leave the capillary at the anodic end during the concentration when the original sample is dissolved in the leading buffer. For such proteins, the sample should be dissolved in the terminator (in this case, part of the population

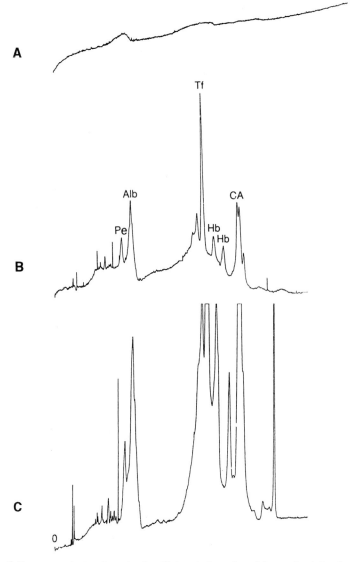

Fig. 6. Free zone electrophoresis of a dilute solution of model proteins following in-tube concentration. The principle is shown in Fig. 5. Sample: A 20-μg/ml concentration of each of the proteins phycoerythrin (Pe), human serum albumin (Alb), human transferrin (Tf), human hemoglobin A (Hb), and carbonic anhydrase (CA). (A) Prior to concentration (a 20-μg/ml concentration of each protein). The width of the applied sample zone was as large as 5 mm. (B) Following concentration by a combination of displacement electrophoresis and a hydrodynamic counterflow. The whole capillary was filled with sample solution (see Fig. 5). *Step I* (concentration by displacement electrophoresis combined with a hydrodynamic

of slow proteins may migrate out of the capillary at its cathodic end). Figure 6 illustrates the efficiency of this concentration technique. All experimental details are given in the caption to Fig. 6. For other concentration methods based on displacement electrophoresis, see Refs. 10, 11, and 39–44. It should be stressed that a change in the length and diameter of the capillary requires a change in the buffer levels in the electrode vessels. Although the difference in buffer levels is not critical, as explained in Ref. 36, it can easily be established by running colored proteins in transparent glass tubes (from Modulohm, Herlev, Denmark). Such experiments show that the width of the final, concentrated sample zone is 0.2–0.4 mm. From this width and the length of the capillary (155 mm) one can calculate that the experiment presented in Fig. 6 represents a concentration factor of 400- to 800-fold.

VIII. Multipoint Detection

The apparatus mentioned in the Introduction for zone electrophoresis in a rotating capillary has the advantage that it permits automatic scanning of the capillary with UV light at any time. One can thus follow the separation and do kinetic studies and see whether anything unexpected happens during a run. The set of electropherograms recorded affords several measuring points, which increases the accuracy in the determination of migration time

[39] C. Schwer and F. Lottspeich, *J. Chromatogr.* **623**, 345 (1992).
[40] F. Foret, E. Szoko, and B. L. Karger, *J. Chromatogr.* **608**, 3 (1992).
[41] D. Kaniansky, J. Marák, V. Madajová, and E. Šimuničová, *J. Chromatogr.* **638**, 137 (1993).
[42] T. Hirokawa, A. Ohmori, and Y. Kiso, *J. Chromatogr.* **634**, 101 (1993).
[43] J. L. Beckers and M. T. Ackermans, *J. Chromatogr.* **629**, 371 (1993).
[44] F. Foret, E. Szökö, and B. L. Karger, *Electrophoresis* **14**, 417 (1993).

counterflow; Fig. 5a–c): Leading buffer in the anode vessel, 0.015 M HCl, titrated to pH 8.5 with Tris; terminating buffer, 0.1 M glycine, titrated to pH 8.5 with NaOH; the buffer level in the anode vessel was 2 cm higher than that in the cathodic vessel; the stationary boundary was located 1–2 cm from the cathodic end of the capillary. Dimensions of the capillary: 155 (145) × 0.05 (i.d.) mm; detection wavelength, 220 nm; voltage, 500 V for 15 min. The sample was dissolved in leading buffer. *Step II* (analysis by free zone electrophoresis following replacement of the glycine–NaOH buffer by the HCl-Tris buffer; Fig. 5d): no difference in the buffer levels in the electrode vessels; voltage, 3000 V for 18 min. (C) Like (B), with the difference that the anode vessel as well as the capillary was filled with sample solution. The experiments were performed in a BioFocus 3000 (Bio-Rad, Hercules, CA). (Reprinted from *J. Chromatogr.* **676**, 421, 1994, with kind permission of Elsevier Science-NL, Sara Burgerhartstraat 25, 1055 KV Amsterdam, The Netherlands.)

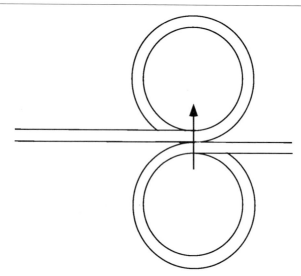

Fig. 7. Multipoint detection with minimum zone broadening. The first, second, and third electropherograms are obtained when the sample zones pass the first, second, and third windows, respectively. The arrow indicates the position of the detecting UV beam.

and peak area. For several reasons, the narrow fused silica tubing is not suitable for such scanning. However, by using a multipoint detection technique one approaches the advantages offered by a true scanning technique.[45] The technique is outlined in Fig. 7. The polyimide coating of the fused silica tubing is burned off at places corresponding to the desired migration distances. The tubing is coiled into a figure eight so that the detection windows become placed side by side or on top of each other in the detecting UV beam. The first electropherogram is obtained as the solutes pass the first window, the second as they pass the second window, and so on. When the solute zone migrates through the first loop it has become tilted, corresponding to a zone broadening of $2\pi\phi$, where ϕ is the diameter of the capillary.[45,46] The same tilting (corresponding to 81°) occurs in the second loop, but in the opposite direction, because the tubing is coiled into a figure eight. Therefore, there is no net tilting after passage of two loops. However, the compensation is not complete if broadening caused by diffusion is also considered. Experiments in gels have shown that the

[45] T. Srichaiyo and S. Hjertén, *J. Chromatogr.* **604**, 85 (1992).
[46] S. Wicar, M. Vilenchik, A. Belenkii, A. S. Cohen, and B. L. Karger, *J. Microcol. Sep.* **4**, 339 (1992).

total zone broadening is smaller when the tubing is coiled into a figure eight instead of a spiral.

Because any curvature in the capillary gives rise to tilting of a zone, i.e., zone broadening, the capillary should be straight in all electrophoresis experiments. Therefore, all of the capillary electrophoresis apparatuses that we have constructed were designed with this in mind.[11] Even a straight non-zone-deforming capillary can be employed for multipoint detection if the capillary following the recording of the first electropherogram is displaced longitudinally to a position where the UV beam passes the second window, etc.[45] For interesting applications of this detection technique, see Ref. 47.

For new information related to this chapter, see reference 48.

Acknowledgments

This work was financially supported by the Swedish Natural Science Research Council, and by the Knut and Alice Wallenberg and Carl Trygger Foundations.

[47] P. Sun and R. A. Hartwick, *J. Liquid Chromatogr.* **17**(9), 1861 (1994).
[48] S. J. Hjertén, *Jpn. J. Electroph.* **39**, 105 (1995).

[14] Micellar Electrokinetic Chromatography

By Norio Matsubara and Shigeru Terabe

I. Overview

Terabe *et al.* first demonstrated the use[1] and principle[2] of micellar electrokinetic chromatography (MEKC) based on the suggestion made by Nakagawa.[3] Since then, diverse applications of this method to various organic compounds have been reported. Although MEKC uses an electric field as the driving force for the migration of analytes in a manner similar to capillary zone electrophoresis (CZE), the difference between MEKC and CZE lies, first, in its separation mechanism. Separation by CZE is based on the difference in electrophoretic mobility of ions. In contrast, MEKC uses the difference in the partition ratio of compounds between aqueous phase and micellar phase (more accurately, pseudostationary phase). Second, MEKC can separate not only ionic compounds but also nonionic

[1] S. Terabe, K. Otsuka, K. Ichikawa, A. Tsuchiya, and T. Ando, *Anal. Chem.* **56**, 111 (1984).
[2] S. Terabe, K. Otsuka, and T. Ando, *Anal. Chem.* **57**, 834 (1985).
[3] T. Nakagawa, *Newsl., Div. Colloid Surf. Chem., Chem. Soc. Jpn.* **6**(1), 1 (1981).

compounds at the same time. Because the migration order of nonionic compounds and, in many cases, ionic compounds is determined by the degree of hydrophobic interaction between micelles and analytes, the separation mechanism is more similar to that of reversed-phase high-performance liquid chromatography (RP-HPLC) than to electrophoresis, except that MEKC achieves high theoretical plate numbers (>100,000) for many compounds routinely. The reason is mainly because there is a difference in the flow profile in a capillary tube from HPLC. MEKC is a powerful tool, especially for the separation of small molecules. In this chapter we concentrate on practical aspects and the knowledge necessary to use MEKC. For more physicochemical and theoretical aspects[4,5] or for a complete list of references,[6,7] readers should refer to the publications cited.

II. Theory

A. Separation Mechanism

Figure 1 shows the basic principle of separation by MEKC. An ionic surfactant dissolved in an aqueous pH buffer solution at a concentration higher than the critical micelle concentration (CMC) forms micelles usually spherical in nature, and the size of the micelles depends on the aggregation number of the individual surfactant. When sodium dodecyl sulfate (SDS), one of the most popular surfactants in MEKC, is employed, 1 micelle consists of about 60 molecules and is negatively charged. The SDS micelle is, therefore, driven toward the positive electrode by electrophoresis. The velocity is indicated by v_{ep}, which is a negative value; the converse direction is designated as the positive direction. Because the inner wall of a fused silica capillary is negatively charged in most cases, the electroosmotic flow is in the direction of the negative electrode, with a velocity of v_{eo}. Shown in Fig. 2, under the conditions of pH >5, the SDS micelle migrates toward the negative electrode at a net velocity of $v_{mc} = (v_{eo} + v_{ep})$ because the absolute value of v_{eo} is larger than v_{ep}.[8] A neutral analyte injected into the

[4] S. Terabe, Micellar electrokinetic chromatography. *In* "Capillary Electrophoresis: Theory, Methodology and Applications" (N. A. Guzman, ed.). p. 65. Marcel Dekker, New York, 1993.

[5] P. D. Grossman and J. C. Colburn (eds.), "Capillary Electrophoresis—Theory and Practice." Academic Press, New York, 1992.

[6] S. F. Y. Li, "Capillary Electrophoresis—Principles, Practice and Applications." Elsevier, Amsterdam, The Netherlands, 1992.

[7] J. Videvogel and P. Sandra, "Introduction to Micellar Electrokinetic Chromatography." Hüthig, Heidelberg, Germany, 1992.

[8] K. Otsuka and S. Terabe, *J. Microcol. Sep.* **1**, 150 (1989).

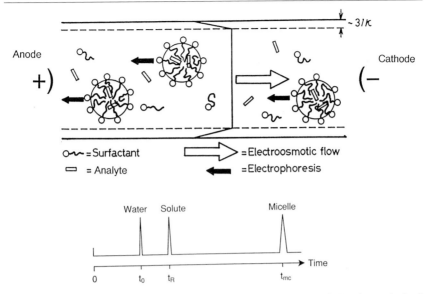

FIG. 1. Separation mechanism in MEKC. (*Bottom:* Reprinted from S. Terabe *et al., Anal. Chem.* **57,** 834. Copyright 1985 American Chemical Society.)

capillary under this condition migrates at a velocity of v_{eo}, with a migration time of t_0 if it barely distributes itself in the micellar phase, and migrates at a velocity of v_{mc}, with a migration time of t_{mc}, if it is totally incorporated into the micelles. An example of the former is methanol, and the latter is an azo dye, Sudan III. Migration times of all electrically neutral compounds, t_R, exist between t_0 and t_{mc}, depending on the distribution coefficients into the micellar phase. The difference, $(t_{mc} - t_0)$, is the time window available for the separation of peaks on the chromatogram.

B. Resolution Equation

Resolution in MEKC, R_s, is given by the following equation[2]:

$$R_s = \frac{\sqrt{N}}{4}\left(\frac{\alpha - 1}{\alpha}\right)\left(\frac{k'_2}{1 + k'_2}\right)\left[\frac{1 - t_0/t_{mc}}{1 + (t_0/t_{mc})k'_1}\right] \tag{1}$$

where N is the theoretical plate number; α is the separation factor, which is equal to k'_2/k'_1; and k'_1 and k'_2 are capacity factors of analytes 1 and 2, respectively. Equation (1) is similar to that of conventional chromatography

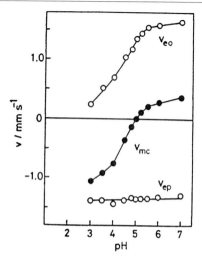

FIG. 2. Dependence of electrophoretic velocities on pH: v_{eo}, electroosmotic velocity; v_{mc}, migration velocity of SDS micelle; v_{ep}, electrophoretic velocity of SDS micelle. Conditions: Micellar solution, 0.10 M SDS; column, 0.05-mm i.d. $\times$ 650 mm; length of the column used for separation, 500 mm; current, 50 μA; applied voltage, about 14.7 kV; detection wavelength, 220 nm. [Reproduced from K. Otsuka and S. Terabe, $J.$ $Microcol.$ $Sep.$ **1,** 150 (1989) with permission of MicroSeparations, Inc.]

except for the last term, which is a function of t_0/t_{mc}. The dependency of R_s on the plate number and the separation factor is the same as that of conventional chromatography. The major difference between MEKC and conventional chromatography in Eq. (1) is that the optimum capacity factor for the separation of analytes depends on the t_0/t_{mc} ratio.

When the peaks of analyte 1 and analyte 2 are very close on the chromatogram, we can assume $k_1' = k_2' = k'$. On this condition, Eq. (1) is described as

$$R_s = \frac{\sqrt{N}}{4}\left(\frac{\alpha - 1}{\alpha}\right)f(k')\qquad(2)$$

where $f(k')$ denotes

$$f(k') = \left(\frac{k'}{1 + k'}\right)\left[\frac{1 - t_0/t_{mc}}{1 + (t_0/t_{mc})k'}\right]\qquad(3)$$

To find the optimum k', k'_{opt}, that gives the highest resolution, we can differentiate Eq. (3) and find the value of k' that makes the first derivative of Eq. (3) equal to zero.[9]

$$k'_{opt} = \sqrt{\frac{t_{mc}}{t_0}} \qquad (4)$$

If the pH of the buffer solution is 7, t_{mc}/t_0 is around 4 for an SDS micelle, thus k'_{opt} is close to 2. The results given by Eq. (4) suggest that extremely large or small capacity factors are not favorable for separation by MEKC. In general, capacity factors between 1 and 5 are recommended for the optimal resolution of compounds. The methods for manipulating capacity factors, mainly to reduce partition coefficients of hydrophobic compounds to the micellar phase in order to enhance resolution, are discussed in Sections III and IV.

III. Basic Elements That Influence Separation Conditions in Micellar Electrokinetic Chromatography

A. Selection of Surfactant

The structure of a surfactant affects the selectivity of separation. Three factors [(1) the length of hydrophobic alkyl chain, (2) the electric charge and the nature of the hydrophilic group, and (3) the chirality of surfactant] have been investigated.

The aggregation numbers and CMCs of anionic, cationic, and nonionic surfactants are listed in Table I. As shown in Table I, surfactants that have an alkyl chain length shorter than 10 generally have very high CMCs and high conductivities[10]; they are therefore not suitable for micelle-forming reagents, but are used mainly as additives to micelles to change selectivity.[11–13] Surfactants that have an alkyl chain length longer than 14, that is, tetradecyl groups, have such high Krafft points that they are seldom used in MEKC. The Krafft point is specific to each surfactant and is described as the breakpoint in the temperature–solubility curve, above which the solubility of surfactant increases sharply and finally exceeds the critical micelle concentration. In other words, micelles are formed only when the temperature is higher than the Krafft point. Because the difference in

[9] J. P. Foley, *Anal. Chem.* **62**, 1302 (1990).
[10] D. E. Burton, M. J. Sepaniak, and M. P. Maskarinec, *J. Chromatogr. Sci.* **25**, 514 (1987).
[11] H. Nishi, N. Tsumagari, and S. Terabe, *Anal. Chem.* **61**, 2434 (1989).
[12] H. Nishi, N. Tsumagari, T. Kakimoto, and S. Terabe, *J. Chromatogr.* **477**, 259 (1989).
[13] R. A. Wallingford, P. D. Curry, Jr., and A. G. Ewing, *J. Microcol. Sep.* **1**, 23 (1989).

TABLE I
ANIONIC, CATIONIC, NONIONIC, AND CHIRAL SURFACTANTS

Surfactant	Critical micelle concentration[a] (mM)	Aggregation number	Krafft point (°C)
Anionic surfactant			
Sodium octyl sulfate	130–155	20	—
Sodium decyl sulfate	31–41	50	8
Sodium dodecyl sulfate (SDS)	8.1	62	16
Sodium tetradecyl sulfate (STS)	2.1 (50°)	138 (0.1 M NaCl)	30
Sodium hexadecyl sulfate	0.4	—	45
Sodium dodecanesulfonate (SDDS)	9.8	54	38
Sodium N-lauroyl-N-methyltaurate (LMT)	8.7	—	<0
Cationic surfactant			
Cetyltrimethylammonium bromide (CTAB)	0.92	61	25
Cetyltrimethylammonium chloride (CTAC)	1.3 (30°)	—	—
Nonionic surfactant			
n-octyl-β-D-glucoside	25	—	—
Polyoxyethylene sorbitan monolaurate (Tween 20)	0.059	—	—
Chiral surfactant			
Sodium N-dodecanoyl-L-valinate (SDVal)	5.7 (40°)	—	—
Bile salts			
Sodium cholate (SC)	13–15	2–4	—
Sodium deoxycholate (SDC)	4–6	4–10	—
Sodium taurocholate (STC)	10–15	5	—
Sodium taurodeoxycholate (STDC)	2–6	—	—

[a] At 25°.

selectivity based on the alkyl chain length is not significant,[2] the dodecyl group is generally the most suitable surfactant, although surfactants with longer alkyl chains give larger distribution coefficients for hydrophobic analytes.

When SDS is chosen as a surfactant for MEKC separation, one thing to keep in mind is the selection of cation in the buffer. Sodium salt is generally preferred to potassium salt because potassium dodecyl sulfate has much less solubility than SDS and is prone to precipitate at room temperature. The nature of the hydrophilic group affects the selectivity of polar analytes. Sodium dodecane sulfonate (SDDS)[2] and sodium N-lauroyl-N-methyltaurate (LMT)[12,14] showed an apparent difference in selectivity compared to SDS.

Cationic micelles such as cetyltrimethylammonium bromide (CTAB)

[14] H. Nishi, N. Tsumagari, T. Kakimoto, and S. Terabe, J. Chromatogr. 465, 331 (1989).

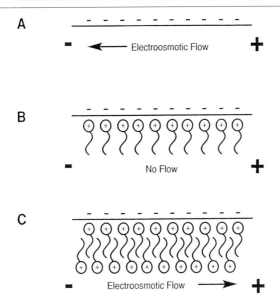

FIG. 3. Effect of cationic surfactant on the surface charge of a fused silica capillary. (A) No surfactant added; electroosmotic flow in normal direction. (B) Electrostatic adsorption of the positively charged surfactant head groups to the negative silanol groups on the silica surface of the capillary inner wall. (C) Admicellar bilayer formation by hydrophobic interaction between the nonpolar chains, resulting in a reversal of the electroosmotic flow. [Reproduced from Å. Emmer et al., J. Chromatogr. **547**, 544 (1991) with permission of Elsevier Science Publishers BV.]

showed substantially different selectivity compared to SDS among nonionic compounds as well as ionic compounds.[15] It should be noted that cationic surfactants are adsorbed electrostatically on the inner wall of a fused silica capillary, to form a bilayer, and turn the apparent surface charge from negative to positive, and thus reverse the direction of electroosmotic flow from the negative to positive electrode. Figure 3 shows the mechanism of such a reversal of surface charge by using cationic micelles. This phenomenon is useful to avoid peak broadening of positively charged proteins, which tend to be adsorbed electrostatically on the silica surface of a capillary under the normal conditions.[16] Care must be taken because halide anions generate corrosive gases at the anode.

Nonionic micelles have been used for the separation of peptides.[17,18]

[15] K. Otsuka, S. Terabe, and T. Ando, J. Chromatogr. **332**, 219 (1985).
[16] Å. Emmer, M. Jansson, and J. Roeraade, J. Chromatogr. **547**, 544 (1991).
[17] S. A. Swedberg, J. Chromatogr. **503**, 449 (1990).
[18] N. Matsubara and S. Terabe, Chromatographia **34**, 493 (1992).

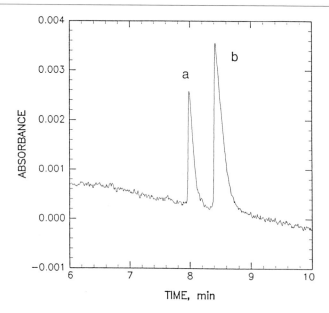

Motilin (Human):   Phe-Val-Pro-Ile-Phe-Thr-Tyr-Gly-Glu-Leu-Gln-
                    Arg-Met-Gln-Glu-Lys-Glu-Arg-Asn-Lys-Gly-Gln

Motilin (KW-5139): Phe-Val-Pro-Ile-Phe-Thr-Tyr-Gly-Glu-Leu-Gln-
                    Arg-Leu-Gln-Glu-Lys-Glu-Arg-Asn-Lys-Gly-Gln

FIG. 4. Separation of (a) [Met[13]]motilin and (b) [Leu[13]]motilin. Electrophoretic solution, 120 mM Tween 20 in 10 mM phosphate solution (pH 1.5); applied voltage, 10 kV; current, 26 μA; capillary inner diameter, 50 μm; column length, 27 cm (effective length, 20 cm). [Reproduced from N. Matsubara and S. Terabe, *Chromatographia* **34,** 493 (1992) with permission of Friedr. Vieweg & Sohn Verlagsgesellschaft mbH.]

One of the nonionic surfactants, Tween 20, has a lower affinity for peptides than does SDS, and allows a successful separation of closely related peptides[18] that are not separated by SDS micelle alone[19] because distribution coefficients to SDS micelle are too high (Fig. 4). However, this method should be included as an extended form of MEKC rather than in our traditional category of MEKC because nonionic micelles have no electrophoretic mobility, in contrast to SDS micelles, which have the electrophoretic mobility in the opposite direction to electroosmotic flow and enable

[19] T. Yashima, A. Tsuchiya, and O. Morita, *Chromatography* **12**(5), 124 (1991).

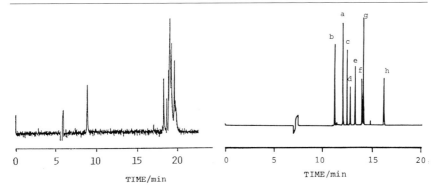

Fig. 5. Separation of eight corticosteroids. *Left:* Mixture of hydrocortisone, hydrocortisone acetate, betamethasone, cortisone acetate, triamcinolone acetonide, and fluocinolone acetonide, dexamethasone acetate, fluocinonide. Capillary, 65 cm × 50-μm i.d. (50 cm to the detector); applied voltage, 20 kV; separation solution, 50 mM SDS in 20 mM borate–20 mM phosphate buffer (pH 9.0). [Reproduced from S. Terabe *et al., J. Chromatogr.* **545,** 359 (1991) with permission of Elsevier Science Publishers.] *Right:* Conditions: buffer, the same as at left, but containing 0.1 M sodium cholate instead of SDS; other conditions are the same as on the left. Solutes: a, hydrocortisone; b, triamcinolone; c, betamethasone; d, hydrocortisone acetate; e, dexamethasone acetate; f, triamcinolone acetonide; g, fluocinolone acetonide; h, fluocinonide. [Reproduced from H. Nishi *et al., J. Chromatogr.* **513,** 279 (1990) with permission of Elsevier Science Publishers BV.]

the separation of nonionic compounds. The use of nonionic micelles is still rare, except for mixed micelles[20] with SDS in MEKC. Zwitterionic micelles are also used for MEKC separation.[17,21]

The use of sodium N-dodecanoyl-L-valinate and bile salt micelles, which are chiral, is reported to separate optically active compounds. Details are discussed in Section VI,B. Another characteristic of bile salts is their low solubilizing power. Hydrophobic compounds are generally difficult to separate by SDS alone because their distribution coefficients to the micellar phase are too high, as is mentioned in Section IV,B and IV,C. Bile salt micelles have been used successfully to solve this problem, taking advantage of low solubilizing power, which, in turn, gave a distribution coefficient for an optimal resolution (Fig. 5).[22,23]

[20] H. T. Rasmussen, L. K. Goebel, and H. M. McNair, *J. Chromatogr.* **517,** 549 (1990).
[21] H. K. Kristensen and S. H. Hansen, *J. Chromatogr.* **628,** 309 (1993).
[22] H. Nishi, T. Fukuyama, M. Matsuo, and S. Terabe, *J. Chromatogr.* **513,** 279 (1990).
[23] R. O. Cole, M. J. Sepaniak, W. L. Hinze, J. Gorse, and K. Oldiges, *J. Chromatogr.* **557,** 113 (1991).

B. Concentration of Surfactant

The relationship between the distribution coefficient K and the capacity factor of the analyte, k', is shown by Eq. (5),[2]

$$k' = K\bar{v}(C_{srf} - \text{CMC}) \qquad (5)$$

where $\bar{v}$ is the partial specific volume of the micelle and C_{srf} is the surfactant concentration. Equation (5) suggests that the capacity factor increases linearly with surfactant concentration when the surfactant concentration is higher than the CMC, so the proper adjustment of this concentration is important to optimize k'. However, an increased conductivity, which induces Joule heating, limits the highest concentration, and the value of the CMC for a given surfactant decides the lowest concentration usable in MEKC. Thus, the concentration range of 20–100 mM for SDS is often used. To increase surfactant concentration the use of small-bore capillary columns rather than a normal column of 50-μm i.d. is required to control Joule heating, but the detection sensitivity is reduced.

C. Effect of pH

Although the nature and concentration of buffer solution do not significantly affect selectivity in MEKC, the pH of the buffer does affect the selectivity of separation of ionic substances. The ionized forms of analytes having the same charge as micelles, e.g, organic acids, are less incorporated into anionic micelles and exhibit decreased distribution coefficients.[24] However, the migration order is not always predictable because ionic substances themselves have electrophoretic mobilities even when they are not incorporated into micelles. The above theory is not applicable to analytes that have the opposite charge to that of micelles, because positive analytes tend to form ion pairs with micelles and comigrate. Another effect of pH is to change the velocity of electroosmotic flow to extend the migration window (see Section II,A).

D. Effect of Temperature

Temperature affects the distribution coefficient and mobility of analytes. Analytes have different temperature dependencies of distribution coefficients,[25,26] therefore temperature also affects selectivity. Raising the temper-

[24] K. Otsuka, S. Terabe, and T. Ando, *J. Chromatogr.* **348**, 39 (1985).
[25] S. Terabe, T. Katsura, Y. Okada, Y. Ishihama, and K. Otsuka, *J. Microcol. Sep.* **5**, 23 (1993).
[26] S. Terabe, T. Katsura, Y. Ishihama, and Y. Okada, "Proceedings of the Fourteenth International Symposium on Capillary Chromatography," p. 515. Foundation for the International Symposium on Capillary Chromatography, Miami, FL, 1992.

ature also results in an increase in v_{eo} and v_{ep} of the micelle owing to a decreased viscosity, with the decrease of migration time without altering selectivity. There is a report in which the number of theoretical plates was increased for some compounds when the temperature was raised from 28 to 39° because of peak sharpening.[27]

IV. Aqueous Phase Modifiers: Optimizing Capacity Factor and Changing Selectivity

The addition of third substances such as organic solvents, urea, ion-pair reagents, or selective complexing reagents (e.g., cyclodextrin) to micellar solvents has been employed in order to optimize the capacity factor, or to change the selectivity of separation. (Refer to Ref. 28 for more details about the manipulation of selectivity.)

A. Cyclodextrin

Cyclodextrins are used in electrokinetic chromatography (EKC) in two different ways. One is the so-called cyclodextrin electrokinetic chromatography (CDEKC). Cyclodextrin is a nonionic compound and has no electrophoretic mobility. If one of the hydroxy groups in cyclodextrin is modified by an ionic group, this compound can be used instead of a micelle. An analyte molecule included in the cavity of the modified cyclodextrin has a different overall charge because of the modified group attached to cyclodextrin, with a change in electrophoretic mobility. Terabe et al.[29] reported the separation of aromatic isomers by 2-O-carboxymethyl-β-cyclodextrin (β-CMCD) on the basis of this principle. However, the low electrophoretic mobility of β-CMCD relative to electroosmosis reduced the time window available for the separation, as discussed in Section II.

The other method is cyclodextrin-modified micellar electrokinetic chromatography (CD-MEKC),[30,31] which uses nonionic cyclodextrin as an additive to SDS–MEKC. Because cyclodextrin barely distributes itself into the micellar phase but has a hydrophobic cavity in its center, the capacity factor of hydrophobic compounds that fit well in the cavity of cyclodextrin is reduced, which changes the selectivity of the separation of specific compounds. For example, separation of polychlorinated benzene congeners by MEKC with or without γ-cyclodextrin is reported.[31] In this case, the addition

[27] A. T. Balchunas and M. J. Sepaniak, *Anal. Chem.* **60,** 617 (1988).

[28] S. Terabe, *J. Pharm. Biomed. Anal.* **10,** 705 (1992).

[29] S. Terabe, H. Ozaki, K. Otsuka, and T. Ando, *J. Chromatogr.* **332,** 211 (1985).

[30] H. Nishi, T. Fukuyama, and S. Terabe, *J. Chromatogr.* **553,** 503 (1991).

[31] S. Terabe, Y. Miyashita, O. Shibata, E. R. Barnhart, L. R. Alexander, D. G. Patterson, B. L. Karger, K. Hosoya, and N. Tanaka, *J. Chromatogr.* **516,** 23 (1990).

of urea was not effective but cyclodextrin improved selectivity remarkably. In the cases of CDEKC and CD-MEKC, the selectivity of both methods was varied by changing the size of the cavity or the modification of the hydroxy groups of cyclodextrin.[28] In general, γ-cyclodextrin is the most effective and 10 mM is enough for this purpose. For highly hydrophobic compounds (usually $k' > 10$), the cyclodextrin concentration must be increased to 40 mM or higher to reduce k', partitioning to micellar phase. Another example of CD-MEKC is the selective separation of optical isomers, which is described in Section VI,B.

B. Organic Solvent

The second way to change the capacity factor is by the addition of organic solvents miscible with water. The capacity factor decreases with an increase in concentration of the organic solvents in the mobile phase. Methanol or acetonitrile has been used for this purpose, but the addition of these solvents to more than 20% in volume may break down micelles or adversely reduce performance. The organic solvent is used to separate compounds that have relatively hydrophobic groups and migrate near or at t_{mc}. Capacity factors of these compounds are reduced and the separation is optimized by the addition of organic solvent. Separation of motilins[19,32] and insulins from various animals[32,33] has been reported on the basis of this principle (Fig. 6). The addition of organic solvent to the micellar solution usually reduces electroosmotic velocity, extending the time window and enhancing resolution[34–36] (see Section II).

C. Urea

The third way to optimize the capacity factor is by addition of urea to the solution. As is well known, a high concentration of urea increases the solubility of some hydrophobic compounds.[37–39] This effect was used in MEKC to separate corticosteroids.[40] Migration times of eight corticoste-

[32] T. Yashima, A. Tsuchiya, O. Morita, and S. Terabe, *Anal. Chem.* **64,** 2981 (1992).
[33] T. Yashima, A. Tsuchiya, and O. Morita, "Proceedings of the 11th Symposium on Capillary Electrophoresis," p. 53. Denki-Eidou Bunseki Kenkyu Kondan-kai (Division of Analytical Electrophoresis), Japan Society for Analytical Chemistry, Tokyo, Japan, 1991.
[34] K. Otsuka, S. Terabe, and T. Ando, *Nippon Kagaku Kaishi* 950 (1986).
[35] J. Gorse, A. T. Balchunas, D. F. Swaile, and M. J. Sepaniak, *HRC-CC, J. High Resolut. Chromatogr. Chromatogr. Commun.* **11,** 554 (1988).
[36] M. M. Bushey and J. W. Jorgenson, *J. Microcol. Sep.* **1,** 125 (1989).
[37] P. Mukerjee and A. Ray, *J. Phys. Chem.* **67,** 190 (1963).
[38] M. J. Schick, *J. Phys. Chem.* **68,** 3585 (1964).
[39] D. B. Wetlaufer, S. K. Malik, L. Stoller, and R. L. Coffin, *J. Am. Chem. Soc.* **86,** 508 (1964).
[40] S. Terabe, Y. Ishihama, H. Nishi, T. Fukuyama, and K. Otsuka, *J. Chromatogr.* **545,** 359 (1991).

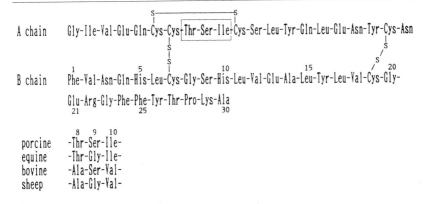

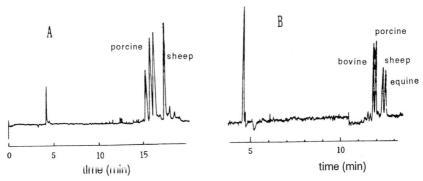

FIG. 6. Separation of insulins by MEKC with acetonitrile. Conditions: capillary, 650 mm ×
0.05-mm i.d., 500 mm to detector; electrophoretic solutions: (A) 50 mM acetate buffer (pH
3.6) containing 10 mM CTAB and 5% acetonitrile (first peak: equine; third peak: bovine);
(B) 50 mM borate buffer (pH 8.5) containing 50 mM SDS and 15% acetonitrile; temperature
ambient; detection wavelength, 215 nm. (Reprinted from T. Yashima et al., Anal. Chem. **64**,
2981. Copyright 1992 American Chemical Society.)

roids in 50 mM SDS solution were very close and mutual separation was
difficult. The addition of 6 M urea to this solution achieved baseline separation of all eight corticosteroids. The reason is mainly because the capacity
factor k' was reduced to an optimal value for the separation of these
hydrophobic compounds. The effect was studied by the measurement of
solubility of the corticosteroids and calculation of the distribution coefficients from the migration times in the solution containing various concentrations of urea. Urea changes not only the capacity factor of hydrophobic
compounds but also the selectivity. Figure 7 shows a comparison of the
separation of phenylthiohydantoin (PTH)–amino acids[40] (a) in the absence

a

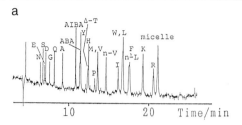

b

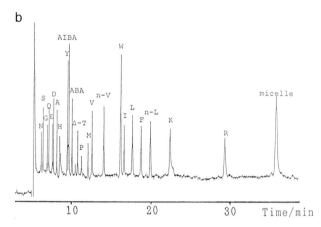

FIG. 7. Electrokinetic chromatogram of a mixture of 23 PTH–amino acids. The peaks are identified with one-letter abbreviations for the amino acids. AIBA, 2-aminoisobutyric acid; ABA, 2-aminobutyric acid; Δ-T, PTH-dehydrothreonine. The micelle is traced with timepidium bromide. (a) Capillary, 50 cm × 52-μm i.d. (30 cm to the detector); separation solution, 50 mM SDS in 100 mM borate–50 mM phosphate buffer; applied voltage, 10.5 kV. (b) Separation solution, 100 mM SDS and 4.3 M urea in the same buffer as in (a); other conditions as in (a). [Reproduced from S. Terabe *et al., J. Chromatogr.* **545,** 359 (1991) with permission of Elsevier Science Publishers BV.]

and (b) in the presence of 4.3 M urea. In Fig. 7b, complete separation of 23 PTH-derivatized amino acids was achieved whereas some of the PTH-derivatized amino acids were not separated in Fig. 7a. However, the mechanism by which the urea changes the selectivity is not well understood.

D. Ion-Pair Reagents

As already mentioned in Section I, MEKC is similar to reversed-phase HPLC for the separation of electrically neutral analytes. However, there is a difference in the separation of ionic analytes. Because micelles are charged, either repulsive or attractive electrostatic forces occur between analyte and micelles. The addition of an ion-pair reagent alters this interaction, and hence selectivity. Anionic analytes such as carboxylates are repelled from SDS micelles by the electrostatic repulsive force. When a tetraalkylammonium salt is added to this solution, it forms ion pairs with carboxylates and partly with SDS micelles and neutralizes the charge of the carboxylates. Thus, an increased migration time of analytes was observed[11] because of greater incorporation of analytes into the micelles. Cationic analytes, conversely, decreased[11] in migration time on addition of tetraalkylammonium salts to a mobile phase containing SDS micelles because ionic interaction of analytes and micelles was suppressed.

V. Instrumentation

A. Operating Conditions

Here, we give typical operating conditions in MEKC for the novice practitioner. The same capillaries, detectors, and high-voltage power supplies as used for CZE, or commercially available CZE instruments, can be used to perform MEKC. However, capillaries must have a reasonable electroosmotic flow because the time window of MEKC and the migration times of samples depend on the electroosmotic flow. Rinsing new capillary columns with 0.1 M sodium hydroxide prior to use is often recommended. Standard conditions are summarized in Table II. When resolution is not satisfactory at these conditions, consider ways to optimize the capacity factor first. If k' is less than 0.5, the concentration of SDS may be increased up to 200 mM. If k' is still too small, the surfactant should be replaced. If k' is too large (i.e., $k' > 10$), reduce the concentration of SDS, although it should be kept higher than the CMC (~5 mM). Other options in the

TABLE II
STANDARD OPERATING CONDITIONS

Parameter	Condition
Running solution	50 mM SDS in 50 mM borate buffer (pH 8.5–9.0)
Capillary	50- to 75-μm i.d. $\times$ 20–50 cm (from the injection end to the detector)
Applied voltage	10–20 kV (current should be kept below 100 μA)
Temperature	25° or ambient
Sample solvent	Water or methanol
Sample concentration	0.1–1 mg/ml
Injection end	Positive or anodic end
Injection volume	Below 2 nl (or less than 1-mm width in the column)
Detection	200–210 nm (depends on the sample)

last case are a change of surfactant, adjustment of pH, and the addition of aqueous phase modifiers. The use of bile salt instead of SDS reduces k', which is more suitable to hydrophobic compounds. At higher pH, organic acids have lower k' values because of ionic repulsion. Various additives have been reported to increase the solubility of hydrophobic compounds in the aqueous phase (see Section IV).

B. Detection Techniques

The transparency of SDS micelles down to 200 nm enables detection of aromatic and other organic compounds that absorb in the ultraviolet (UV) range. This is still the most popular detection technique in MEKC. Other detection techniques used in MEKC are laser-induced fluorescence (LIF) measurements, indirect fluorescence measurements, and electrochemical detection. Each method has its advantages and disadvantages. Of the three, LIF is the most sensitive method and suitable for the detection of trace amounts of compounds; however, derivatization to label the sample with a fluorescent probe is often necessary. (For examples of LIF detection, see [19] in this volume.[40a]) Indirect fluorescence is universal for all compounds that have no fluorescence. The detection is based on the exclusion of fluorophore (quinine) from micelles by analytes, shown as negative peaks on the chromatograms.[41] However, the detection limit is only comparable to that of the UV absorption technique because of baseline noise that originates from high background fluorescence. Amperometric detection is sensitive and selective. In addition, unlike optical detectors, the sensitivity is independent of the capillary diameter, so it is possible to enhance resolution with a narrower capillary tube. The weak point of this approach is that few organic compounds show an oxidation or reduction current; also, isolation of the detector electrode from the electric field used for electrophoresis is necessary at the capillary–detector electrode interface. Typical applications include detection of catechols and catecholamines (see Section VI,D).

VI. Applications

A. Positional (Ring) Isomers

Historically, the compounds first separated by MEKC were ring isomers of phenolic derivatives.[1] Successful separation of phenol, three cresols,

[40a] T. T. Lee and E. S. Yeung, *Methods Enzymol.* **270**, Chap. 19, 1996 (this volume).
[41] L. N. Amankwa and W. G. Kuhr, *Anal. Chem.* **63**, 1733 (1991).

three chlorophenols, and six xylenols was performed within 20 min. The theoretical plate numbers calculated from the chromatogram were 210,000 for phenol, 260,000 to 350,000 for cresols and chlorophenols, and 300,000 to 400,000 for xylenols and p-ethylphenols, corresponding to a theoretical plate height of 1.9–3.6 μm. These results proved the power of MEKC. Later, the application was extended to the separation of 20 polychlorinated phenols.[24]

B. Optical Isomers

Although the optical isomer category largely overlaps with the previous descriptions of chiral bile salt micelles in Section III,A and cyclodextrin in Section IV,A, we introduce some typical examples for this specific field.

Optical resolution was first reported[42] by CZE (or EKC) using additives of a Cu(II)–L-histidine complex, which was in equilibrium in a capillary with analyte amino acids, making diastereomers with DL-amino acids on complexation. In the field of MEKC, there are three approaches to the separation of optical isomers: the first is to use chiral micelles, the second is to add chiral additives to a micellar solution, and the third is to convert chiral isomers into diastereomers by chemical derivatization. Dobashi et al.[43,44] demonstrated the use of N-dodecanoyl-L-valinate as a chiral surfactant and applied it to the separation of esters of DL-amino acids. The use of a chiral bile salt micelle was reported by Terabe et al.[45] for dansyl amino acids, and by Nishi et al.[46,47] for drugs such as carboline derivatives and diltiazem in order to separate optical isomers.

The second approach is also popular. The use of a copper complex in CZE for the chiral separation of amino acids[42] was also applied to MEKC.[48] In this case, L-aspartyl-L-phenylalanine (aspartame) was the additive to sodium tetradecyl sulfate (STS) micelles. Cyclodextrin, added to SDS micelles, also gave an excellent selectivity for optical isomers. The chiral separation of some dansylated amino acids in 100 mM SDS and 60 mM γ-cyclodextrin was achieved (Fig. 8).[49] Other chiral additives and mixed

[42] E. Gassmann, J. E. Kuo, and R. N. Zare, Science 230, 813 (1985).

[43] A. Dobashi, T. Ono, S. Hara, and J. Yamaguchi, Anal. Chem. 61, 1984 (1989).

[44] A. Dobashi, T. Ono, S. Hara, and J. Yamaguchi, J. Chromatogr. 480, 413 (1989).

[45] S. Terabe, M. Shibata, and Y. Miyashita, J. Chromatogr. 480, 403 (1989).

[46] H. Nishi, T. Fukuyama, M. Matsuo, and S. Terabe, J. Microcol. Sep. 1, 234 (1989).

[47] H. Nishi, T. Fukuyama, M. Matsuo, and S. Terabe, J. Chromatogr. 515, 233 (1990).

[48] P. Gozel, E. Gassmann, H. Michelsen, and R. N. Zare, Anal. Chem. 59, 44 (1987).

[49] Y. Miyashita and S. Terabe, "Applications Data DS-767." Beckman Instruments, Fullerton, CA, 1990.

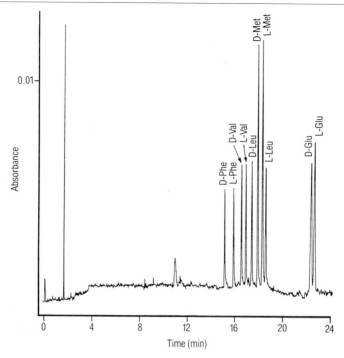

FIG. 8. Chiral separation of five dansylated DL-amino acids. Separation was performed with 60 mM γ-cyclodextrin in a 100 mM SDS solution (pH 8.3) at 12 kV (47 μA), capillary temperature of 20°, and detection wavelength of 200 nm. (Reproduced from Y. Miyashita and S. Terabe, "Applications Data DS-767." Beckman Instruments, Fullerton, CA, 1990.)

micelles with SDS investigated in MEKC are copper complexes of N,N-didecyl-L-alanine,[50] digitonin,[51] and saponins.[52]

An example of the third approach is shown in Fig. 9. A mixture of racemic amino acids was derivatized with optically active 2,3,4,6-tetra-O-acetyl-β-D-glucopyranosyl isothiocyanate (GITC), which was followed by separation with SDS–MEKC.[53] The diastereomers of GITC derivatives had similar electrophoretic mobilities and were not separated by CZE, but they were successfully separated by SDS–MEKC. Marfey's reagent (1-fluoro-2,4-dinitrophenyl-5-L-alanineamide) was also used in the third approach.[54]

[50] A. S. Cohen, A. Paulus, and B. L. Karger, *Chromatographia* **24**, 15 (1987).
[51] K. Otsuka and S. Terabe, *J. Chromatogr.* **515**, 221 (1990).
[52] Y. Ishihama and S. Terabe, *J. Liq. Chromatogr.* **16**, 933 (1993).
[53] H. Nishi, T. Fukuyama, and M. Matsuo, *J. Microcol. Sep.* **2**, 234 (1990).
[54] A. D. Tran, T. Blanc, and E. Leopold, *J. Chromatogr.* **516**, 241 (1990).

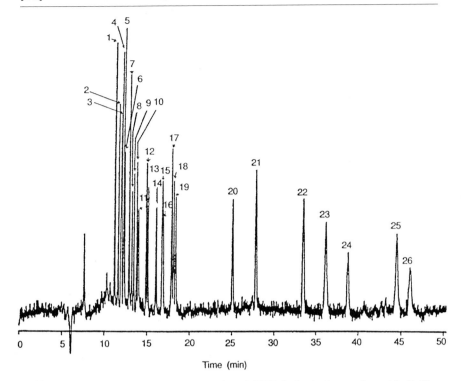

FIG. 9. Micellar EKC separation of 14 pairs of GITC-derivatized DL-amino acids. Buffer: 0.02 *M* phosphate–borate buffer solution, pH 9.0, containing 0.23 *M* SDS. Applied voltage; 20 kV. Peaks: 1, L-Pro + L-Thr; 2, L-Ala; 3, D-Pro; 4, L-His; 5, D-Ala + L-Val; 6, D-Thr; 7, D-His; 8, L-Tyr; 9, L-Met; 10, D-Val; 11, L-Ile; 12, L-PheG; 13, L-Leu; 14, D-Met; 15, D-Tyr; 16, D-Ile; 17, D-PheG; 18, L-Phe; 19, D-Leu; 20, D-Phe; 21, L-Trp; 22, D-Trp; 23, L-Lys; 24, L-Arg; 25, D-Lys; 26, D-Arg. [Reproduced from H. Nishi *et al.*, *J. Microcol. Sep.* **2**, 234 (1990) with permission of MicroSeparations, Inc.]

C. Pharmaceuticals

The application of MEKC in this area has been attracting great interest among analytical chemists. The area covers wide ranges of compounds, e.g., vitamins, barbiturates, steroids, antibiotics, anesthetics, and antipyretics. The main reasons MEKC is preferred to CZE in this area are, first, that MEKC can separate both ionic and nonionic compounds simultaneously and, second, that MEKC can separate derivatives with the same charge and a similar structure, which is often required in drug analysis. The binding of SDS to proteins is a problem in protein analysis by MEKC, which was used, in reverse, to detect small molecules in the matrix of human

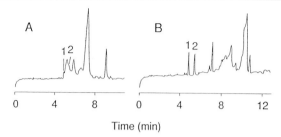

Fig. 10. Separation of cefpiramide (peak 2) and antipyrine (peak 1) in human plasma achieved in (A) 50 mM phosphate buffer at pH 8.0, and in (B) 10 mM SDS added to (A). Capillary, 60 cm × 50-μm i.d. (40 cm to the detector) (Scientific Glass Engineering, Inc.); applied voltage, 15 kV; current, 11 μA; detection wavelength, 280 nm. [Reproduced from T. Nakagawa *et al.*, *Chem. Pharm. Bull.* **36**, 1622 (1988) with permission of the Pharmaceutical Society of Japan.]

plasma (Fig. 10).[55] Sodium dodecyl sulfate allowed the direct injection of plasma with the effect of solubilizing proteins, releasing of protein-bound solutes, and delaying of the elution of proteins that often cause strong interference. The use of bile salt[56] or the addition of γ-cyclodextrin[57] to SDS micelles was also effective for the analysis of hydrophobic drugs, as mentioned in Sections III,A and IV,A.

Refer to Section VI,B for a description of the chiral separation of drugs and to another review article[58] for a more comprehensive description of the separation of pharmaceuticals by MEKC.

D. Catechols and Catecholamines

An important characteristic of catechols and catecholamines is that they are electrochemically active and are detected by amperometry both selectively and sensitively (Fig. 11).[59] However, because the amines have positive charge at neutral pH, they tend to be adsorbed on the negatively charged silica surface, which causes slightly tailed peaks.[60]

E. Amino Acid Derivatives

Although Jorgenson and Lukacs[61] reported the separation of dansyl derivatives of amino acids in the earliest stage of CZE, the peaks were

[55] T. Nakagawa, Y. Oda, A. Shibukawa, and H. Tanaka, *Chem. Pharm. Bull.* **36**, 1622 (1988).
[56] H. Nishi, T. Fukuyama, M. Matsuo, and S. Terabe, *J. Chromatogr.* **498**, 313 (1990).
[57] C. P. Ong, C. L. Ng, H. K. Lee, and S. F. Y. Li, *J. Chromatogr.* **547**, 419 (1991).
[58] H. Nishi and S. Terabe, *Electrophoresis* **11**, 691 (1990).
[59] R. A. Wallingford and A. G. Ewing, *J. Chromatogr.* **441**, 299 (1988).
[60] R. A. Wallingford and A. G. Ewing, *Anal. Chem.* **60**, 258 (1988).
[61] J. W. Jorgenson and K. D. Lukacs, *Anal. Chem.* **53**, 1298 (1981).

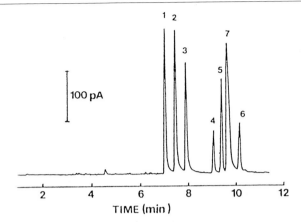

FIG. 11. Electrokinetic separation of catechols as borate complexes: 10 mM dibasic sodium phosphate–6 mM sodium borate at pH 7 with 10 mM SDS; separation capillary, 64.3 cm; detection capillary, 1.7 cm; separation potential, 20 kV (7 μA); injection, 4 sec at 20 kV. Peaks: 1, norepinephrine; 2, epinephrine; 3, 3,4-dihydroxybenzylamine; 4, dopamine; 5, L-3,4-dihydroxyphenylalanine (Dopa); 6, catechol; 7, 4-methylcatechol. [Reproduced from R. A. Wallingford and A. G. Ewing, *J. Chromatogr.* **441,** 299 (1988) with permission of Elsevier Science Publishers BV.]

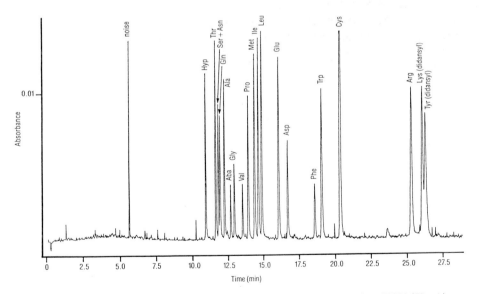

FIG. 12. Analysis of 21 dansyl amino acids. Separation was performed at 15 kV (52 μA), capillary temperature of 20°, and detection wavelength of 200 nm. Separation solution: 100 mM SDS in 100 mM borate (pH 8.3). (Reproduced from Y. Miyashita and S. Terabe, "Applications Data DS-767." Beckman Instruments, Fullerton, CA, 1990.)

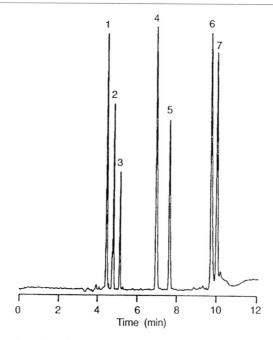

FIG. 13. Separation of peptides by MECC (micellar electrokinetic capillary chromatography). Samples: 1, des-Tyr^1-Met-enkephalin; 2, Met-enkephalin; 3, Leu-enkephalin; 4, [Val^5]angiotensin II; 5, angiotensin II; 6, angiotensin I; 7, angiotensin III. Conditions: effective capillary length, 40 cm; applied voltage, 275 V/cm ($I = 48 \mu$A); detection wavelength, 215 nm; injection, by electromigration for 2 sec; buffer, 10 mM borate–boric acid, 50 mM SDS, pH 8.5. [Reproduced from A. Wainwright, *J. Microcol. Sep.* **2,** 166 (1990) with permission of MicroSeparations, Inc.]

close together in three groups of basic, acidic, and neutral amino acids. The amino acids in the last group were rather difficult to separate by CZE. Later, improved resolution of dansylated amino acids was reported by Miyashita and Terabe (Fig. 12).[49] Similarly, PTH–amino acids were separated by both Otsuka *et al.*[15] and Terabe *et al.*,[40] and *o*-phthaldialdehyde (OPA)–amino acids were separated by Liu *et al.*,[62] all by MEKC. MEKC is suitable for the separation of compounds that have the same or similar charge, such as amino acid derivatives. Chiral separation of amino acid derivatives is described separately in Section VI,B.

[62] J. Liu, K. A. Cobb, and M. Novotny, *J. Chromatogr.* **468,** 55 (1988).

F. Peptides

When analyte peptides have different chain lengths or different overall charge, they can be separated by CZE because they have different electrophoretic mobilities. However, if two or more peptides have the same chain length and the same total charge, they usually have such close electrophoretic mobilities that they need to be separated by MEKC. The separation of angiotensin derivatives (biologically active peptides that have 7–10 amino acid residues) and related compounds has been reported by several groups.[17,18,63,64] Figure 13 shows one example. Larger peptides such as motilins and insulins have also been separated by SDS–MEKC.[32]

G. Proteins

The applications of MEKC to proteins are not yet popular because charged micelles bind irreversibly to the backbone of proteins and alter the overall charge and broaden their peaks. There are two reports that describe effective use of surfactants in CZE of proteins. One describes the addition of nonionic surfactant in a surface-coated capillary column.[65] Performance of separation of proteins was enhanced because hydrophobic interaction of proteins and octadecyl groups on the coated capillary wall was reduced. Another paper reported the use of a fluorinated cationic surfactant for the separation of basic proteins, already mentioned in Section III,A.[16] Size fractionation of proteins based on SDS–capillary gel electrophoresis is discussed in chapter 13 in Volume 271 of this series.[65a]

H. Nucleic Acids

Separation of nucleotides and bases by MEKC has been reported by some groups,[66,67] but sequencing of oligonucleotides and DNA fragments, which is of primary interest in this field, is left to capillary gel electrophoresis.

[63] A. Wainright, *J. Microcol. Sep.* **2**, 166 (1990).

[64] J. Liu, K. A. Cobb, and M. Novotny, *J. Chromatogr.* **519**, 189 (1990).

[65] J. K. Towns and F. E. Regnier, *Anal. Chem.* **63**, 1126 (1991).

[65a] B. L. Karger, F. Foret, and J. Berka, *Methods Enzymol.* **271**, Chap. 13, 1996.

[66] A. S. Cohen, S. Terabe, J. A. Smith, and B. Karger, *Anal. Chem.* **59**, 1021 (1987).

[67] K. H. Row, W. H. Griest, and M. P. Maskarinec, *J. Chromatogr.* **409**, 193 (1987).

[15] Structure–Mobility Relationships in Free Solution Zone Electrophoresis

By STAN MICINSKI, METTE GRØNVALD, and BRUCE JON COMPTON

Introduction

The correlation of the structure of a solute with electrophoretic mobility is of interest for numerous reasons. Mobility in free solution gives a direct measure of the net effective charge of the solute, which in turn can be a sensitive and direct measure of many solution phase interactions. Among these are stability, activity, and solubility, which for solutes such as proteins, the subject of this work, are all influenced by solution pH and the charge state of the protein. From a bioanalytical perspective, a knowledge of how solute structure affects mobility aids in the optimization of high-resolution separations. In general, correlation of solute structure with mobility, or the equivalent charge state of a solute, is fundamental to an understanding of how proteins behave in solution.

With this in mind, it is noteworthy that capillary electrophoresis (CE) conducted in free solution is an ideal means of determining the effective net charge of proteins. Presented here are the results of investigations that have shown that mobility predictions of peptides and proteins as a function of solution pH are attainable if the amino acid content of the solute is known.[1,2] As an introduction to this subject, the body of this work demonstrates that the easily calculated valence–pH titration curves for proteins are directly proportional to mobility–pH titration curves. The Appendix at the end of this chapter extends the discussion, emphasizing calculated and actual protein valence, and demonstrating how equations derived by inspection in the body of the work are arrived at fundamentally.

It should be mentioned that an extrapolation of this work is affirmation that solute mobility in free solution is determined more by protein valence than size. Thus, because free solution separations are primarily charge based, the observation that background buffer pH plays the predominant role in determining resolution in free solution CE is given a quantitative basis.

---

[1] B. J. Compton, *J. Chromatogr.* **559**, 357 (1991).
[2] B. J. Compton and E. A. O'Grady, *Anal. Chem.* **63**(22), 2597 (1991).

METHODS IN ENZYMOLOGY, VOL. 270

Electrophoretic Mobility Modeling

Despite the fact that a theoretical basis for predicting solute electrophoretic mobility as a function of charge and size has been available for over half a century,[3] most analyses are developed empirically. This is understandable in light of the fact that much of the previous work could not be easily reduced to practice. The work described here sacrifices generality for practicality in that it is limited in scope to proteins.

The motion of an unbound charged solute in an external electrical field is determined by the nature of the solute itself (size, shape, magnitude of the net charge, etc.) and the external environment of the surrounding medium (viscosity, dielectric strength, pH, etc.). All of these influence the resulting magnitude of the net force on the solute, which in turn determines electrophoretic mobility.

Empirical Relationships

It should be noted that the complexity of the subject has resulted in numerous empirical attempts at describing mobility of peptides.[4,5] While these efforts are meritorious in that they successfully describe the relationships under study, they lack generality and hence predictive value because they apply in a quantitative sense only to the exact experimental conditions under which they were derived. Also, none of these studies have accommodated the roles of background buffer ionic strength and pH. This is a serious omission because optimization of CE separations is principally achieved through control of these two important attributes of the background buffer. As will be seen, buffer ionic strength defines how solute size influences mobility while pH determines directly the charge state of the solute.

Notwithstanding the limitations of this previous work, its findings must be incorporated into any generalized model of mobility. In summary, the empirical relationships have indicated that solute mobility is, with one exception, a direct function of net charge (Z) and inversely dependent on molecular weight (M) to a power of 1/3 to 2/3. The exception regards a charge relationship that can be explained by examining the differences between calculated (Z_c) and actual or effective charge (Z_a) on a protein and the role that electrostatic charge reduction has in rationalizing these two values.[1,2] It should be noted that valence (Z) and charge (i) are equated by the constant e, the elementary charge ($i = Ze$). Also, by convention

[3] D. C. Henry, *Proc. R. Soc. London* **A133,** 106 (1931).
[4] E. C. Rickard, M. M. Strohl, and R. G. Nielsen, *Anal. Biochem.* **197,** 197 (1991).
[5] R. E. Offord, *Nature (London)* **211,** 591 (1966).

electrophoretic mobility (μ) is defined in unit electric field (1 V/m), hence the following treatment assumes field strength is unity.

With this taken into consideration, the empirical relations can be summarized by the expression

$$\mu = CZ_a/M^n \tag{1}$$

where C is a system-dependent constant, and n varies from 1/3 to 2/3 depending on both the magnitude of M and as shown in the Appendix. It should be mentioned that n approaches 1/3 for small solutes such as simple ions in low ionic strength buffers and 2/3 for large solutes such as colloids and particles in high ionic strength buffers. Peptides and proteins fall within a broad molecular weight range and are studied under a variety of solution ionic strengths, giving rise to the variable molecular weight dependencies seen in the past.

Semiempirical Mobility Modeling

The theoretical (calculated) net valence (Z_c) of a protein is simply a sum of the individual valences and, as a function of pH, can be calculated directly from the protein's amino acid content from sequence information, using the familiar Henderson–Hasselbalch equation. Posttranslational modifications to the protein, if known or suspected, can likewise be accounted for by proper consideration of ionizable substituents. The result of this effort is a Z_c–pH titration curve that is directly proportional to the corresponding μ–pH titration curve.

It is important to note that the reason Z_c and μ are proportional over a wide range of pH is because Z_c, again the calculated valence, is related to actual solute valence (Z_a) by a simple constant f_z (where $f_z = Z_a/Z_c$) that is pH independent. This rather remarkable pH independence of f_z allows determination of solute valence or equivalent mobility at one pH to be extrapolated over an entire pH range of interest. Recent work indicates the approximate magnitude of f_z may be determined *a priori*, as further discussed in the Appendix.[1,2]

Z_c–pH Titration Curves via Henderson–Hasselbalch Equation

The theoretical net charge, Z_c, as a function of pH, is calculated using the familiar Henderson–Hasselbalch equation in the following form[6]:

$$Z_c = \sum_{n=1,4} \frac{P_n}{1 + 10^{pH - pK(P_n)}} + \sum_{n=1,5} \frac{N_n}{1 + 10^{pK(N_n) - pH}} \tag{2}$$

[6] A. Sillero and J. M. Ribeiro, *Anal. Biochem.* **179**, 319 (1989).

TABLE I
Intrinsic Ionization Potentials for Some Common
Amino Acids and Terminal Groups and
Corresponding Amino Acid Contents of Bovine
Serum Albumin and Recombinant Human FXIII

Amino acid type	Terminal group	pK_a	BSA	rhFXIII
P1	His	6.4	17	87
P2	tNH$_2$	8.2	1	0
P3	Lys	10.4	59	38
P4	Arg	12	23	45
N1	tCOOH	3.2	1	1
N2	Asp	4	41	87
N3	Glu	4.5	59	75
N4	Cys	9	35	9
N5	Tyr	10	20	29

P_n and N_n refer to the respective positively and negatively charged amino acids listed in Table I. Solving for Z_c as a function of pH is accomplished by substitution of the appropriate number of charged amino acid residues. The ionization constants (pK) used in this work are given in Table I and are the intrinsic constants for free amino acids. The actual ionization constants differ depending both on solution conditions (ionic strength and dielectric constant, temperature, etc.) and on the actual location of a charged residue in a protein.

Experimental Method

By way of example, two proteins have been chosen to illustrate charge-based separations in free solution zone CE. Bovine serum albumin (BSA, estimated pI 5.73, molecular weight 69,000) from Sigma Chemical Co. (St. Louis, MO) and recombinant human factor XIII (rhFXIII, estimated pI 5.67, molecular weight 168,000) produced at ZymoGenetics, Inc. (Seattle, WA) were chosen because of their availability. These proteins are maintained in solution containing 0.5 mM EDTA, 10 mM glycine, 2% (w/v) sucrose, and 0.04% (w/v) formamide, pH 7.2. Free zone capillary electrophoresis for protein mobility determinations is performed using an untreated 50-mm i.d. × 27 cm capillary from Polymicro Technologies (Phoenix, AZ). A detection window 7 cm from the end is made by cleaning off the polyimide coating, using a flame, giving a capillary length of 20 cm. A P/ACE System 2050 (Beckman, Palo Alto, CA) with a detection wavelength of 214 nm is used for all measurements.

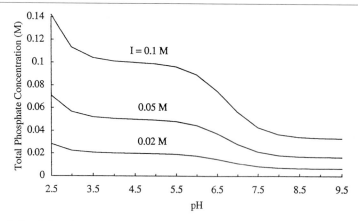

FIG. 1. Total phosphate concentration in various constant ionic strength phosphate buffers as a function of pH, illustrating how ionic strength varies with pH.

Separations are conducted in 50 mM phosphate (pH range 2.0 to 3.5) or borate buffers (pH range 7.0 to 9.0, sodium salts). These pH ranges represent conditions under which these buffers have relatively constant ionic strength (see, for example, Fig. 1 for orthophosphate). The expected influence of ionic strength (I) on μ is relatively weak because $\mu = f(I^{1/2})$. Specific ionic interactions are ignored in this study, although both phosphate and borate can potentially interact in a specific fashion with these proteins.

All sample concentrations are 0.4 mg of protein per milliliter. Sample loading is by pressure for 2 sec, giving a calculated load volume of 0.13 nl. All experiments are conducted at 12 kV at a temperature of 25.0°. Higher voltages result in current fluctuations and irreproducible results attributed to capillary heating. The capillary is purged with 0.1 M NaOH and running buffer prior to each analysis.

Amino acid data used for calculating the BSA protein charge are taken from the Protein Information Resource (PIR) data bank while for rhFXIII amino acid composition data comes from ZymoGenetics sequence data. These data are also summarized in Table I. Mobility results are corrected by accounting for mobility contributions from endosmotic flow, as previously mentioned, based on the mobility of formamide.[1,2]

At each pH for each component, determination of the migration time is done in replicates of at least three. The mean standard deviation for the measurements is 3.5% or lower. Only at a pH of 7.0 are the deviations greater (5%), probably because at this pH rhFXIII has limited solubility.

Calculating Z_c and μ–pH Titration Curves

The following steps are followed to calculate Z_c and μ–pH titration curves.

1. Amino acid data such as seen in Table I are substituted into Eq. (2) and Eq. (2) is solved for the desired pH range, for example, pH 2.0–11.0. These calculations can be accomplished using personal computer spreadsheet application programs.
2. One measurement of solute mobility at a convenient pH, for instance pH 2.0, is made. Electrophoretic mobility is calculated from CE data by the following expression:

$$\mu = L_1 L_2 / Vt \qquad (3)$$

   where L_1 and L_2 are the capillary length in meters from end to end and from end to detection window, respectively, V is the applied voltage, and t is the time in seconds a solute zone spends migrating from the capillary end to the detection window.
      Calculations for example: Using the experimental conditions described in the previous section ($L_1 = 0.27$ m, $L_2 = 0.20$ m, $V = 12$ kV) and Eq. (3), for a solute requiring 900 sec to be detected a mobility of 5×10^{-9} m^2/Vs is found.
3. Converting the Z_c–pH curve to a μ–pH curve requires a scale change. This is accomplished by determining a scaling factor, μ/Z_c, where the values for μ and Z_c are for the same pH.
4. This scaling factor is then applied to the entire pH range by multiplying respective Z_c values by the scaling factor.
5. For optimization of CE separations, the μ values from the μ–pH curve are substituted into Eq. (3) and solved for t as a function of pH. The optimum pH is arrived at by inspection.

Results and Discussion

As mentioned previously, conversion of the Z_c–pH curve to a μ–pH titration curve explicitly assumes that a pH-independent factor such as f_z exists because the calculated scaling factor, based on Eq. (1) and the definition of Z_c, is equivalent to M^n/Cf_z. The great diversity of proteins and their properties makes it unlikely that f_z will be absolutely constant throughout a broad pH range because changes in protein conformation and secondary interactions between the protein and background buffer components or the protein itself (aggregation) will likely change the microenvironment of charged residues enough to alter the fundamental state of the protein.

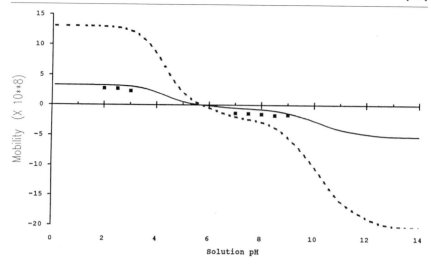

FIG. 2. Plot of the calculated (μ_c, ---), measured ($\mu_{a,m}$, discrete data points), and corrected ($\mu_{a,c}$, —) mobilities of BSA as a function of solution pH. Mobilities at the low and high pH ranges were determined in phosphate and borate buffers, respectively. The measured mobilities were corrected for endosmotic flow. The calculated mobilities are from the Appendix. The corrected mobilities were scaled to the measured mobilities by using μ measured at pH 2.0.

A demonstration that this approach is feasible and that f_z is relatively constant over a wide pH range is shown in Figs. 2 and 3. In these examples, μ and Z_c were determined at a pH of 2.0 and applied to the rest of the pH range shown. Both Figs. 2 and 3 show that the actual titration curve ($\mu_{a,m}$, the discrete data points) is coincident with the scaled calculated titration curve ($\mu_{a,c}$). A detailed calculation of the theoretical (μ_c) curve (see Appendix for details), is also shown to illustrate how μ_c and $\mu_{a,c}$ are related by f_z.

In the case of BSA and rhFXIII the values of f_z were found to be 3.1 and 3.3, respectively. The physical interpretation of these two values is that the maximum total valence on the protein at any given pH (Z_c) is approximately three times the actual valence (Z_a). The discrepancy is due to charge suppression, which occurs to some extent whenever a protein is not in solution at its isoelectric point. It is noteworthy that while CE measurement of protein valence gives Z_a, an actual acid–base titration of a protein will result in Z_t, which has the same magnitude as Z_c.[6,7]

The μ–pH titration curve can be used to directly predict optimum pH

---

[7] P. C. Hiemenz, "Principles of Colloid and Surface Chemistry." Marcel Dekker, New York, 1986.

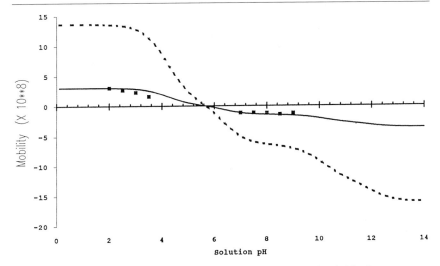

Fig. 3. A plot of rhFXIII results, similar to that for BSA in Fig. 2.

conditions for the separation of complex mixtures. Direct substitution of experimentally determined μ into Eq. (3) gives mobility time (t) for a given solute under defined experimental conditions. While the examples presented here were an oversimplification for pedagogical purposes, in theory suspected isoelectrotypes of a heterogeneous protein sample can be resolved by this approach of pH optimization. An alternative use of this approach is to confirm, at least in theory, the identity of bands in an electropherogram. The separation of isoelectrotypes of a monoclonal antibody was previously performed in free solution CE.[1] In this application it was postulated that the heterogeneity of the antibody was due to deamidation. A comparison of the theoretical titration curves for hypothetical deamidated forms of the antibody indicated they could be resolved at a pH approaching neutrality. This was verified experimentally, lending credence to the deamidation hypothesis as well as resulting in an analytical tool useful for measuring the heterogeneity of the antibody.

It is expected that the maximum valences of solutes such as proteins occur at the extremes of pH but at these extremes, proteins as a class tend to have similar valences. This is illustrated here for BSA and rhFXIII, two very different proteins yet with similar mobilities. One explanation for this is that these biomolecules are limited in the maximum attainable charge density (valence per unit surface area). This limitation can be hypothesized to be due to the balance between the cohesive forces (ionic, van der Waals, and hydrophobic) that determine the higher order structure of a protein, the coulombic forces (attractive and repulsive) on the surface, and the ionic and dielectric state of the background buffer that tends to minimize the ionic surface interactions.

Thus, while separations at the extremes of solution pH are currently experimentally attractive in CE with regard to charge reduction at capillary walls (at low pH) and solute–wall repulsion (at high pH), these extremes of pH limit the selectivity of the method. This leads one to conclude that the potential general usefulness of CE currently resides in its high efficiency, while the ability to predict (and control) solute mobility accurately at intermediate pH will greatly increase its usefulness in the future.

Appendix: Detailed Treatment and Discussion of Solute Electrophoretic Mobility as Applied to Proteins

Electrophoretic Mobility of Proteins Based on Debye–Hückel–Henry Theory

Proteins are usually characterized by their primary, secondary, and tertiary structures, which translate into the three-dimensional conformation that determines form and function. This structure indirectly affects the charge on a protein at a given pH because the equilibrium constants of the ionizable amino acids are influenced by the local environment and in general are not the same as those in free solution. However, in general charged groups will lie on the surface of the protein with the possible exception of residues in the cleft of the active site.

The forces acting on the discrete nature of the surface charges on a protein are well known and include electrostatic forces, van der Waals forces, hydrogen bonding, and hydrophobic interactions, all of which influence the conformation of the protein itself as well as its interaction with the medium, for example, by adsorption of buffer ions and by hydration effects. While computer models exist that account for all of these interactions, they are far too complex for general usage in that they require a high-resolution coordinate map of every amino acid in the protein. Simpler models that treat the continuum nature of the particle and medium such as the Debye–Hückel–Henry theory of charged particles lend sufficient detail and yet are generally applicable. Where such theory fails to give an accurate numerical result, a semiempirical approach is taken. Such is the case, for example, when the Debye–Hückel–Henry theory fails to predict the extent of the charge suppression due to the counterion adsorption and hydration at the surface which is needed to predict *a priori* the effective surface charge inside the shear interface (or zeta potential at the shear interface). The semiempirical approach presented below accounts for these problems and is generally applicable and easily implemented.

A reasonable starting point for discussing the electrophoretic mobility of a charged particle is the classic Debye–Hückel–Henry[3] theory of electrophoresis, expressed by

$$\mu = \frac{Z_a e}{6\pi \eta r_s} \frac{\Phi(\kappa r_s)}{(1 + \kappa r_s)} \qquad (4)$$

$$\kappa = \left(\frac{2e^2 N}{\varepsilon k T} I \right)^{1/2} \qquad (5)$$

where Z_a is the actual charge, η is the viscosity of the medium (0.000895 kg·m^{-1}sec^{-1} for water, 25°), κ is the Debye parameter, which is a measure of the thickness of the ionic double layer surrounding the protein, r_s is the Stokes radius of the protein, I is the solution ionic strength (mol/m^3), ε is the dielectric constant of water (80 at 25°), N is Avogadro's number, k is Boltzmann's constant (1.38×10^{-23} J), and $\Phi(\kappa r_s)$ is Henry's function (1.0–1.5), which takes into account the distortion of the electrical field due to the presence of the charged particle. For reference purposes κ has a value of approximately $1 \times 10^{-8} I^{1/2}$, with units of m$^{1/2}$mol$^{-1/2}$ for 25° in water. When I in this expression is taken into account the units for κ are m^{-1}.

In this case the Debye–Hückel restriction for small values of the zeta potential applies as determined from the inequality $(\zeta e/kT) < 1$, where ζ is the zeta potential, e is the elementary charge (1.6×10^{-19} C), and T is the absolute temperature.[7] Henry's function $\Phi(\kappa r_s)$ varies in a sigmoidal fashion between the values of 1.0 and 1.5 for intermediate values of κr_s near unity and it bridges the gap between small ions and large particles, i.e., particles in the colloidal size range.[8] The parameter $\Phi(\kappa r_s)/(1 + \kappa r_s)$ is often considered a retardation factor to the simpler Debye–Hückel equation for finite-sized particles and small values of the zeta potential.

Charge Effects

Proteins are amphoteric macromolecules whose net charge depends on the ionization potential of its ionic groups. The net valence, Z, is simply the sum of the individual charges that result from those amino acid groups that are ionized.[8] Calculation of the theoretical charge of a protein is usually accomplished by the Henderson–Hasselbalch equation [Eq. (2)], which gives the net valence (Z_c) as a function of solution pH.

[8] J. T. G. Overbeek, in "Advances in Colloid Science" (H. Mark and E. J. W. Verwey, eds.), p. 97. Wiley-Interscience, New York, 1950.

Such a titration curve is easily generated, as shown in the main body of this work, given the pK values of the nine ionizable amino acids and the terminal amine and carboxylic acid groups (Table I). This general expression can be applied to other ionizable groups such as those attributed to posttranslational modification of proteins (glycosylation, phosphorylation, etc.), by treating these groups in a fashion analogous to that of the amino acids. Such theoretical titration curves are shifted from that obtained from membrane potentials or by electrophoresis determinations for several reasons. The most important cause is the formation of the first peptide bond, which induces changes in the neighboring amino and carboxylic acid groups.[4] The overall peptide bond sequence is, of course, the backbone of the primary structure of the protein. Amino acid groups that may be distal from the ionizable one of interest according to the primary structure may exert intramolecular forces on the free charge if they are spatially close together in the three-dimensional conformation. In other words, the secondary and tertiary structures give the local microenvironment that affects the ionization potential of the amino acid. If the group is hidden in the hydrophobic interior such as may occur in the active site of an enzyme, the low dielectric constant will oppose ionization. Those that are exposed to the aqueous exterior are influenced by the ionic strength of the medium.

In summary, there are many structural influences that will shift the acid and base equilibrium constants of a protein from those of the free amino acid. Because these effects are not likely to be quantitated without experimental validation for each protein of interest or sophisticated modeling of the intramolecular forces in three dimensions, the Henderson–Hasselbalch equation must be used as a first approximation along with an empirical approach to account for the charge shielding. Given the other uncertainties due to the electrical double layer discussed below, such an approximation is warranted.

There are secondary effects that reduce the theoretical charge of a protein below that which is predicted on the basis of the pK values of the free amino acids. Such a reduction in the effective charge is due to the charge suppression of the surface charges by the inner portion of the electrical double layer, the so-called Stern layer. Within the Stern layer specific adsorption of the counterion (or possibly even a coion) occurs that reduces the net charge, and because this adsorption occurs within the shear interface, the adsorbed charges are assumed fixed. Specific adsorption usually lowers the surface potential and is functionally dependent on the type of ions present in solution and their concentration (ionic strength). Specific adsorption in electrocapillary studies is well documented and follows the lyotropic series according to ionic valance and size. Also, the dielectric constant of water within the Stern layer is probably considerably less than in the bulk solution owing to the exceptionally high electrical field in this region.[7] Such effects are difficult to quantitate because of the complex nature of the adsorption, which is both electrostatic and chemical in nature, and because of the difficulties in estimating the thickness of the shear layer. The resultant charge suppression leads to a discrepancy between the theoretical charge, based on the amino acid content of the protein, and the measured charge by either electrophoresis or membrane potentials.[7,8]

The usual approach to circumventing these difficulties is to measure indirectly the actual charge of the protein within the Stern layer via membrane potential measurements. The measured potential assumed to be the zeta potential at the sruface of shear (the effective surface of the molecule) can be equated to the actual charge from the expression $Z_a = 4\pi\varepsilon\zeta r_s(1 + \kappa r_s)$ for spherical particles, where ε is the medium dielectric constant. The Debye–Hückel restriction that the potential be small, and consequently not distort the Stern layer, applies. As is discussed below, the modern use of capillary electrophoresis via widely available commercial units makes the direct calculation of electrophoretic mobilities from first principles an easily applied technique. From Eq. (4) the actual charge can be determined and will, of course, depend on the product of the particle radius and the Debye parameter (κr_s), because this determines the retardation factor to be used in Eq. (4).

Size and Shape Effects

Molecular size and shape are important in calculating the mobility of a protein because they are determinants in the Stokes radius in Eq. (4). The size of particles such as proteins, which lie at the lower end of the colloidal size range, is difficult to measure directly except by specialized techniques such as light scattering. However, because the intent of a generally useful theory is one that can easily be reduced to practice, more amenable methods are sought. As discussed by Oncley,[9] the radius of a molecule can be related to the molecular weight (M) by

$$r_e = (3Mv/4\pi N)^{1/3}(f/f_0) \tag{6}$$

where r_e is the equivalent radius, v is the protein partial specific volume (approximately constant for most proteins, 0.70–0.75), and f/f_0 is the frictional ratio. Thus, this simple expression contains both the size parameter M and the shape parameter f/f_0. The molecular weight is, of course, easily calculated from the amino acid sequence of a protein.

[9] J. L. Oncley, *Ann. N.Y. Acad. Sci.* **41**, 121 (1941).

The shape of a protein is important from the standpoint of frictional drag. The shape of the protein will influence the orientation so as to minimize the drag force while in motion. The fixed charges, if not randomly and evenly distributed, will impart the equivalent of a dipole moment on the protein, causing it to assume a fixed orientation when subject to an electrical field. Because proteins are large enough relative to the solute molecules with diffusion coefficients on the order of 10–100 times less, the random interaction of the molecular diffusion is not expected to smooth out the effects of this orientation.

The frictional ratio f/f_0 accounts for the asymmetry as well as solvation, is equal to unity for unsolvated spheres, and can be determined experimentally from sedimentation or diffusion data because $f = kT/D$, where D is the molecular diffusion coefficient. Typical values for f/f_0 range from 1.0 to 1.7 for globular proteins.[9]

The equivalent radius may be taken as the Stokes radius in Eq. (4) because the radius of a solvated protein is roughly equivalent to that at the shear interface. It is also important to note that the protein radius appears in two terms in Eq. (4), giving a molecular weight dependency by substitution with Eq. (6) of $M^{-1/3}$ and $M^{-2/3}$ as shown in the final section. Such a dependence shown in a single expression unifies earlier empirical models that show dependencies of $-1/3$, $-1/2$, and $-2/3$ power depending on the assumptions used in the derivations and the media over which the electrophoresis was carried out.[4,5]

Medium and Buffer Effects

The mobility of a protein is affected by the external environment as well as by the physicochemical characteristics of the protein itself. The viscosity of the medium is inversely related to the mobility as shown by Eq. (4). The dielectric strength of the medium plays a much more complex role. As shown by Eq. (4) the dielectric constant affects the Debye parameter by changing the magnitude of the electrical field surrounding the protein as well as the solute ions. A decrease in the dielectric constant by, for example, adding organic solvent to the medium acts to increase the electrical field and decrease the thickness of the ionic atmosphere ($1/\kappa$) surrounding the charge. The dielectric constant has an even more subtle effect on the mobility than the direct influence of this parameter on the Debye parameter. Changing the dielectric constant also affects the dissociation constants of the weakly ionizable amino acids of the protein. A lower dielectric constant favors the neutral species and leads to a net lowering of the overall protein charge. For most protein applications, however, mobility and protein structure are studied in aqueous buffered systems so that the viscosity and dielectric constant are taken to be that of pure water. It is the number and type of ions that are most variable because these are used in controlling the solution pH and ionic strength.

The role of ionic strength in determining mobility is important for several reasons. The primary effect of ionic strength as shown by Eq. (4) is in the $\Phi(\kappa r_s)/I^{1/2}$ dependency, which is most apparent for large values of $\kappa r_s \gg 1$. Thus, for a given value of the Henry function, which ranges only from 1.0 to 1.5, the ionic strength will have at worst an $I^{-1/2}$ dependency. This effect on mobility is well established as shown by the pioneering work of Henry.[3] Secondary but perhaps more influential factors include the effect of ionic strength on the adsorption of buffer ions onto the protein surface (charge suppression) and on the equilibrium constants of the ionizable groups of the protein. Increasing the ionic strength is expected to cause more charges to adsorb onto the protein surface in a Langmuir-type isotherm, up to a surface coverage that corresponds to saturation.[9] Saturation at high ionic strength corresponds to maximum charge suppression and a minimum in the effective charge on the protein. This lends additional support to the need to conduct mobility measurements at constant ionic strength when pH effects are being studied. Also, as mentioned, the intrinsic adsorbability of buffer ions onto the protein surface is dependent on the types (valence and size) present in solution. These secondary effects are difficult to quantitate owing to the complex nature of the interactions. However, such effects are negligible for pH studies done at constant ionic strength and when the buffer ions are essentially unchanged.

Numerous studies of the effect of pH on protein mobility by changing the net charge of the protein have been reported. Abramson[10] and Longsworth[8] both examined the influence of pH on mobility. The work by Abramson did not attempt to fix the ionic strength because the work was over a sufficiently small pH range to diminish this effect. Longsworth used weakly dissociated acids and bases (phosphates and acetates) to vary pH over a wide range while adding NaCl to control ionic strength.

As shown in Fig. 1 for phosphate-buffered systems the ionic strength varies with pH at constant total phosphate (the usual method for preparing phosphate buffer). These results were calculated for ionization constants of $pK_1 = 2.12$, $pK_2 = 7.21$, and $pK_3 = 12.0$ and taking into account all charged species in the determination of ionic strength. Thus, phosphate-buffered systems near physiological pH show the greatest variation in ionic strength while ionic strength changes very little at low pH (3–6), where

[10] H. A. Abramson, M. H. Gorin, and L. S. Moyer, *Chem. Rev.* **24,** 345 (1939).

the ionic strength is equal to the molar concentration of total phosphate. Such equilibration curves may need to be considered unless large amounts of added salts (relative to the pH controlling acids and bases) are used to control ionic strength.

Calculating Charge Suppression

As previously discussed, the theoretical charge of a protein, Z_c, as calculated by Eq. (3), and actual charge, Z_a, as calculated from mobility measurements using Eqs. (4)–(6), are not coincident. Therefore, the calculated charge is of limited usefulness unless it can be equated to the actual charge. Several investigators[6,7] have postulated that a constant of proportionality exists between the charge of a protein as measured by titration, presumed to measure the theoretical charge Z_c, and the charge as measured by electrophoresis, presumed to be equated to the actual charge Z_a, within the shear interface. The constant of proportionality is independent of pH so long as the ionic strength of the medium is held constant. Overbeek[8] used the data of Longsworth to show that for egg albumin the charges by titration are 1.7 times those calculated from electrophoresis over a pH range for 3 to 12 when the ionic strength was fixed at 0.1 M. Thus, for a given ionic strength, a charge suppression factor as defined by $f_z = Z_a/Z_c$ can be experimentally determined independent of pH and is expected to be valid over the entire pH range. Such constancy of f_z does, of course, assume that the ionic strength is held constant by ions of the same chemical specificity for the protein surface and that the protein is stable over the pH range of interest. Finally, because f_z is independent of pH, it can be used to calculate the mobility of a protein along the entire pH curve from the theoretical charge of a protein. Given two proteins whose amino acid content is known, f_z for each protein can be experimentally determined at a single pH and their resultant μ–pH curves used to optimize their resolution over the entire pH range as further discussed in the body of this work.

Mobility from Capillary Zone Electrophoresis

Electrophoresis is performed in many different modes and in complex media. However, an as yet barely discovered use for capillary zone electrophoresis is for determining electrophoretic mobilities and solute diffusion coefficients, which together will give solute molecular weight.

Because the type of buffer and ionic strength affect the mobility of solutes, the conditions of the electrophoresis operation should be the same as those for which the mobility analysis is sought. If multiple runs are to be made at various pH values, it is desirable to maintain a relatively constant ionic strength either by the addition of a salt or by considering the equilibrium relationships of the buffer as was shown in Fig. 1 for phosphate-buffered systems.

The initial investigation of protein should be conducted at a pH far from the pI of all proteins studied because such conditions will lend greater accuracy in the determination of the charge suppression factor, f_z. A low pH is favored so long as the protein is stable and soluble, for several reasons. (1) Below the pK_a of the silanol groups of the typical uncoated glass capillary (pH < 2.5), the equilibrium favors the neutral species and protein adsorption is minimized, giving more accurate mobility data; (2) the suppression of silanol charges on the capillary wall also acts to dampen out any electroosmotic flow. Because the applied voltage is directly proportional to the number of theoretical plates, high voltages are favored until Joule heating becomes significant, especially for large-diameter capillaries. The maximum voltage will depend on the conductivity of the buffer system. Analysis of the voltage-to-current relationship from several capillary runs will give an indication of the degree of heating. A linear relationship indicates that no excessive heating has occurred.

Once the μ–pH curves have been generated for each protein then their resolution can be optimized by selection of the pH at which their mobility difference is the greatest. This can be done with graphical aids by inputting μ into Eq. (3) and plotting t against pH. As previously shown, such an approach can be used to optimize the separation of immunoglobulin G (IgG) isoelectrotypes.[1]

Because Eqs. (4)–(6) are of fundamental value but offer little direct aid in predicting mobility of proteins and Eq. (1) has been shown to be empirically useful and consistent with a variety of experimental systems, it is important to show that these sets of equations are self-consistent.

Combining Eqs. (4)–(6) gives, with simple algebraic manipulation, the following:

$$\mu = \frac{C_1 Z_a}{C_2 M^{1/3} + C_3 M^{2/3} I^{1/2}} \tag{7}$$

where the constants C_1–C_3 are complex constants from Eqs. (4)–(6). Substitution into these constants results in $C_1 = 9.5 \times 10^{-18}$ m^{3}V^{-1}sec^{-1}, $C_2 = 6.62 \times 10^{-10}$ mol$^{1/3}$kg$^{-1/3}$m, and $C_3 = 4.55 \times 10^{-11}$ m$^{5/2}$mol$^{1/6}$kg$^{-2/3}$, where it is assumed that $\Phi(\kappa r_s)$ and (f/f_0) are unity. Substitution into Eq. (7) of Z_c from Eq. (2) over an extended pH range results in a μ_c–pH titration curve as shown for BSA and rhFXIII in Figs. 2 and 3. Measuring actual mobility at one pH (in the examples presented here, solution pH was 2.0) and taking the ratio of actual mobility, as calculated by Eq. (3), to μ_c gives f_z. This constant is then applied to the entire pH range to give the resultant $\mu_{a,c}$ curves shown in Figs. 2 and 3.

Equation (7) also shows that the dependency of μ on M varies from $M^{-1/3}$ and $M^{-2/3}$, depending on both the magnitude of M and medium ionic strength I. Equation (7) reduces directly to Eq. (1), where C represents an aggregate of C_1–C_3. Thus, a physical basis for the empirical studies that were used to arrive at Eq. (1) is established.

[16] Capillary Isoelectric Focusing

By TIM WEHR, MINGDE ZHU, and ROBERTO RODRIGUEZ-DIAZ

Introduction

Capillary isoelectric focusing (cIEF) combines the high resolving power of conventional gel isoelectric focusing with the advantages of capillary electrophoresis (CE) instrumentation. In this technique, proteins are separated according to their isoelectric points in a pH gradient formed by ampholytes under the influence of an electric potential. When performed in the capillary format, the use of small-diameter capillaries with efficient dissipation of Joule heat permits the application of high field strengths for rapid focusing. The separations can be performed in free solution, without the need for gels. The use of ultraviolet (UV)-transparent fused silica capillaries enables direct on-tube optical detection of focused protein zones, without the requirement for staining. In some commercial CE systems, the cIEF process can be performed automatically, allowing unattended analysis of multiple samples.

As in conventional IEF, the high protein-resolving power in capillary IEF depends on the focusing effect of the technique. At steady state, the distribution of ampholytes forms a stable pH gradient within which proteins become focused at the position where their net charge is zero, i.e., where pH equals pI. Diffusion of a protein toward the anode will result in acquisition of positive charge, resulting in return to the focused zone. Similarly, diffusion toward the cathode will result in acquisition of negative charge, causing back-migration to the zone. As long as the field is applied, electrophoretic migration thus counters the effects of diffusion. Because of this focusing effect, cIEF typically generates extremely narrow zones, with resolving power potentially higher than other modes of capillary electrophoresis.

---

[1] Deleted in proof.
[1a] Deleted in proof.

METHODS IN ENZYMOLOGY, VOL. 270

Because CE instruments use on-tube detection at a fixed point along the capillary, cIEF must include a means of transporting the focused zones past the detection point. Three approaches have been used to mobilize focused zones. In chemical mobilization, changing the chemical composition of the anolyte or catholyte causes a shift in the pH gradient, resulting in electrophoretic migration of focused zones past the detection point. In hydrodynamic mobilization, focused zones are transported past the detection point by applying pressure or vacuum at one end of the capillary. In electroosmotic mobilization, focused zones are transported past the monitor point by electroosmotic pumping. Each of these approaches requires different instrument configurations and different strategies for optimizing the cIEF separation. Each of these methods is discussed, with emphasis on the technique of chemical mobilization.

Capillary Isoelectric Focusing with Chemical Mobilization

Capillary isoelectric focusing with chemical mobilization, first described by Hjertén and Zhu,[1b] is a three-step process. In the first step, sample is mixed with the ampholyte and introduced into the capillary by pressure or vacuum. Once the entire capillary is filled with the sample plus ampholyte mixture, the capillary ends are immersed in ampholyte and catholyte solutions. The second (focusing) step commences with application of high voltage. During this stage, ampholytes migrate under the influence of the electric field to generate a pH gradient; the range of the gradient in the capillary is defined by the composition of the ampholyte mixture. At the same time, protein components in the sample migrate until, at steady state, each protein becomes focused in a narrow zone at its isoelectric point. Focusing is achieved rapidly (typically a few minutes in short capillaries) and is accompanied by an exponential drop in current. The focusing process can be monitored by the movement of nascent protein zones past the detection point. When focusing is complete (as determined by attainment of a minimum current value or minimum rate of current decrease), the final step (mobilization) begins with substitution of the anolyte or catholyte solutions with a suitable mobilization solution.

Principle of Chemical Mobilization

The theoretical basis of chemical mobilization was described by Hjertén et al.[2] At steady state, the electroneutrality condition in the capillary during

[1b] S. Hjertén and M. Zhu, *J. Chromatogr.* **347,** 265 (1985).
[2] S. Hjertén, J.-L. Liao, and K. Yao, *J. Chromatogr.* **387,** 127 (1987).

focusing is

$$C_{H^+} + \Sigma\, C_{NH_3^+} = C_{OH^-} + \Sigma\, C_{COO^-} \tag{1}$$

where C_{H^+}, C_{OH^-}, $C_{NH_3^+}$, and C_{COO^-} are the concentrations of protons, hydroxyl ions, and positive and negative groups in the ampholytes, respectively. In anodic mobilization, addition of a nonproton cation X^{n+} to the anolyte introduces another term to the left side of the equation:

$$C_{X^{n+}} + C_{H^+} + \Sigma\, C_{NH_3^+} = C_{OH^-} + \Sigma\, C_{COO^-} \tag{2}$$

Because $C_{H^+} C_{OH^-}$ is constant, migration of the nonproton cation into the capillary will result in a reduction in proton concentration, i.e., an increase in pH. Similarly, addition of a nonhydroxyl anion Y^{m-} to the catholyte in cathodic mobilization yields a similar expression:

$$C_{H^+} + \Sigma\, C_{NH_3^+} = C_{OH^-} + \Sigma\, C_{COO^-} + C_{Y^{m-}} \tag{3}$$

indicating that migration of a nonhydroxyl anion into the capillary results in a reduction in hydroxyl concentration, i.e., a decrease in pH. Progressive flow of nonproton cations (anodic mobilization) or nonhydroxyl anions (cathodic mobilization) will therefore cause a progressive pH shift down the capillary, resulting in mobilization of proteins in sequence past the monitor point. Because high voltage is applied during the mobilization process, proteins remain focused in the moving zones, with no loss in resolution. The position of zones relative to each other does not change across the major portion of the gradient during mobilization, so that mobilization times can be correlated with isoelectric points.

Capillary Selection

To obtain good resolution when performing capillary IEF with chemical mobilization, it is essential that electroosmotic flow (EOF) be reduced to a very low level. In the presence of significant levels of EOF, attainment of stable focused zones is prevented, resulting in band broadening. Therefore, the use of coated capillaries is necessary for this technique. A viscous polymeric coating is recommended for greatest reduction in EOF, and use of neutral, hydrophilic coating materials reduces protein interactions. Both adsorbed and covalent coatings have been used for cIEF, but covalent coatings have the advantage of enhanced stability. The most commonly employed coating chemistry has been that described by Hjertén.[3] In this procedure, a bifunctional silane such as γ-methacryloxypropyltrimethoxysi-

---

[3] S. Hjertén, *J. Chromatogr.* **347,** 191 (1985).

lane is reacted with silanol groups on the internal surface of the capillary. After covalent attachment of this reagent, the acryl group is reacted with acrylamide in the presence of N,N,N',N'-tetramethylethylenediamine (TEMED) and ammonium persulfate without any cross-linking agent to form a monolayer coating of linear polyacrylamide covalently attached to the surface. Capillaries coated with this type of procedure exhibit as much as a 40-fold reduction in EOF, and reduced protein adsorption.[4] It has been reported[5] that the quality of fused silica tubing varies among suppliers and from batch to batch, and that attaining a stable coating requires optimized capillary pretreatment and coating procedures. Capillaries coated with linear polyacrylamide are commercially available from Bio-Rad Laboratories (Hercules, CA).

The low level of EOF in coated tubes permits separations to be carried out in very short capillaries. Earlier work using chemical mobilization was performed using capillaries as short as 11 cm with internal diameters up to 200 μm.[6] More recently, 12- to 17-cm capillaries with internal diameters of 25 μm have been used.[7] While use of the small inner diameter capillaries improves resolution through better heat dissipation, the high surface-to-volume ratio increases the potential for protein–wall interactions. These effects are greatly diminished by use of an inert capillary coating.[3] Although use of small inner diameter capillaries lowers detection sensitivity by reduction in the detector light path, this is generally not a concern because zone concentrations in cIEF are extraordinarily high. For example, focusing of a protein into a 0.5- to 1.0-mm zone results in a 170- to 340-fold increase in protein concentration relative to the sample.

Sample Preparation and Injection

Sample preparation for cIEF includes adjustment of sample salt levels, selection of the appropriate ampholyte composition, and dilution of the sample to the proper protein concentration. The ionic strength of the sample should be as low as possible, preferably lower than 50 mM. Excessive sample ionic strength due to the presence of buffer, salts, or ionic detergents will interfere with the isoelectric focusing process, requiring very long focusing times and causing peak broadening during mobilization. Elevated current due to the presence of salt can increase the risk of precipitation as proteins become concentrated in focused zones. Samples with salt concen-

[4] M. Zhu, R. Rodriguez, D. Hansen, and T. Wehr, *J. Chromatogr.* **516,** 123 (1990).
[5] S. Hjertén, and K. Kubo, *Electrophoresis* **14,** 390 (1993).
[6] S. Hjertén and M. Zhu, *J. Chromatogr.* **346,** 265 (1985).
[7] M. Zhu, R. Rodriguez, and T. Wehr, *J. Chromatogr.* **559,** 479 (1991).

trations of 50 mM or greater should be desalted by dialysis, gel filtration, or ultrafiltration.

The ampholyte composition should be selected based on the desired separation range. For separating multicomponent samples containing proteins with widely different isoelectric points, or for estimating the pI of an unknown protein, a wide-range ampholyte blend should be selected, e.g., pH 3–10. The final ampholyte concentration should be between 1 and 2%. To detect proteins at the basic end of the gradient during cathodic mobilization, it is necessary that the gradient span only the effective length of the capillary, e.g., the distance from the capillary inlet to the detection point. In cases in which the total capillary length is much greater than the effective length, many sample components may focus in the "blind" segment distal to the monitor point and be undetected during mobilization. In this case, a basic compound such as TEMED can be used to block the distal section of the capillary. As a rule of thumb, the ratio of TEMED concentration (%, v/v) to ampholyte concentration should be approximately equal to the ratio of the "noneffective" capillary length to total length. For example, if the effective length of a 20-cm capillary is 15 cm, a 0.5% concentration of TEMED should be added if the final ampholyte concentration is 2%. Note that use of TEMED is required only if the sample contains proteins focusing at the basic end of the gradient in cathodic mobilization (e.g., proteins with pI values above 8 when using pH 3–10 ampholytes). Use of TEMED for cIEF of proteins with pI values below 8 will have no beneficial effect, and in fact will decrease resolution because a steeper pH gradient is generated in the effective length of the capillary. In the case of anodic mobilization of acidic proteins (pI less than 5), a similar strategy can be considered, e.g., addition of an acid compound such as glutamic or aspartic acid to the ampholyte mix.

In situations in which enhanced resolution of proteins with similar pI values is desired, the use of narrow-range ampholyte mixtures may be considered. Narrow-range ampholyte mixtures generating gradients spanning 1–3 pH units are available from several commercial sources. However, our experience with this approach to high-resolution cIEF has been disappointing, perhaps owing to the limited number of ampholyte species in narrow-range "cuts." One proposed solution has been to blend narrow-range ampholytes from several manufacturers.[8]

The final protein concentration in the sample plus ampholyte mixture will depend on sensitivity requirements and the solubility of the protein components under focusing conditions. As an approximation, a final con-

---

[8] S. Hjertén, in "Capillary Electrophoresis" (P. D. Grossman and J. C. Colburn, eds.), p. 205. Academic Press, San Diego, CA, 1992.

centration of 0.5 mg/ml per protein should provide adequate sensitivity and satisfactory focusing plus mobilization performance. However, many proteins may precipitate during focusing at this starting concentration, because the final protein concentration in the focused zone may be as high as 200 mg/ml. Immunoglobulins, membrane proteins, and high molecular weight or hydrophobic proteins in general have a high risk of precipitation in cIEF. In such cases, use of very dilute protein solutions may be required. Prior to injection, the prepared sample should be centrifuged for 2–3 min in a microcentrifuge to remove any particulate material and to degas the solution.

Once prepared, the sample plus ampholyte mixture is introduced into the capillary. This can be accomplished by pressure injection or by vacuum. For good quantitative precision, sufficient volume should be loaded into the capillary to ensure that the tube contains a homogeneous mixture of sample. Therefore the capillary should be injected with at least three to five tube volumes of sample plus ampholytes.

Focusing

The focusing step begins with the immersion of the capillary in the anolyte (dilute phosphoric acid) and catholyte (dilute sodium hydroxide) solutions, followed by application of high voltage. Typically the catholyte solution is 20–40 mM NaOH, and the anolyte is half the catholyte molarity, e.g., 10–20 mM H_3PO_4. It is important that the catholyte be prepared fresh, as sodium hydroxide solutions will gradually take up carbon dioxide from the atmosphere. The presence of carbonate salts in the catholyte will interfere with the focusing process. Use of higher NaOH concentrations (e.g., 40 mM) minimizes this problem. For narrow-bore capillaries (e.g., 25-μm i.d.), field strengths of 300–900 V/cm or greater can be used, but our experience indicates that 600 V/cm is optimal. At greater field strengths, shorter analysis times can be achieved at the expense of resolution.

On application of high voltage, the charged ampholytes migrate in the electric field, with basic ampholytes migrating toward the cathode and acidic anolytes migrating toward the anode. A pH gradient begins to develop, with low pH toward the anode and high pH toward the cathode. Because the pH of the anolyte (catholyte) is lower (higher) than any of the ampholyte species, migration of ampholytes out of the capillary is prevented and, at steady state, a pH gradient across the capillary is formed with each ampholyte (and protein component in the sample) focused at a pH equal to its pI. During this process the initial current is high and drops as the number of charge carriers diminishes during focusing. Focusing is usually considered to be complete when the current has dropped to a level approximately 10%

of its initial value and the rate of change approaches zero. It is generally not advisable to prolong focusing beyond this point, as resistive heating increases the risk of protein precipitation. Also, loss of ampholytes at the acidic or basic end of the gradient can give rise to anodic or cathodic drift.[9] Anodic drift can be minimized by increasing the phosphoric acid concentration of the anolyte.[10]

Proteins begin to form into nascent zones at the margins of the capillary and, as focusing progresses, the zones migrate past the monitor point. The detection profile generated during this step may not exhibit well-resolved peaks, but it is reproducible and characteristic of the sample. In cases in which protein precipitation prevents acquisition of reliable data during mobilization, the focusing profile can sometimes yield useful information about the sample.[11]

Mobilization

At the completion of the focusing step, high voltage is turned off and the anolyte or catholyte is replaced by the mobilization reagent. High voltage is again applied to begin mobilization. As in focusing, field strengths of 300–900 V/cm can be used for mobilization, with optimum separations achieved in small inner diameter capillaries using a field strength of about 600 V/cm. The choice of anodic vs cathodic mobilization and the composition of the mobilization reagent depend on isoelectric points of the protein analytes, and the goals of the separation. Because the majority of proteins have isoelectric points between 5 and 9, cathodic mobilization is most often used.

The most common chemical mobilization method is addition of a neutral salt such as sodium chloride to the anolyte or catholyte; sodium ion serves as the nonproton cation in anodic mobilization and chloride functions as the nonhydroxyl anion in cathodic mobilization. A suggested cathodic mobilization reagent is 80 mM NaCl in 40 mM NaOH. At the beginning of mobilization, current initially remains at the low value observed at termination of focusing, but gradually begins to rise as chloride ions enter the capillary. Later in mobilization, when chloride is present throughout the tube, a rapid rise in current signals the completion of mobilization (Fig. 1A).

Ideally, mobilization should cause focused zones to maintain their relative position during migration, i.e., zones should be mobilized as a train past the monitor point. In practice, movement of ions into the capillary

[9] R. A. Mosher and W. Thormann, *Electrophoresis* **11,** 717 (1990).
[10] J. R. Mazzeo and I. S. Krull, *J. Microcolumn Sep.* **4,** 29 (1992).
[11] F. Kilár, *in* "A Handbook of Capillary Electrophoresis: A Practical Approach" (J. Landers, ed.), p. 97. CRC Press, Boca Raton, FL, 1996.

causes a pH change at the capillary end that gradually progresses deeper into the capillary. The rate of change depends on the amount of coion moving into the capillary, the mobility of the coion, and the buffering capacity of the carrier ampholytes. The actual slope of the pH gradient changes across the capillary, becoming shallower in the direction opposite to mobilization (Fig. 2). Neutral and basic proteins are efficiently mobilized toward the cathode with sodium chloride, and mobilization times correlate well with pI (Fig. 3). However, acidic proteins at the far end of the capillary are mobilized with lower efficiency and may exhibit zone broadening or be undetected.

Use of zwitterions is an alternative approach that provides more effective mobilization of protein zones across a wide pH gradient.[7] For example, cathodic mobilization with a low-pI zwitterion enables efficient mobilization of proteins with pI values ranging from 4.65 to 9.60. The proposed mechanism for zwitterion mobilization couples a pH shift at the proximal end of the tube with a displacement effect at the distal end as the zwitterion forms an expanding zone within the gradient at its isoelectric point. Effective zwitterion mobilization depends on selection of the appropriate mobilization reagent. For example, cathodic mobilization requires a zwitterion with an isoelectric point between the pH of the anolyte and the pI of the most acid analyte protein. The current level during zwitterion mobilization is lower than that observed in salt mobilization (Fig. 1B).

Detection

The strong absorbance of ampholytes at wavelengths below 240 nm (Fig. 4) makes detection of proteins in the low-UV region impractical. Therefore 280 nm is generally used for absorbance detection in cIEF. This results in a loss in detector signal of as much as 50-fold relative to detection at 200 nm, but the high protein concentrations in focused peaks more than compensate for the loss in sensitivity imposed by 280-nm detection. Because ampholyte species vary in abundance and UV extinction, the low-UV profile of ampholytes detected during cIEF can provide some information about the ampholyte distribution in the capillary. When a multiwavelength or scanning UV detector is used, correlation of the migration position of ampholyte peaks with protein standards could allow the ampholyte profile to be used for internal standardization.

Troubleshooting Capillary Isoelectric Focusing Separations

Protein precipitation is the major source of difficulty in capillary isoelectric focusing. In the focusing process, proteins become highly concentrated at zero net charge, conditions that promote aggregation and loss of solubil-

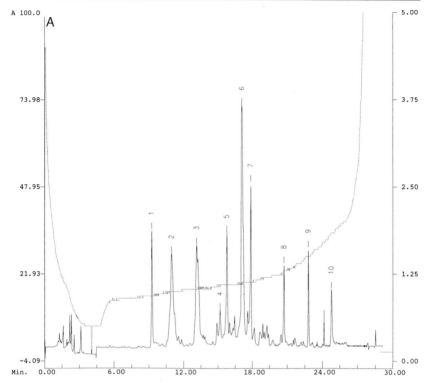

Fig. 1. Current levels during focusing and cathodic mobilization with sodium chloride (A) or zwitterion (B). Conditions: capillary, 17 cm × 25 μm (coated); focusing and mobilizing anolyte, 20 mM H₃PO₄; focusing catholyte, 40 mM NaOH; mobilizing catholyte, 40 mM NaOH plus 80 mM NaCl (A) or zwitterion (B); polarity, positive to negative; focusing conditions, 15 kV for 240 sec; mobilizing voltage, 15 kV; capillary temperature, 20°; detection, 280 nm; sample, Bio-Rad IEF protein standard diluted 1:24 in 2% Bio-Lyte 3–10 ampholytes (Bio-Rad, Hercules, CA). Solid trace, focusing and mobilization electropherogram; dotted trace, current in microamperes. Peak identification: 1, cytochrome c; 2–4, lentil lectins; 5, contaminant; 6, human hemoglobin C; 7, equine myoglobin; 8, human carbonate dehydratase; 9, bovine carbonate dehydratase; 10, β-lactoglobulin; 11, phycocyanin.

ity. Precipitation in cIEF is manifested by current fluctuation or loss, by variations in peak heights or migration patterns, and by spikes in the electropherogram as particulates transit the detection point. Protein precipitation can be minimized by reducing the focusing time (which may, however, reduce resolution) or by reducing protein concentration (which will reduce sensitivity). The most effective means of reducing protein precipitation is by addition of protein-solubilizing agents such as organic modifiers and surfactants to the sample plus ampholyte mixture. Organic modifiers such

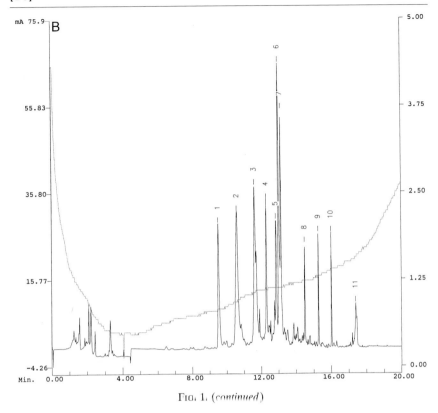

FIG. 1. (continued)

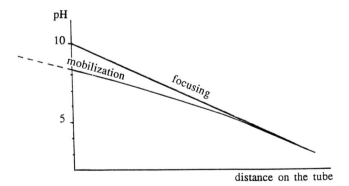

FIG. 2. Variation of pH with distance along the capillary during focusing and cathodic mobilization.

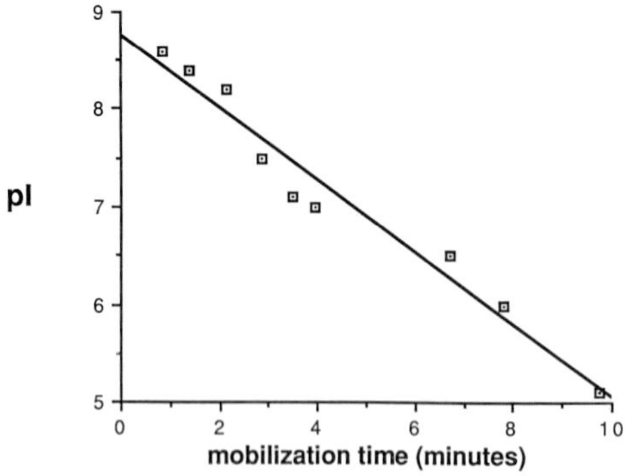

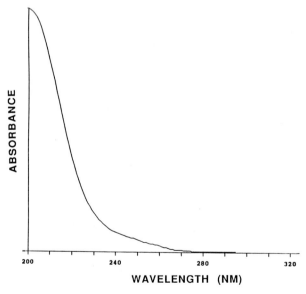

FIG. 3. Isoelectric point vs mobilization time using cathodic mobilization with sodium chloride (conditions as described in Fig. 1A).

as ethylene glycol (5–40%) serve to reduce hydrophobic interactions that promote aggregation.[2] We have found nonionic detergents such as Triton X-100 to be quite effective in reducing precipitation; however, the reduced form (available from Sigma Chemical Co., St. Louis, MO) should be used to minimize UV interference. Brij 35 (Fluka Biochemika Buchs, Switzerland) is another UV-transparent nonionic detergent that should prove use-

FIG. 4. Ultraviolet spectrum of pH 3–10 ampholytes.

ful. We have on occasion used chaotropic agents such as 7 M urea to help solubilize proteins. However, chaotropes and aggressive surfactants can change the migration behavior of proteins.

Variations in viscosity within the tube due to variations in protein concentration or in the composition of the sample may cause variations in migration times although the relative peak positions remain the same. For this reason, use of internal standards is strongly recommended in cIEF, particularly when the technique is to be used to estimate the isoelectric point of an unknown protein.

Variable electroosmotic flow can contribute to poor reproducibility in cIEF. This may be caused by loss of the capillary coating after prolonged use, particularly at the cathodic end where the coating is continually exposed to high-pH conditions. Increased EOF can be recognized by delayed appearance of peaks during the focusing step, and early migration of peaks during mobilization, accompanied by loss of resolution. Well-coated capillaries should have lifetimes of up to several hundred analyses, and poor separation due to protein adsorption may be misdiagnosed as coating failure.

Gradual adsorption of protein to the capillary wall can cause deterioration of capillary performance in cIEF. This can be minimized by purging the capillary between each analysis with one or more wash reagents, such as dilute acid (e.g., 10 mM H_3PO_4) or surfactants [e.g., 1% (w/v) Brij or 1% (w/v) sodium dodecyl sulfate (SDS)]. A capillary with degraded performance can often be regenerated with extended washes with these same solutions. In cases of serious performance deterioration, a brief wash with 1% SDS in basic solution (NaOH, pH 10) or washing with 0.1% (v/v) trifluoroacetic acid in 40:60 (v/v) water–acetonitrile may regenerate the capillary. The capillary should be washed extensively with water following exposure to aggressive wash solvents. Best lifetime for coated capillaries is achieved by washing the capillary with dilute acid and water after usage, followed by purging with dry nitrogen for 5 min before storage.

Capillary Isoelectric Focusing with Hydrodynamic Mobilization

Capillary isoelectric focusing with hydrodynamic mobilization was first described by Hjertén and Zhu.[6] Mobilization was accomplished by displacing focused zones from the capillary by pumping anolyte solution into the capillary using a high-performance liquid chromatography (HPLC) pump equipped with a T connection to deliver a flow rate into the capillary of 0.05 μl/min. Voltage was maintained during mobilization, and on-tube detection using a UV detector was employed.

Hydrodynamic mobilization by vacuum and using on-tube detection has been described by Chen and Wiktorowicz.[12] In this approach, a four-step vacuum-loading procedure was used sequentially to introduce segments of catholyte [20 mM NaOH plus 0.4% (w/v) methylcellulose], ampholytes plus methylcellulose, sample solution, and a final segment of ampholytes plus methylcellulose from the anodic end of the capillary. Following loading of the capillary, focusing was carried out for 6 min at a field strength of about 400 V/cm, then mobilization of focused zones toward the cathode was performed by applying vacuum at the capillary outlet with high voltage simultaneously maintained to counteract the distorting effects of laminar flow. A dimethylpolysiloxane (DB-1)-coated capillary (J&W Scientific, Folsom, CA) combined with addition of methylcellulose to the catholyte and ampholyte solutions served to suppress EOF. Relative mobility values for proteins were calculated by normalizing zone migration times to the migration times of the catholyte–ampholyte and ampholyte–anolyte interfaces. A plot of relative mobility values vs pI values of protein standards was linear over a pH range of 2.75–9.5.

Capillary Isoelectric Focusing with Electroosmotic Flow Mobilization

Isoelectric focusing using capillaries with significant levels of EOF is a two-step process, with focusing occurring while sample proteins are being transported toward the detection point by electroosmotic flow. This technique has been used with both uncoated capillaries and with capillaries coated to reduce (but not eliminate) EOF.

Two approaches have been reported, one in which the sample plus ampholyte mixture was introduced as a plug at the inlet of a capillary prefilled with catholyte, and another in which the entire capillary was prefilled with sample plus ampholyte mixture. In the first approach, described by Thormann et al.,[13] 75-μm i.d. × 90 cm uncoated capillaries were filled with catholyte [20 mM NaOH plus 0.06–0.3% (w/v) hydroxypropylmethylcellulose (HPMC)] and a 5-cm segment of sample in 2.5–5% (w/v) ampholytes was injected at the inlet (anodic) of the capillary by gravity. After immersion of the capillary inlet in anolyte (10 mM H$_3$PO$_4$), high voltage was applied at a field strength of 220 V/cm. Formation of the pH gradient and focusing of proteins into zones occurred as the sample segment was swept toward the monitor point at the distal end of the capillary. The addition of HPMC to the catholyte served to dynamically coat the fused

---

[12] S.-M. Chen and J. E. Wiktorowicz, *Anal. Biochem.* **206,** 84 (1992).

[13] W. Thormann, J. Caslavska, S. Molteni, and J. Chmelík, *J. Chromatogr.* **589,** 321 (1992).

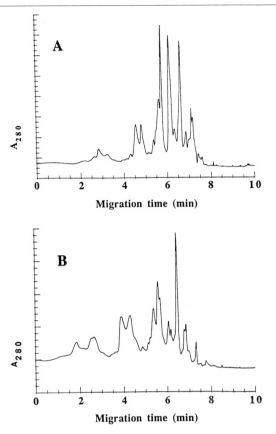

FIG. 5. Capillary IEF patterns of (A) type I recombinant tissue plasminogen activator and (B) type II recombinant tissue plasminogen activator. Conditions: 17 cm × 25 μm coated capillary, focusing for 2 min at 12 kV, mobilization at 8 kV, detection at 280 nm. The ampholyte solution contained 2% ampholytes (pH 6–8), 2% CHAPS, and 6 M urea. The microheterogeneity observed in these pure preparations of rtPA is expected based on the known glycoforms of rtPA. [Reproduced from K. W. Kim, *J. Chromatogr.* **559,** 401 (1991), with permission.]

silica wall, thereby reducing protein adsorption and EOF. Successful application of this technique depends on optimization of the HPMC concentration, ampholyte concentration, and sample load to minimize protein adsorption and to modulate the EOF level so that focusing approaches completion before the detection point is reached.

A capillary IEF method using EOF mobilization in which the entire effective length of the capillary was filled with ampholyte plus sample

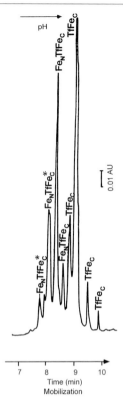

FIG. 6. Capillary IEF of iron-complexed transferrin. Conditions: 14 cm × 100 μm coated capillary, focusing for 6 min at 4 kV, mobilization at 4 kV, sample in 2% ampholytes (pH 5–7), detection at 280 nm. Peaks represent various isoforms of transferrin with iron bound at the C terminal-binding site (TfFe$_C$) or at both the C terminal- and N terminal-binding sites (Fe$_N$TfFe$_C$). [Reproduced from F. Kilár and S. Hjertén, *Electrophoresis* **10**, 23 (1989), with permission.]

was reported by Mazzeo and Krull.[14,15] In initial studies using uncoated capillaries, methylcellulose was added to modulate EOF and TEMED was used to block the detector–distal capillary segment. This approach was successful only for neutral and basic proteins owing to variations in the rate of EOF during the separation. As the separation progressed, the drop in average pH due to mobilization of the basic segment of the pH gradient into the catholyte resulted in diminished EOF. This in turn caused peak broadening and poor resolution for acidic proteins. Improved mobilization

[14] J. R. Mazzeo and I. S. Krull, *Anal. Chem.* **63**, 2852 (1991).
[15] J. R. Mazzeo and I. S. Krull, *J. Chromatogr.* **606**, 291 (1992).

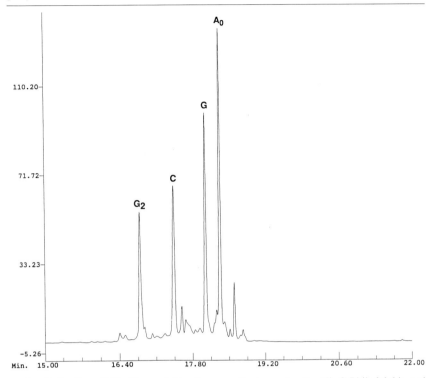

Fig. 7. Capillary IEF of hemoglobins from a patient carrying the Hb G Philadelphia and Hb C mutations. Conditions: 17 cm × 25 μm coated capillary, focusing for 5 min at 10 kV, mobilization at 10 kV, sample in 2% ampholytes (pH 3–10). Peaks represent normal human hemoglobin A_0 and hemoglobin variants carrying the C, G, or C and G mutations. [Reproduced from M. Zhu, T. Wehr, V. Levi, R. Rodriguez, K. Shiffer, and Z. A. Cao, *J. Chromatogr. A* **652**, 119 (1993), with permission.]

of acidic proteins was achieved using commercial C_8-coated capillaries (Supelco, Inc., Bellefonte, PA) in which EOF varied less with pH.[16] However, pH-dependent variation of EOF was still significant enough that plots of p*I* vs migration time were not linear over broad pH ranges. Use of multiple internal standards was recommended for accurate p*I* determination with this method.

Summary

Capillary isoelectric focusing is a useful analytical technique for characterization of protein mixtures and determination of protein isoelectric

---

[16] J. R. Mazzeo, J. A. Martineau, and I. S. Krull, "Methods: A Companion to Methods in Enzymology" Vol. 4, p. 205. Academic Press, San Diego, CA, 1992.

points. It is particularly useful in separation of protein glycoforms (Fig. 5),[17] characterizing protein microheterogeneity (Fig. 6),[18] and resolution of charge variants (Fig. 7).[19] The capillary focusing process is analogous to conventional isoelectric focusing in gels, while the requirement for zone mobilization is unique to the capillary format with on-tube detection. A variety of mobilization methods have been described, and the selection of the mobilization method for a particular application depends on the capillary type, the instrument configuration, and the type of proteins to be analyzed. Capillary IEF is generally successful for proteins with a molecular weight up to about 150,000 that exhibit good solubility in aqueous buffers, but may be unsatisfactory for large or hydrophobic proteins. Because of precipitation and variation in mobilization efficiencies, use of internal standards is recommended in most applications.

Capillary IEF can be compared to conventional gel IEF in terms of sample throughput and sensitivity. Conventional gels require approximately 4–6 hr to cast, run, and stain the gel, depending on whether silver or Coomassie staining is used. A typical gel contains 10 sample lanes, yielding a throughput of 25–35 min/sample. Capillary IEF separations (including focusing and mobilization) are typically 15–20 min. The mass sensitivity of conventioanl gel IEF is 36–47 ng for Coomassie staining and 0.5–1.2 ng for silver staining. In capillary IEF, sensitivity will depend on the volume of sample injected; assuming a capillary with a volume of 100 nl is completely filled with sample prior to focusing, the limit of detection will be approximately 1 μg/ml or 0.1 ng injected. Thus capillary IEF compares favorably with conventional gel IEF in terms of detectivity and analysis time, and has the additional benefit of complete automation of the process including separation and data reduction.

[17] K. W. Kim, *J. Chromatogr.* **559**, 401 (1991).
[18] F. Kilár and S. Hjertén, *Electrophoresis* **10**, 23 (1989).
[19] M. Zhu, T. Wehr, V. Levi, R. Rodriguez, K. Shiffer, and Z. A. Cao, *J. Chromatogr. A* **652**, 119 (1993).

[17] Isotachophoresis

By LUDMILA KŘIVÁNKOVÁ, PETR GEBAUER, and PETR BOČEK

Concise Theory of Isotachophoresis

Analytical isotachophoresis (ITP) is one of the modes of capillary electrophoresis and is suitable for analyzing mixtures of ionogenic substances in solutions.[1,2] It is a fast separation method: By using voltages of 5–20 kV and capillaries of 0.2- to 0.8-mm i.d. and 10–50 cm in length, filled typically with 0.01 M electrolytes, the analysis times are 5–40 min. The precision of quantitative analysis is usually 1–3 relative percent. By using ultraviolet (UV) absorbance detection, the usual detection limit lies at the nanogram level, which corresponds to the concentration on the parts per million (ppm) level, assuming sample volumes of tens of microliters.

The parameter decisive for the separation is the electrophoretic mobility (u) of the analyte, which is defined as the velocity (v) of an analyte due to a unit electric field strength (E):

$$u = v/E \tag{1}$$

For species present in solution in various forms (e.g., a weak acid present as an anion or nondissociated molecule), the effective mobility ($\bar{u}$) is defined as the weighted average of the mobilities of the various forms of the substance present in the solution (the mobility of an uncharged particle is obviously zero):

$$\bar{u} = \sum_{i=0}^{n} x_i u_i \tag{2}$$

where $i = 0, 1, \ldots, n$ denotes the neutral molecule and related dissociated forms, and x is their molar fraction. Equation (2) shows how to control the separation by controlling the chemical equilibria in the solution, e.g., by selecting a proper pH.

Isotachophoresis is performed in a discontinuous buffer system consisting of a leading electrolyte containing a high-mobility ion and of a terminating electrolyte containing a low-mobility ion. The sample solution is introduced between these two electrolytes and on passage of electric

[1] F. M. Everaerts, J. L. Beckers, and T. P. E. M. Verheggen, "Isotachophoresis. Theory, Instrumentation and Applications." Elsevier, Amsterdam, 1976.
[2] P. Boček, M. Deml, P. Gebauer, and V. Dolník, "Analytical Isotachophoresis." VCH Verlagsgesellschaft, Weinheim, Germany, 1988.

METHODS IN ENZYMOLOGY, VOL. 270

current the sample substances separate from the original mixed sample zone into pure zones of individual substances, owing to the differences in their electrophoretic mobilities. After the separation is completed, a steady state exists: The system is formed by a stack of adjacent zones between the leading and terminating zones, each containing only one substance. The zones migrate in the order of decreasing mobilities of their substances but all with the same velocity:

$$v_{ITP} = iu_L/\kappa_L = iu_A/\kappa_A = iu_B/\kappa_B = \cdots = iu_T/\kappa_T \qquad (3a)$$
$$v_{ITP} = E_L u_L = E_A u_A = E_B u_B = \cdots = E_T u_T \qquad (3b)$$

where $i[Am^{-2}]$ is the electric current density, u_i, κ_i, and E_i are the mobility, specific conductivity and electric field strength of component/zone i, respectively, and the subscripts L, A, B, and T relate to the leading substance, sample substances, and the terminating substance, respectively. The scheme of such a migrating zone stack is found in Fig. 1a.

The ITP steady state has three special features. The first is the absence of any background electrolyte in the separated zones. A zone of an analyte contains the analyte itself and a counterion (migrating from the leading zone and common to all zones). Beside these species, there are no other ions that significantly contribute to the electric conductivity in the zone in question. Obviously, the buffering of pH can be controlled by the selection of the counterionic system. The second feature is the self-sharpening effect, which manifests itself in that the boundaries between the migrating sample zones are permanently sharp. Under the usual analysis conditions (see above), the boundary width is on the order of 10^{-6} m. The third feature is the concentrating effect, which may be characterized as follows: (1) the concentration in each sample zone is constant throughout this zone (see the concentration profiles in Fig. 1b); and (2) the value of this concentration is a function of the mobility of this sample and is directly proportional to the concentration of the leading substance. In effect, the concentrations in the sample zones are adjusted to the concentration of the leading electrolyte such that the conductivities of the zones fit with Eq. (3a). This indicates that the lower the mobility of the zone constituent, the lower the conductivity of the zone (see Fig. 1c). The concentrating effect is important in isotachophoretic practice as it allows the concentration of minor sample components into pure ITP zones.

In the course of separation, the resolution of a two-component sample increases linearly from zero to a maximum value that corresponds to complete separation (the steady state) and then remains unchanged. The speed of the separation is proportional to the selectivity between the two components defined as the relative difference between effective mobilities. The speed of the process, however, must always be considered in relation to

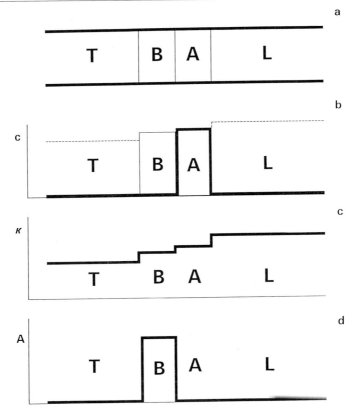

FIG. 1. Diagram of an ITP system of sample zones A and B migrating between the zones of the leader L and terminator T (a) and their concentration (b), conductivity (c), and UV absorbance (d) profiles.

the total analysis time, which depends on the electric current, the concentration of the leading electrolyte, and the capillary geometry.

From the above-cited characteristics, some general rules can be concluded on how to compose electrolyte systems for ITP. They are summarized in Table I. Table II presents an overview of recommended electrolyte systems for anionic and cationic analysis. The systems must always be selected so that the operating pH region ensures sufficient ionization of the analytes.

Practice of Isotachophoresis

Figure 2 gives a phenomenological view of one possible means of performing ITP analysis. Figure 2a shows the isotachopherograph being filled

TABLE I
SELECTION OF ELECTROLYTE SYSTEMS IN ISOTACHOPHORESIS[a]

	Cationic ITP	Anionic ITP
Suitable leading ion	K^+, NH_4^+, (Na^+)	Cl^-
Terminator	H^+	OH^-
Alternative terminator	Weak base	Weak acid
	Effective mobility approx. $5-10 \times 10^{-9}$ m^2V^{-1} sec^{-1}	Effective mobility approx. $5-10 \times 10^{-9}$ m^2V^{-1} sec^{-1}
Counterion	Weak acid	Weak base
Condition of ionization of analytes	$pH \leq pK_{BH} + 1$	$pH \geq pK_{HA} - 1$

[a] Reprinted from P. Boček, M. Deml, P. Gebauer, and V. Dolník, "Analytical Isotacho-phoresis," p. 188. VCH Verlagsgesellschaft, Weinheim, Germany, 1988, with permission.

with the working electrolytes. The two electrolyte chambers LC and TC are filled with a leading and a terminating electrolyte, respectively. These two chambers are connected by the separation capillary, which can be closed by the valve V. The capillary is connected through a semipermeable membrane M, which (together with valve V) closes the system against hydrodynamic flow during the analysis to the leading electrode chamber LC. The capillary is further equipped with a drain O and with a septum S serving for the introduction of the sample (Fig. 2b). Figure 2c shows the situation after a high-voltage supply (HV) of constant electric current has been connected across the capillary via the leading and terminating electrodes (leading and terminating electrolytes). The mixed zone of the sample components (two in Fig. 2) resolves into individual zones of pure components. Figure 2d shows a later point in time, when the sample is completely separated. All zones migrate with constant velocity. The concentration of substances in their zones is independent of the introduced amount. The lengths of the sample zones are given by the amount of the sample introduced. When this moving stack of sample zones passes through detector D, it can be recorded (Fig. 2e). Quantitation is performed simply by measuring zone lengths (step lengths on the record).

The basic instrumental arrangement of the ITP analysis can be modified in different ways to meet special requirements. A common problem in the analyses of biological fluids is the separation of minor sample constituents in the presence of a bulk amount of some other substance (e.g., chloride or sodium). For this, a system of two coupled columns may be used (Fig. 3). In the first (preseparation) column of large diameter, the effective concentration of the minor sample components and their separation from the major component are performed; in the second (analytical) column

TABLE II
RECOMMENDED BASIC ELECTROLYTE SYSTEMS[a]

Counterion	pH	Terminating electrolyte	Analyzed compounds
A. Anionic analysis[b]			
β-Alanine	3.1–4.1	Glutamic acid, propionic acid, caproic acid	Nucleotides, carboxylic acids, sulfo acids, phosphono acids, peptides
ε-Aminocaproate	4.1–5.1	Glutamic acid, MES, propionic acid, pivalic acid	Carboxylic acids, acidic amino acids, polyphosphates, phosphono acids, uric acid
Creatinine	4.5–5.5	Pivalic acid, MES, $NaHCO_3$	Carboxylic acids, acidic amino acids
Histidine	5.5–6.5	MES, HEPES	Carboxylic acids, peptides, acidic amino acids
Imidazole	6.6–7.6	MES, veronal	Antibiotics, some zwitterions, selected proteins and peptides
Tris	7.6–8.6	Glycine + $Ba(OH)_2$, β-alanine + $Ba(OH)_2$, asparagine + $Ba(OH)_2$	Purines, pyrimidines, amino acids, peptides, proteins
Ammediol	8.3–9.3	Glycine + $Ba(OH)_2$, β-alanine + $Ba(OH)_2$, EACA + $Ba(OH)_2$, GABA + $Ba(OH)_2$	Purines, pyrimidines, peptides, amino acids, proteins, phenols
Ethanolamine	9.0–10.0	EACA + $Ba(OH)_2$	Proteins, amino acids, phenols, thiols
B. Cationic analysis[c]			
Acetate	4.2–5.2	Acetic acid, GABA	
Propionate	4.4–5.4	Acetic acid, creatinine	
MES	5.7–6.7	Histidine, imidazole	
Veronal	6.9–7.9	Imidazole, triethylenediamine	
Asparagine	8.3–9.3	Triethanolamine, Tris	
Glycine	9.1–10.1	Triethanolamine, NH_3	
β-Alanine	9.8–10.8	NH_3, ethanolamine	

[a] GABA, γ-Aminobutyric acid; EACA, ε-aminocaproic acid; MES, morpholinoethanesulfonic acid; Tris, tris(hydroxymethyl)aminomethane. Reprinted from P. Boček, M. Deml, P. Gebauer, and V. Dolník, "Analytical Isotachophoresis," p. 191. VCH Verlagsgesellschaft, Weinheim, Germany, 1988, with permission.
[b] Leading electrolyte: 0.002–0.02 M HCl (leading ion: Cl^-) + counterion + additive [0.05–0.2% poly(vinyl alcohol), 0.05–0.5% hydroxymethyl-, hydroxypropyl-, hydroxypropylmethylcellulose, 0.05–0.5% Triton X-100].
[c] Leading electrolyte: 0.005–0.05 M KOH (leading ion: K^+) + counterion.

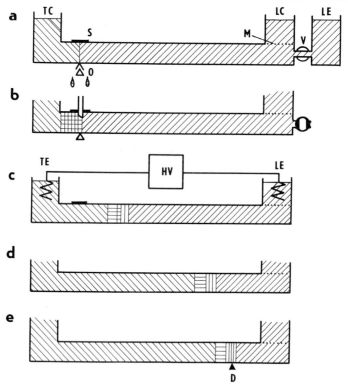

FIG. 2. Diagram of an ITP analysis. (a) filling the column and capillary with operating electrolytes; (b) injection of the sample; (c) application of high voltage; (d) separation; (e) detection. TC, Terminating electrode chamber; TE, terminating electrolyte; LE, leading electrolyte; LC, leading electrode chamber; S, septum; V, valve; M, semipermeable membrane; O, outlet; HV, high voltage; D, detector. (Reprinted from P. Boček, M. Deml, P. Gebauer, and V. Dolník, "Analytical Isotachophoresis," p. 40. VCH Verlagsgesellschaft, Weinheim, Germany, 1988, with permission.)

having a small diameter, the final separation of the minor components and their sensitive detection proceed. This procedure can be combined by using different leading electrolytes in the two capillaries, and moreover the ITP preseparation can be followed by a zone-electrophoretic step instead of an ITP step in the analytical capillary.

Two kinds of detectors are used in ITP. Universal detectors measure electric field strength in the zones or their conductivity and provide a nonzero detection response to any zone. The detection record has a steplike character (Fig. 1c). Specific detectors respond only to substances having a given property. The UV absorption detector, e.g., detects only zones

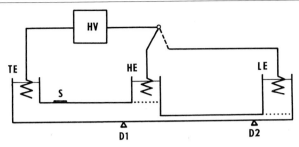

Fig. 3. Diagram of the column-coupling system in the column-switching mode. TE, terminating electrolyte–terminating electrode chamber; S, septum; HE, helping (auxiliary) electrode
chamber; LE, leading electrolyte–leading electrode chamber; HV, high-voltage power supply;
D1, detector in the preseparation capillary; D2, detector in the analytical capillary. After the
sample has passed detector D1, the current is switched over from TE–HE to TE–LE and
the analytes are detected by detector D2.

containing UV-absorbing species (Fig. 1d). In any case, the ideal detection
record of a sample zone represents a rectangular step that is characterized
by two parameters: the length, which is a direct measure of quantity, and the
height, which is a qualitative parameter. All usual techniques of quantitative
analysis are applicable, e.g., the internal or external standard technique
and the standard addition technique.

Applications of Isotachophoresis

Several features of the principle of capillary isotachophoresis (CITP)
and of the instruments employed make this method highly applicable in
the field of biochemistry. The simplicity of performance of the simultaneous
determination of both strong and weak acids or bases, high sensitivity
and separation efficiency, short separation times, low running costs, and
(usually) lack of requirement for sample pretreatment are the main advantages of ITP separations, especially in complex matrices of biological samples. There are three main areas where ITP has already proved to be a
suitable tool for biochemical analyses.[3]

1. The separation of small ions up to 1000 Da is the main application
   for ITP. Because many reactions in enzymology are monitored by
   observing concentration changes of small ions, e.g., organic acids,
   ITP can be used.

[3] P. Gebauer, V. Dolník, M. Deml, and P. Boček, in "Advances in Electrophoresis" (A.
Chrambach, M. J. Dunn, and B. J. Radola, eds.), Vol. 1, p. 283. VCH Verlagsgesellschaft,
Weinheim, Germany, 1987.

2. ITP can also successfully separate larger molecules including peptides and proteins up to 100,000 Da. Here, the concentrating capabilities of ITP are of great advantage. The operation in the column-coupling mode enables sophisticated separations to be performed, providing valuable information.

3. Its concentrating capabilities make ITP a useful preseparation technique for other separation methods that can be operated in an on-line arrangement as well. The on-line combination of ITP and capillary zone electrophoresis (CZE) must be mentioned in particular. It can be performed using the same experimental arrangement as shown in Fig. 3, where the first (ITP) step provides a preseparation of analytes and their concentration into narrow zones. In the following (CZE) step the effective resolution and sensitive detection of the analytes take place.[4]

Representative examples of reported analyses of each of the application areas mentioned in the preceding list are presented below. Although a number of them have been performed using special commercial instrumentation for ITP that is no longer available, the described procedures can be applied without any change, using present-day commercial instruments for capillary zone electrophoresis that are also able to operate in the ITP mode. All that must be done is to select a correct combination of a capillary with a suitable inner surface (with respect to electroosmosis), voltage polarity, and driving current.

Isotachophoresis of Small Ions in Enzymology

Urine or plasma is a typical example of a complex matrix. Simultaneous analysis of anions or cations in untreated urine has been described by Pep and Vonderschmitt.[5,6] For the separation of anions, complexation with Cd^{2+} included in the leading electrolyte as the counterion is employed. The separation of cations is performed in the presence of a complex-forming nonionic detergent, Triton X-100 (Fig. 4). The leading electrolyte is 10 mM tetramethylammonium hydroxide, 12 ml/liter acetic acid, 0.12 g/liter hydroxypropylmethylcellulose (HPMC), and 8 ml/liter Triton X-100, pH 5.62. The terminating electrolyte is 30 mM $Cd(NO_3)_2$, pH 6. Both electrolytes are prepared in 98% (v/v) methanol. The current during separation is 200 μA and, when the zones are close to the detector, the current is

---

[4] L. Křivánková, P. Gebauer, W. Thormann, R. A. Mosher, and P. Boček, *J. Chromatogr.* **638**, (1993).

[5] P. Pei and D. J. Vonderschmitt, *J. Clin. Chem. Clin. Biochem.* **25**, 253 (1987).

[6] P. Pei and D. J. Vonderschmitt, *J. Clin. Chem. Clin. Biochem.* **26**, 91 (1988).

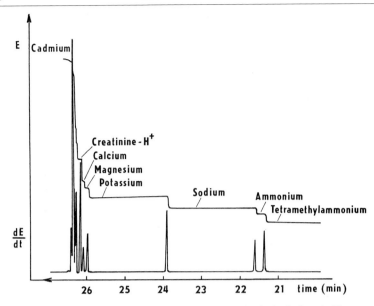

FIG. 4. ITP analysis of the most abundant cationic species in 1 μl of pooled human urine. [Reprinted from P. Pei and D. J. Vonderschmitt, *J. Clin. Chem. Clin. Biochem.* **26,** 91 (1988).]

decreased to 75 μA. For determination of calcium, anions forming complexes with calcium are precipitated with lead acetate, The mean deviation from 100% recovery of chloride, sulfate, and orthophosphate and ammonium, sodium, potassium, magnesium, calcium, and creatinine added to urine samples is reported to be 3.1%.

The column-coupling system of a CS Isotachophoretic Analyzer (URVJT, Spišská Nová Ves, Czechoslovakia) enables the separation of thiodiacetic acid, a metabolite of vinyl chloride, in the presence of a 10^4-fold excess of chloride ions in urine[7] (see Fig. 5). Different leading electrolytes are used in the preseparation and analytical capillaries. The preseparation is performed in 10 m*M* HCl plus β-alanine, 0.2% (w/v) hydroxypropylcellulose, pH 3.4, as leading electrolyte, and in this step, chlorides and phosphates are driven out of the preseparation capillary to the auxiliary electrode. After the portion of the sample containing thiodiacetic acid reaches the end of the preseparation capillary, the current is switched across the leading and terminating electrodes and the analysis continues in the leading electrolyte composed of 10 m*M* HCl plus β-alanine, 0.2% (w/v) hydroxypropylcellulose, pH 4.3. The driving current during detection is 45

[7] L. Křivánková, E. Samcová, and P. Boček, *Electrophoresis* **5,** 226 (1984).

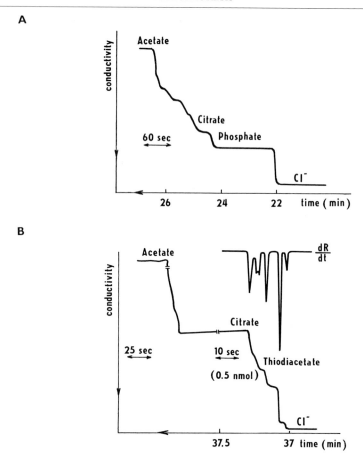

Fig. 5. ITP determination of thiodiacetate in 5 μl of urine from a person exposed to vinyl chloride. (A) The record from the preseparation capillary; (B) the record from the analytical capillary. *dR/dt* is the derivation of the conductivity signal. [Reprinted from L. Křivánková, E. Samcová, and P. Boček, *Electrophoresis* **5**, 226 (1984), with permission.]

μA. Untreated urine (5 μl) is injected and the lowest detectable concentration of thiodiacetic acid is about 6×10^{-6} *M*. The reproducibility of analyses in the range of $2–15 \times 10^{-5}$ *M* is about 3%.

As an example of the separation of metabolites in plasma, the determination of glycolic, glyoxylic, oxalic, and formic acids, the metabolites found in the plasma of humans poisoned with ethylene glycol, can be mentioned.[8] Heparinized plasma (1–15 μl) is injected into an LKB (Bromma, Sweden) 2127 Tachophor equipped with both conductivity and UV detectors. The leading electrolyte is 3 mM HCl plus 2 mM NaCl, pH 2.52, and the terminating electrolyte is 10 mM acetic acid. To the leading electrolyte, 0.2–0.4% HPMC is added. The driving current of 200 μA is decreased to 100 μA prior to detection.

The simultaneous analysis of pyruvate, acetoacetate, lactate, and 3-hydroxybutyrate in the plasma of patients with diabetes mellitus is shown in Fig. 6.[9] Heparinized plasma (5 μl) is injected into a CS Isotachophoretic Analyzer equipped with a column-coupling system and conductivity detectors. For preseparation, 10 mM HCl plus β-alanine, 0.2% (w/v) hydroxypropylcellulose (HPC), pH 4.2, is used as leading electrolyte, the analytical capillary is filled with 5 mM HCl plus glycine, 0.2% (w/v) HPC, pH 3.0, and 20 mM nicotinic acid serves as terminating electrolyte. The driving current for detection is 20 μA. The method is used for observation of metabolism of patients during various types of physical training.

Extracts of microorganisms present another complex medium that can be directly analyzed by CITP. Cyclic 2,3-diphosphoglycerate, a metabolite exclusively present in methanogenic bacteria, is separated from other phosphate-containing compounds in an extract of *Methanobacterium thermoautotrophicum* in an electrolyte system composed of leading electrolyte: 10 mM HCl plus 6-amino-n-hexanoic acid, 0.05% poly(vinyl alcohol), pH 4.5, and terminating electrolyte: 10 mM acetic acid, pH 4.8 (Fig. 7).[10] One microliter of the cell-free extract containing about 30 μg of dry weight of the cell material is injected into an LKB 2127 Tachophor equipped with a conductivity detector. The driving current is 500 μA, and the total analysis lasts 15 min. Results of analyses are in good agreement with spectrophotometric assay; however, the detection limit of the ITP method is lower by two orders of magnitude and the performance of the analysis is less time consuming. The same apparatus and a similar electrolyte system are used for the analysis of carboxylic acids in bacterial fermentation products.[11]

The determination of alkaloids of *Fumaria parviflora* and *Fumaria capreolata* can be presented here as an example of separations of complex

[8] S. Øvrebø, D. Jacobsen, and O. M. Sejersted, *J. Chromatogr.* **416**, 111 (1987).
[9] L. Křivánková and P. Boček, *J. Microcol. Sep.* **2**, 80 (1990).
[10] L. G. M. Gorris, J. Korteland, R. J. A. M. Derksen, C. van der Drift, and G. D. Vogels, *J. Chromatogr.* **504**, 421 (1990).
[11] J. S. van der Hoeven, H. C. M. Franken, P. J. M. Camp, and C. W. Dellebarre, *Appl. Environ. Microbiol.* **35**, 17 (1978).

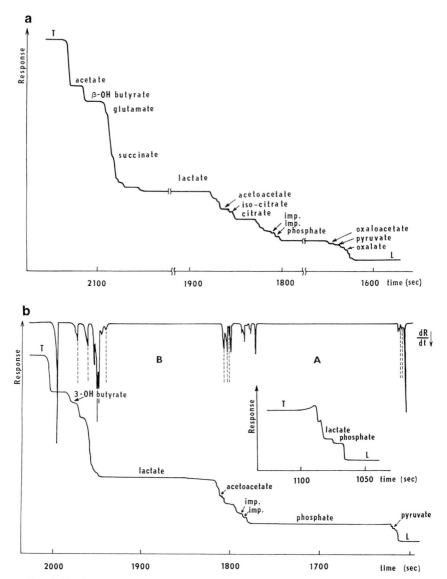

FIG. 6. (a) Identification of some acids in human plasma by ITP. Twenty-five microliters of diluted (1:4) heparinized plasma was enriched with acetate, 3-hydroxybutyrate, glutamate, succinate, acetoacetate, isocitrate, citrate, oxaloacetate, pyruvate, and oxalate so that their concentrations were about 0.05–0.1 mM and analyzed. [Reprinted from L. Křivánková and P. Boček, *J. Microcol. Sep.* **2**, 80 (1990), with permission.] (b) ITP analysis of 5 μl of untreated heparinized plasma from a diabetes mellitus patient. (A) The record from the preseparation capillary; (B) the record from the analytical capillary. dR/dt is the derivation of the conductivity signal. [Reprinted from L. Křivánková and P. Boček, *J. Microcol. Sep.* **2**, 80 (1990), with permission.]

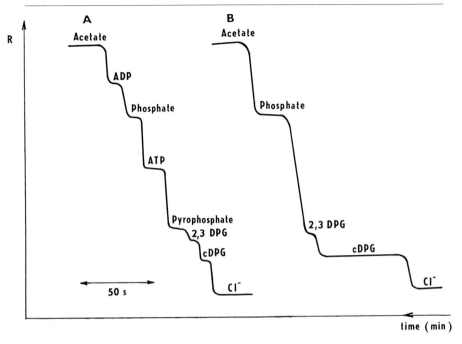

FIG. 7. ITP analysis of phosphate-containing metabolites. (A) One microliter of a reference mixture containing cyclic 2,3-diphosphoglycerate (cDPG) and 2,3-diphosphoglycerate (2,3-DPG) at 1 mM and other ions at 2 mM concentrations; (B) 1 μl of a cell-free extract prepared from *Methanobacterium thermoautotrophicum* containing about 30 μg dry weight of cell material. Detection was performed with a conductivity detector. [Reprinted from L. G. M. Gorris, J. Korteland, R. J. A. M. Derksen, C. van der Drift, and G. D. Vogels, *J. Chromatogr.* **504**, 421 (1990), with permission.]

plant matrices.[12] The methanol extract of dried, pulverized plants is concentrated, acidified with acetic acid, and washed with diethyl ether. The pH of the aqueous phase is adjusted to 8 with Na_2CO_3 and extracted with diethyl ether, giving fraction I. The aqueous layer is then adjusted to pH 12 with NaOH and extraction with diethyl ether yields fraction II (Fig. 8). Acidification of the aqueous phase to pH 5 with HCl, addition of KI, and extraction with chloroform result in fraction III. Aliquots of the individual fractions are dissolved in diluted HCl (1 mg in 1 ml of 5 mM HCl) and 5 or 10 μL of this solution is injected for the cationic ITP, using laboratory-made equipment with coupled columns resembling the CS Isotachophoretic Analyzer. The leading electrolyte is 10 mM potassium acetate buffer, pH 5, with 0.5 g/liter of poly(vinyl alcohol). Acetic acid (10 mM) serves as the

[12] I. Válka and V. Šimánek, *J. Chromatogr.* **445**, 258 (1988).

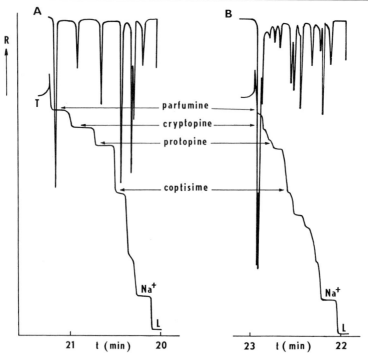

FIG. 8. ITP analysis of alkaloids. (A) Ten microliters of a standard mixture containing a 0.1 m*M* concentration of each alkaloid; (B) 5 μl of an extract of *Fumaria parviflora* (2 mg/ml). A conductometric detector was used. [Reprinted from I. Válka and V. Šimánek, *J. Chromatogr.* **445**, 258 (1988), with permission.]

terminating electrolyte. The driving current in the analytical capillary is 50 μA. The results are comparable to those of HPLC analyses. As a further example of ITP application in plant biochemistry, the separation of flavonoids in the extract from *Calamagrosis sachalinensis* can be mentioned.[13]

Nucleotides released from rat liver during the extraction with 0.7 *M* perchloric acid and 20% (v/v) methanol (7 ml/g liver), prepared very quickly to avoid metabolic changes, are analyzed directly in the extract neutralized with ethanolamine buffer, pH 7.2, and KOH.[14] Sample (10 μl) is injected into an LKB 2127 Tachophor and analyzed. The leading electrolyte consists of 5 m*M* HCl adjusted to pH 3.9 with β-alanine, and 0.25–0.75% HPMC. The terminating electrolyte is 5 m*M* caproic acid. The driving current during

[13] A. Hiraoka, K. Yoshitama, T. Hine, T. Tateoka, and T. N. Tateoka, *Chem. Pharm. Bull.* **35**, 4317 (1987).
[14] G. Eriksson, *Anal. Biochem.* **109**, 239 (1980).

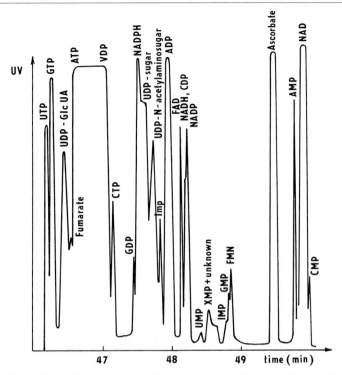

FIG. 9. ITP analysis of nucleotides in a liver extract of rat. Eight microliters of liver extract and 2 μl of nonabsorbing spacers at 0.5 mM concentrations were injected. [Reprinted from G. Eriksson, *Anal. Biochem.* **109**, 239 (1980), with permission.]

detection is 37.5 μA. Nonabsorbing spacers are added to the sample to distinguish zones having similar extinction coefficients (Fig. 9). The described working conditions also work for other nucleotide separations, e.g., in the analysis of nucleotides released from several strains of gram-positive and gram-negative bacteria and the yeast *Candida albicans* treated by high-voltage pulse and ultrasound.[15]

Enzymatic reactions can be followed by ITP to determine the activity of an enzyme or to identify products of the reaction. Oerlemans *et al.*[16] employ enzymatic reactions to identify purines and pyrimidines in urine. One hundred microliters of diluted (1:5) urine is incubated for various periods of time at 37° with 3 μl of the enzyme. For the incubation with

---

[15] H. Hülsheger, S. Husmann-Holloway, E. Borriss, and J. Potel, *in* "Analytical and Preparative Isotachophoresis" (C. J. Holloway, ed.), p. 157. Walter de Gruyter, Berlin, 1984.

[16] F. Oerlemans, C. de Bruyn, F. Mikkers, T. Verheggen, and F. Everaerts, *J. Chromatogr.* **225**, 369 (1981).

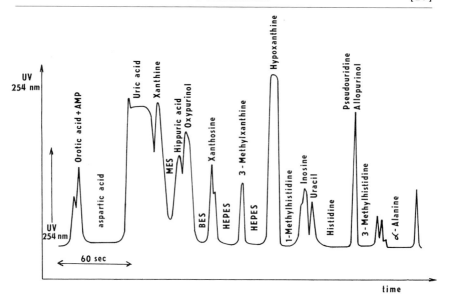

FIG. 10. ITP analysis of urine from a Lesch–Nyhan patient under allopurinol treatment. Three microliters of diluted urine (1:5) and 1 μl of the spacer mixture (1–4 mM) were injected. [Reprinted from F. Oerlemans, C. de Bruyn, F. Mikkers, T. Verheggen, and F. Everaerts, *J. Chromatogr.* **225**, 369 (1981), with permission.]

purine-nucleoside phosphorylase, 0.5 mM ribose 1-phosphate is used as cosubstrate; for adenine phosphoribosyltransferase, phosphoribosylpyrophosphate is used. Furthermore, xanthine oxidase and uricase are employed. For the analysis, 3 μl of the diluted urine is injected directly into an LKB 2127 Tachophor equipped with a UV detector. The operational system consists of the leading electrolyte: 5 mM HCl plus ammediol (2-amino-2-methyl-1,3-propanediol) to pH 8.55 and 0.3% hydroxypropylcellulose, and the terminating electrolyte: 20 mM β-alanine plus Ba(OH)₂, pH 10.4–10.5. A set of spacers is added to the sample to be able to distinguish compounds possessing close extinction coefficients (Fig. 10).

The possibility to follow simultaneously both substrates and reaction products is widely employed in enzymology. As examples, the transformation of pyruvate to succinate by calf heart mitochondrial enzymes,[17] the conversion of pyruvate into lactate by lactate dehydrogenase from pig heart,[18] and the degradation of hyaluronate oligosaccharides by β-N-acetyl-

[17] A. Kopwillem, *J. Chromatogr.* **82**, 407 (1973).
[18] A. J. Willemsen, *J. Chromatogr.* **105**, 405 (1975).

glucosaminidase or bovine testicular hyaluronidase[19] can be mentioned. Figure 11 shows an ITP analysis of the incubation mixture for hydrolysis of UDPglucuronic acid by pyrophosphatase from the microsomal fraction of mammalian liver homogenate.[20] The primary products, UMP and glucuronic acid 1-phosphate, are further cleaved by phosphatase to uridine and phosphate, and glucuronic acid and phosphate, respectively. The analysis is carried out using an LKB 2127 Tachophor equipped with a UV detector. The leading electrolyte is 5 mM HCl plus β-alanine, pH 3.89, 0.25% HPMC, and the terminating electrolyte is 5 mM caproic acid. The current during detection is 50 μA. Incubations of UDPGA with microsomal enzyme preparations are carried out in 100 mM Tris-HCl buffer at pH 8 at 25°, and the protein concentration is adjusted to about 1 mg/ml. Aliquots (10 μl) removed at different time intervals after initiation of incubation are injected directly into the analyzer.

In the study of an interferon-induced 2-5A system, ITP can effectively substitute for the assays of 2-5A phosphodiesterase activity using radiolabeled compounds.[21] The 2-5A system, which forms 2-5 oligoadenylates from ATP and then splits their 2-5 phosphodiester bonds, involves 2-5A synthetase, ribonuclease L, and 2-5A phosphodiesterase. As the substrate, the 2-5A trimer core is used, reaction products of its degradation are identified (Fig. 12), and the degradation is followed in time using an LKB Tachophor 2127 analyzer filled with 10 mM HCl plus β-alanine plus 0.3% methylcellulose, pH 3.65, as the leading electrolyte and 10 mM caproic acid as the terminating electrolyte. Five microliters of the incubation mixture is usually injected. The current during detection is 75 μA.

Isotachophoresis of Peptides and Proteins

Isotachophoresis is also helpful in peptide analyses. It can be used to follow procedures of peptide synthesis, purification, or hydrolysis, with respect both to small ions as well as larger peptides and their side or degradation products. Figure 13 presents ITP records of different purification steps of Asp-Gln.[22] Asp-Gln is synthesized by the N-carboxy anhydride method and purified by gel filtration on Sephadex G-10. Ten microliters of the sample (5.49 μg) is injected into an LKB 2127 Tachophor. Ultraviolet detection is performed at 206 nm. The leading electrolyte is 10 mM HCl

[19] P. Nebinger, *J. Chromatogr.* **354,** 530 (1986).
[20] C. J. Holloway, S. Husmann-Holloway, and G. Brunner, *J. Chromatogr.* **188,** 235 (1980).
[21] G. Bruchelt, M. Buedenbender, K. H. Schmidt, B. Jopski, J. Treuner, and D. Niethammer, *J. Chromatogr.* **545,** 407 (1991).
[22] P. Stehle and P. Fürst, *J. Chromatogr.* **346,** 271 (1985).

A

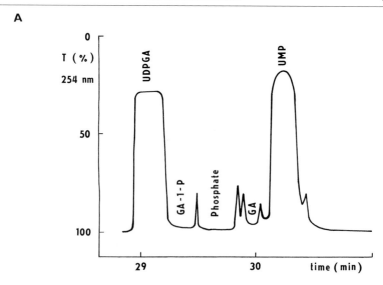

B

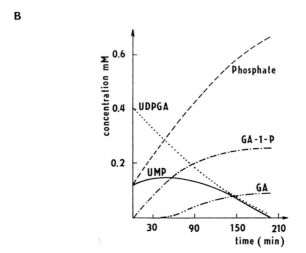

FIG. 11. Assay of the enzymatic hydrolysis of UDPglucuronic acid by ITP. (A) Composition of the incubation mixture at a later stage of hydrolysis. UDPGA, UDPglucuronic acid; GA-1-P, glucuronic acid 1-phosphate; GA, glucuronic acid. (B) Kinetic profile of UDPGA hydrolysis derived from the ITP analyses. [Reprinted from C. J. Holloway, S. Husmann-Holloway, and G. Brunner, *J. Chromatogr.* **188,** 235 (1980), with permission.]

A

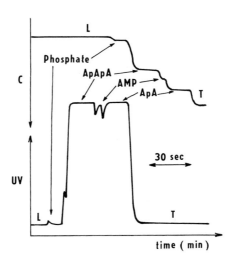

B

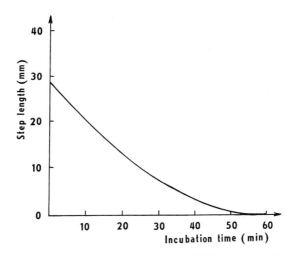

Fig. 12. ITP determination of 2-5A phosphodiesterase activity. (A) The analysis of degradation products of the reaction mixture of 2-5A trimer core and snake venom phosphodiesterase. (B) Time course of 2-5A trimer core (ApApAp) degradation. Ten microliters of 3 mM ApApAp was incubated with 10 μl of snake venom phosphodiesterase (0.01 IU/ml). [Reprinted from G. Bruchelt, M. Buedenbender, K. H. Schmidt, B. Jopski, J. Treuner, and D. Niethammer, *J. Chromatogr.* **545**, 407 (1991), with permission.]

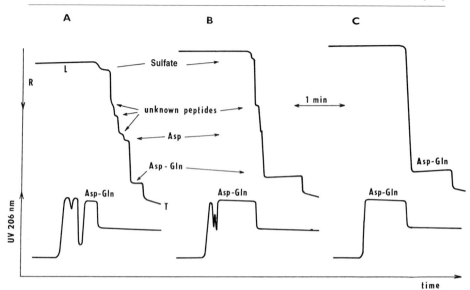

FIG. 13. ITP analyses before (A) and after (B and C) different purification steps. Ten microliters of the solution containing 5.49 μg of the material was analyzed. R, Response of a conductivity detector. [Reprinted from P. Stehle and P. Fürst, *J. Chromatogr.* **346,** 271 (1985), with permission.]

plus Bis–Tris, 0.4% HPMC, pH 6, and the terminating electrolyte is 10 m*M* morpholineethanesulfonic acid (MES) plus Tris, pH ~6. The current during detection is 60 μA.

Useful information on the purity determination of peptides can be obtained when ITP is applied in addition to reversed-phase high-performance liquid chromatography (RP-HPLC) as these methods are based on different physicochemical parameters. Complementary results of both RP-HPLC and ITP of insulin, adrenocorticotropin hormone (ACTH), β-endorphin, vasopressin, and cholecystokinin are demonstrated in the study of Janssen *et al.*[23] Figure 14 shows the ITP separation of five closely related basic fragments of ACTH. Experiments are performed in a home-made apparatus equipped with a polytetrafluoroethylene separation capillary (0.2-mm i.d., 320 mm in length) with UV (254 nm) and resistance detection. The leading electrolyte is 0.01 M sodium acetate, pH 4.5, with 0.2% HPMC, and the terminating electrolyte is 0.01 M β-alanine with acetic acid added to pH 5.0.

[23] P. S. L. Janssen, J. W. van Nispen, M. J. M. van Zeeland, and P. A. T. A. Melgers, *J. Chromatogr.* **470,** 171 (1989).

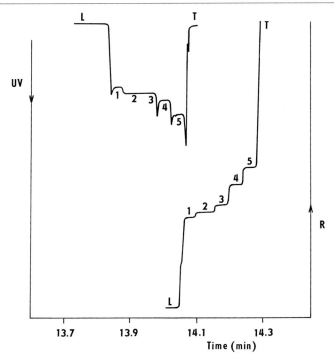

FIG. 14. ITP separation of closely related ACTH fragments. 1, ACTH(1–18), p*I* 11.5; 2, ACTH(1–17), p*I* 10.9; 3, ACTH(1–16)NH$_2$, p*I* 10.9; 4, ACTH(1–16), p*I* 10.4; 5, ACTH(1–14), p*I* 10.3. [Reprinted from P. S. L. Janssen, J. W. Van Nispen, M. J. M. Van Zeeland, P. A. T. A. Melgers, *J. Chromatogr.* **470,** 171 (1989), with permission.]

For the isolation of lysozyme from commercial ovalbumin,[24] preparative ITP is applied and CITP serves both as a predictor for establishment of the optimum operational conditions and to check the purity of the resulting product. Figure 15 shows the isotachopherograms of a blank (Fig. 15A), of a commercial ovalbumin (Fig. 15B), and of the pure lysozyme product (Fig. 15C). The sample in the latter two analyses is spiked with low molecular weight non-UV-absorbing spacers. This, together with the use of fast-scanning multiwavelength UV detection, provides a clear picture of the composition of the samples. CITP performed both in a commercial LKB 2127 Tachophor using a 0.5-mm i.d. polytetrafluoroethylene capillary with a conductivity and single-wavelength UV detection and in a homemade setup with a 75-mm i.d. silica capillary with multiwavelength UV detection provides identical results, which demonstrates how a procedure for ITP

---

[24] J. Caslavska, P. Gebauer, and W. Thormann, *J. Chromatogr.* **585,** 145 (1991).

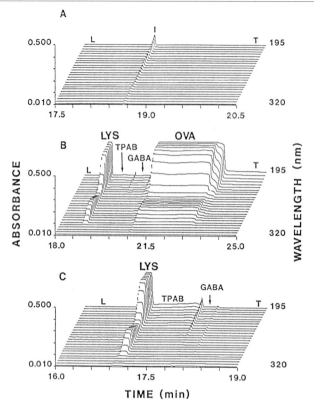

Fig. 15. ITP control of purification of lysozyme (LYS) from commercial ovalbumin (OVA). (A) blank; (B) analysis of commercial ovalbumin spiked with tetrapentylammonium bromide (TPAB) and γ-aminobutyric acid (GABA); (C) analysis of a fraction of purified lysozyme together with the two spacers. [Reprinted from J. Caslavska, P. Gebauer, and W. Thormann, *J. Chromatogr.* **585,** 145 (1991), with permission.]

analysis may be easily transferred between instruments of very different parameters.

An analysis of human serum proteins is demonstrated here, using a Shimadzu IP-2A (Kyoto, Japan) apparatus adapted for fully automated operation.[25] The leading electrolyte solution is 5 mM HCl, 9.3 mM 2-amino-2-methyl-1-propanol, pH 9.9, and the terminating electrolyte is 50 mM tranexamic acid and KOH, pH 10.8. The 230-mm long polytetrafluoroethylene (PTFEP) capillary of 0.5-mm i.d. is coated with HPMC. Figure 16a shows a record of normal human serum proteins. Four microliters of serum

[25] T. Manabe, H. Yamamoto, and T. Okuyama, *Electrophoresis* **10,** 171 (1989).

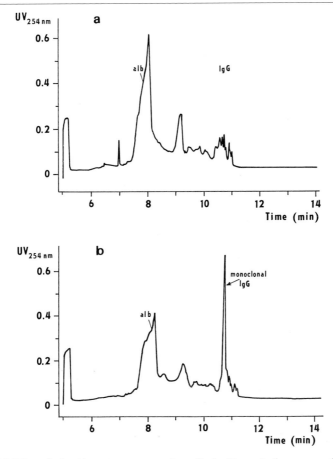

F<small>IG</small>. 16. ITP analysis of human serum proteins spiked with ampholine spacers. (a) Pattern of serum from a normal person; (b) pattern of serum from a monoclonal gammopathy patient. [Reprinted from T. Manabe, H. Yamamoto, and T. Okuyama, *Electrophoresis* **10,** 172 (1989), with permission.]

is diluted with a solution containing 400 μl of a 16-fold diluted stock solution containing 4% Ampholine carrier ampholytes (pH 3.5–10) and 0.1% sodium azide. Five microliters of the sample solution is introduced into the capillary. Figure 16b shows the record of human serum proteins from monoclonal gammopathy patients, clearly demonstrating the differences in mobility and in quantity between the two monoclonal immunoglobulin G (IgG) species. Reproducibility is about 3%, and the total time needed for one analysis cycle is not longer than 26 min.

Another interesting ITP application in the serum protein analysis is the

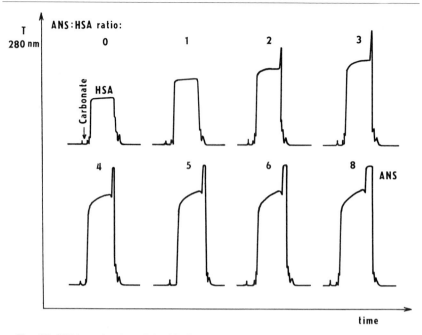

FIG. 17. ITP investigation of the binding of 8-anilino-1-naphthalenesulfonic acid (ANS) to human serum albumin (HSA). Analyses were performed with 5 μl of 0.1 mM HSA and 0–0.8 mM ANS mixtures. [Reprinted from C. J. Holloway and G. Bulge, *J. Chromatogr.* **234**, 454 (1982), with permission.]

biochemical and functional characterization of apolipoprotein B-containing lipoproteins.[26] For an analytical ITP run, 2 μl of the mixture of serum incubated with Sudan Black B for 30 min at 4°C and spacers (2:1) are introduced into an LKB 2127 Tachophor instrument. Phosphoric acid (5 mM), 20 mM ammediol, pH 9.2, and 0.25% HPMC are used as the leading electrolyte, and 0.1 M valine and 20 mM ammediol, pH 9.4, constitute the terminating electrolyte. During detection the current is 50 μA. The same electrolytes are also used for a preparative subfractionation of lipoproteins carried out in an Elphor VAP 22 (Bender & Hobein, Munich, Germany).

Isotachophoresis provides highly informative results in the study of interactions between proteins and ligands. An illustrative example is its contribution to the determination of the number and types of binding sites for 8-anilino-1-naphthalenesulfonic acid (ANS) to albumin.[27] In Fig. 17

[26] G. Nowicka, T. Brüning, B. Grothaus, G. Kahl, and G. Schmitz, *J. Lipid Res.* **31**, 1173 (1990).
[27] C. J. Holloway and G. Bulge, *J. Chromatogr.* **234**, 454 (1982).

isotachopherograms (measured with an LKB Tachophor) of 5-μl volumes of mixtures of ANS and human serum albumin (HSA) incubated at various stoichiometric ratios (0.1 mM HSA, 0.1–0.8 mM ANS) are shown. The results indicate that at least three molecules of the dye are bound tightly to the protein and the resulting complex shows a changed extinction coefficient, as well as an altered step height in the record. At ratios greater than 3:1, a further increase in UV absorption occurs, indicating that further molecules are bound to the protein and, simultaneously, the step due to the free dye appears in the record. Similarly, three or four high-affinity binding sites for bilirubin and one binding site for biliverdin were found in the study of their interactions with bovine serum albumin.[28] Again, an LKB 2127 Tachophor is suited for this analysis, which is performed in 5 mM HCl plus 10 mM 2-amino-2-methyl-1-propanol with 0.2% HPMC, pH 9, as the leading electrolyte and 5 mM 6-amino-n-caproic acid adjusted to pH 10.65 with Ba(OH)$_2$ as the terminating electrolyte.

Isotachophoresis as Preseparation Technique

The on-line combination of ITP and CZE is advantageous not only in separations of minor sample components in a complex matrix,[29,30] but also in the improvement of detection in peptide and protein analyses. Relatively large volumes of diluted peptide samples containing an excess of other ions can be injected and after preseparation and cleanup in the ITP stage a qualitative and quantitative analysis is performed in the CZE stage. An example of an ITP–CZE separation of a protein mixture is shown in Fig 18.[31] A laboratory-made coupled-column system is filled with the leading electrolyte: 10 mM ammonium acetate–acetic acid, pH 4.8, with 1% Triton X-100, and the terminating electrolyte: 20 mM ε-aminocaproic acid–acetic acid, pH 4.4, for the ITP stage, and the terminating electrolyte is also used here as the background electrolyte in which the zone electrophoresis is performed. A 10^{-8} M protein mixture (12 μl) is injected and separated first in a 200 mm $\times$ 0.4-mm i.d. PTFE preseparation capillary and then analyzed and detected in a fused silica capillary (390 mm $\times$ 0.075-mm i.d.) coated with linear polyacrylamide. Compared to a single CZE analysis, a 1000-fold improved detection level is reported.

The reproducibility of the ITP–CZE combination technique is satisfactory, as confirmed in Ref. 32. With repeated injections of plasma samples

[28] P. Oefner, A. Csordas, G. Bartsch, and H. Grunicke, *Electrophoresis* **6**, 538 (1985).
[29] L. Křivánková, F. Foret, and P. Boček, *J. Chromatogr.* **545**, 307 (1991).
[30] D. Kaniansky, J. Marák, V. Madajová, and E. Šimuničová, *J. Chromatogr.* **638**, 137 (1993).
[31] F. Foret, E. Szoko, and B. L. Karger, *J. Chromatogr.* **608**, 3 (1992).
[32] D. S. Stegehuis, U. R. Tjaden, and J. van der Greef, *J. Chromatogr.* **591**, 341 (1992).

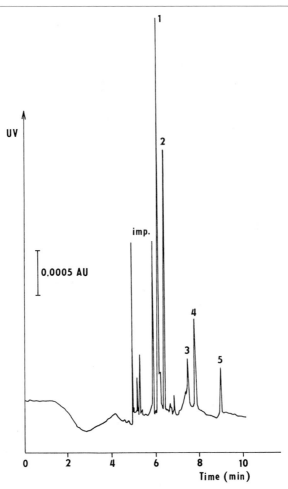

Fig. 18. CZE separation of a protein mixture after on-line ITP preconcentration. 1, Lyso-zyme; 2, cytochrome *c*; 3, trypsin; 4, ribonuclease A; 5, α-chymotrypsinogen A. [Reprinted from F. Foret, E. Szoko, and B. L. Karger, *J. Chromatogr.* **608,** 3 (1992), with permission.]

spiked with a fluorescein isothiocyanate derivative of the peptide angioten-sin III, a reproducibility of 3.5% RSD is reported. When using only single-step CZE, a rapid decrease in performance of the analysis is observed.

From other published reports of ITP used with other techniques, the combination with mass spectrometry[33] should be mentioned. Using an elec-

---

[33] R. D. Smith, J. A. Loo, C. J. Barinaga, C. G. Edmonds, and H. R. Udseth, *J. Chromatogr.* **480,** 211 (1989).

trospray ionization interface, the capabilities of this combination for the analysis of peptides and proteins are demonstrated. The concentrating effect of ITP combined with the sensitivity of mass spectrometry detection provide, e.g., for myoglobin and cytochrome c, detection limits in the range of hundreds of femtomoles.

Future of Isotachophoresis

Isotachophoresis has proved to be a useful method for quantitative analysis of low molecular weight substances, especially in diluted samples. It offers some advantages for the analysis of high molecular weight substances, particularly in terms of its effective separation capabilities. Here we can expect a continuation of the present spectrum of applications. The use of ITP may be of special value as an on-line preseparation technique for capillary zone electrophoresis, where the concentrating (stacking) power of ITP is combined with the detection sensitivity of CZE and thus the full resolving power of CZE can be utilized. Isotachophoresis can be used in the first stage, thus creating sharp zones of accumulated analytes, which can serve then as a sampling stack for CZE.

[18] Two-Dimensional Liquid Chromatography–Capillary Electrophoresis Techniques for Analysis of Proteins and Peptides

By ALVIN W. MOORE, JR., JOHN P. LARMANN, JR., ANTHONY V. LEMMO, and JAMES W. JORGENSON

Introduction

The probability of resolving all of the components in a sample decreases as the number of components in that sample increases. Mathematical models set forth by Giddings[1,2] predict that as sample complexity increases, complete separation of component peaks can be achieved only with high-efficiency separations. Even the most efficient one-dimensional (1D) techniques have insufficient peak capacity for the separation of highly complex biological samples. Two-dimensional (2D) separations have peak capacities

[1] J. C. Giddings, *Anal. Chem.* **56,** 1258 (1984).
[2] J. C. Giddings, *HRC-CC, J. High Resolut. Chromatogr. Commun.* **10,** 319 (1987).

approximately equal to the product of the peak capacities of the component 1D separations, and are therefore preferable for these samples. Until recently, planar 2D techniques such as isoelectric focusing–polyacrylamide gel electrophoresis (EIF–PAGE) have been the primary method for the analysis of complex samples.[3] However, planar techniques tend to be labor intensive and are more prone to problems with reproducibility. Also, sample recovery and identification can be problematic. A coupled-column design is appealing because of the potential for automation and decreased analysis time with online detection.

Coupled columns are often used in "heartcutting" configuration[4,5] in liquid and gas chromatography (LC and GC). In these systems, the goal is to isolate a single compound or group of compounds in a complex mixture. The first column separates interfering components from the compounds of interest. Only a small portion of the first column effluent containing the compounds of interest is collected and shunted to the second column for further separation. For example, in GC a first-dimension column may be used for class separations (alkanes, alkenes, aromatics). The portion of the effluent containing the compounds of interest, say the alkenes, is shunted into a second column of a different chemistry, which separates the different alkenes from each other for identification and quantitation. A detector may be placed after the first column to determine when effluent should be diverted to the second column, or for a routine analysis the necessary timing may be known in advance.

Note that in these heartcutting types of analysis there is interest in only a small number of the components eluting from the first column. Most components are never diverted to the second column and thus are not fully separated in a 2D format. If the user is interested in all of the sample components, such as in peptide or protein maps, then the coupled column design must be altered. All of the effluent from the first-dimension column must be analyzed by the second-dimension column, to generate 2D information for all peaks in the analysis.

Figure 1a is a schematic diagram of a comprehensive 2D system. Sample is injected onto the first dimension and separated. All effluent from the first separation passes through an interface where it is sampled into the second dimension. After separation in the second dimension, sample components pass with the second dimension effluent into a detector. Detection

[3] P. H. O'Farrell, *J. Biol. Chem.* **250,** 4007 (1975).

[4] W. Bertsch, *in* "Multidimensional Chromatography" (H. J. Cortes, ed.), p. 74. Marcel Dekker, New York, 1990.

[5] H. J. Cortes and L. D. Rothman, *in* "Multidimensional Chromatography" (H. J. Cortes, ed.), p. 219. Marcel Dekker, New York, 1990.

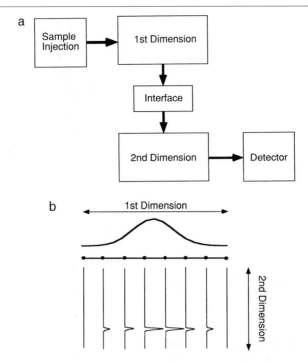

Fig. 1. (a) A comprehensive 2D separation system. (b) Sampling considerations in two dimensions.

is usually only done at the end of the second dimension, to avoid any band-broadening effects that might be caused by a detector placed in-line between the first and second dimensions.

In a heartcutting scheme, the relative analysis times for the first and second dimensions are not critical. The second dimension only analyzes the effluent from a brief period of the first-dimension separation. If, however, all of the first-dimension effluent is to be sampled into the second, the time required for the second separation limits the sampling of the first. If detection is done only at the end of the second dimension, as in Fig. 1a, then each entire analysis in the second dimension samples only a single "point" from the first. This is shown graphically in Fig. 1b. Characterization of the first-dimension separation is thus inherently limited by the speed of the second-dimension analysis relative to the first.

In 1990, Bushey and Jorgenson[6] introduced a 2D separation scheme as described previously. A coupled-column design was used that employed

[6] M. M. Bushey and J. W. Jorgenson, *Anal. Chem.* **62,** 978 (1990).

reversed-phase liquid chromatography (RPLC) as the first dimension and capillary zone electrophoresis (CZE) as the second, to give a 2D RPLC–CZE system. This method was termed *comprehensive,* as all effluent from RPLC analysis was analyzed by CZE, so that peaks are specified by both an RPLC retention time and a CZE migration time. Data from this system resemble that of traditional 2D slab gel electrophoresis already mentioned.

In this chapter we present three approaches to 2D coupled-column designs. Each approach is comprehensive as defined, and each uses a form of liquid chromatography as the first dimension with capillary zone electrophoresis as the second. Liquid chromatography and CZE are complementary techniques because their solvent systems are compatible but their separation mechanisms are different (see [2] in this volume[6a]). The interface between the two systems is also simple: either an automated valve or a fluid tee. Detection is by ultraviolet (UV) absorption or laser-induced fluorescence (LIF). The first approach couples microcolumn size-exclusion chromatography (SEC) with CZE for the separation of proteins. The second approach combines RPLC and CZE for the separation of tryptic peptides. The last approach couples RPLC with Fast-CZE, a variation of CZE, enabling fast 2D separations.

Two-Dimensional Size-Exclusion Chromatography–Capillary Zone Electrophoresis

Experimental

Reagents

Human serum and buffer reagents are purchased from Sigma (St. Louis, MO). Formamide is obtained from Fisher Scientific (Pittsburgh, PA). All chemicals are used as received. The buffer used for both the SEC and CZE separations is 10 mM Tricine, 25 mM Na$_2$SO$_4$, 0.005% (w/v) sodium azide, adjusted to pH 8.23 with NaOH. A Barnstead Nanopure system with a 0.2-μm pore size nylon membrane filter (Alltech Associates) is used to generate deionized water for buffer solutions.

Sample Preparation

Human serum (lyophilized powder) is reconstituted in 500 μl of buffer containing 8% (w/v) formamide (1 ml is the original volume according to

[6a] S.-L. Wu and B. L. Karger, *Methods Enzymol.* **270**, Chap. 2, 1996 (this volume).

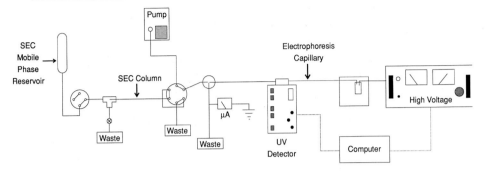

Fig. 2. Instrumental diagram for 2D SEC–CZE. See text for detailed description.

Sigma). Rather than filtering to remove any undissolved particulate matter, the serum sample is subjected to centrifugation at 16,000 g for 2 min. After centrifugation, the supernatant is drawn off and stored at 4°.

Column-Packing Procedure

The size-exclusion column is a 105 cm by 250-μm i.d. fused silica capillary (Polymicro Technologies), slurry packed in our laboratory with 6 μm, spherical, Zorbax GF450 particles (Rockland Technologies, Inc.). Columns are typically packed in 8–10 hr and allowed to settle overnight under pressure. Before use, the column is flushed with water for 4 hr, then flushed overnight with mobile phase. The packing procedure has been described in detail previously.[7]

Instrumentation

Figure 2 is a schematic diagram of the entire instrumental setup, described in detail as follows.

Chromatographic System. The chromatographic injection system has been modified from a static split injection system that had been developed for open tubular LC in our laboratory.[8] Injections are made with a pressurized reservoir containing a small vial of sample. The microcolumn is removed from the static split injection tee and inserted into the pressurized reservoir, where head pressure of 7 atm (100 psi) is applied for 8 min to inject the desired volume of sample. After the injection is complete, the microcolumn is returned to the static split injection tee, where 41 atm (600 psi) of head pressure is applied to begin chromatography. This approach

[7] R. T. Kennedy and J. W. Jorgenson, *J. Microcol. Sep.* **2**, 120 (1990).
[8] E. J. Guthrie and J. W. Jorgenson, *J. Chromatogr.* **255**, 335 (1983).

is taken rather than performing a static split injection, which requires several hundred microliters of sample per injection.

Electrophoresis System. A ±30-kV high-voltage power supply (Spellman High Voltage Electronics Corp.) is used in the negative high-voltage mode. Capillary electrophoresis is performed in an untreated fused silica capillary with an inner diameter of 50 μm. The capillary is 38 cm overall, 20 cm to the detection window. Capillary zone electrophoresis conditions include 5-sec electromigration injection at -3 kV and 4-min runs at -8 kV.

SEC–CZE Interface. The SEC microcolumn is interfaced to the CZE system through an electrically actuated six-port valve (Valco Instruments). The valve is fitted with a 300-nl external collection loop made from 15 cm of 50-μm i.d./350-μm o.d. fused silica. The microcolumn is interfaced to the valve by connection with a 3-cm long section of 50-μm i.d./350-μm o.d. fused silica capillary. This avoids exposing the frit at the end of the microcolumn to the valve fittings. The union between these capillaries is made by a "sleeve" of 0.007-in. i.d./1-16-in. o.d. Teflon tubing. This approach increases column lifetime and allows easy insertion and removal of the SEC microcolumn from the valve.

The electrophoresis capillary is coupled to the valve through a Delrin tee (Alltech Associates), which is connected to the valve by a 5-cm section of 0.005-in. i.d./1/16-in. o.d. PEEK tubing (Upchurch Scientific). The end of the capillary is held in place near the center of the tee. A 20-cm section of 0.040-in. i.d. stainless steel tubing is also inserted into the Delrin tee. This tubing serves as both a waste line and as the ground electrode (anode) for the CZE system. A microammeter is placed between the tubing and electrical ground to monitor CZE current. Hydrodynamic flow within the CZE capillary is minimized by keeping the height of this stainless steel waste line level with the cathodic buffer reservoir. The operation and timing of the valve have been previously described.[6] A Hewlett-Packard (Palo Alto, CA) 1050 pump provides a flush rate of 100 μl/min.

Detection. A Linear Instruments model 200 variable wavelength UV–Vis detector with an on-column capillary flowcell is used for CZE detection. Detection is done at 214 nm.

Instrument Control. A Hewlett-Packard Vectra 386/25 computer is used to control the valve, high-voltage power supply, and data collection system. The computer is equipped with a Labmaster multifunction data acquisition board (Scientific Solutions). Software written in-house with QuickBasic 4.5 (Microsoft Corp.) provides control over experimental parameters and allows for data processing and analysis. Spyglass Transform and Spyglass Format (Spyglass, Inc.), raster imaging software for the Macintosh provide gray-scale images of the 2D data.

Results and Discussion

Figure 3 is a gray-scale image of the separation of protein components in reconstituted human serum. Signal intensity is presented in logarithm (base 10) scale to bring out subtle, low-intensity peaks. The large band observed is due to the high concentration of serum albumin constituents. These albumin components make up the majority of the total concentration of protein present in human serum. There are other protein components present in human serum, but many of these species are present at concentration levels below the detection limit of the current system.

The system shown here represents the first comprehensive scheme to couple two capillary separation techniques that both use liquid mobile phases. In general, separation of protein samples is notoriously problematic. Protein species tend to undergo strong adsorption to silica surfaces in both chromatographic and electrophoretic separations. Further difficulty arises from the lack of sensitive detection schemes applicable to a wide range of proteins. The combined use of microcolumn techniques for protein separation results in greater overall resolving power than independent use of either technique. Future work is directed toward a new interfacial approach

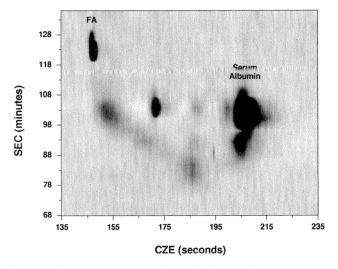

FIG. 3. Analysis of human serum sample by 2D SEC–CZE. SEC injection was 8 min at 7 atm (100 psi). Head pressure of 41 atm (600 psi) was applied to generate a flow rate of 275 nl/min for the chromatographic separation. Electrophoresis capillary was 38 cm overall, 20 cm to the detection window. CZE conditions: 5-sec electromigration injection at −3 kV and 4-min runs at −8 kV. Data collection: 2 points/sec.

that will allow the use of smaller inner diameter SEC microcolumns operated at near-optimal flow rates. This will greatly improve the separation
efficiency of the SEC separation. This improvement in separation efficiency
will be necessary in order to increase the overall peak capacity and resolving
power of the 2D system. In addition, implementation of techniques available
for reducing protein–fused silica interactions in CZE[9–11] will be investigated. Employment of these techniques often leads to dramatic improvements in CZE separation efficiency, and thus would also enhance the performance of the 2D system.

Two-Dimensional Reversed-Phase Liquid Chromatography–Capillary Zone Electrophoresis

Experimental

Materials. Triethylamine (TEA), trifluoroacetic acid (TFA), 2-mercap-
toethanol, iodoacetic acid, porcine thyroglobulin, and trypsin are purchased
from Sigma. Dimethyl sulfoxide (DMSO) is purchased from Mallinckrodt
(St. Louis, MO). Tetramethylrhodamine isothiocyanate isomer-5 (TRITC)
is purchased from both Molecular Probes (Eugene, OR) and Research
Organics. Acetonitrile (Optima grade) is purchased from Fisher Scientific
and water is purified for HPLC with a Barnstead Nanopure system. Sodium
phosphate diabasic is purchased from EM Science. All reagents are used
as received.

Sample Preparation. The porcine thyroglobulin is first irreversibly reduced using Canfield's procedure.[12] Thyroglobulin (10 mg) is dissolved in
1 ml of a 10 M urea–0.1 M boric acid solution with the pH adjusted to 8.5.
Twenty microliters of 2-mercaptoethanol is added, the solution is flushed
with nitrogen, and then held at 37° for 4 hr. The protein is precipitated on
the addition of 7 ml of absolute ethanol–2% (v/v) HCl and centrifuged.
The protein pellet is added to 50 μg of iodoacetic acid in 1 ml of 8 M
urea–0.1 M boric acid, pH 8.5. After 10 min, 250 μl of 2-mercaptoethanol
is added to react with the excess iodoacetic acid. The protein is precipitated
by the addition of 7 ml of absolute ethanol. It is centrifuged and resuspended
in ethanol two times.

The reduced thyroglobulin is digested with trypsin (30:1, w/w) for 24
hr at 37° in 0.1 M boric acid buffer at pH 8.4. After enzymatic digest, the

[9] R. M. McCormick, *Anal. Chem.* **60,** 2322 (1988).
[10] J. K. Towns and F. E. Regnier, *Anal. Chem.* **63,** 1126 (1991).
[11] M. M. Bushey and J. W. Jorgenson, *J. Chromatogr.* **480,** 301 (1989).
[12] R. E. Canfield, *J. Biol. Chem.* **238,** 2691 (1963).

sample is filtered (0.2-μm pore size; Gelman). The fluorescent derivatizing reagent TRITC is dissolved in DMSO at 10 mg/ml. A 25-μl aliquot of this solution is combined with 75 μl of digested thyroglobulin (1 mg/ml). The reaction is allowed to proceed at room temperature for 4 hr in the dark. For all tagging reactions, TRITC is in threefold molar excess to the tryptic peptides, with TRITC having a final concentration of 5.6 mM in the reaction mixture. This mixture is diluted 100-fold into the initial RPLC mobile phase prior to injection.

Instrumentation. The 2D instrumental configuration is diagrammed in Fig. 4. The first-dimension RPLC column is a Zorbax 2.1 × 150 mm 300 SB-C$_8$ column. Sample is injected onto this column with a Rheodyne 7125 manual valve fitted with a 20-μl loop. A Hewlett-Packard model 1050 HPLC solvent delivery system is used to generate the RPLC elution gradient. Effluent from the RPLC column is collected in an eight-port electrically actuated valve fitted with two 10-μL loops (Valco Instruments). The comprehensive coupling of liquid chromatography to capillary zone electrophoresis with a six-port electrically actuated valve has been previously reported by our group.[6]

The second-dimension CZE runs are performed as follows (Fig. 4). While the LC effluent fills loop A in the eight-port valve, the contents of loop B are injected and run on the CZE capillary. Once loop A is filled, the valve is switched so that the loop A contents are injected onto the CZE

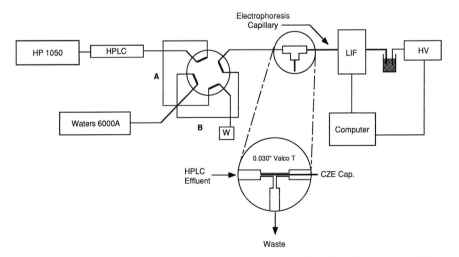

FIG. 4. RPLC–CZE instrumental configuration. A and B, Two 10-μl loops in the eight-port electrically actuated valve. LIF, Laser-induced fluorescence; W, waste; HV, high-voltage power supply. See text for further explanation.

capillary, with loop B now being filled by LC effluent. Each successive CZE run must take place in the time it takes to fill the 10-μl loops, with any overflow of these loops going to waste. The entire RPLC run is essentially collected and analyzed by the second dimension, CZE.

The CZE capillary is interfaced to the valve with a 1/16-in. stainless steel tee with a 0.030-in. i.d. through-hole (Valco Instruments). The eight-port valve is connected to the tee with 3 cm of 0.007-in. stainless steel tubing. The fused silica capillary (PolyMicro Technologies) is brought in the other side of the tee, through the 0.030-in. inner spacing of the tee, and positioned as close as possible to the 0.007-in. stainless tubing from the valve. The fused silica capillary cannot slide into this piece of stainless as it has an outer diameter of 360 μm (0.014 in.). The third side of the tee is fitted with 0.030-in. PEEK tubing (Upchurch Scientific) and connected to waste. When the valve switches, the flush pump (Waters 6000A) forces the contents of the newly filled loop into the interface tee. The contents of the loop flush past the CZE capillary and out to waste. By adjusting the voltage, sample is electromigrated onto the CZE capillary. Note that most of the volume (99.9%) in the loop flows to waste, but the second dimension effectively samples all of the loop contents. In addition to pumping the loop contents into the tee, the flush pump also continuously baths the end of the capillary with running buffer during the CZE run.

The entire valve is held at electrical ground and serves as the anode for CZE. Current is monitored in electrophoresis by placing a microammeter between the valve and electrical ground. A Spellman $\pm$30-kV high-voltage power supply is operated in a negative polarity mode. Negative high voltage (-13.5 kV) is applied to the detection side of the capillary in a cathodic buffer vial. The fused silica capillary is 15-μm i.d./360-μm o.d. and 14 cm in length from the tee to the high-voltage reservoir. In this experiment, electroosmotic flow carries compounds from the tee at ground to negative high voltage. The eight-port valve and high voltage are under computer control (Fig. 5) to synchronize the CZE injections. The eight-port valve alternates between the two loops on a 50% duty cycle. When the valve is actuated, the high voltage is lowered to the injection voltage in order to electromigrate sample onto the CZE capillary. Voltages less than the run voltage are necessary to avoid overloading the CZE capillary with sample. Before returning to the run voltage, the power supply is set to 0 V. If the voltage is not set to zero, peaks in the CZE runs are tailed. This voltage programming reduces the effect of poor washout of sample from the 10-μl loops. The choices of injection voltage, injection time, and zero voltage time are dependent on the flow rate of the flush pump. Injection voltage is always 10% of the run voltage for 2 sec, followed by 2 sec of zero voltage. The flush pump operates at 0.8 ml/min or 13 μl/sec. This flow rate flushes

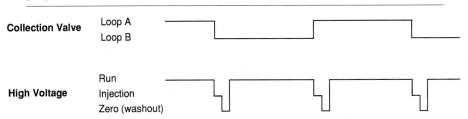

FIG. 5. RPLC–CZE timing diagram. The eight-port valve is run on a 50% duty cycle. While one loop is being injected onto the second dimension, the other loop is collecting effluent from the first dimension. In this manner, no sample is lost in the interface between injections. When the valve switches to make an injection, the run voltage is lowered to the injection voltage for 2 sec. Prior to returning to the run voltage, there is a period of 2 sec at 0 V. This discriminates against exponential washout of the collection loops.

the 10-μl loops more than five times prior to returning to the run voltage. Because the injection sequence requires 4 sec and the CZE run time is 15 sec, the second-dimension CZE runs are repeated every 19 sec.

Detection is accomplished on the capillary 5 cm from the tee with the total length of the capillary at 14 cm (9 cm from detection point to negative high voltage). A high-sensitivity, low-cost laser-induced fluorescence (LIF) detection system utilizing a 1.5-mW green helium–neon laser is used.[13] The excitation spectrum of TRITC is an excellent match for the 543.5-nm line on this laser, and this combined with its high molar absorptivity and photostability make it an exceptional tag. Fluorescence is detected at 90° from excitation with a photomultiplier tube (PMT).

Data collection and timing of the valve and high-voltage power supply are all under computer control. A Macintosh IIcx equipped with a multifunction data acquisition board from National Instruments (NB-MIO-16X) runs under software written in-house using LabView (National Instruments). Data are presented with Spyglass software.

Separation Conditions. For RPLC, linear gradients are run from 20 to 70% B in 50 min, where A is 0.1% TFA (v/v) in water and B is 0.1% TFA (v/v) in acetonitrile. The flow rate is 70 μl/min. The CZE run buffer in the negative high-voltage reservoir and in the flush pump is 10 mM Na_2HPO_4–20 mM TEA, pH 11.0.

Results and Discussion

Figure 6 is a gray-scale image of an RPLC–CZE run of TRITC-labeled tryptic peptides of porcine thyroglobulin. The number of tryptic peptides

[13] D. Y. Chen, H. P. Swerdlow, H. R. Harke, J. Z. Zhang, and N. J. Dovichi, *J. Chromatogr.* **559**, 237 (1991).

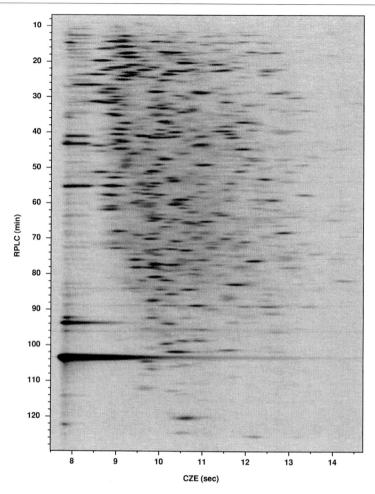

Fig. 6. Two-dimensional chromatoelectropherogram of TRITC-labeled tryptic peptide fragments of reduced porcine thyroglobulin.

for such a protein number in the hundreds. Incomplete enzymatic digestion, which is always a reality, makes the sample even more complex. Either RPLC or CZE alone would prove inadequate for the complete separation of such a complex digest. Therefore, the enhanced peak capacity of a 2D method such as RPLC–CZE is necessary for these separations. The spot capacity for this system is estimated at 3000. As the number of compounds reaches approximately 10% of this capacity, saturation of the separation space occurs, and there will begin to be significant overlapping of peaks.

Thus, this technique should be capable of resolving digests containing several hundred peaks.

Note the large peak in the chromatoelectropherogram at an RPLC retention time of 104 min. This is the excess TRITC from the derivatization reaction. With the resolving power of this system it is possible to use excess reagent and not compromise the separation by peak overlaps. In fact, 1D CZE with excess TRITC is nearly impossible as all electrophoretic peaks will be hidden by the reagent peak from 7.7 to 9.6 sec. Excess derivatizing agent allows all amines in the sample to be labeled regardless of the derivatization reaction kinetics.

Presently this technique is being used for the analysis of variants of cytochrome *c*. There is evidence that it may be useful for elucidating post-translational modifications such as deamidation. In addition, this technique has possibilities for ultratrace analysis in complex samples as it combines high separation efficiencies with the low detection limits of laser-induced fluorescence detection.

Two-Dimensional Reversed-Phase Liquid Chromatography–Fast Capillary Zone Electrophoresis

Experimental

Reagents. A Barnstead Nanopure system is used to generate deionized water for all samples and buffers. All chemicals are used as obtained. Buffers and solutions are filtered with 0.2-μm pore size membrane filters (Alltech) before use. Aqueous buffers contain 0.005% (w/v) sodium azide as an antimicrobial agent. Aqueous buffer for RPLC and CZE is 0.01 *M* NaH$_2$PO$_4$, adjusted to pH 6.85 with NaOH. This is the CZE buffer used in all experiments. For the RPLC gradient runs, solvent A is CZE buffer, solvent B is 30% acetonitrile (Fisher Optima)–70% CZE buffer (v/v).

Samples are filtered with 0.2-μm pore size Acrodisc syringe filters (Gelman, Inc.) prior to use in the CZE or RPLC systems. Water-soluble sodium fluorescein (Aldrich, Milwaukee, WI) in CZE buffer is used to check the optical alignment in the Fast-CZE system. Fluorescein isothiocyanate (FITC) is obtained from Sigma. Acetone and pyridine used in tagging reactions are from Fisher.

Sample Preparation. Horse heart cytochrome *c* is digested with trypsin (30:1, w/w) for 24 hr at 37° in 0.1 *M* boric acid buffer, pH 8.4. After enzymatic digest, the sample is filtered (0.2-μm pore size; Gelman). To tag digest components with FITC, a portion of the digest solution is diluted into 200 μl of 0.1 *M* boric acid, pH 8.4. FITC is dissolved in 1 ml of 80:20 (v/v) acetone–pyridine solution. The concentrations of digest solution and

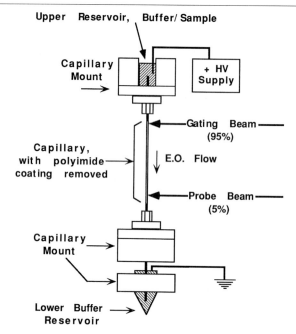

Fɪɢ. 7. Fast-CZE instrumental diagram. See text for explanation.

FITC solution are adjusted to give a threefold molar excess of the tryptic peptides over FITC in the sample. A 50-μl aliquot of FITC solution is added to 200 μl of digest solution and allowed to react in darkness of 24 hr at room temperature. The final concentration of fluorescent tag in tagged digest samples is 4.9 mM. Samples are diluted 1000-fold into CZE buffer before analysis by RPLC–Fast-CZE.

Fast-CZE. Our present Fast-CZE instrument is as described previously[14] with the following modifications. A mechanical shutter (Newport Optical Corp.) has been substituted for the acoustooptic modulator used in our previous work because the high speeds possible with the modulator are unnecessary and the optical setup with the shutter is simpler. Significantly, the jointed capillary is no longer used. Modifications to the capillary mount have enabled us to use a short length of a single capillary and still maintain the high electric field necessary for Fast-CZE.

Figure 7 is a schematic diagram of the present Fast-CZE instrument. The capillary is mounted vertically, with buffer reservoirs at the top and bottom for contact with the high-voltage power supply. The capillary mount

[14] C. A. Monnig and J. W. Jorgenson, *Anal. Chem.* **63,** 802 (1991).

is constructed of electrically insulating material to make this arrangement possible. Gravitationally driven flow is not a concern because the capillary has an inner diameter of only 10 μm. When doing only Fast-CZE, the lower buffer reservoir is connected to electrical ground (through a microammeter, not shown, to monitor capillary current) and the upper reservoir is connected to positive high voltage. Electroosmotic flow is from top to bottom, toward the more negative electrode. A high-power argon ion laser (maximum power, 1 W) is split into two beams that are focused at different heights on the capillary. The upper beam contains more than 95% of the total laser power and is the gating (or bleach) beam, while the lower beam contains the remaining 5% of the laser power and is the probe (detection) beam.

In Fast-CZE, the upper buffer reservoir is also the sample reservoir. Sample is continuously electromigrated into the capillary as long as voltage is applied. "Injections" of samples for individual CZE runs are done with the gating beam of the laser. The gating beam is normally on. Because of the high power of the gating beam, all sample passing through it is photodegraded into nonfluorescent species. As long as the gating beam is on, practically no fluorescent species reach the probe beam further down the capillary. Only a low-level background fluorescence is seen at the probe beam. To make an injection, the gating beam is momentarily blocked by closing the shutter. The shutter is closed for only a few milliseconds, but in this time a small slug of unbleached material passes into the region of the capillary between the beams. This "optically injected" sample is separated between the beams by differential electromigration of its components as with standard CZE, and the components are then detected as they pass through the probe beam.

Instrumental. A diagram of the 2D RPLC–Fast-CZE system is shown in Fig. 8. The RPLC pump is a Hewlett-Packard HP1050 model quarternary gradient pump. The pump flow rate for the 2D runs is 250 μl/min. The linear gradient for the elution of tryptic digest samples is from 100% solvent A to 100% solvent B over 15 min, then 100% solvent B for 5 min. Injections are made onto the RPLC column via a Rheodyne 7125 manual valve with a 20-μl sample loop. The analytical column is a 2.1 × 150 mm Vydac (The Separations Group, Hesperia, CA) protein and peptide C_{18} column. The entire effluent from the RPLC column passes through a connecting capillary of untreated silica, 75-μm i.d./350-μm o.d. by 75 cm, into the interface tee for the Fast-CZE system.

For the 2D system, the upper buffer reservoir in Fast-CZE is replaced with a 0.010-in. i.d. stainless steel tee (Valco, Inc.) with 1/16-in. standard fittings. The remaining port on the tee is connected to waste through large-bore (1-mm i.d.) PEEK tubing. The tee makes contact with electrical ground

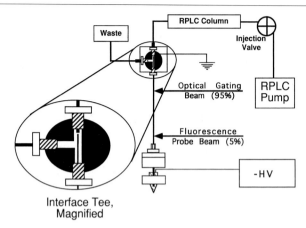

FIG. 8. RPLC–Fast-CZE 2D instrumental diagram. The inset shows a close-up of the interface tee and the end of the CZE capillary. All effluent from the RPLC column flows across the end of the CZE capillary and out to waste. A fraction of all RPLC effluent is drawn into the CZE capillary by electromigration, but individual Fast-CZE injections are controlled by the gating beam.

and the polarity of the high-voltage supply is reversed so that negative high voltage is applied to the lower buffer reservoir. Electroosmotic flow is toward the more negative electrode, and thus is still from top to bottom as described earlier. The CZE capillary is 12 cm by 10-μm i.d./145-μm o.d. The applied voltage for the 2D analyses is 20 kV, so that the applied electric field is 1.7 kV/cm. The distance between the point of injection and the point of detection (gating beam to probe beam) is 1.2 cm. This is the effective capillary length over which separation occurs.

For a 2D analysis, effluent from the RPLC column flows through the connecting capillary into the tee, across the end of the CZE capillary, and out the waste line. The large-bore waste line helps ensure that there is minimal pressure-induced flow in the small-bore capillary. Constant high voltage is applied to the CZE capillary throughout the analysis, so that a small fraction of the RPLC effluent is continuously electromigrated into the CZE capillary. No interface valve is needed because individual CZE injections are done with the optical injection system based on the gating beam from the laser. As sample components elute from the RPLC column, they are further separated by Fast-CZE and detected by fluorescence at the probe beam. Fluorescence signal is collected with a PMT and computerized data acquisition system as described in our earlier work.[14] Gray-scale images of the 2D data are generated with Spyglass software.

Results and Discussion

A typical Fast-CZE analysis is shown in Fig. 9. The lower trace is the fluorescence signal seen at the probe beam, the actual electropherogram, while the upper trace represents laser power. To make an optical injection, the gating beam of the laser is momentarily blocked, so that an unbleached slug of material passes into the separation region between the beams. This is shown in Fig. 9 by the large dip in laser power in the upper trace. The shutter is actually placed before the beam splitter in the optical path, so that both beams of the laser are blocked when the shutter is closed. Thus for each injection there is an accompanying dip in the background fluorescence signal seen at the probe beam. This serves as a useful injection marker for each CZE run. The sample peaks appear seconds later, after separation of the sample components by differential electromigration between the beams. Here the large peak at about 1.5 sec is water-soluble fluorescein and the four smaller peaks are FITC-tagged amino acids. The amount of sample injected is controlled by the length of time the shutter is closed as well as by the voltage applied to the capillary.

Figure 10 is a 2D RPLC–Fast-CZE analysis of an FITC-tagged tryptic digest of horse heart cytochrome *c*. For these data, the data acquisition rate was 150 Hz. The RPLC elution gradient for this analysis was only 20 min long, and Fast-CZE runs were done every 2.5 sec. Only a fraction of the entire 2D data set is shown in Fig. 10, to focus attention on the area

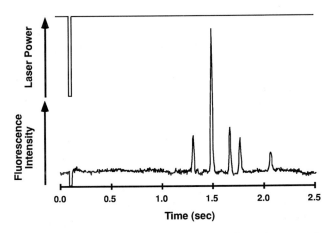

FIG. 9. Fast-CZE analysis timing diagram. This diagram shows the temporal relationship between laser power and observed fluorescence signal. The injected sample is a mixture of water-soluble fluorescein and four FITC-tagged amino acid standards.

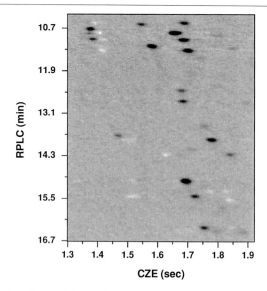

Fɪɢ. 10. 2D RPLC–Fast-CZE analysis of tryptic digest of horse heart cytochrome *c*. Gray-scale image is a portion of the entire 2D data set for this analysis.

containing the majority of the peaks. This small part of the entire 2D data set shows the inherent high peak capacity of the 2D method. The RPLC peaks in this run are typically 20 sec wide at the base. Notice that even with such sharp RPLC peaks, Fast-CZE runs done every 2.5 sec ensure that the RPLC dimension is well sampled. Using Fast-CZE as the second dimension makes 2D analysis possible in the time normally taken to do the RPLC analysis alone. In addition, the RPLC elution gradient need not be as complex as if RPLC were used alone because the second dimension does part of the "work" of the separation.

This method has applications for sample "fingerprinting" techniques such as tryptic mapping and purity analysis. As Fig. 10 shows, the 2D system allows for specification of a number of different peptide fragments by both CZE migration time and LC elution time. To identify individual peptide peaks, known standards could be run individually to match up these migration and elution times, but this would obviously be labor intensive. Sample fingerprinting, however, would be much less so. Even if none of the peaks are specifically identified, the 2D pattern can be used for comparison of sample identity and for confirmation of peak purity. The same sample analyzed under the same conditions should give the same pattern, and sample impurities should appear as spurious peaks in the pattern.

Such tryptic mapping is routinely done by RPLC, but often requires development of complex gradient methods specific to each sample. The same can be said of some methods of purity confirmation. The advantage of a 2D system is the inherent high peak capacity. If the peak capacity of an analysis method is sufficient for complex samples of hundreds of components, the method can be readily used for less complex samples with very little sample-specific method development.

Acknowledgments

This work has been supported by the National Science Foundation (Grant CHE-9215320) and the National Institutes of Health (Grant GM39515). A. W. M., Jr. received support from the U.S. Department of Education. The authors thank Hewlett-Packard for the gift of the HP 1050 solvent delivery systems aned the Vectra 386/25 computer. J. P. L., Jr. and A. V. L. thank Rockland Technologies for the donation of the Zorbax RPLC column and the GF450 stationary phase.

[19] Capillary Electrophoresis Detectors: Lasers

By THOMAS T. LEE and EDWARD S. YEUNG

Introduction

The miniaturization of electrophoresis in the capillary form (CE) has advanced separation science to new levels of performance. However, improved separation power is not sufficient unless coupled to sensitive detection. Because only 1 pmol or less of an analyte is involved in the typical analysis, preparative applications of CE have been difficult so far and, hence, the greatest value of the technique lies in the analysis of complex mixtures through fractionation. Even though numerous approaches can be adopted to enhance the resolution of a wide spectrum of analytes, the detection of the separated analytes in the capillary tubing remains the stumbling block to successful analysis in many situations. To this end, the narrow bandwidths and low analyte quantities in CE render most of the detector technologies originally developed for liquid chromatography (LC) inappropriate and pose new challenges to the designers of detectors for CE. This is especially valid when one is dealing with biological matrices, where the determination of components present in trace quantities is often necessary. Hence, the urgent need for the sensitive detection of various types of analytes in CE is obvious.

While it is not the intention of this chapter to review all aspects of detection in CE, a brief discussion of the pros and cons of other detection schemes is needed in order to present a coherent perspective on where laser-based methods stand among the plethora of detection methodologies in existence. The interested reader may also refer to other reviews on CE detection methods elsewhere.[1–10]

Non-Laser-Based Detectors for Capillary Electrophoresis

The on-column UV–visible absorbance detector utilizing incoherent light sources accompanies all commercial CE instruments and is by far the most widely used. With variable wavelength selectivity, the sensitivity of the detector can be tuned to optimize detectability for most types of analytes with limits of detection (LOD) at the micromolar level.[11] Exceptions are analytes lacking chromophores that absorb appreciably at wavelengths longer than around 200 nm (e.g., carbohydrates, polyamines). In such instances, indirect absorbance detection, where the displacement of chromophores in the running buffer by the presence of nonabsorbing analytes, may be used to obtain LODs of about 10 μM.[12,13] However, when high sensitivity is necessary, the performance of conventional absorbance detectors becomes inadequate. Even with the use of U- or Z-shaped flow cells[14,15] or rectangular capillaries,[16] the LODs for most analytes remain unsatisfactory.

[1] S. F. Y. Li, "Capillary Electrophoresis." Elsevier, Amsterdam, 1992.
[2] W. G. Kuhr, *Anal. Chem.* **62**, 403R (1990).
[3] W. G. Kuhr and C. A. Monnig, *Anal. Chem.* **64**, 389R (1992).
[4] M. V. Novotny, K. A. Cobb, and J. Liu, *Electrophoresis* **11**, 735 (1990).
[5] E. S. Yeung and W. G. Kuhr, *Anal. Chem.* **63**, 275A (1991).
[6] P. Jandik and W. R. Jones, *J. Chromatogr.* **546**, 431 (1991).
[7] Z. Deyl and R. Struzinsky, *J. Chromatogr.* **569**, 63 (1991).
[8] E. C. Huang, T. Wachs, J. J. Conboy, and J. D. Henion, *Anal. Chem.* **62**, 713A (1990).
[9] M. J. F. Suter, B. B. DaGue, W. T. Moore, S. N. Lin, and R. M. Caprioli, *J. Chromatogr.* **553**, 101 (1991).
[10] E. R. Schmid, *Chromatographia* **30**, 573 (1990).
[11] Y. Walbroehl and J. W. Jorgenson, *J. Chromatogr.* **315**, 135 (1984).
[12] S. Hjertén, K. Elenbring, F. Kilar, J. Liao, A. J. Chen, C. J. Siebert, and M. Zhu, *J. Chromatogr.* **403**, 47 (1987).
[13] F. Foret, S. Fanali, L. Ossicini, and P. Bocek, *J. Chromatogr.* **470**, 299 (1989).
[14] G. J. M. Bruin, G. Stegeman, A. A. C. Van, X. Xu, J. C. Kraak, and H. Poppe, *J. Chromatogr.* **559**, 163 (1991).
[15] J. P. Chervet, R. E. J. Van Soest, and M. Ursem, *J. Chromatogr.* **543**, 439 (1991).
[16] T. Tsuda, J. V. Sweedler, and R. N. Zare, *Anal. Chem.* **12**, 2149 (1990).

For certain applications, electrochemical detectors may be advantageous. Impressive LODs of 10–100 nM have been demonstrated for several inorganic cations with potentiometry.[17] The universal conductivity detector offers an LOD of about 1 μM, but at the expense of having to use low concentrations of buffering electrolytes.[18-21] For easily oxidized or reduced analytes, amperometric detectors are able to deliver LODs at the 10–100 nM level.[22,23] For nonelectroactive analytes, indirect amperometric[24] or potentiometric[25] detection can be used. Limits of detection of around 1 μM have been reported with indirect amperometry.[24] Although versatile and inexpensive, most electrochemical detectors suffer from gradual diminution in response stemming from electrode contamination and, therefore, stringent means of reproducibly regenerating the electrode surface are often necessary. This problem is especially severe when the sample matrix contains high molecular weight species.

Two CE detection schemes based on radioactivity have also been reported.[26,27] Both designs respond to high-energy radiation emitted by radioisotopes present in analyte molecules and are, hence, highly selective. Limits of detection of between 0.1 and 1 nM have been demonstrated for ^{32}P-labeled nucleotides[26] and a ^{99}Tc-chelated tumor-imaging agent.[27] Because of their unique selectivity and high sensitivity, radioisotope detectors are especially suitable for the determination of trace species in complex biological matrices. However, the requirement that the analyte be prelabeled may render the technique incompatible with some applications, including the analysis of natural systems, screening of food preparations or pharmaceutical formulations, etc.

The use of mass spectrometers as detectors for CE opens up the feasibility of positive analyte identification and structural elucidation in complex mixtures. As with LC/mass spectrometry (MS), the issues of sensitivity, reliability, and cost are central to the success of CE/MS. However, LODs

[17] C. Haber, I. Silvstri, S. Roosli, and W. Siman, *Chimia* **45**, 117 (1991).

[18] X. Huang, J. A. Luckey, M. J. Gordon, and R. N. Zare, *Anal. Chem.* **61**, 766 (1989).

[19] F. Mikkers, F. Everaerts, and T. Verheggen, *J. Chromatogr.* **169**, 11 (1979).

[20] M. Deml, F. Foret, and P. Bocek, *J. Chromatogr.* **320**, 159 (1985).

[21] X. Huang, R. N. Zare, S. Sloss, and A. G. Ewing, *Anal. Chem.* **63**, 189 (1991).

[22] R. A. Wallingford and A. G. Ewing, *Anal. Chem.* **59**, 678 (1987).

[23] C. E. Engstrom-Silvermann and A. G. Ewing, *J. Microcol. Sep.* **3**, 141 (1991).

[24] T. Olefirowicz and A. G. Ewing, *J. Chromatogr.* **499**, 713 (1990).

[25] A. Manz and W. Siman, *Anal. Chem.* **59**, 74 (1987).

[26] S. Pentoney, R. Zare, and J. Quint, *J. Chromatogr.* **480**, 259 (1989).

[27] K. D. Altria, C. F. Simpson, A. K. Bharij, and A. E. Theobald, *Electrophoresis* **11**, 732 (1990).

ranging from 1 μM to 1 nM for macromolecules such as proteins have already been demonstrated.[28-32] More detailed discussions of CE/MS are given elsewhere in this volume.

Fluorescence detectors equipped with incoherent light sources provide sensitivity one to two orders of magnitude higher than absorbance detectors.[33] Because of their low cost and commercial availability, lamp-based fluorescence detectors are preferred over their counterparts relying on absorbance for analytes that fluoresce appreciably. On the other hand, various labeling schemes that incorporate highly fluorescent molecules onto naturally nonfluorescent analytes may be adopted to take advantage of the good sensitivity, linearity, and selectivity of lamp-based fluorescence detectors.

Laser-Based Detectors for Capillary Electrophoresis

At present, use of the laser as a light source represents the state of the art in CE detector technologies as far as sensitivity is concerned. The most valuable attribute of the laser in CE detection is perhaps the monochromaticity and coherent nature of laser light, which allow the tight focusing of a large amount of light into a small volume of space confined by the detection region. Significant enhancement in the sensitivity of various spectroscopic detection schemes including absorbance, fluorescence, refraction, and Raman scattering has been realized through the replacement of conventional light sources with the laser. It is to this diverse aspect of CE detection that the remainder of this chapter is devoted.

Fluorescence

When trace to ultratrace levels of analytes are involved, laser-induced fluorescence (LIF), which offers higher sensitivity than any other detection scheme in CE, is usually the method of choice. With fluorescein- or rhodamine-labeled analytes, LODs on the order of picomolar or a few hundred

[28] R. Smith, C. Barinaga, and H. Udseth, *Anal. Chem.* **60,** 1948 (1988).
[29] R. D. Smith, J. A. Loo, C. Barinaga, C. Edmonds, and H. Udseth, *J. Chromatogr.* **480,** 211 (1989).
[30] M. Moseley, L. Deterding, K. Tomer, and J. Jorgenson, *Rapid Commun. Mass Spectrom.* **3,** 87 (1989).
[31] E. D. Lee, W. Mueck, J. D. Henion, and T. R. Covey, *Biomed. Environ. Mass Spectrom.* **18,** 87 (1989).
[32] R. Caprioli, W. Moore, M. Martin, B. DaGue, K. Wilson, and S. Moring, *J. Chromatogr.* **480,** 247 (1989).
[33] J. S. Green and J. W. Jorgenson, *J. Chromatogr.* **352,** 337 (1986).

molecules are possible. At present, five LIF detection schemes of significance in CE have been demonstrated. Their experimental configurations, merits, drawbacks, and applications are outlined below.

Laser-Induced Fluorescence with Orthogonal Geometry. Because laser light is well collimated and monochromatic, it can be focused to a beam diameter approaching the diffraction limit of light. Thus, a high photon flux can be delivered to the interior of a capillary to allow efficient excitation of fluorescence. More important, however, is the fact that a laser beam can be focused relatively "cleanly" into the interior of the capillary (i.e., without significant scattering from the capillary wall) so that scattered light does not give rise to a high level of background during measurement. With the capillary tilted at an angle to the excitation beam and the emitted fluorescence collected in a direction perpendicular to the plane of the capillary and excitation beam, the background from scattered light is minimized after appropriate spatial and spectral filtering.[34-36]

As shown in Fig. 1, an on-column LIF detector with orthogonal geometry can be simple and rugged.[37] It consists of eight main components: a capillary holder (CH), 1-cm focal length lens (L), ×20 microscope objective (MO), microscope objective holder (MOH), mirrors (M), mirror mounts (MM), photomultiplier tube (PMT) with filter (F), Plexiglas box (PB), and two light shields (LS). The laser beam enters the otherwise light-tight PB through a 3-mm hole. The L is rigidly mounted to the PB so that the focal point of the excitation beam is uniquely defined and used as a reference point for all the other components. With a small section of its coating removed, the capillary is mounted on a two-dimensional stage (CH) capable of 10-μm resolution. Two short pieces of quartz capillary (350-μm o.d. × 250-μm i.d.) glued to the CH serve to guide the separation capillary through the optical region. This can hold the common 150-μm o.d. capillary tubing. Alternatively, 350-μm o.d. capillaries can be inserted into the holder without the guides. The mounted capillary is at an angle of 20° with respect to the laser beam.

With this design, linear dynamic ranges of at least four orders of magnitude and an LOD of 3 pM of fluorescein injected have been documented for both a 50- and 75-μm i.d. and 145-μm o.d. capillary.[37,38] Despite the low cost and user-friendly features of the above-described setup, its sensitivity is comparable to the other more instrumentally elaborate designs based on

[34] E. Guthrie, J. Jorgenson, and P. Dluzneski, *J. Chromatogr.* **22,** 171 (1984).
[35] P. Gozel, E. Gassmann, H. Michelsen, and R. N. Zare, *Anal. Chem.* **59,** 44 (1987).
[36] W. G. Kuhr and E. S. Yeung, *Anal. Chem.* **60,** 1832 (1988).
[37] T. T. Lee and E. S. Yeung, *J. Chromatogr.* **595,** 319 (1992).
[38] E. S. Yeung, P. Wang, W. Li, and R. W. Giese, *J. Chromatogr.* **608,** 73 (1992).

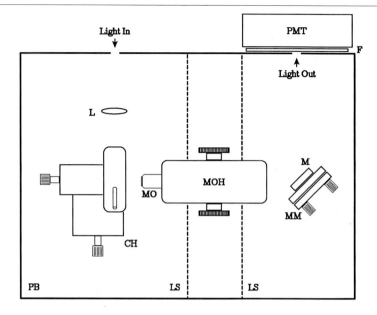

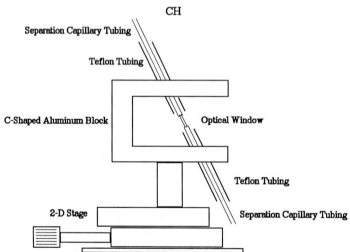

FIG. 1. Experimental setup of orthogonal LIF detector. See text for details. [Reproduced with permission from T. T. Lee and J. Yeung, *J. Chromatogr.* **550,** 831 (1991).]

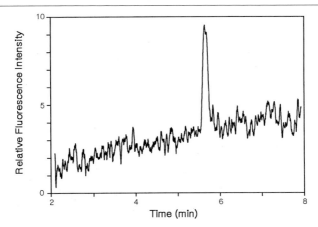

FIG. 2. Electropherogram of 5×10^{-10} M conalbumin injection. [Reproduced with permission from T. T. Lee and J. Yeung, *J. Chromatogr.* **550**, 831 (1991).]

the sheath-flow cuvette,[39] charge-coupled device (CCD) detector with axial illumination,[40] and epilumination fluorescence microscope.[41] However, with the on-column orthogonal geometry design, it has been found that when capillaries with inner diameter values of less than 50 μm are used, the LOD degrades by roughly 10 times, a consequence of increased scattering and diminished signal size.[42] When excited in the ultraviolet (UV) region, the signal-to-noise ratio (S/N) also suffers when capillaries filled with viscous media such as gels or polymer solutions, which normally contain high concentrations of luminescent impurities and scattering centers, are used. Nonetheless, when mass detection sensitivity as well as medium luminescence and scattering are not a concern, the above-described on-column LIF detector offers state-of-the-art concentration sensitivity at low cost and with ease of construction and operation. With deep ultraviolet excitation (275 nm) from an argon ion laser, tryptophan-containing native proteins can be detected at the 0.1 nM level at injection (Fig. 2). Limits of detection of 0.1 and 0.2 nM for, respectively, conalbumin and bovine serum albumin in CE have been reported.[37] When labeled with 3-(4-carboxybenzoyl)-2-quinolinecarboxaldehyde, the separated fragments from the hydrolysate of a 10 nM poly(galacturonic acid) solution can be detected.[43]

[39] Y.-F. Cheng and N. J. Dovichi, *Science* **242**, 562 (1988).
[40] J. V. Sweedler, J. B. Shear, H. A. Fishman, R. N. Zare, and R. H. Scheller, *Anal. Chem.* **63**, 496 (1991).
[41] L. Hernandez, J. Escalona, N. Joshi, and N. Guzman, *J. Chromatogr.* **559**, 183 (1991).
[42] T. Lee, E. S. Yeung, and M. Sharma, *J. Chromatogr.* **565**, 197 (1991).
[43] J. Liu, O. Shirota, and M. V. Novotny, *Anal. Chem.* **64**, 973 (1992).

Laser-Induced Fluorescence with Sheath Flow Cuvette. As the inner diameter of the capillary decreases to below 50 μm, flicker noise from the wall-scattered excitation light for the on-column orthogonal geometry LIF format becomes significant and the *S/N* begins to degrade. However, to further lower the quantity of an analyte that is amenable to analysis by CE, it is necessary to decrease the inner diameter of the capillary tubing used for separation. An ingenious way of solving this problem is through the attachment of a sheath flow cuvette to the detection end of the capillary.[39]

As depicted in Fig. 3, the steady sheath flow supplied by an LC pump or sustained by gravity concentrically encloses the effluent from the capillary, thereby conferring spatial stability on the postcapillary detection region. This is essentially the same technology that is employed in flow cytometry. By varying the sheath flow rate, one can control the diameter of the effluent stream confined by the sheath flow, with the optimal effluent stream diame-

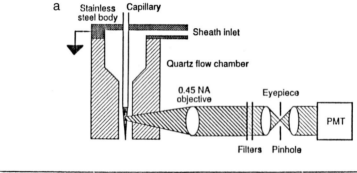

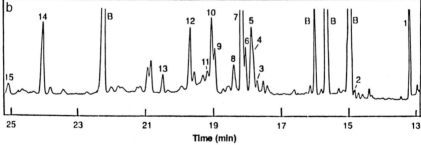

Fig. 3. (a) Schematic diagram of LIF detector with sheath-flow cuvette. (b) Separation of between 2 and 7 amol of 18 amino acids. The separation is driven by a 25-kV potential, and a pH 10 buffer is used for both the separation and the sheath stream. Injection was for 10 sec at 2 kV. Amino acids identified: peak 1, Arg; 2, Lys; 3, Leu; 4, Ile; 5, Trp; 6, Met; 7, Phe, Val, His, and Pro; 8, Thr; 9, Ser; 10, Cys; 11, Ala; 12, Gly; 13, Tyr; 14, Glu; and 15, Asp. B, Peaks associated with the reagent blank. (Reproduced with permission from Ref. 39.)

ter being about 10 μm. The electrical circuit is completed by grounding the stainless steel sheath inlet. Up to 50 mW of laser power may be used for a laser beam orthogonal to the effluent stream and collection optics focused to a beam diameter of 10 μm within the cuvette before photobleaching of fluorescein-labeled analytes negate signal-to-noise enhancement.[44] Because the cuvette is of excellent optical quality, flicker noise from the wall-scattered light is reduced to below the shot noise (photon counting statistics) level. Hence, LIF detection in CE with the sheath flow cuvette is shot noise limited. As few as 600 molecules injected from a 1 pM solution of an amino acid labeled with tetramethylrhodamine isothiocyanate (TRITC) can be detected in CE.[45] Similar results have been obtained with tetramethylrhodamine-labeled DNA fragments during one spectral channel sequencing.[46] Unlike the on-column geometry, CCD with axial illumination and epilumination formats, the problems with impurity fluorescence in the UV region and scattering from the gel matrix are nullified with the use of postcolumn detection using the sheath flow cuvette.

Laser-Induced Fluorescence with Charge-Coupled Device and Axial Illumination. An alternative to the above-described LIF detection schemes is to couple laser light axially into the exit end of the capillary with the resulting fluorescence over a 2-cm long region of observation collected by a cylindrical mirror. The collected fluorescence is then directed into a spectrograph where the light is dispersed and registered on a CCD detector two-dimensionally (Fig. 4).[40] Hence both spatial and spectral information are obtained without loss of sensitivity. To reduce the quantity of data produced and the read-out time, the CCD is operated in the time-delayed integration (TDI) mode. In TDI, the rate of read-out from the serial register is synchronized with the migration rate of an analyte band in the region of observation. Because the dominant source of noise in a low-light level measurement with the CCD is from data read-out, TDI lowers the number of read-outs and improves the S/N. Besides, the signal from an analyte band is integrated over its entire residence time in the region of observation. The S/N can be improved further be binning several rows of pixels together prior to read-out in such a way that the narrow width of an analyte band is not distorted significantly. By inserting a long-pass cut off filter before the spectrograph slit, stray light from Rayleigh scattering is decreased and fluorescein isothiocyanate (FITC)-labeled amino acids present at picomolar concentrations can be detected.[40] Disadvantages of LIF detection with axial

[44] S. Wu and N. J. Dovichi, *J. Chromatogr.* **480,** 141 (1989).

[45] J.-Y. Zhao, D.-Y. Chen, and N. J. Dovichi, *J. Chromatogr.* **608,** 117 (1992).

[46] H. Swerdlow, J. Z. Zhang, D. Y. Chen, H. R. Harke, R. Grey, S. Wu, N. J. Dovichi, and C. Fuller, *Anal. Chem.* **63,** 2835 (1991).

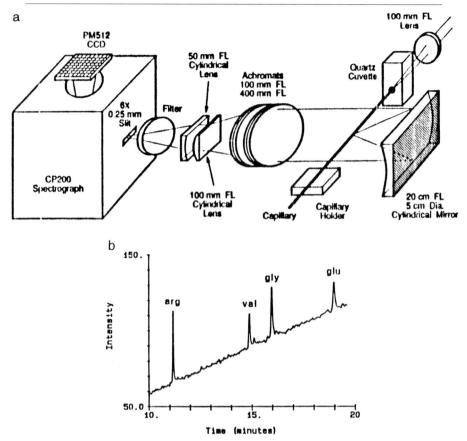

FIG. 4. (a) Experimental setup of LIF detector with CCD and axial illumination. (b) TDI electropherogram of four FITC–amino acids, with the concentration of injected FITC–arginine, FITC–valine, and FITC–glycine at 7×10^{-11} M and FITC–glutamate at 8×10^{-11} M. The fluorescence intensity is obtained by using a spectral window between the Rayleigh and major Raman band. The sloping base line is caused by a continually increasing integration time corresponding to a decreasing shift rate. (Reproduced with permission from Ref. 40, Copyright 1991, American Chemical Society.)

illumination and CCD include the necessity of using low excitation powers to minimize photobleaching of analytes while avoiding shadowing, high cost of the CCD detector, and instrumental complexity.

Laser-Induced Fluorescence with Epilumination. The employment of an epilumination fluorescence microscope with laser excitation offers LODs comparable to the best detection schemes in CE.[41] In the collinear geometry of excitation and collection, a high-numerical aperture (NA) lens can be used to allow the tight focusing of light into the detection region while increasing the solid angle for light collection (Fig. 5). Because confocal excitation and detection are used, the depth of field of the optical system is sufficiently small that only the interior of the capillary is probed. Stray light is rejected in the same manner as high spatial resolution is achieved in confocal microscopy. As shown in Fig. 5, laser light is directed into the detection region within the capillary subsequent to reflection from a dichroic mirror and focusing through a 4.5-mm (0.75 NA) lens. The emitted fluorescence is then collected by the same lens and directed back to the dichroic mirror, which selectively transmits the long-wavelength fluorescence while reflecting the Rayleigh scattered component. After passing through a set of long-pass, spatial and notch filters, the fluorescence is detected with a photomultiplier tube.

Impressive LODs of picomolar solutions of FITC-labeled amino acids at injection and linear dynamic ranges of over five orders of magnitude have been demonstrated.[41] In addition to being an on-column detection scheme, an advantage of the setup is the ease of operation of the device. Because the microscope already possesses the optics for visually checking the quality of alignment of the focused light within the capillary detection region and a precise three-dimensional displacement device for spatial manipulation, the switching of capillaries is easily accomplished.

The collinear geometry of excitation and collection also opens up the feasibility of performing separations in multiple capillaries simultaneously, thereby increasing the throughput of CE. By laying multiple capillaries in parallel on a translatable stage, DNA-sequencing runs have been performed (Fig. 6).[47] With the employment of 2-color fluorescence detection and a 2-dye labeling protocol, a raw sequencing rate of 100,000 bases in about 2 hr using 200 capillaries may be possible in the future.[48]

Laser-Induced Fluorescence with Fiber Optic Coupling. The best performance for LIF requires careful alignment to match the laser beam to the core of the separation capillary to avoid stray light. For the nonexpert, such alignment may not be an easy task. To provide a more rugged system for the routine user, fiber optics was employed to deliver the laser light in the first commercial version of a CE–LIF detector. An elliptical reflector then collected the fluorescence and guided it to a phototube. Filters are used throughout to further isolate the emission from the excitation beam. Because the dimension of the fiber is large, mechanical alignment is relatively easy. This system provided performance that is somewhat poorer than the other state-of-the-art LIF schemes described above, but with little

[47] X. C. Huang, M. A. Quesada, and R. A. Mathies, *Anal. Chem.* **64,** 967 (1992).
[48] X. C. Huang, M. A. Quesada, and R. A. Mathies, *Anal. Chem.* **64,** 2149 (1992).

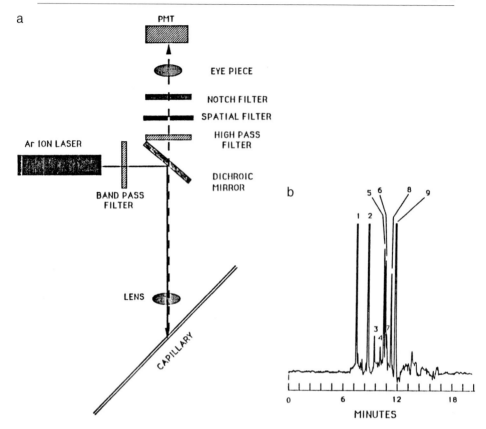

FIG. 5. (a) Schematic diagram of epilumination LIF detector. (b) Electropherogram of a mixture of six FITC–amino acids. Laser power, 4 mW; spatial filter, 2-mm diameter; objective, 0.75 NA; ocular, ×10. The analyte concentration was $2.1 \times 10^{-9} M$, and the injection volume was 1 nl. The amount injected was 2.1 amol. Peaks: 1, arginine; 2, blank (probably unreacted FITC); 3, blank; 4, impurity of reagent; 5, leucine; 6, methionine; 7, cysteine; 8, alanine; 9, glycine. FITC–arginine and FITC–glycine peaks were clipped. [Reproduced from L. Hernandez, J. Escalona, N. Joshir, and N. Guzman. *J. Chromatogr.* **559,** 183 (1991), with permission.]

user intervention. The main reason is inadequate rejection of stray light and inefficient utilization of the laser output due to irradiation of the entire capillary rather than the liquid core alone. This is not a serious concern with excitation in the visible region, but can become a critical limitation for ultraviolet excitation.

Other Considerations. The picomolar detection sensitivity reported for all of the above-described LIF schemes in CE was obtained with analytes

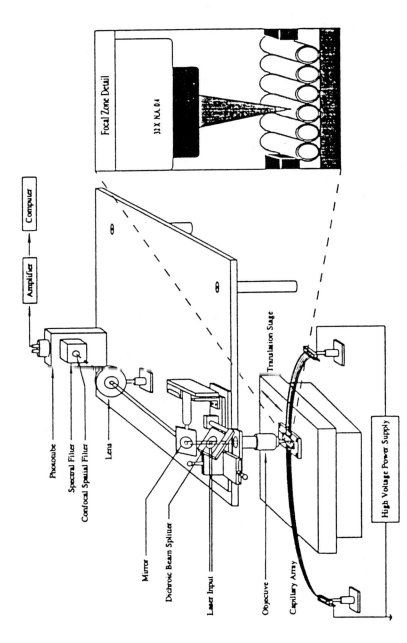

FIG. 6. DNA sequencing by scanning a capillary array. (Reproduced with permission from X. C. Huang, M. A. Quesada, and R. A. Mathies, *Anal. Chem.* **64**, 967. Copyright 1992, American Chemical Society.)

labeled with highly fluorescent derivatives (e.g., fluorescein, rhodamine). In practice, because most analytes do not fluoresce appreciably naturally, to detect them with LIF would require derivatization. For biological macromolecules, an important exception is proteins, which can be detected with 0.1 nM sensitivity when excited at 275 nm.[37] However, the derivatization of macromolecules in CE is a nontrivial task.

For precolumn labeling, variable extents of label incorporation into each macromolecule give rise to components with different mobilities and, hence, render quantitation difficult. For instance, the precolumn labeling of myoglobin resulted in multiple peaks in the electropherogram.[49] Because most labeling reagents (e.g., fluorescein isothiocyanate [FITC], naphthalene-2,3-dicarboxaldehyde [NDA]) were designed for operation at concentration levels higher than nanomolar to micromolar, it is unclear whether their use in situations in which the analyte is present at picomolar levels is suitable. The impressive detection limits reported for LIF of labeled biomolecules have been based exclusively on diluting the samples after derivatization at higher concentrations and with larger amounts of analytes. This problem is especially severe when interference from the sample matrix is a concern. Even when adverse effects accompanying labeling efficiency and matrix interference are not serious, one would need to devise a separation condition under which the large unreacted and hydrolyzed reagent peaks are well resolved from the analyte peaks of interest. This is because the high fluorescence background from peaks associated with the reagent easily obscures the presence of the small analyte peaks. With nanoliter-sized samples, traditional schemes of derivatization necessitate dilution and, hence, decrease the effective detectability of LIF.

Because all fluorogenic reagents are, in reality, at least slightly fluorescent in their native and hydrolyzed forms, on- and postcolumn derivatization schemes are unable to offer sensitivity close to precolumn labeling with LIF, as a high reagent-to-analyte ratio is often necessary for efficient reaction. An additional difficulty with on-column labeling is that the reagent must exhibit fast kinetics of binding with the analyte. Otherwise, peak broadening from electrodiffusion due to slow interspecies conversion would become significant.[50] As only nanoliter-sized analyte zones are involved, the instrumental requirement of postcolumn derivatization is complex in order to avoid peak broadening resulting from mixing.[49]

From the discussion above, it is obvious that although absolute detection sensitivity is crucial to LIF, it is by no means the only critical facet in the practical implementation of LIF detection in CE. Innovative approaches

[49] B. Nickerson and J. Jorgenson, *J. Chromatogr.* **480,** 157 (1989).
[50] D. F. Swaile and M. J. Sepaniak, *J. Liq. Chromatogr.* **14,** 869 (1991).

to derivatization and native fluorescence detection methods are vital to future success of LIF detection in CE.

Absorbance

In light of the difficulty inherent in derivatization, sensitive absorbance measurements represent a potentially powerful solution to the problem of detection in CE. This is because most native analytes absorb appreciably in some parts of the electromagnetic spectrum, in contrast to the scarcity of natural fluorophors. In other words, absorbance is a much more general method of detection than fluorescence. As noted earlier, conventional absorbance detectors fail to deliver high sensitivity and offer an LOD of only 10^{-3} to 10^{-4} absorbance unit (AU), which corresponds to about micromolar levels of analyte. Use of the laser, however, has made possible the development of a whole new generation of methods for the sensitive measurement of absorbance in CE.

Thermal–Optical Detection. By far, one of the most sensitive absorbance detection techniques in CE is thermal–optical (TO) detection. An impressive LOD of 3×10^{-7} AU or a 40 nM concentration of 4-dimethylaminoazobenzene-4'-sulfonyl- (DABSYL) arginine in a 50-μm i.d. capillary has been reported.[51] In TO detection, light from a pump laser is focused into the detection region confined by the wall of a cylindrical capillary (Fig. 7). A probe laser beam, focused into the detection region so that it intersects the pump beam, is positioned slightly off center on the photodiode. In the presence of an absorbing analyte, nonradiative relaxation of the resulting excited states gives rise to the release of heat into the solvent medium. The ensuing elevation in temperature decreases the density of the solvent and, hence, changes the refractive index of the medium in the vicinity of the pump beam. The change in refractive index is registered through the deflection of the probe beam in the plane of the pump and probe beams. The extent of deflection is, in turn, proportional to the concentration of the analyte. Similar to fluorescence measurements but unlike conventional absorbance measurements, TO absorbance detection is essentially a low-background technique. With phase-sensitive detection through modulation of the pump beam intensity, TO absorbance detection is highly sensitive. A linear dynamic range exceeding four orders of magnitude has been demonstrated.[52]

At present, CE with TO absorbance detection has been applied exclusively to the determination of labeled amino acids.[51,52] One such important

[51] M. Yu and N. J. Dovichi, *Appl. Spectrosc.* **43**, 196 (1989).
[52] K. C. Waldron and N. J. Dovichi, *Anal. Chem.* **64**, 1396 (1992).

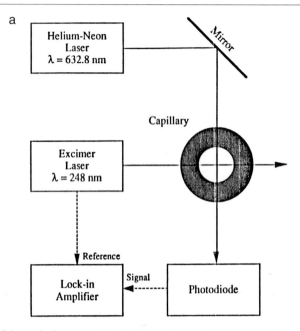

FIG. 7. (a) Schematic diagram of TO absorbance detector. (b) High-speed micellar capillary electrophoresis separation of PTH–amino acids. A 50-μm i.d., 39-cm long capillary was used for the separation; the distance from the injector to the detector was 34 cm. The separation proceeded at 8 kV. Injection was for 5 sec at 500 V. A 12.5 mM, pH 7.0 borate–phosphate buffer, containing 35 mM sodium dodecyl sulfate, was used for the separation. Threonine produced a small peak that overlapped that of asparagine. (Reproduced with permission from Ref. 52, Copyright 1992, American Chemical Society.)

application is the separation of 20 phenylthiohydantoin–amino acids present in subfemtomole amounts, which, in conjunction with developments of miniaturized reactors, carries the potential of lowering the quantity of protein required for sequencing via the Edman degradation reaction to the femtomolar level. The detection of native analytes at low concentrations in CE with TO absorbance detection utilizing ultraviolet laser excitation promises to address many significant issues in analysis in the future.

However, the determination of analytes at such low concentrations with absorbance detection may suffer from severe interference from matrix components via absorbance and refractive index effects, especially when one is dealing with biological samples. Besides, the alignment of two tightly focused laser beams in a picoliter intersection volume within the capillary is no trivial task. The detector is also not rugged, as the sensitivity of TO absorbance detection varies dramatically with relative beam locations and,

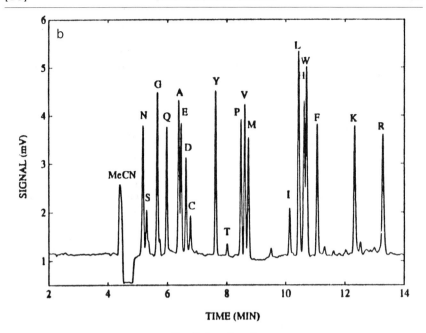

FIG. 7. (*continued*)

in fact, situating the CE system and TO detection optics on separate tables has been cited as a reason for degradation in performance.[51] Finally, the high cost of employing two lasers is a deterrent to practical use as well.

Laser-Induced Capillary Vibration. Rather than probing the effects of heat on the refractive index of the solution in the detection region within the capillary, the resulting local tension of the capillary is monitored in laser-induced capillary vibration (CVL).[53] In CVL detection, the periodic generation of heat subsequent to the absorption of light from an intensity-modulated pump laser beam induces a corresponding periodic change in the local temperature in the detection region of the capillary, which in turn causes the local tension on the capillary to oscillate. This oscillation results in the vibration of the whole section of the capillary between two fixed points. The magnitude of the vibration is sensed by the periodic deflection of a probe laser beam. With CVL, an absorbance detection sensitivity comparable to TO detection in CE has been reported.[53]

The experimental setup of CVL detection in CE is shown in Fig. 8. A pump laser beam is focused into the detection region that is located halfway between a pair of holders that prevent those two points on the capillary

[53] J. Wu, T. Odake, T. Kitamori, and T. Sawada, *Anal. Chem.* **63,** 2216 (1991).

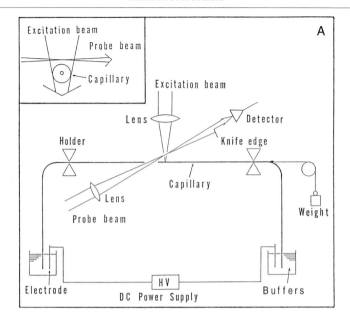

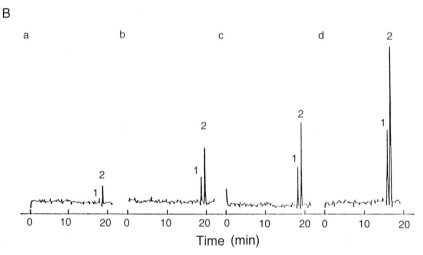

FIG. 8. (A) Experimental setup of CVL detector. (B) Electropherograms of tryptophan (peak 1) and phenylalanine (peak 2). Introduction amounts: (a) Trp, 100 fmol; Phe, 1 pmol; (b) Trp, 300 fmol; Phe, 3 pmol; (c) Trp, 600 fmol; Phe, 6 pmol; (d) Trp, 1.2 pmol; Phe, 12 pmol. (Reproduced with permission from Ref. 53, Copyright 1991, American Chemical Society.)

from moving. A suspended weight is used to stretch the same section of the capillary to maintain tension. Focused in an orthogonal direction to the pump beam and capillary axis so that its beam waist just misses the body of the capillary, the probe laser beam senses the position of the capillary by way of its deflection by the capillary. With lock-in detection, the extent of the deflection of the probe beam is extracted and found to be proportional to the concentration of the absorbing analyte in the detection region. A linear dynamic range of three orders of magnitude has been documented.[53]

Unlike TO absorbance detection, CVL should be relatively insensitive to the effects of changes in the refractive index (RI) of the solution in the detection region. This should render CVL detection less prone to baseline drifts or fluctuations from sample matrix effects and gradient schemes than TO detection. In addition, because the probe beam is not focused into the capillary, CVL detection should be compatible with very small inner diameter capillaries, as the optical quality of the deflected probe beam is always preserved. With ultraviolet excitation, native tryptophan can be detected at the 10 μM level in a 30-μm i.d. capillary.[53]

Axial-Beam Detection. In the cross-beam arrangement, the small inner diameter of the capillary in CE limits the sensitivity of conventional absorbance detectors. However, the path length can be increased by directing light through the axis of the capillary so that the width of the analyte band becomes the effective path length for light attenuation. Because the typical width of an analyte band at the detection region is about 3 mm whereas a common inner diameter of the capillary is 50 μm, a 60-fold increase in the absorbance path length in switching from the cross- to axial-beam arrangement is realized.

Because of the monochromaticity and directional nature of laser light, it can easily be focused into the capillary axially. As depicted in Fig. 9, the injection end of the capillary is immersed in a buffer vial constructed from a cuvette to allow preservation of the optical quality of the beam.[54] The capillary is enclosed in a cylindrical glass support to minimize the effects of vibration on background noise. The exit end of the capillary is then terminated in a second cuvette with the light emerging from the capillary collected with a microscope objective that focuses the transmitted light onto a photodiode.

Figure 10 depicts the electropherogram procured with the above-described setup. After differentiation, the familiar form of the electropherogram is obtained and no information is lost. Unfortunately, for an aqueous buffer–fused silica capillary system, a large fraction of the transmitted light

[54] J. A. Taylor and E. S. Yeung, *J. Chromatogr.* **550,** 831 (1991).

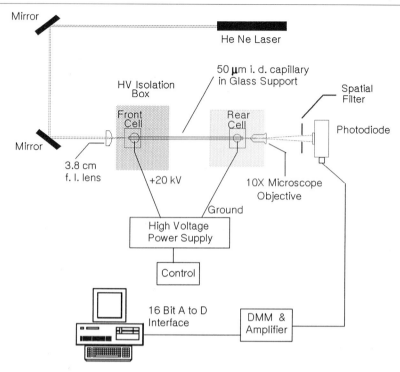

FIG. 9. Schematic diagram of axial-beam absorbance detector. [Reproduced with permission from J. A. Taylor and E. S. Yeung, *J. Chromatogr.* **550,** 831 (1991).]

traverses through not only the solution phase, but also the fused silica wall of the capillary. This increases the noise on the transmitted light resulting from capillary vibration and renders the sensitivity of the system no better than with the conventional cross-beam format. Nonetheless, by using poly-tetrafluoroethylene (PTFE) tubing and a 50% (v/v) ethylene glycol–aqueous buffer system containing sodium dodecyl sulfate (SDS), total internal reflection of the injected light is actualized and noise from capillary vibration is reduced. With this arrangement, an LOD of 3×10^{-5} AU for a 50-μm i.d. capillary has been documented.[54]

The major drawback of axial-beam absorbance detection is its incompatibility with aqueous-based buffers in bare fused silica capillaries, as PTFE tubing possesses poor heat-dissipating capabilities and an overwhelming majority of CE separations have been performed in 100% aqueous-based buffers. A second disadvantage is that when a highly absorbing constituent in the sample is present, the signal from the transmitted light may be swamped out by the shot noise of the detector and, hence, the detectability

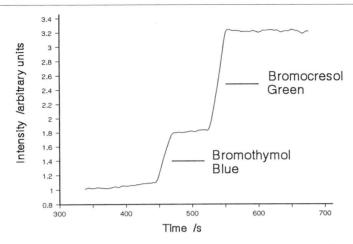

FIG. 10. Electropherogram acquired with axial-beam absorbance detection. [Reproduced with permission from J. A. Taylor and E. S. Yeung, *J. Chromatogr.* **550**, 831 (1991).]

of other analytes in the same run is adversely affected. The small linear dynamic range of the detection scheme is also problematic. Axial-beam absorbance detection may eventually gain popularity in the future with concurrent development of new capillary coatings and/or nonaqueous-based buffers.

Multireflection Cell. Another means by which the effective path length in an absorbance measurement in CE can be increased is through the multiple reflection of light within the detection region.[55] As shown in Fig. 11, by coating the outer surface of the capillary with a layer of silver, it is possible to make a beam of light pass through the detection region within the capillary many times before the transmitted light is detected. Once again, the directional nature of laser light is critical to such a design. A large number of reflections and, therefore, a large effective path length can be obtained by reducing the angle of incidence. But reduction of the angle of incidence to below 5° results in a large decrease in transmitted intensity. Hence, further gains in sensitivity are not possible. On the other hand, the number of reflections can be increased by lengthening the distance between the two windows on the silver-coated cell (D2). However, the high efficiency of CE separations limits D2 to below 0.9 mm. Even with the above constraints, an LOD of 2×10^{-5} AU for a 75-μm capillary (in the absence of

[55] T. Wang, J. H. Aiken, C. W. Huie, and R. A. Hartwick, *Anal. Chem.* **63**, 1372 (1991).

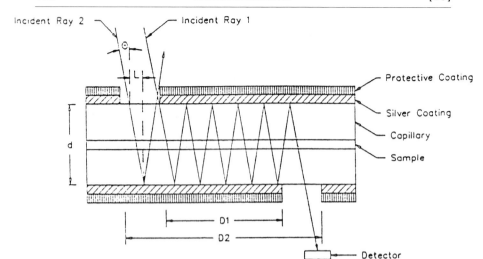

Fig. 11. Diagram of multireflection cell. (Reproduced with permission from T. Wang, J. H. Aiken, C. W. Huie, and R. A. Hartwick, *Anal. Chem.* **63,** 1372. Copyright 1991, American Chemical Society.)

an applied field) has been reported, representing a slight improvement over the conventional cross-beam design for CE monitoring.[56]

As with all laser-based absorbance detection schemes, the paucity of laser wavelengths and high cost of lasers are disadvantages of the multireflection cell. The two-order of magnitude linear dynamic range is also not practical for many applications. In light of the limitations of laser-based absorbance detectors and the generality of absorbance detection in CE, the need for innovative means of improving the sensitivity of absorbance measurements in CE is both obvious and urgent.

Refractive Index

When the analyte of interest does not possess the appropriate chromophores or fluorophores, and when indirect detection (see below) is not convenient, detection schemes based on refractive index (RI) effects become valuable. Because of their universality, these detectors can be used to screen unknown samples for unidentified analytes with unspecified physical properties. Even in the monitoring of previously identified samples, the occurrence of unexpected impurities can be detected with such detection schemes. As in the case with absorbance, the small inner diameter of the capillary in CE imposes an upper limit on the optical path length and

[56] A. E. Bruno, B. Krattiger, F. Maystre, and H. M. Widmer, *Anal. Chem.* **63,** 2689 (1991).

renders traditional RI detection schemes inappropriate. Therefore, detectors relying on unconventional designs have been developed.

Interference Fringe Sensing. To detect small changes in the RI of the solution within the picoliter detection volume in CE, changes in the location of an interference fringe produced by the diffraction of laser light in passing through the detection region of the capillary are measured.[56] When a laser beam is perpendicularly focused into the detection region of the capillary with a cylindrical lens whose focal plane is orthogonal to both the beam direction and capillary axis, interference of light results in the formation of dark and bright fringes on planes surrounding the capillary axis. For a given capillary, the location of each fringe depends on the refractive index of the medium in the detection region. As a consequence, it is possible to measure the concentration of an analyte in the detection region by noting the change in the location of a given fringe.

As shown in Fig. 12, the main components of such an RI detector include a laser with focusing optics, Peltier-cooled sample cell, and position-sensitive diode. Because of the cylindrical shape of the capillary, the ensuing interference pattern with an air-filled sample cell is too complicated to interpret. Hence, the detection region of the capillary is immersed in an index-matching fluid, which has an RI that matches that of fused silica and is confined in a cylindrical sample cell, so as to simplify the interference pattern and allow the selection of the best fringe for use. With an analyte flowing past the detection region by pressure, an LOD of 3×10^{-8} refractive index unit and a four-order of magnitude linear dynamic range have been reported. However, the LOD degrades by roughly 10 times during electrophoresis, a consequence of temperature fluctuations from Joule heating. Hence, the sensitivity of RI detection in CE is limited by temperature instability during separation. Besides, long-term temperature drifts is also problematic. With this detection scheme, the CE separation of five native saccharides and an LOD of $10 \mu M$ for saccharose have been demonstrated.[56]

Refractive Index Gradient Detection. One way of discarding the unfavorable effects of temperature drifts is through the employment of RI gradient detection. Although modulation of the separation voltage has been used to reduce the effects of temperature drifts, its adverse effects on separation efficiency as well as instrumental complexity and expense render the approach incompatible with practical CE.[57] Notwithstanding, RI gradient detection based on Schlieren optics has been shown to be advantageous for certain applications.[58-60]

[57] C.-Y. Chen, T. Demana, S.-D. Huang, and M. D. Morris, *Anal. Chem.* **61**, 1590 (1989).
[58] T. McDonnell and J. Pawliszyn, *Anal. Chem.* **63**, 1884 (1991).
[59] J. Wu and J. Pawliszyn, *Anal. Chem.* **64**, 219 (1992).
[60] J. Wu and J. Pawliszyn, *Anal. Chem.* **64**, 224 (1992).

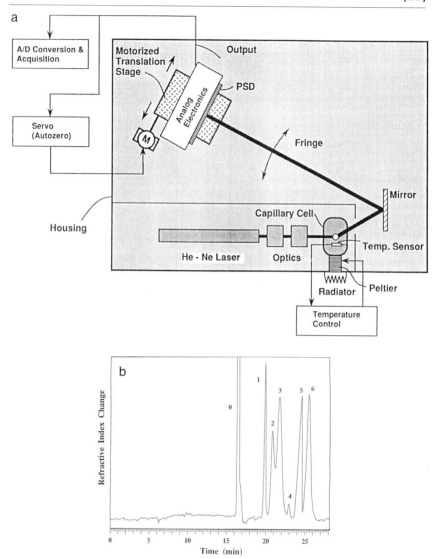

FIG. 12. (a) Experimental setup of interference-fringe sensing RI detector. (b) Electropherogram of a mixture of five underivatized saccharides. The peaks are identified as buffer (0), saccharose (1), N-acetylglucosamine (2), cellubiose (3), impurity (4), N-acetylgalactosamine (5), and lactose (6). Conditions: each 1% except for saccharose, 0.5%; buffer, 100 mM tetraborate, pH 9; capillary length, 70 cm (55 cm to detector), 50-μm i.d.; CE voltage, 14 kV; current, 50 μA; injection time, 7 sec at 12 kV; thermocooler temperature, 27°; interference fringe, $n - 2$. (Reproduced with permission from Ref. 56, Copyright 1991, American Chemical Society.)

In RI gradient detection with schlieren optics, a laser beam is focused into the detection region in the capillary in a direction perpendicular to the capillary axis (Fig. 13). The transmitted light is imaged onto a position-sensitive diode so that the output from the diode in the absence of an RI gradient is extinguished. In the presence of an RI gradient (or an analyte concentration gradient), however, the transmitted light is deflected from its original position and, hence, an output from the diode is produced. After the passage of the analyte band, the transmitted light returns back to the position of extinction on the diode and no signal is sent out. Because modern position-sensitive diodes possess low dark noise and are sensitive and linear, slight deflections of the transmitted light can be detected sensitively and accurately. Because the detector responds to the gradient in RI, slow drifts in the relative location of the capillary and temperature during separation are not recorded. Instead, a differentiated version of the ordinary electropherogram is obtained.

The visualization of proteins present at 0.1 μM levels and fractionated by capillary isoelectric focusing (CIEF) can be achieved with RI gradient detection.[59] This is made possible by the good heat-dissipating capability of the capillary, which gives rise to sharp bands in CIEF. In addition, the peak height in the differentiated electropherogram is approximately proportional to the concentration of the analyte, albeit only a one-order of magnitude linear dynamic range is found. The use of schlieren optics for the whole column detection of protein bands in CIEF without mobilization promises to increase the speed of analysis of CIEF in future.[60]

Other than its low cost and universality, RI gradient detection with schlieren optics in CE is characterized by ease of construction and operation. However, its incompatibility with the more popular forms of CE [such as capillary zone electrophoresis (CZE), micellar electrokinetic chromatography (MEKC), and capillary gel electrophoresis (CGE)] has not been demonstrated. Refractive index gradient detection has also been applied to capillary isotachophoresis (CITP). Despite what has been claimed theoretically,[58] the resolving power of CITP remains inferior to the capabilities of other forms of CE.

Indirect Detection

Given the low sensitivity of RI detection and the incompatibility of RI gradient detection with the more popular forms of CE, indirect detection remains the only viable and general means to universal detectability in CE. Although indirect detection based on electrochemical detectors has been shown to be quite sensitive,[24] its lack of long-term stability due to electrode contamination is detrimental to routine analysis. The use of indirect fluo-

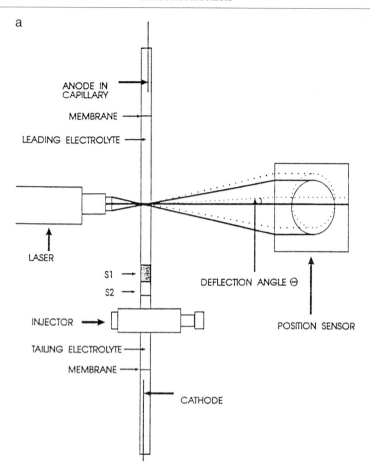

a

ANODE IN
CAPILLARY

MEMBRANE

LEADING ELECTROLYTE

LASER

S1
S2

DEFLECTION ANGLE Θ

INJECTOR

POSITION SENSOR

TAILING ELECTROLYTE

MEMBRANE

CATHODE

FIG. 13. (a) Experimental setup of RI gradient detector. (b) Isotachopherogram of biodegradation mixture. Chloride ion (LE) and MES (TE) were at 8 mM. Applied potential was 3 kV. (Reproduced with permission from Ref. 58, Copyright 1991, American Chemical Society.)

rometry bypasses such problems and defines the state-of-the-art in indirect detection in CE today,[61] as shown in Fig. 14.

    With the exception that an intensity-stabilized excitation beam is required, the experimental arrangement of indirect fluorometry is identical to that of direct LIF with orthogonal geometry.[61] The intensity stabilization is necessary, as the source flicker noise on the large fluorescence background in the electropherogram limits the ultimate detectability achievable with

[61] W. G. Kuhr and E. S. Yeung, *Anal. Chem.* **60**, 2642 (1988).

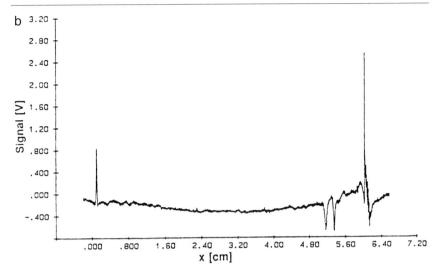

FIG. 13. (*continued*)

indirect fluorometry. As discussed earlier, the indirect signal originates from the displacement of charged fluorophores in the running buffer by the presence of analytes possessing the same polarity of charge. The corresponding decrease in the fluorescence intensity thus constitutes the detection signal. To optimize sensitivity, background fluorophores with mobilities close to that of the analyte are utilized to maximize the sharpness and preserve the symmetry of the indirect peak. An LOD of around 0.1 μM for various inorganic anions and a linear dynamic range exceeding two orders of magnitude in CE have been documented.[62] Several neutral alcohols resolved by MEKC have also been detected with indirect fluorometry, but with LODs at only the 10 μM level.[63]

In spite of its universal detectability and good sensitivity, practical implementation of indirect fluorometry in CE is limited by the constraints that detection places on separation. This is a consequence of the low background fluorophore concentrations necessary for the maintenance of optimal sensitivity and the difficulty in selecting well-behaved fluorophores with appropriate mobilities and pK_a values. Moreover, the reproducibility of CE separations with indirect fluorometric detection is extremely sensitive to the composition of the sample. If indirect fluorometric detection is to be imple-

[62] L. Gross and E. S. Yeung, *J. Chromatogr.* **480,** 169 (1989).
[63] L. N. Amankwa and W. G. Kuhr, *Anal. Chem.* **63,** 1733 (1991).

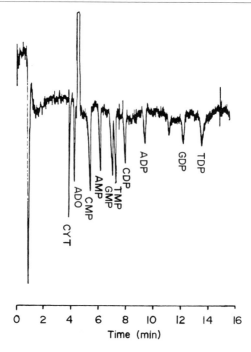

Fig. 14. Separation of nucleotides and nucleosides. Indirect detection of 150 fmol each of nucleosides (adenosine, ADO; cytidine, CYT) and nucleotides (adenosine 5′-monophosphate, AMP; cytidine 5′-monophosphate, CMP; guanosine 5′-monophosphate, GMP; thymidine 5′-monophosphate, TMP; adenosine diphosphate, ADP; cytidine diphosphate, CDP; guanosine diphosphate, GDP; and thymidine diphosphate, TDP) in 0.5 mM salicylate. A 1-sec, 30-kV injection of 20 μM each in salicylate buffer at pH 3.5 was followed by electrophoresis at 30 kV on a 70-cm (25-μm i.d., 150-μm o.d.) untreated column. (Reproduced with permission from Ref. 61, Copyright 1988, American Chemical Society.)

mented in routine CE analyses, the problems just outlined will need to be addressed.

There is of course the related technique of indirect photometric detection in CE.[12,13] As one goes to smaller and smaller capillaries, the absorption path length decreases rapidly to render indirect photometry insensitive. Fluorescence, however, can maintain useful sensitivities down to 5 μm capillaries. This is why indirect fluorescence outperforms indirect photometry for detection in CE but not necessarily in liquid chromatography. Such an application to the study of single human erythrocytes has been demonstrated,[64] as shown in Fig. 15.

[64] B. L. Hogan and E. S. Yeung, *Anal. Chem.* **64,** 2841 (1992).

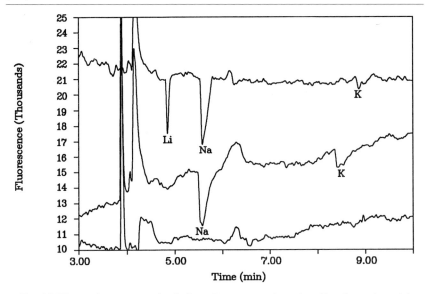

Fig. 15. Electropherograms using indirect fluorescence detection. *Top:* 2-sec electroinjection of 45 μM standards. Li (11.7 fmol injected); Na (10.7 fmol); K (6.5 fmol). *Middle:* One human erythrocyte-injected. *Bottom:* Blank of extracellular matrix. The calibration curve for electroinjected LiCl standards was linear from 5.0×10^{-6} to 8.4×10^{-4} M with an r^2 of 0.996 ($n = 20$). We determined the peak area precision of 2-sec 60 μM LiCl standard injections to be 2.0% RSD with a retention time precision of 0.22% RSD ($n = 7$). (Reproduced with permission from Ref. 64, Copyright 1992, American Chemical Society.)

Future Outlook on Capillary Electrophoresis Detectors

While the first decade in the history of CE has witnessed the transformation of an academically appealing idea into full commercialization, the successful integration of CE into various aspects of analysis in the real world in the decade pending would depend on the usefulness of its capabilities over existing methodologies. To the extent that innovative separation strategies and new applications continue to materialize, inadequate detection sensitivity remains the major obstacle to the acceptance of CE by the practical analyst. The transfer of LC absorbance detector technologies to CE has served the purpose of introducing CE to the analytical community and conventional absorbance detectors will continue to be an integral part of CE in the future. However, the present detection sensitivity of commercial absorbance detectors is a far cry from what is required for practical analyses on numerous occasions. As a result, the successful permeation of CE into the real world of analysis would rely on the development of reasonably sensitive, inexpensive, and general detectors. Further improvements in con-

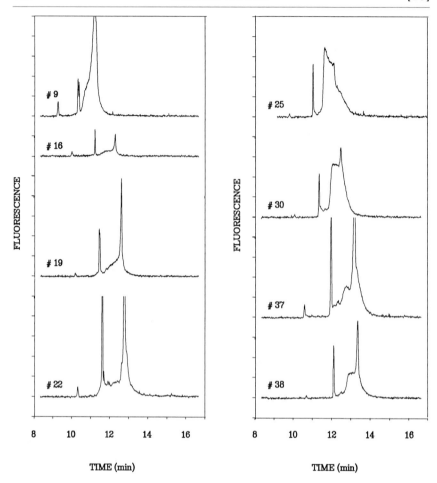

FIG. 16. Electropherograms of proteins in several individual human erythrocytes. Numbers refer to consecutive run numbers over the entire series of 39 trials. A common scale factor is used throughout. (Reproduced with permission from Ref. 65, Copyright 1992, American Chemical Society.)

ductivity, absorbance, RI, and indirect fluorometric detectors are thus necessary.

Even though financially well-endowed laboratories may now procure LIF detectors commercially, the implementation of CE for the purpose of trace to ultratrace analysis is not yet straightforward. Concurrent developments of selective and highly fluorescent labeling reagents and reliable derivatization protocols are needed to realize the full potential of LIF in

CE. The impressive mass sensitivity of CE/LIF is thus far merely a scholarly illusion, until clever means of microsample manipulation are devised. A promising possibility to unambiguous but sensitive detection is the exploitation of the native fluorescence of analytes through ultraviolet excitation. An example[65] is shown in Fig. 16. Unfortunately, the high cost of such lasers remains a major deterrent to practical use. To this end, the "blue-shifting" of semiconductor lasers (which, in fact, are stil in the red) together with the utility of nonlinear optics will almost certainly render LIF detectors more accessible in the future.

The powerful combination of fractionation and structural elucidation through CE/MS could significantly advance the analysis of biomolecules in complex mixtures. As with LC/MS, the issues of cost, reliability, sensitivity, and ease of operation will always be central to the implementation of CE/MS. However, given the continued and committed effort in both academia and industry in CE/MS research, the next decade could witness the initial payoffs of such technology.

While the explosive growth in CE detector technologies since the 1980s is continuing, the inadequate performance of existing CE detectors remains unsatisfactory. This is by no means a reflection of the exhaustion of new ideas. Instead, it indicates that the era of the direct transfer of LC detector technologies to CE is over. For the true admirers of revolutionary innovations, exciting developments are yet to come.

Acknowledgment

The Ames Laboratory is operated for the U.S. Department of Energy by Iowa State University under Contract No. W-7405-Eng-82. This work was supported by the Director of Energy Research, Office of Basic Energy Sciences, Division of Chemical Sciences, and the Office of Health and Environmental Research.

---

[65] T. T. Lee and E. S. Yeung, *Anal. Chem.* **64,** 3045 (1992).

Section III

Mass Spectrometry

Instrumentation
Articles 20 through 23

[20] Fast Atom Bombardment Mass Spectrometry

By William E. Seifert, Jr. and Richard M. Caprioli

Introduction

One of the most important pieces of experimental data in the analysis of the molecular structure of biological compounds is the determination of the mass of the molecule. While a number of methods are used to determine the molecular mass of peptides, oligosaccharides, and oligonucleotides [e.g., gel-permeation chromatography, native and sodium dodecyl sulfate–polyacrylamide gel electrophoresis (SDS–PAGE)], the most accurate method is mass spectrometry. An enormous impact has been made on the biochemical research community by the introduction of desorption ionization techniques for mass spectrometric analysis. These desorption ionization methods permit the analysis of macromolecules that in the past could have been analyzed only by extensive cleavage and derivatization. These so-called "soft" ionization techniques include fast atom bombardment, secondary ion, field desorption, desorption chemical ionization, californium-252 plasma desorption, laser desorption, thermospray, and electrospray. Fast atom bombardment (FAB) mass spectrometry has found a particular niche in the mass analysis of biological molecules up to approximately 5000 Da.

Fast atom bombardment mass spectrometry (FABMS) was developed by Barber and co-workers[1] in an attempt to alleviate the surface charging associated with static secondary ion mass spectrometry. However, the technique made a major impact when the same group[2,3] published the novel procedure of introducing the analyte in a viscous liquid matrix. Since that time, FAB has been used to analyze a wide variety of molecules ranging from oligopeptides to organometallics and buckminsterfullerenes. Owing to its ability to introduce polar or charged molecules into the gas phase with concomitant ionization, FABMS has been used extensively in the analysis of compounds of biological origin.

This chapter reviews the principles of FABMS and continuous-flow

---

[1] M. Barber, R. S. Bordoli, R. D. Sedgwick, and A. N. Tyler, *J. Chem. Soc. Chem. Commun.* 325 (1981).

[2] M. Barber, R. S. Bordoli, G. V. Garner, D. B. Gordon, R. D. Sedgwick, L. W. Tetler, and A. N. Tyler, *Biochem. J.* **197**, 401 (1981).

[3] M. Barber, R. S. Bordoli, G. J. Elliott, R. D. Sedgwick, and A. N. Tyler, *Anal. Chem.* **54**, 645A (1982).

(CF) FABMS, discusses methods for obtaining optimal spectral results, and surveys some applications to biochemical research. It is beyond the scope of this chapter to review all of the work that has been accomplished using FAB, and the reader is referred to other works for further references and details.[4–14]

Static Fast Atom Bombardment

In FAB mass spectrometry, high-energy atomic particles are used to bombard the surface of a liquid sample. These atoms are usually inert gases such as argon or xenon that have been given translational energies of 5–10 keV. As pointed out by Gaskell,[15] FAB is properly considered a variant of secondary ion mass spectrometry (SIMS), which uses a beam of high-energy ions to bombard a sample target. Although a liquid matrix was first used with FAB, Aberth et al.[16] demonstrated that similar results can be obtained if a primary beam of Cs^+ ions is used to bombard the sample in a liquid matrix (liquid secondary ion mass spectrometry, LSIMS) instead of a beam of fast argon or xenon atoms. Thus, the charge of the primary bombarding beam is apparently of little importance in the production of sample ions. What is of major importance in either FABMS or LSIMS is the momentum of the bombarding particle.[15] The momentum can be varied either by increasing the mass of the bombarding particle (e.g., using xenon instead of argon) or by increasing the acceleration energy of the particle. For the purposes of this chapter, FAB and LSIMS will be considered synonymous, although it should be noted that to increase sensitivity and

[4] A. L. Burlingame, D. S. Millington, D. L. Norwood, and D. H. Russell, *Anal. Chem.* **62,** 268R (1990).

[5] A. L. Burlingame, T. A. Baillie, and D. H. Russell, *Anal. Chem.* **64,** 467R (1992).

[6] K. L. Busch, *in* "Mass Spectrometry of Peptides" (D. M. Desiderio, ed.), p. 173. CRC Press, Boca Raton, FL, 1991.

[7] R. M. Caprioli, *in* "Biologically Active Molecules" (U. P. Schluneggar, ed.), p. 39. Springer-Verlag, Berlin, 1989.

[8] R. M. Caprioli and W. T. Moore, *Methods Enzymol.* **193,** 214 (1990).

[9] R. M. Caprioli, *Biochemistry* **27,** 513 (1988).

[10] E. De Pauw, *Mass Spectrom. Rev.* **5,** 191 (1986).

[11] E. De Pauw, *Adv. Mass Spectrom.* **11,** 383 (1989).

[12] E. De Pauw, A. Agnello, and F. Derwa, *Mass Spectrom. Rev.* **10,** 283 (1991).

[13] C. Fenselau and R. J. Cotter, *Chem. Rev.* **87,** 501 (1987).

[14] J. L. Gower, *Biomed. Mass Spectrom.* **12,** 191 (1985).

[15] S. J. Gaskell, *in* "Mass Spectrometry in Biomedical Research" (S. J. Gaskell, ed.), p. 3. John Wiley & Sons, Chichester, 1986.

[16] W. Aberth, K. M. Straub, and A. L. Burlingame, *Anal. Chem.* **54,** 2029 (1982).

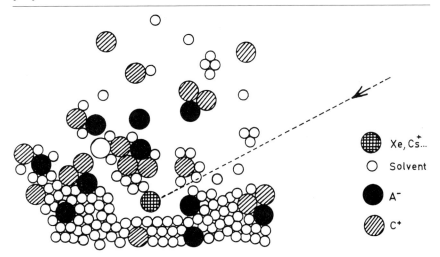

Fig. 1. Schematic representation of the sputtering of a solution of an electrolyte by a fast atom or fast ion. (From *Mass Spectrom. Rev.* E. DePauw, **5,** 199. Copyright © 1986, John Wiley & Sons, Inc. Reprinted by permission of John Wiley & Sons, Inc.)

decrease the source pressure many investigators are converting from FAB to LSIMS.[6,17]

The procedure for obtaining an FAB mass spectrum is relatively simple. A sample solution of approximately 1 μl is mixed with 1–2 μl of glycerol, or some other viscous liquid, and a 2-μl aliquot of this sample–matrix solution is placed on the target, usually the tip of a direct insertion probe. On being inserted into the ion source of the mass spectrometer, the sample is bombarded with atoms or ions having translational energies of 6 to 8 keV. As the high-energy bombarding particles hit the liquid sample, molecules from the surface of the liquid are "sputtered" into the gas phase, as shown in Fig. 1. Both positively and negatively charged ion species as well as neutral molecules are desorbed from the surface and enter the electric fields of the mass spectrometer ion source. Depending on the experiment being performed, the appropriately charged ions are directed into the mass analyzer where the mass-to-charge ratio is measured.

The mechanism for the production of gaseous ions in FAB/LSIMS is complex and still the subject of some discussion. In part, it undoubtedly involves the sputtering of preformed ions from the sample solution. However, because ions are observed from species that are not ionic in the liquid phase, ion–molecule reactions in the high-pressure region above the sample

---

[17] S. P. Markey and M.-C. Shih, *in* "Continuous-Flow Fast Atom Bombardment Mass Spectrometry" (R. M. Caprioli, ed.), p. 45. John Wiley & Sons, Chichester, 1990.

target (the selvedge) must play an important role.[15] Whatever the detailed mechanism for ionization, the bombardment process results in the production of cationized adducts, e.g., $[M + H]^+$, $[M + Na]^+$, and $[M + K]^+$, as well as negatively charged ions, primarily $[M - H]^-$. The mass spectra obtained from FAB are characterized by these ionized molecular species along with significantly less abundant fragment ions. Depending on the analyte and the conditions of ionization, the amount of fragmentation may provide enough data to obtain structural information about the molecule.

Liquid Sample Matrix

One of the most important characteristics of FAB is that the analysis involves the use of a liquid sample. A viscous liquid, termed the *matrix,* is required to minimize evaporation and prolong the liquid state in the high vacuum of the mass spectrometer. The matrix is usually composed of high concentrations (50–95%) of a viscous organic compound with low volatility, such as glycerol, in water. Owing to the fluid nature of the liquid sample, the surface is constantly refreshed with new molecular material from the interior of the sample droplet.

The ability to analyze liquid samples by mass spectrometry enables the biochemist to obtain direct molecular weight and structural data on many polar and charged molecules that previously could not be analyzed without chemical modification. In addition, the use of this liquid matrix allows the direct analysis of biochemical reactions as they occur.

The use of a viscous organic matrix also introduces some significant difficulties. One of the noticeable features of an FAB mass spectrum is that there is a peak at every mass-to-charge value. This chemical background is mainly the result of radiation damage of the organic components of the matrix during the bombardment process and significantly increases the noise level and in effect decreases sensitivity. Matrix cluster ions are also produced in the bombardment process (see Fig. 2), which, depending on the compound of interest, can interfere with the ions produced from the analyte. Because the FAB process sputters molecules from the surface of the liquid sample, the use of liquid solvents or matrices of different physical and chemical properties or the presence of salts can produce spectra having distributions of ions that may not be representative of the original sample. Also, hydrophobic and hydrophilic interactions between the compounds of interest affect their relative concentrations in the surface layers of the sample droplet compared to their bulk concentrations. Finally, because the process occurs in the high-vacuum environment of the mass spectrometer, evaporation of the liquid sample changes the concentration over time, and

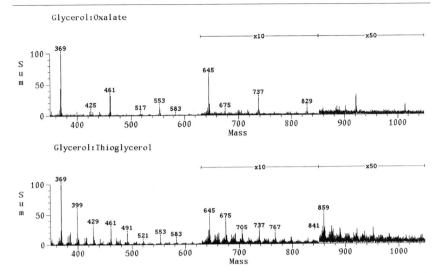

FIG. 2. Fast atom bombardment mass spectra of two common matrices, showing the matrix-derived cluster ions. *Top:* Glycerol acidified with oxalic acid (40 mM oxalic acid in 90% aqueous glycerol). *Bottom:* Glycerol–thioglycerol (1 : 1, v/v).

hence the relative intensities of the peaks in the FAB mass spectra can also change with time.

Surface concentration is a major factor in FABMS because only the surface layers are sputtered in the bombardment process. Therefore, those factors that affect the surface concentration of analyte molecules can lead to a discrimination of the signal produced for each species. For example, Fig. 3 displays the $[M + H]^+$ region of the FAB mass spectrum obtained from the analysis of an equimolar mixture of seven heptapeptides differing in sequence only in the middle three residues. Table I presents the sequence of these peptides along with the hydrophilic index and the charge on the peptides at pH 1.[18] The differences in their hydrophilicities and net charges in solution prevent the peptides from occupying the surface in proportion to their solution concentrations. Hence, the intensities of the $[M + H]^+$ ions in the FABMS of this mixture vary greatly. It was also shown that addition of peptides that are more hydrophilic to a solution of a selected peptide tends to increase its intensity, while adding more hydrophobic peptides can decrease the intensity of the selected peptide.[18]

[18] R. M. Caprioli, W. T. Moore, G. Petrie, and K. Wilson, *Int. J. Mass Spectrom. Ion Proc.* **86,** 187 (1988).

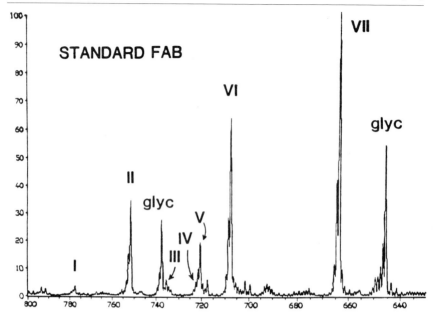

FIG. 3. FAB mass spectrum of a mixture of seven peptides. Peaks designated by "glyc" are ions from the glycerol matrix. [Reproduced with permission from R. M. Caprioli *et al.*, *Int. J. Mass Spectrom. Ion Proc.* **86,** 187 (1988).]

TABLE I

HYDROPHILIC INDICES AND CHARGES OF PEPTIDES IN MIXTURE USED FOR FAST ATOM BOMBARDMENT ANALYSES[a]

Compound	Sequence	$[M + H]^+$	Hydrophilic index[b]	Charge at pH 1
I	Ala-Phe-Lys-Lys-Ile-Asn-Gly	777.4	37	+3
II	Ala-Phe-Asp-Asp-Ile-Asn-Gly	751.3	80	+1
III	Ala-Phe-Lys-Ala-Lys-Asn-Gly	735.4	331	+3
IV	Ala-Phe-Lys-Ala-Asp-Asn-Gly	722.3	353	+2
V	Ala-Phe-Lys-Ala-Ile-Asn-Gly	720.4	59	+2
VI	Ala-Phe-Asp-Ala-Ile-Asn-Gly	707.3	80	+1
VII	Ala-Phe-Ala-Ala-Ile-Asn-Gly	663.3	80	+1

[a] Reproduced with permission from R. M. Caprioli *et al.*, *Int. J. Mass Spectrom. Ion Proc.* **86,** 187 (1988).

[b] Calculated by the method of S. Naylor, A. F. Findeis, B. W. Gibson, and D. H. Williams, *J. Am. Chem. Soc.* **108,** 6359 (1986) from the data of H. B. Bull and K. Breese, *Arch. Biochem. Biophys.* **161,** 665 (1974).

DePauw and co-workers[10-12] and Gower[14] have reviewed the use of different matrices for FABMS and LSIMS. Gower[14] listed three major requirements for choosing a matrix.

1. It must dissolve the compound to be analyzed, allowing the molecule to diffuse easily to the surface so that it can replenish molecules that have been sputtered and ionized.

2. It should be relatively low in volatility so that evaporation does not significantly limit the time in which spectra can be obtained.

3. The matrix should not chemically react with the compound being analyzed, or if it does, it should be in a reproducible and predictable way.

For peptides, the matrices most commonly used are glycerol, thioglycerol, or glycerol–thioglycerol and dithiothreitol–dithioerythritol mixtures, sometimes with appropriate cosolvents or pH modifiers such as trifluoroacetic or oxalic acids. Thiodiethyleneglycol has been reported[19] to be an efficient matrix for peptides that apparently decreases the suppression effects observed with the use of glycerol. Carbohydrates have been successfully analyzed using amine matrices[20,21] in addition to glycerol and thioglycerol, while triethylene glycol and hexamethylphosphoric triamide–triethylene glycol were superior matrices for the negative ion FABMS of gangliosides, particularly those with multiple sialic acid residues.[22] Oligonucleotides have been successfully analyzed with glycerol, thioglycerol, or one of the amine matrices. The aromatic compound m-nitrobenzyl alcohol (NBA) was found by Barber *et al.*[23] to be a useful matrix for the analysis of certain cyclic peptides and organometallic compounds such as cobalamin. NBA is used in the FABMS analysis of C_{60}- and C_{70}-fullerenes, as well as nonpolar compounds such as porphyrins. It has been reported[24] that the sensitivity of mononucleotides is increased by two orders of magnitude when they are first derivatized with a trimethylsilylating reagent and analyzed with NBA as the matrix.

The presence of salts in the matrix can have either beneficial or deleterious effects on a FABMS analysis. Some investigators have increased the ion signal obtained from certain nonpolar compounds by adding either

[19] F. De Angelis, R. Nicoletti, and A. Santi, *Org. Mass Spectrom.* **23**, 800 (1988).
[20] K. Harada, F. Ochiai, M. Suzuki, Y. Numajiri, T. Nakata, and K. Shizukuishi, *Mass. Spectrosc.* **32**, 121 (1984).
[21] L. M. Mallis, H. M. Wang, D. Loganathan, and R. J. Linhardt, *Anal. Chem.* **61**, 1453 (1989).
[22] R. Isobe, R. Higuchi, and T. Komori, *Carbohydr. Res.* **233**, 231 (1992).
[23] M. Barber, D. Bell, M. Eckersley, M. Morris, and L. Tetler, *Rapid Commun. Mass Spectrom.* **2**, 18 (1988).
[24] Q. M. Weng, W. M. Hammargren, D. Slowikowski, K. H. Schram, K. Z. Borysko, L. L. Wotring, and L. B. Townsend, *Anal. Biochem.* **178**, 102 (1989).

ammonium or alkali metal salts to the sample. On the other hand, the uncontrolled presence of inorganic salts and the adduct ions they produce can confuse the identification of the molecular species, dilute the signal produced by the molecular species over several peaks, or in the case of too much salt, suppress the ion current produced from the organic sample. This is a significant problem when dealing with biological samples, because of the ubiquitous presence of salts, especially those of sodium. To decrease the levels of salt in these samples, it is recommended that desalting procedures such as reversed-phase liquid chromatography (LC), solid phase extraction, or gel filtration be used when appropriate. Simple procedures for desalting samples have been published by several authors.[25-28] Figure 4 shows the difference in the FABMS obtained from a plant glycoside following two different desalting procedures using a solid phase extraction column.[27] As the salt is removed from the sample, the signal for the molecular species, which is spread over four peaks in the untreated sample ($[M + H]^+$, $[M + Na]^+$, $[M - H + 2Na]^+$, and $[M - 2H + 3Na]^+$), is consolidated into a single ion corresponding to the protonated molecule ($[M + H]^+$).

Applications of Static Fast Atom Bombardment Mass Spectrometry

Molecular Mass Determination. One of the primary uses of FABMS is for the determination of molecular mass of compounds within a biological sample. Advances in analyzer technology have enabled magnetic instruments to scan beyond m/z 20,000 and quadrupole instruments are produced that will scan up to m/z 4000. However, the FAB ionization process is limited by the low ion yields obtained from samples having higher molecular masses. Thus, there is a practical limit of about m/z 6000 for routine FABMS measurements of peptides and other biological compounds. The sensitivity of these instruments can be in the picomolar range for specific compounds analyzed under optimal conditions; however, it can be much less depending on the chemical nature of the compound (especially its surface activity), its molecular mass, the matrix used, and certain instrumental parameters such as the mass range scanned and the rate of scan. For example, for a peptide of approximately 2000 Da, FABMS spectra can usually be obtained on a 5- to 10-pmol sample dissolved in 2 μl of 90% (v/v) glycerol or thioglycerol that yield a signal-to-noise ratio of about 10:1 on the $[M + H]^+$ ion. However, more than 1 nmol may be required to obtain

[25] K. I. Harada, K. Masuda, K. Okumura, and M. Suzuki, *Org. Mass Spectrom.* **20,** 533 (1985).
[26] D.-C. Moon and J. A. Kelley, *Biomed. Environ. Mass Spectrom.* **17,** 229 (1988).
[27] Q. M. Li, L. Dillen, and M. Claeys, *Biol. Mass Spectrom.* **21,** 408 (1992).
[28] T.-F. Chen, H. Yu, and D. F. Barofsky, *Anal. Chem.* **64,** 2014 (1992).

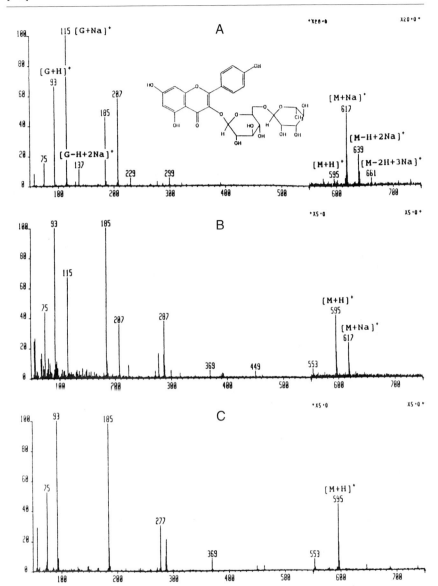

FIG. 4. FAB mass spectra obtained for plant kaempferol-3-O-rutinoside after three methods of sample preparation: (A) no desalting; (B) desalting using deionized distilled water; (C) desalting using 0.02 *M* HCl. (From *Biolog. Mass Spectrom.* Q. M. Li *et al.*, **21,** 408. Copyright © 1992 John Wiley & Sons, Inc. Reprinted by permission of John Wiley & Sons, Inc.)

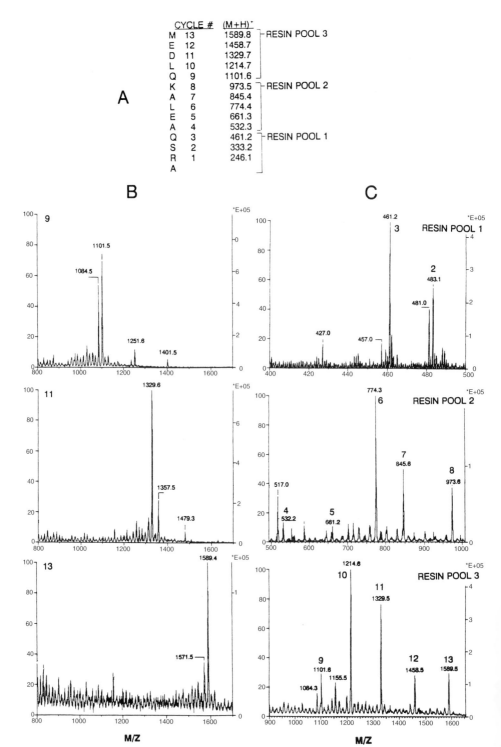

	CYCLE #	(M+H)*	
M	13	1589.8	RESIN POOL 3
E	12	1458.7	
D	11	1329.7	
L	10	1214.7	
Q	9	1101.6	
K	8	973.5	RESIN POOL 2
A	7	845.4	
L	6	774.4	
E	5	661.3	
A	4	532.3	
Q	3	461.2	RESIN POOL 1
S	2	333.2	
R	1	246.1	
A			

A

B

C

the same signal intensity from an underivatized oligosaccharide analyzed under the same conditions.

A valuable application of FABMS has been in the verification of the molecular mass of synthetic peptides produced by automated solid phase synthesis. The main feature that makes FABMS such an attractive method for assessment of peptide structure is its ability to determine precisely the molecular mass of the peptide to within less than 1 Da. Thus, amino acid insertions, deletions, or chemical modifications to the peptide can readily be discerned. Moore and Caprioli[29] have devised a method that can monitor the progress of solid phase peptide synthesis. The technique uses a micro-TFMSA (trifluoromethanesulfonic acid) cleavage and deblocking method that releases sufficient peptide for analysis from a 1-mg aliquot of peptide–resin. Figure 5 shows the results obtained from the stepwise assessment of the solid phase synthesis of a 14-amino acid peptide. With this method, each reaction can be monitored or the peptide–resin aliquots can be pooled to give a retrospective synthetic history. The use of this methodology allows for the quick and unambiguous location of the position of a deletion in the sequence of amino acids, even when more than one residue of a specific amino acid is in the sequence.

Reaction Monitoring. One of the positive features of using a liquid sample to obtain a FABMS spectrum is the ability to directly monitor reactions occurring in aqueous solutions. It has been shown[30] that FABMS can be used to determine the kinetic parameters for the tryptic cleavage of polypeptide substrates in the molecular mass range of 500–3500 Da. The measurements are made by removing timed aliquots from the enzymatic reaction mixture and obtaining the FABMS of the mixture. The increase in the intensity of product (cleavage) peptides as well as the decrease in intensity of the native peptide is then measured in each of the aliquots.

[29] W. T. Moore and R. M. Caprioli, *in* "Techniques in Protein Chemistry II" (J. J. Villafranca, ed.), p. 511. Academic Press, San Diego, CA, 1991.
[30] R. M. Caprioli, *Mass. Spectrom. Rev.* **6**, 237 (1987).

FIG. 5. Stepwise assessment of the automated solid phase synthesis of a 14-amino acid peptide by micro-TFMSA/CF-FABMS analysis. (A) The peptide sequence, synthetic cycle numbers, and the expected $[M + H]^+$ values for the products; (B) the mass spectra of the products derived following cycles 9, 11, and 13; (C) the mass spectra obtained from the analysis of the three peptide-resin pools indicated in (A). The larger, boldfaced numbers above the ion signals represent cycle numbers; the smaller numbers represent observed mass assignments. (Reproduced with permission from W. T. Moore and R. M. Caprioli.[29])

This is shown in Fig. 6 for the tryptic hydrolysis of oxidized bovine insulin B-chain. Two possible trypsin cleavages are possible with insulin B-chain, but the hydrolysis of the Arg^{22}–Gly^{23} bond is considerably faster than the Lys^{29}–Ala^{30} bond. The kinetic constants obtained from the FABMS monitoring of the tryptic hydrolysis of the Arg^{22}–Gly^{23} bond are consistent with data obtained from other sources.[30]

Shao et al.[31] have utilized FABMS to identify and quantify the products of enzymatic reactions involving glycan processing during glycoprotein synthesis. FABMS spectra were obtained after Golgi enzymes processed either glycans or glycans linked to a protein. The glycans had a general structure R-Man_x, where R is 6-(biotinamido)hexanoyl-Asn-$GlcNAc_2$. By using glycans of different structure or the presence or absence of substrate donors or inhibitors, the discrete steps in the processing of the oligosaccharides were identified.

Fast Atom Bombardment Tandem Mass Spectrometry. The scarcity of fragmentation in FABMS is a distinct advantage in the determination of molecular masses, but it is also a disadvantage when trying to obtain structural information. Tandem mass spectrometry (MS/MS) can be utilized, however, to obtain structural information from FAB mass spectra. The use of MS/MS has been reviewed by Busch et al.,[32] in a volume that includes the principles and concepts of MS/MS as well as the various types of instrumentation and a wide variety of analytical applications. The basic principle of MS/MS is the measurement of the mass-to-charge ratios of ions before and after undergoing a collision reaction within the mass spectrometer.[32] There are two major modes of MS/MS operation: *product ion scanning* and *precursor ion scanning*. With the product ion scan, the first stage of mass analysis is set only to transmit ion of a selected m/z. Additional energy is then transferred to these ions, usually by collision with a neutral gas such as helium or argon, which causes further fragmentation in a process called *collision-induced dissociation* (CID). The newly formed fragment (product) ions are then analyzed in the second mass analysis stage. In a *precursor ion scan,* the second mass analyzer is set only to transmit ions of a selected m/z emerging from the CID cell. The first mass analyzer is scanned to determine which precursor ions give rise to the selected product ion.

One area in which FAB MS/MS has been successfully utilized has been in the determination of the amino acid sequence of peptides and proteins.

[31] M.-C. Shao, C. C. Q. Chin, R. M. Caprioli, and F. Wold, *J. Biol. Chem.* **262**, 2973 (1987).
[32] K. L. Busch, G. L. Glish, and S. A. McLuckey, "Mass Spectrometry/Mass Spectrometry: Techniques and Applications of Tandem Mass Spectrometry." VCH Publishers, New York, 1988.

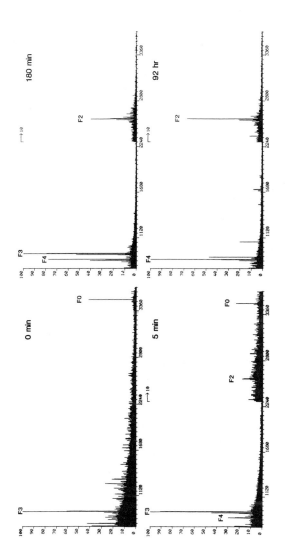

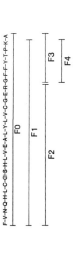

FIG. 6. Time-course mass spectra obtained from the tryptic hydrolysis of oxidized bovine insulin B-chain. Tryptic fragments are indicated below the sequence and above the observed ions.

The application of MS/MS to peptide sequencing has been reviewed for both high-energy[33] and low-energy[34] CID analysis. The general procedure involves the enzymatic digestion of the protein by proteases such as trypsin or chymotrypsin followed by the LC isolation of the product peptides. The peptides are then subjected to FAB MS/MS analysis, whereby the sequence is deduced from the fragment (product) ion patterns. This important area has been reviewed in detail[33,34] and the reader is referred to these works.

Fast atom bombardment MS in combination with peptide N-glycosidase F (PNGase-F), an enzyme that hydrolyzes the β-aspartylglycosylamine bond of N-linked oligosaccharides, has been utilized to identify the sites of N-glycosylation in glycoproteins of known sequence.[35] Gonzalez et al.[36] have modified this procedure to make the identification of N-glycosylation sites easier so that it could be utilized with glycoproteins of unknown sequence. The glycoprotein is treated with PNGase-F in a buffer containing 40 atom% $H_2^{18}O$, causing the oligosaccharide-linked asparagine residues to be converted to aspartate residues containing 40% ^{18}O in the β-carboxyl group. Following digestion with endoproteases and isolation of the peptides by LC, the ^{18}O-labeled peptides are identified by FABMS. The appropriate peptides then undergo CID analysis, giving rise to product ions that are used to ascertain the amino acid sequence. The method is schematically presented in Fig. 7. A comparison of the CID spectra obtained from a deglycosylated pentapeptide prepared in 40 atom% $H_2^{18}O$ and in normal water is shown in Fig. 8. Those fragment ions that contain an ^{18}O-labeled aspartate residue are easily recognized by the doublet of peaks separated by two mass units having an intensity ratio of approximately 6:4.

Tandem mass spectrometry is particularly useful in obtaining information on complex mixtures of compounds because of the additional mass separation step prior to the final analysis. Kayganich and Murphy[37] have used negative ion FAB MS/MS along with mild acid hydrolysis to distinguish between three subclasses of glycerophosphoethanolamines (GPEs) and to identify those that contain arachidonate in human neutrophils without the need for derivatization and additional chromatographic separation steps. Tandem mass spectrometry is required because isobaric ions (those having the same nominal m/z) are produced from several different molecu-

[33] K. Biemann, *Methods Enzymol.* **193,** 455 (1990).

[34] D. F. Hunt, T. Krishnamurthy, J. Shabanowitz, P. R. Griffin, J. R. Yates III, P. A. Martino, A. L. McCormack, and C. R. Hauer, *in* "Mass Spectrometry of Peptides" (D. M. Desiderio, ed.), p. 139. CRC Press, Boca Raton, FL, 1991.

[35] S. A. Carr and G. D. Roberts, *Anal. Biochem.* **157,** 396 (1986).

[36] J. Gonzalez, T. Takao, H. Hori, V. Besada, R. Rodriquez, G. Padron, and Y. Shimonishi, *Anal. Biochem.* **205,** 151 (1992).

[37] K. A. Kayganich and R. C. Murphy, *Anal. Chem.* **64,** 2965 (1992).

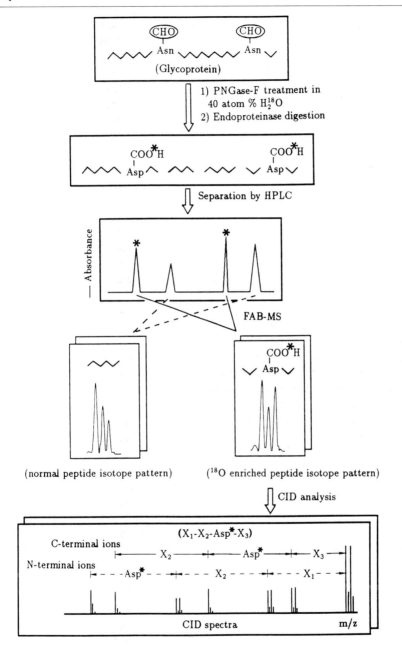

F1G. 7. Schematic diagram of the method for the determination of the specific N-glycosylation sites in a glycoprotein using FAB mass spectrometry and CID analysis. (Reproduced
with permission from J. Gonzalez et al.[36])

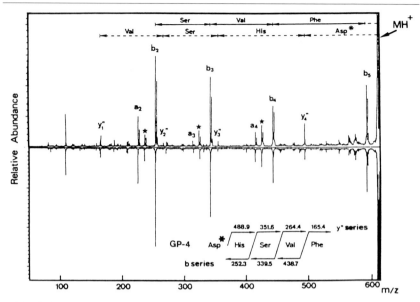

FIG. 8. Comparison of the CID FABMS spectra of the peptide GP-4 prepared in 40 atom% $H_2^{18}O$ (top) and normal water (bottom). (Reproduced with permission from J. Gonzalez et al.[36])

lar species, even though there are significant structural differences between the subclasses of GPE. Arachidonate-containing GPE species give rise to a strong CID fragment ion at m/z 303. Therefore, a precursor ion scan for m/z 303 identifies all arachidonate-containing GPE species. The authors took advantage of the lability of plasmalogens to mild acid hydrolysis to help distinguish the plasmalogen arachidonate-containing GPEs from those isobaric 1-O-alkylacyl species that contain double bonds within the alkyl chain. Figure 9A shows the precursor ion scan for all molecular species giving rise to the arachidonate-indicating fragment at m/z 303. After subjecting the sample to mild acid hydrolysis, the precursor ion scan presented in Fig. 9B shows only the nonplasmalogen species giving rise to the arachidonate fragment. The results obtained from the FAB MS/MS method were not statistically different from the results published using more conventional methods, but the MS/MS method was significantly faster and less labor intensive.

Continuous-Flow Fast Atom Bombardment

Continuous-flow FAB (CF-FAB) was devised as a means of introducing aqueous solutions of nonvolatile, charged and polar compounds into the

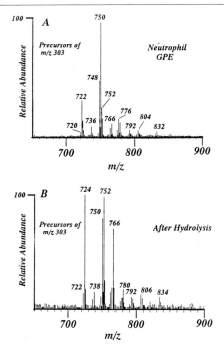

FIG. 9. Negative ion FAB MS/MS analysis of human neutrophil GPE showing the precursor ion scans for m/z 303, indicating the arachidonic acid-containing species before (A) and after mild acid hydrolysis (B). (Reprinted with permission from K. A. Kayganich and R. C. Murphy, *Anal. Chem.* **64**, 2968. Copyright 1992, American Chemical Society.)

FAB mass spectrometer. It was designed to take advantage of the analytical capabilities of FAB, while eliminating or significantly reducing the drawbacks of the technique.[38–40] CF-FAB makes use of a flowing carrier solution, containing significantly reduced amounts of matrix, to refresh the target surface continuously. The result is a technique that has been used in a wide variety of applications including the analysis of polypeptides, mixtures arising from protease digestion, oligosaccharides, nucleotides, prostanoids, and many other biological compounds, including drugs and drug metabolites. A similar technique, termed frit-FAB, uses a fine mesh or frit as the target surface and has also been used with success in the analysis of biological samples. In general, the term "continuous-flow" FAB has been used to describe the overall procedure of introducing a constant flow of liquid

[38] R. M. Caprioli, T. Fan, and J. S. Cottrell, *Anal. Chem.* **58**, 2949 (1986).
[39] R. M. Caprioli and T. Fan, *Biochem. Biophys. Res. Commun.* **141**, 1058 (1986).
[40] R. M. Caprioli, *Anal. Chem.* **62**, 477A (1990).

to the target surface, whether a frit or mesh is used. This convention is used here and either arrangement is referred to as CF-FAB. CF-FAB has been the subject of a monograph concerned with the operational aspects, applications, and experiences of a number of investigators using the technique.[41]

Design and Operational Considerations

The basic CF-FAB probe interface is usually a stainless steel tube through which a fused silica capillary having an internal diameter of 50–100 μm is passed. The capillary terminates at the sample target (probe tip). A high-vacuum seal is fitted within the probe and a pump-out port is provided near the target end. Depending on whether the probe is used for magnetic sector (i.e., high voltage) or quadrupole (low voltage) operation, the probe tip may be electrically isolated from the stainless steel shaft by means of an insulator, usually made of Vespel. The fused silica capillary tubing also serves to isolate the probe from the high voltages in the source of a magnetic sector instrument. The size of the fused silica capillary, while not critical, does affect the dynamics of the carrier flow. For example, a 1-m length of capillary having an internal diameter of 75 μm operated at a flow rate of 10 μl/min will have a very low back pressure at the pump (less than 10 psi).[42] However, the use of chromatographic columns, additional microvalves, and fittings, as well as flow restrictors and splitters, will significantly affect the pressure required to maintain this flow.

The basic principle of CF-FAB operation involves the delivery of an aqueous sample solution to the sample target and the ensuing removal of the liquid at approximately the same rate, leaving only a thin film of liquid on the sample target for fast atom bombardment. A continuous flow of carrier solution is usually provided by means of a syringe-type or microbore-LC pump. Samples can be introduced into the mass spectrometer either by injection into the flow of carrier solution, included in the carrier solution itself (direct infusion), or as eluents from an LC or capillary electrophoresis (CE) instrument. The carrier solution typically is composed of a 5–10% aqueous glycerol solution delivered at a flow rate of 5–10 μl/min. To maintain stability on the sample target and increase compatibility with chromatographic sources, the carrier solution can be modified with low concentrations of volatile organic solvents such as methanol or acetonitrile.

[41] R. M. Caprioli (ed.), "Continuous-Flow Fast Atom Bombardment Mass Spectrometry." John Wiley & Sons, Chichester, 1990.
[42] R. M. Caprioli, *in* "Continuous-Flow Fast Atom Bombardment Mass Spectrometry" (R. M. Caprioli, ed.), p. 1. John Wiley & Sons, Chichester, 1990.

Stability. To obtain high-quality results from a CF-FAB analysis, stable operation of the interface must be maintained. Instability of the ion beam produced in the FAB process can occur for a number of reasons, but three parameters have been found to have a significant effect on CF-FAB stability. The first of these is efficient removal of the carrier solution from the CF-FAB target. Delivery and removal of carrier solution to and from the probe tip should be balanced in order to maintain a thin film of liquid on the target. When too much liquid is allowed to accumulate on the probe tip, instability of the ion beam occurs due to surges in source pressure resulting from disruptive boiling of the sample droplet. Accumulation of carrier solution on the target also decreases the performance of the CF-FAB analysis by broadening chromatographic peaks and decreasing sensitivity.

Early versions of the CF-FAB probe attempted to balance the evaporation of the carrier solution to its delivery with gentle heating, a process that can be difficult to reproduce especially if the temperature of the probe tip cannot be accurately monitored or if a variety of solvent systems are used for the carrier solutions. An alternative method for ensuring efficient removal of the carrier solution is the use of capillary action to draw the liquid from the target onto a filter paper pad[43] or other wick-type material.[44,45] When an absorbant wick is used to draw the liquid from the probe tip, evaporation of the liquid takes place at a more even rate owing to the increased surface area afforded by the wick. Balancing the delivery of the carrier solution and the amount of heat applied to evaporate the liquid becomes less critical, and stability is achieved and maintained with little difficulty. Under typical CF-FAB conditions in which an aqueous or mixed solvent system is used containing 1–5% glycerol at a flow rate of 5 μl/min, stable operation can be maintained for 4–6 hr.

Another method used to achieve stable CF-FAB conditions is the use of a stainless steel frit as the target and termination point of the capillary. This strategy was first used as an interface for LC/FABMS.[46,47] The use of the frit causes the carrier solution to be dispersed over the target and provides for a more stable evaporation of the solvent from the target surface. The original design of this frit-FAB probe allowed for flow rates of only 1–2 μl/min. Several commercial designs now use a stainless steel screen, which being more coarse than the frit, allows for higher flow rates.

[43] R. M. Caprioli, "Proc. 36th ASMS Conf. Mass Spectrom., June 5–10, San Francisco, CA," p. 729. 1988.
[44] W. E. Seifert, Jr., A. Ballatore, and R. M. Caprioli, *Rapid Commun. Mass Spectrom.* **3**, 117 (1989).
[45] M.-C. Shih, T.-C. Lin Wang, and S. P. Markey, *Anal. Chem.* **61**, 2582 (1989).
[46] Y. Ito, T. Takeuchi, D. Ishii, and M. Goto, *J. Chromatogr.* **346**, 161 (1985).
[47] T. Takeuchi, S. Watanbe, N. Kondo, D. Ishii, and M. Goto, *J. Chromatogr.* **435**, 482 (1988).

The actual choice of an open capillary/wick system or a frit system is probably dictated more by which system is provided for a particular instrument from the manufacturer. We have used both systems in our laboratory and, in our hands, the open capillary/wick system is preferable in that it yields greater sensitivity and generally gives sharper chromatographic peaks.

A second factor having a major effect on stability is the presence or absence of liquid backflow from the target to the probe shaft.[42] It is important to prevent the flow of carrier solution from the probe tip through the space between the fused silica capillary and the probe tip aperture and into the shaft. If liquid gets into this space, it can rapidly vaporize and erupt outward and disrupt the sample surface. We have found that the best way to prevent this is to pass the capillary through a septum placed in the target tip as a device to stop the backflow, and to minimize the dead volume within the target tip, particularly the space between the fused silica capillary and the hole in the probe tip.

High pressures in the source can also cause instability of the FAB gun itself. This becomes more of a problem when volatile organic solvents are used in the carrier solution. We have remedied this condition by replacing the outer aluminum cathode disk with one fabricated with a shaft, about 1 cm long, that fits inside the gun barrel.[42] The aperture in the redesigned cathode is also slightly smaller in diameter. Both of these modifications have the effect of reducing the conductance of gas back into the FAB gun assembly. Cesium ion guns are preferable in this regard because they are less affected by source pressure increases.

Memory Effects. Continuous introduction of samples into any analytical instrument always brings with it the possibility of carryover of sample from one analysis to the next. With CF-FAB, careful attention to operating conditions will minimize the possibility of contamination of one sample with a previously analyzed compound. The conditions that usually give rise to memory effects are also those conditions that give rise to instability of the ion signal, i.e., build-up of carrier solution due to a flow rate that is too rapid, or inadequate evaporation, or accumulation of glycerol or other viscous matrix component on the probe tip due to a restricted flow rate or a too-high probe temperature. These conditions yield poor CF-FAB performance resulting in either very broad peaks from flow injection and chromatographic analyses, or contamination of subsequent mass spectra with previously analyzed samples. If the CF-FAB conditions are maintained properly, then memory effects can be kept to less than 0.1% for samples injected at 2-min intervals.[48]

---

[48] R. M. Caprioli, *in* "Biologically Active Molecules" (U. P. Schluneggar, ed.), p. 59. Springer-Verlag, Berlin, 1989.

Performance Considerations

Advantages. By decreasing the amount of viscous organic matrix from 80 to 90% as is the case in static-FAB to 5–10% with CF-FAB, certain performance advantages are observed. One of the most obvious effects is the decrease in background ions contributed by the organic matrix components. Structural features in the mass spectrum that are obscured in the static-FAB spectrum are often easily discerned in a CF-FAB spectrum.[39] Several investigators have reported increases in sensitivity with CF-FAB in comparison to static-FAB.[39,49,50] In some cases, the sensitivity increase can be in excess of two orders of magnitude. This increase in sensitivity can be attributed to a large extent to the decrease in chemical noise from the organic matrix, although there is some increase in sensitivity resulting from a higher yield of ions derived from the sample owing to the higher water content of the matrix.[39] Absolute sensitivity is highly compound dependent, but in general, a peptide having a molecular weight of 1500–2000 can usually be determined using as little as 0.5–5 pmol.

The use of CF-FAB also results in a significant decrease in ion suppression effects.[18,51] This can have important consequences when analyzing complex mixtures. The occurrence of ion suppression effects in static-FAB has meant that when analyzing a complex mixture such as those obtained from the proteolytic digestion of proteins, only 50–80% of the expected peptides can be observed. When CF-FAB is utilized, the ion suppression effects are minimized, enabling the investigator to observe most, if not all, of the peptides in the mixture.

The combination of decreased chemical noise and ion suppression effects with CF-FAB can have dramatic results in comparison to static-FAB. In some cases, a peak that could not be discerned above background when analyzed by static-FAB is easily observed when analyzed by CF-FAB. Figure 10 presents a comparison of the static-FAB mass spectrum of an octapeptide containing a C-terminal homoserine lactone residue with that obtained by CF-FAB.[52] The spectrum obtained by static-FAB on 30 pmol of peptide did not contain a peak above background corresponding to the $[M + H]^+$ ion of the peptide. The CF-FAB mass spectrum obtained from 20 pmol of the peptide, on the other hand, had a prominent $[M + H]^+$ ion with a signal-to-noise ratio of approximately 20:1. Tomer *et al.*[53] have shown that the use of a coaxial CF-FAB interface can increase the sensitivity

[49] T.-C. Lin Wang, M.-C. Shih, S. P. Markey, and M. W. Duncan, *Anal. Chem.* **61**, 1013 (1989).

[50] J. A. Page, M. T. Beer, and R. Lauber, *J. Chromatogr.* **474**, 51 (1989).

[51] R. M. Caprioli, W. T. Moore, and T. Fan, *Rapid Commun. Mass Spectrom.* **1**, 15 (1987).

[52] R. M. Caprioli and M. J.-F. Suter, *Int. J. Mass Spectrom. Ion Proc.* **118/119**, 449 (1992).

[53] K. B. Tomer, J. R. Perkins, C. E. Parker, and L. J. Deterding, *Biol. Mass Spectrom.* **20**, 783 (1991).

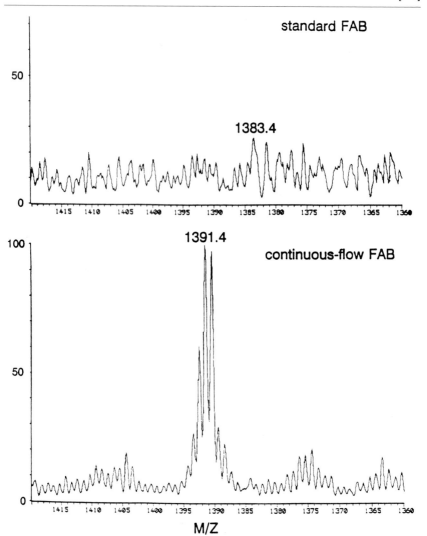

FIG. 10. Mass spectra obtained with 30 pmol of an octapeptide under static-FAB (top) and CF-FAB conditions (bottom). [Reproduced with permission from R. M. Caprioli and M. J.-F. Suter, *Int. J. Mass Spectrom. Ion Proc.* **118/119,** 449 (1992).]

of the analysis of peptides having an approximate molecular weight of 3500 by as much as an order of magnitude over static-FAB. The coaxial CF-FAB interface developed by Tomer and co-workers[54-56] separately de-

[54] J. S. M. de Wit, L. J. Deterding, M. A. Moseley, K. B. Tomer, and J. W. Jorgenson, *Rapid Commun. Mass Spectrom.* **2,** 100 (1988).

livers the matrix solution and the solution carrying the analyte to the probe tip. It consists of a length of fused silica capillary, through which the buffer and sample flow, surrounded by a wider bore capillary that serves as the conduit for the matrix solution. They attribute some of the increase in sensitivity to the coaxial design of the interface. which constrains the analyte to the center of the probe tip, thus possibly presenting a higher concentration of the analyte at the center of the FAB beam.

Disadvantages. While the advantages to using CF-FAB are significant, the dynamic nature of delivering an aqueous solution to the source of a mass spectrometer does present the analyst with some difficulties as well. One of these is the time necessary to attain stable operation following insertion of the CF-FAB probe into the mass spectrometer. As noted previously, this requires balancing the delivery of the carrier solution with its removal by evaporation. Owing to the constant fluctuation of the surface of the liquid film, the signals obtained from CF-FAB analyses can have more variation than those obtained from static-FAB. Reproducible CF-FAB spectra are obtained by averaging spectra over the injection or chromatographic peak. Another disadvantage, particularly for magnetic sector instruments that operate at high accelerating voltages, is the higher source pressures produced by CF-FAB due to the introduction of primarily aqueous solutions. Continuous flow rates of 5–10 μl/min can produce source pressures as high as 2×10^{-4} torr. A magnetic sector mass spectrometer must be properly grounded to avoid damage to the electronics by high-voltage arcs that can (and do) occur at elevated source pressures. Also, because substantial volumes of water and glycerol are being evaporated and/or sputtered into the source, the performance of quadrupole mass analyzers can be decreased due to contamination of the source and quadrupole rods. This also requires that the rough pump oil be changed on a more frequent basis.

Modes of Continuous-Flow Fast Atom Bombardment Operation

The CF-FAB interface may be set up to analyze biological samples in any of three different modes of operation. The samples may be directly infused into the mass spectrometer in a constant flow analysis, injected into the carrier flow by means of a microinjector valve (flow-injection analysis), or introduced via a separation process such as LC or CE.

[55] M. A. Moseley, L. J. Deterding, J. S. M. de Wit, K. B. Tomer, R. T. Kennedy, N. Bragg, and J. W. Jorgenson, *Anal. Chem.* **61**, 1577 (1989).

[56] R. M. Caprioli and K. B. Tomer, *in* "Continuous-Flow Fast Atom Bombardment Mass Spectrometry" (R. M. Caprioli, ed.), p. 93. John Wiley & Sons, Chichester, 1990.

In cases in which it is advantageous to monitor the course of a reaction constantly, the constant-flow mode of operation can be used. In this procedure, the reaction solution is constantly pumped into the mass spectrometer by means of either atmospheric pressure or a syringe pump. When this method is used, either a small concentration of glycerol (5%) is added to the reaction mixture or added through a make-up tee prior to the solution entering the mass spectrometer.[54] The constant-flow analysis is used primarily for reactions in which the concentrations of reactants or products are rapidly changing or when taking individual aliquots is not possible.[57] An example is the enzymatic hydrolysis of ribonuclease S-peptide by a mixture of carboxypeptidases P and Y.[30] Mass spectra were constantly taken as the reaction mixture was continuously infused into the mass spectrometer at a rate of 4 μl/min. The CF-FAB matrix was added to the continuous flow via a low dead-volume tee using a 30% glycerol–0.3% trifluoroacetic acid (TFA) solution in water. Instead of trying to identify the low molecular weight amino acids that were released, the truncated molecular species of the S-peptide were monitored. In all, 14 molecular species were identified in this analysis, corresponding to the removal of 13 amino acids from the C terminus of the polypeptide. In this way, sequence information was obtained on the basis of the mass difference between molecular species, and not primarily on the rate of release of the amino acid cleaved. Time shifts are observed for the intensity maxima of the peptide molecular species in an analysis of this type, and can be used to clear up ambiguities in the analysis, especially when substrates are present in small quantities.

Operating the CF-FAB interface in the flow-injection mode provides a convenient and efficient method of batch sampling. It allows for the injection of sample solutions into the mass spectrometer as often as once every 2 min. To achieve this rate of sample processing, only small volumes (1 μl or less) are injected, making sure that all plumbing dead volumes are minimized or eliminated altogether. This method of analysis can be particularly advantageous when a large number of samples must be analyzed, e.g., quality control of synthetic peptides following cleavage from the resin and time-course analysis of enzymatic hydrolyses.

Flow-injection CF-FAB analysis has another advantage over static-FAB in that quantitative estimates can be made more easily owing to the presence of an "injection" peak. This allows the areas under selected ion chromatograms of the appropriate molecular species to be integrated and backgrounds to be effectively subtracted. Kinetic constants can be easily measured from this type of analysis. For example, the kinetic constants for

[57] R. M. Caprioli, in "Continuous-Flow Fast Atom Bombardment Mass Spectrometry" (R. M. Caprioli, ed.), p. 63. John Wiley & Sons, Chichester, 1990.

the tryptic hydrolysis of the tetrapeptide substrate Met-Arg-Phe-Ala were determined by flow-injection CF-FAB.[9] Time-course aliquots were taken from the reaction mixtures of trypsin and five different concentrations of the substrate, ranging from 0.25 to 1.25 mM. The areas under the peaks of the substrate were measured for each concentration and the rate for that particular reaction was calculated. A Lineweaver–Burk double-recip-rocal plot gave the value of the Michaelis constant, K_m, of 1.85 mM, which was in good agreement with a previously published value of 1.90 mM.[58]

Owing to the mass specific information obtained from a CF-FABMS experiment, rate information can be obtained from the action of a single enzyme on a multiple substrate mixture. Figure 11a shows the time course of the tryptic hydrolysis of one peptide (physalaemin) in an 11-peptide substrate mixture. The peptides in the mixture ranged in molecular weight from 772 to 1758. Figure 11a presents the selected ion chromatogram of the [M + H]$^+$ ion for physalaemin at m/z 1265. Figure 11b compares the rates of reaction of three peptides in this mixture, showing the wide range of affinities of trypsin for these peptide substrates. Seifert *et al.*[59] used this multiple-substrate technique to assess the effect of substrate primary structure on trypsin activity and were able to conclude that it could be used as a survey tool in this type of investigation. In the case of the angiotensin II analogs that were used in this study, those that had an acidic residue on the C-terminal side of the bond being hydrolyzed were cleaved with difficulty, whereas a hydrophobic residue at that site was hydrolyzed with ease.

Negative-ion high resolution CF-FAB has been used to detect and quan-tify benzo[a]pyrene sulfates (BP-SO$_4$) in culture medium.[60] The presence of BP-SO$_4$ is a direct measure of carcinogenic polycyclic aromatic hydrocarbon metabolism and an indication of exposure. The flow-injection mode of operation was used to measure the levels of BP-SO$_4$ isolated from the medium of human hepatoma cells (HepG2) cultured with exposure to benzo[a]pyrene (BP). High resolution ($m/\Delta m = 5000$) was used with single-ion monitoring to completely separate the molecular anion of BP-SO$_4$ from the background, increase the signal-to-noise ratio, and decrease the detection limit. By monitoring m/z 347.0378, the [M$^-$] ion of BP-SO$_4$, the authors were able to obtain a signal-to-noise ratio of 8 on as little as 1.5 pg injected (Fig. 12). The standard curve obtained by injecting different quantities of BP-3-SO$_4$ was linear in the range of 15 to 625 pg. Analysis of the medium from cultured HepG2 cells treated with BP found that it

[58] R. M. Caprioli and L. Smith, *Anal. Chem.* **58,** 1080 (1986).
[59] W. E. Seifert, Jr., W. T. Moore, A. Ballatore, and R. M. Caprioli, "Proc. 38th ASMS Conf. Mass Spectrom., June 3–8, Tucson, AZ," p. 1329. 1990.
[60] Y. Teffera, W. M. Baird, and D. L. Smith, *Anal. Chem.* **63,** 453 (1991).

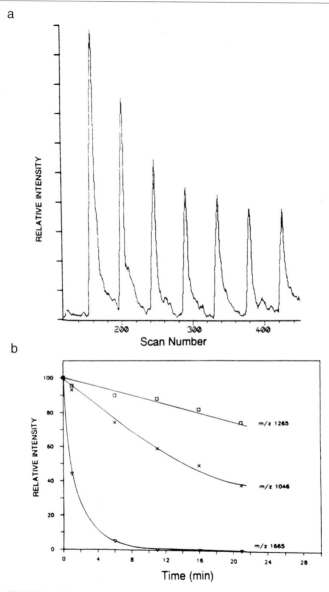

FIG. 11. CF-FAB on-line reaction monitoring of the tryptic hydrolysis of a substrate mixture containing 11 peptides. (a) Selected ion chromatogram for the flow injection analysis of one substrate in the mixture, physalaemin ($[M + H]^+ = 1265$), and (b) rate data for the hydrolysis of three of the substrates. (From R. M. Caprioli *in* "Continuous-Flow Fast Atom Bombardment Mass Spectrometry" (R. M. Caprioli, ed.), p. 63. Copyright 1990 John Wiley & Sons, Ltd. Reprinted by permission of John Wiley & Sons, Ltd.)

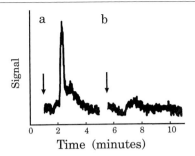

Fɪɢ. 12. CF-FABMS single-ion recording of m/z 347.0378 for the flow-injection analysis at a resolution of 5000 of (a) 1.5 pg of BP-3-SO$_4$ and (b) blank injection. (Reprinted with permission from Y. Teffera *et al.*, *Anal. Chem.* **63**, 453. Copyright 1991, American Chemical Society.)

contained an average of 23 ng/ml, demonstrating that the cells metabolized 3% of the BP into BP-SO$_4$.

Combining Liquid Phase Separations Techniques with Continuous-Flow Fast Atom Bombardment

Interfacing high-performance liquid phase separations techniques with mass spectrometry has some important advantages over other types of detectors. The major advantage is the mass-specific information obtained from a mass spectrometric analysis. The ability to transfer and analyze subpicomole amounts of sample and the time and labor saved in sample handling along with the elimination of the need for derivatization procedures (in most cases) are also noteworthy advantages. However, the requirement of the interface to deal with the high gas load produced by the eluate in the high-vacuum environment of the mass spectrometer ion source is a major obstacle. In the case of the CF-FAB interface, this has largely been dealt with by limiting the amount of flow entering the mass spectrometer.

Three different liquid phase separations techniques have been coupled to CF-FAB including high-performance liquid chromatography (LC), capillary electrophoresis (CE), and microdialysis (MD). A brief discussion of these follows, although the scope of this chapter does not permit an in-depth review of these methods. The reader is referred to several reviews[41,52,61] for a more thorough discussion of examples from the literature.

Liquid Chromatography/Continuous-Flow Fast Atom Bombardment. The first report of coupling liquid chromatography to an FAB mass spectrometer was by Ito *et al.*,[46] who used a packed fused silica capillary column

[61] W. E. Seifert, Jr., W. T. Moore, and R. M. Caprioli, *in* "Mass Spectrometry of Peptides" (D. M. Desiderio, ed.), p. 201. CRC Press, Boca Raton, FL, 1991.

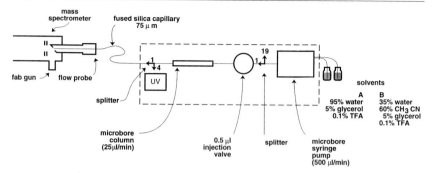

Fig. 13. Schematic diagram for the instrumental setup for microbore LC/CF-FABMS. (Reproduced from the *Journal of Chromatographic Science*, by permission of Preston Publications, A Division of Preston Industries, Inc.)

as the separations device to separate and analyze bile salts. Since that report, CF-FAB has been coupled with conventional-bore LC, microbore LC, and both open-tubular and packed capillary columns to effect LC/MS analyses on a variety of compounds. With conventional-bore LC,[62] because the flow rates are so high (1 ml/min), significant splitting of the effluent stream must be accomplished before it enters the mass spectrometer. The flow rates of microbore and capillary LC are much better suited for CF-FAB because they minimize or eliminate postcolumn splitting, ensuring maximal sensitivity.

A typical microbore LC/CF-FABMS setup is shown schematically in Fig. 13.[63] A commercially available microbore LC that includes a UV detector with a microcuvette was modified by adding a precolumn flow splitter between the syringe pump and the injector. This allows the rapid (500 μl/min or more) formation of eluent gradients prior to the injector. The precolumn splitter allows a flow rate of approximately 25 μl/min to enter the microbore column. A postcolumn splitter is also used to reduce the flow eluting from the column to 5 μl/min before entering the CF-FAB interface. When capillary columns are used as the separations device, the flow rates are low enough (2 μl/min or less) that postcolumn splitting is not necessary.

Glycerol is added to the eluent buffers at a concentration of 3–5% (v/v), although in some cases this may have deleterious effects on chromatographic performance.[64] Tomer and co-workers[53–56] have used their coaxial CF-FAB interface with capillary LC to introduce the matrix to the target

[62] D. E. Games, S. Pleasance, E. D. Ramsey, and M. A. McDowall, *Biomed. Environ. Mass Spectrom.* **15**, 179 (1988).

[63] R. M. Caprioli, B. B. DaGue, and K. Wilson, *J. Chromatogr. Sci.* **26**, 640 (1988).

[64] S. Pleasance, P. Thibault, M. A. Moseley, L. J. Deterding, K. B. Tomer, and J. W. Jorgenson, *J. Am. Soc. Mass Spectrom.* **1**, 312 (1990).

separately from the separations process, thus eliminating glycerol from the chromatographic buffer. Use of this coaxial interface can result in sharper chromatographic peaks and shorter elution times.[64]

The combined LC/CF-FAB technique has been shown to permit the efficient chromatographic separation and on-line mass analysis of mixtures of small peptides, mixtures of peptide hormones with similar molecular weights, and complex mixtures of peptides found in the enzymatic digests of proteins. Its ability to produce $[M + H]^+$ ions (or other cation adducts when salts are present) allows peptides to be easily identified. One example of its utility is shown in the results obtained from a sample of oxytocin that had been treated by autoclaving.[52] Oxytocin is a peptide containing eight amino acids having the sequence

$$\text{Cys-Tyr-Ile-Gln-Asn-Cys-Pro-Leu-Gly(NH}_2)$$
$$\underline{\qquad\text{S}-\text{S}\qquad}$$

The UV chromatogram obtained from the microbore LC/CF-FABMS analysis of the autoclaved oxytocin is presented in Fig. 14. It shows a multiplicity of peaks, with the main oxytocin peak eluting at approximately 17 min. The isotope envelope for the $[M + H]^+$ ion of oxytocin will have the following distributions: m/z 1007.4, 100% intensity; m/z 1008.4, 55%; m/z 1009.4, 25%; and m/z 1010.4, 8% (calculated from the normal isotope distribution). The mass spectra obtained at the beginning and end of the peak are shown in Fig. 15. It is seen that the isotope distribution varies dramatically. Analysis of the mass spectra obtained across the main chromatographic peak at 17 min indicated the presence of several molecular forms of the peptide, i.e., mono- and dideamidated, reduced, and mono- and dioxygenated analogs of oxytocin. The wealth of information obtained from this analysis is a result of the simplicity of CF-FAB mass spectra and the high sensitivity of the LC/CF-FAB combination.

Capillary Electrophoresis/Continuous-Flow Fast Atom Bombardment. Capillary electrophoresis is now an established technique for nanoscale separation of compounds on the basis of their electrophoretic mobilities. The principles and applications of CE are discussed thoroughly. The low flow rates of CE together with the high number of theoretical plates that can be reached with this technique make it a likely candidate for coupling with CF-FAB. This was first shown by Minard *et al.*,[65] who used an open liquid-junction interface in order to preserve the separation efficiency of CE, complete the electrophoretic circuit, and still allow the introduction of both FAB matrix (at 5 μl/min) and CE effluent (at 100 nl/min) into the

[65] R. D. Minard, D. Chin-Fatt, P. Curry, and A. G. Ewing, "Proc. 36th ASMS Conf. Mass Spectrom., June 5–10, San Francisco, CA," p. 950. 1988.

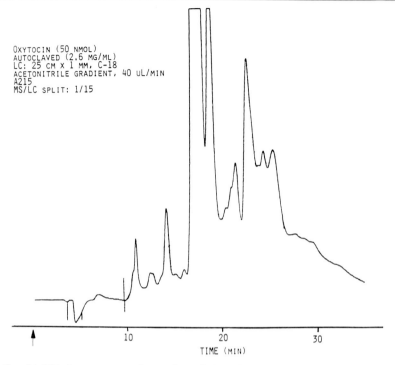

OXYTOCIN (50 NMOL)
AUTOCLAVED (2.6 MG/ML)
LC: 25 CM X 1 MM, C-18
ACETONITRILE GRADIENT, 40 UL/MIN
A215
MS/LC SPLIT: 1/15

TIME (MIN)

FIG. 14. UV chromatogram of a gradient elution obtained from the microbore LC/CF-FABMS analysis of autoclaved oxytocin, using a C_{18} reversed-phase column and water–acetonitrile–TFA gradients. (Reprinted by permission from R. M. Caprioli and M. J.-F. Suter, *Int. J. Mass Spectrom. Ion Proc.* **118/119,** 449. Copyright 1992 by Elsevier Science Publishing Co., Inc.)

mass spectrometer. Moseley *et al.*[66] described a modified coaxial CF-FAB interface in which the electrophoretic capillary was terminated at the CF-FAB probe tip with the matrix solution being pumped to the tip in an outer sheath capillary. This interface was designed so that the electrophoretic transport could be preserved all the way to the FAB probe tip. Using this interface, Moseley *et al.*[67] were able to achieve high separation efficiencies (up to 480,000 theoretical plates) in the analysis of femtomolar quantities of bioactive peptides, including basic peptides such as bradykinin. Figure 16 compares the schematic representations of the liquid-junction and coaxial CE/CF-FAB interfaces.

[66] M. A. Moseley, L. J. Deterding, K. B. Tomer, and J. W. Jorgenson, *Rapid Commun. Mass Spectrom.* **3,** 87 (1989).
[67] M. A. Moseley, L. J. Deterding, K. B. Tomer, and J. W. Jorgenson, *Anal. Chem.* **63,** 109 (1991).

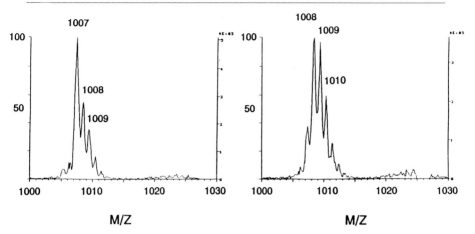

FIG. 15. Mass spectra of oxytocin. *Left:* Recorded at the beginning of the chromatographic peak (16.6 min in Fig. 14). *Right:* Recorded at the end of the peak (17.6 min). (Reprinted by permission from R. M. Caprioli and M. J.-F. Suter, *Int. J. Mass Spectrom. Ion Proc.* **118/119,** 449. Copyright 1992 by Elsevier Science Publishing Co., Inc.)

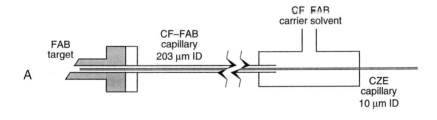

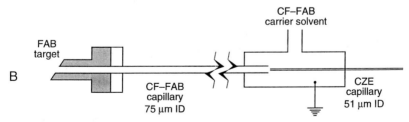

FIG. 16. Schematic representations of the (A) coaxial and (B) liquid-junction interfaces for CE/CF-FABMS. [Reproduced with permission from M. J.-F. Suter and R. M. Caprioli, *J. Chromatogr.* **480,** 247 (1989).]

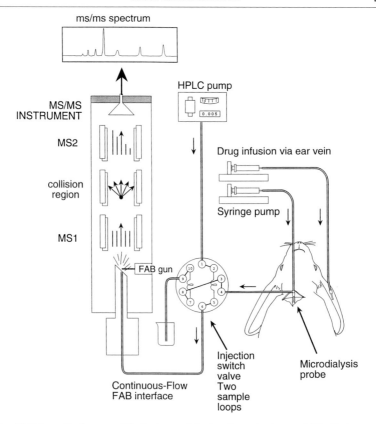

FIG. 17. Schematic diagram of the instrumental setup for an *in vivo* penicillin G pharmacokinetic study in rabbit. (From R. M. Caprioli *in* "Continuous-Flow Fast Atom Bombardment Mass Spectrometry" (R. M. Caprioli, ed.), p. 63. Copyright © 1990 John Wiley & Sons, Ltd. Reprinted by permission of John Wiley & Sons, Ltd.)

Caprioli and co-workers[68,69] have designed an interface that combines some of the features of the liquid-junction and coaxial interfaces. In this interface, the effluent end of the CE capillary and the inlet end of the CF-FAB capillary meet in a short segment of Teflon tubing, which acts as an alignment sleeve. A discontinuous buffer system, in which alkali metal salts were included in the CE buffer but not in the CF-FAB carrier solvent, allowed for high-efficiency separations without significantly compromising CF-FAB detection.

[68] R. M. Caprioli, W. T. Moore, M. Martin, K. B. Wilson, and S. Moring, *J. Chromatogr.* **480,** 247 (1989).
[69] M. J.-F. Suter and R. M. Caprioli, *J. Am. Soc. Mass Spectrom.* **3,** 198 (1992).

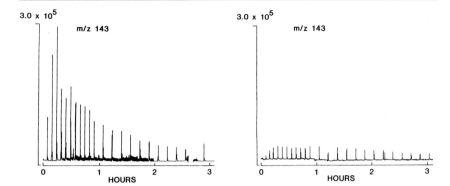

FIG. 18. Selected ion chromatogram for the ion at m/z 143 $[M + H]^+$ obtained from the simultaneous *in vivo* microdialysis/CF-FABMS analysis of valproic acid in rat blood (left) and rat brain (right). Peak intensities were normalized to the most intense peak in the chromatogram of the blood analysis.

Microdialysis/Continuous-Flow Fast Atom Bombardment. Dialysis is a separations technique that is based on the ability of a solute to pass across a semipermeable membrane from an area of higher concentration to an area of lower concentration. Microdialysis makes use of a small probe containing a dialysis membrane to collect dialyzable compounds from biological tissues. The dialysis membrane will pass molecules having molecular weights less than approximately 20,000. Compounds of interest dialyze into the inner chamber of the probe and are then swept into the detector by an aqueous perfusion solution flowing at a rate of about 5 μl/min.

Microdialysis has been coupled to CF-FAB and used in the *in vivo* pharmacokinetic analysis of penicillin G in the rat[70] and rabbit.[57] The experimental setup for this *in vivo* pharmacokinetic study is shown in Fig. 17. Penicillin G passing across the microdialysis membrane from the blood of the animal was monitored using negative-ion CF-FAB MS/MS. The pharmacokinetic curve obtained from this analysis compared favorably with those obtained from that of dog and human by classic off-line non-MS techniques.

Microdialysis CF-FAB has also been used in the study of the disposition of valproic acid in the rat.[71] In this study, the pharmacokinetics of valproic acid in rat brain and blood were compared by *in vivo* microdialysis/CF-FAB following an intravenous dose of 200 mg/kg. Valproic acid concentrations were continuously and simultaneously monitored in both brain and blood over a period of 5 hr. Figure 18 compares the selected ion chromatograms obtained for the brain and blood analyses over the first 3 hr. The

[70] R. M. Caprioli and S.-N. Lin, *Proc. Natl. Acad. Sci. U.S.A.* **87,** 240 (1990).
[71] S.-N. Lin, J. M. Slopis, P. Andrén-Johansson, S. Chang, I. J. Butler, and R. M. Caprioli, "Proc. 39th ASMS Conf. Mass Spectrom., May 19–24, Nashville, TN," p. 583. 1991.

peak levels observed in the rat blood samples were at least an order of magnitude greater than those found in the brain. However, the rate of disappearance from the brain was much slower than that in blood. The results from these experiments demonstrate that changes in drug levels can be measured in the individual live animal.

Conclusion

Fast atom bombardment offers the analyst a sensitive mass spectrometric ionization technique with which to analyze polar and charged molecules up to a molecular weight of about 6000. For higher molecular weight molecules, other desorption ionization techniques such as electrospray MS or matrix-assisted laser desorption MS may be applied effectively to determine the molecular mass. Continuous-flow FAB improves the capabilities of static-FAB by allowing introduction of aqueous solutions into the mass spectrometer, significantly increasing sensitivity, and decreasing the effects of ion suppression that are common with FAB. Liquid phase separation methods such as microbore or capillary LC or CE can be effectively interfaced to a mass spectrometer using CF-FAB, thus expanding the areas of investigation to which FAB can be applied. Continuous-flow FAB also provides a means by which mass specific information can be obtained online from either *in vitro* or *in vivo* experiments in which samples are obtained by microdialysis. Fast atom bombardment has been and continues to be utilized in a wide variety of applications involving biological processes and systems by providing the biochemist with a mass-specific tool to measure molecular weights of compounds isolated from biological materials, and assess the many chemical alterations that can occur in metabolic processes.

[21] Electrospray Ionization Mass Spectrometry

By J. FRED BANKS, JR. and CRAIG M. WHITEHOUSE

Introduction

Electrospray (ES) ionization interfaced with mass spectrometry (MS) has evolved into a powerful tool for the analysis of proteins, peptides, nucleic acids, carbohydrates, glycoproteins, drug metabolites, and other biologically active species.[1-8] The success of ES in this area is largely due

[1] C. M. Whitehouse, R. N. Dreyer, M. Yamashita, and J. B. Fenn, *Anal. Chem.* **57,** 675 (1985).
[2] J. B. Fenn, M. Mann, C. K. Meng, S. F. Wong, and C. M. Whitehouse, *Science* **246,** 64 (1989).

to its ability to extract these fragile chemical species from solution intact, ionize them, and transfer them into the gas phase, where they may be subjected to mass analysis. A unique and powerful characteristic of the ES ionization source is its additional ability to generate ions carrying many charges, whereas most ionization methods generate ions with a single charge. The majority of mass spectrometers are limited to analyzing ions with mass-to-charge ratios (m/z) of only a few thousand, but by dramatically increasing the number of charges (z) through ES ionization, compounds with masses up to several million daltons can be analyzed by MS.[9] The high sensitivity and specificity of ES/MS also make it a useful technique for the analysis of smaller compounds, such as those encountered in environmental and drug metabolite studies, which generally produce singly charged ions.[10–13]

ES/MS hardware can be configured to operate with continuous sample flow introduction, discrete sample injection as in flow-injection analysis (FIA), or on-line sample separation systems such as liquid chromatography (LC) and capillary electrophoresis (CE). Also, because ES is easily configured for these on-line sample introduction schemes, the utility of electrospray mass spectrometry can be enhanced by combining it with complementary analytical techniques. In particular, enzymes can be used to cleave a protein selectively, and the resulting peptide mixture is separated by either LC or CE, which is interfaced on-line to ES/MS. Electrospray and assisted forms of ES can produce sample ions from solution flow rates ranging from 25 μl/min to more than 1 ml/min. Hence submicroliter per minute flow rates encountered with CE separations up to 1-ml/min flow rates from

---

[3] R. D. Smith, J. A. Loo, R. R. Ogorzalek Loo, M. Busman, and H. R. Udseth, *Mass Spectrom. Rev.* **10**, 359 (1991).

[4] J. B. Fenn, M. Mann, C. K. Meng, S. F. Wong, and C. M. Whitehouse, *Mass Spectrom. Rev.* **9**, 37 (1990).

[5] K. L. Duffin, J. K. Welply, E. Huang, and J. D. Henion, *Anal. Chem.* **64**, 1440 (1992).

[6] T. Covey, R. F. Bonner, B. I. Shushan, and J. D. Henion, *Rapid Commun. Mass Spectrom.* **2**, 249 (1988).

[7] E. Huang and J. D. Henion, *J. Am. Soc. Mass Spectrom.* **1**, 158 (1990).

[8] D. F. Hunt, R. A. Henderson, J. Shabanowitz, K. Sakaguchi, M. Hanspeter, N. Sevilir, A. L. Cox, E. Appella, and V. H. Englehard, *Science* **255**, 1261 (1992).

[9] T. Nohmi and J. B. Fenn, *J. Am. Chem. Soc.* **14**, 3241 (1992).

[10] H.-Y. Lin and R. D. Voyksner, *Anal. Chem.* **65**, 451 (1993).

[11] D. Gartiez, "Quantitative and Structural Analysis of Singly Charged Ions from Small Molecules." Presented at the ASMS Fall Workshop on Electrospray Ionization, Boston, MA, 1992.

[12] C. Dass, J. J. Kusmierz, D. M. Desiderio, S. A. Jarvis, and B. N. Green, *J. Am. Soc. Mass Spectrom.* **2**, 149 (1991).

[13] R. D. Voyksner, *Nature (London)* **356**, 86 (1992).

conventional 4.6-mm LC columns can be accommodated on-line with an electrospray ion source interfaced to a mass spectrometer.

In addition to producing intact parent molecular weight information, ES/MS may selectively provide fragment or daughter ion information from a collision-induced dissociation (CID) process, which can occur either in the ES source itself[14–16] or with the use of multiple mass spectrometer stages (MS/MS).[17,18] When in-source CID occurs and more than one parent species is present, it may be difficult to assign the CID fragment ions to individual parent species. For this reason, the CID capability of the ES source is most effectively employed when a single ion species is present. This requirement can be accomplished with complex sample mixtures if an efficient separation technique such as LC or CE is used in-line prior to mass analysis. If mixtures of ions do simultaneously enter the source, then triple quadrupole mass spectrometers can be used in the MS/MS mode, where single components can be m/z selected and exposed to CID conditions within the mass spectrometer itself. Thus, many compound mixture analysis applications requiring triple quadrupole MS/MS capability with CID can be conducted instead using the ES CID capability with a single quadrupole MS in conjunction with a separation technique that fully resolves and separates individual chemical species before they enter the ES source.

Fundamentals

An electrospray ionization source operating at atmospheric pressure was first interfaced to a mass spectrometer in Fenn's laboratory[1] at Yale University (New Haven, CT) with inspiration taken from earlier ion-stopping potential studies using electrospray ionization conducted by Mack *et al.*[19] The term *electrospray* has evolved to describe collectively a basic set of processes encompassing (1) the formation of electrically charged micronsized liquid droplets created from a flowing liquid sample, (2) the extraction of gas phase ions from these same droplets under a high electric field, and (3) the subsequent transport of these ions into a vacuum suitable for mass spectrometric analysis.

[14] R. D. Voyksner and T. Pack, *Rapid Commun. Mass Spectrom.* **5,** 263 (1991).
[15] J. F. Banks, S. Shen, C. M. Whitehouse, and J. B. Fenn, *Anal. Chem.* **66,** 406 (1994).
[16] A. M. Starrett and G. C. Didonato, *Rapid Commun. Mass Spectrom.* **7,** 7 (1993).
[17] J. A. Loo, C. G. Edmonds, and R. D. Smith, *Anal. Chem.* **65,** 425 (1993).
[18] S. A. McLuckey, G. J. Van Berkel, and G. L. Glish, *J. Am. Soc. Mass Spectrom.* **3,** 60 (1991).
[19] L. L. Mack, P. Kralic, A. Rheude, and M. Dole, *J. Chem. Phys.* **52,** 4977 (1970).

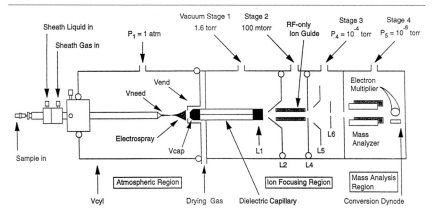

FIG. 1. ES/MS system with quadrupole mass analyzer.

Figure 1 is a schematic of an electrospray ion source as interfaced to a quadrupole mass spectrometer. Although the ES source has been successfully interfaced to magnetic sector,[20-25] Fourier transform,[26-29] ion trap,[10,30] and time-of-flight[31] mass spectrometers, the quadrupole is currently the most common mass spectrometer used with ES and is thus used for illustrative purposes in this chapter. Referring to Fig. 1, sample liquid is introduced through a tube and exits at a conductive needle tip (Vneed) which is maintained typically at 2 to 7 kV relative to the surrounding electrodes: the cylinder (Vcyl), end plate (Vend), and capillary entrance (Vcap). For the ES source illustrated, the liquid introduction needle is operated at ground potential. In conventional or unassisted electrospray, the electro-

[20] C. K. Meng, C. N. McEwen, and B. S. Larsen, *Rapid Commun. Mass Spectrom.* **4,** 151 (1990).

[21] C. Meng, C. N. McEwen, and B. S. Larsen, *Rapid Commun. Mass Spectrom.* **4,** 147 (1990).

[22] R. T. Gallagher, J. R. Chapman, and M. Mann, *Rapid Commun. Mass Spectrom.* **4,** 369 (1990).

[23] B. S. Larsen and C. N. McEwen, *J. Am. Soc. Mass Spectrom.* **2,** 205 (1991).

[24] Y. Wada, J. Tamura, B. D. Musselman, D. B. Kassel, T. Sakari, and T. Matsuo, *Rapid Commun. Mass Spectrom.* **6,** 9 (1992).

[25] R. B. Cody, J. Tamura, and B. D. Musselman, *Anal. Chem.* **64,** 1561 (1992).

[26] S. C. Beu, M. W. Senko, J. P. Quinn, F. M. Wampler III, and F. W. McLafferty, *J. Am. Soc. Mass Spectrom.* **4,** 557 (1993).

[27] K. D. Henry, E. R. Williams, B. H. Wang, F. W. McLafferty, J. Shabanowitz, and D. F. Hunt, *Proc. Natl. Acad. Sci. U.S.A.* **86,** 9075 (1989).

[28] K. D. Henry, J. P. Quinn, and F. W. McLafferty, *J. Am. Chem. Soc.* **113,** 5447 (1991).

[29] J. A. Loo, J. P. Quinn, S. I. Ryu, K. D. Henry, M. W. Senko, and F. W. McLafferty, *Proc. Natl. Acad. Sci. U.S.A.* **89,** 286 (1992).

[30] G. J. Van Berkel, G. L. Glish, and S. A. McLucky, *Anal. Chem.* **62,** 1284 (1990).

[31] J. G. Boyle and C. M. Whitehouse, *Anal. Chem.* **64,** 2084 (1992).

static field is maintained at a sufficiently high level by setting the appropriate voltages on Vcyl, Vcap, and Vend to draw out a Taylor cone from which a thin liquid filament extends. The filament then breaks up into charged droplets primarily due to mechanical Rayleigh fluid instability propagation. The electrosprayed droplets shrink in size until the Rayleigh limit is reached, at which point smaller droplet emission occurs from the primary droplet. Eventually, through this stepwise process of evaporation and droplet breakup, the diameter of the droplet is reduced to the point at which ion emission spontaneously occurs from the droplet surface. The ions produced are then driven by the electrostatic field toward Vend and Vcap, through an orifice, and thence into a vacuum. This movement opposes the drying gas flow of heated nitrogen, which helps to evaporate the droplets. The dielectric capillary in Fig. 1 serves as the orifice into vacuum and has conductive electrodes coated on each end. Positive charged droplets leaving the ground potential liquid introduction needle are driven by the electrostatic field to the Vend and Vcap counterelectrodes, which are set typically at -2 to -7 kV. The ions produced from the charged droplets are swept into the capillary orifice and pulled against the electrostatic field between the capillary entrance and exit electrodes by the gas flowing into vacuum. The ions then enter vacuum at the capillary exit potential, approximately 50–400 V for positive ions with quadrupole mass spectrometers. Because the glass capillary acts as an electrical insulator, the negative kilovolt potentials in the ion production region at atmospheric pressure can be optimized independently of the more modest potentials in the ion-focusing region under vacuum. For conductive metal capillary and nozzle orifices, the orifice potential (exit and entrance) is again typically 80–400 V for positive ions, so that the liquid introduction needle must then be operated at positive kilovolt potentials.

The ions exiting the capillary tube in Fig. 1 are transported through a free jet expansion into vacuum, where a series of dynamic and electrostatic lenses focuses the ions into the mass spectrometer for mass analysis. As illustrated in Fig. 1, highly efficient multipole ion guides operated in the radio frequency (RF)-only mode are often incorporated into the vacuum lens system. The neutral background gas, usually nitrogen, is removed through one or more vacuum-pumping stages between the orifice exit and the mass analyzer chamber. A total of four vacuum stages is shown in the ES/MS quadrupole system in Fig. 1.

The mechanism by which ions are produced from charged liquid droplets is not well understood; however, two models exist that help to describe the ion emission process. In one model initially proposed by Mack et al.,[19] the charged droplets continue their breakup and evaporation until only one molecule is left in each droplet. The charge left behind as the solvent evaporates is then attached to the sample molecule, forming an ion. Work by Fenn[32] has indicated that this ion production mechanism may apply only to larger molecules having molecular weights over 500,000 Da, and that a second model, commonly referred to as the *ion evaporation model* and proposed by Iribarne and Thomson,[33] more accurately describes the mechanism of ion emission from charged droplets for molecules having a molecular weight below 500,000. This second model postulates that through a series of droplet breakups and evaporation, the shrinking radius of the evaporating charged droplet reaches a size and surface charge density at which the electrostatic field at the droplet surface is sufficiently strong to overcome the energy of solvation of an ion on the surface, causing it to "lift-off" into the gas phase.

Variations in the ES source design have emerged, in which electrospray chamber source components are heated and heated metal capillary tubes or nozzles are used as the vacuum orifices.[34,35] When a heated capillary or nozzle is incorporated into an ES source, the charged droplets produced in the atmospheric pressure region may only partially evaporate in the atmospheric pressure chamber and continue to evaporate as they are transported into vacuum. With this drying method, solvent from the evaporating charged droplets is swept into the vacuum along with solute ions and nitrogen carrier gas so that recondensation of the solvent on the ions must be prevented in the rapidly cooling free jet expansion. This solvent recondensation can be avoided by supplying sufficient heating to the expanding gas and setting the voltage differentials on the electrostatic lenses high enough so that ions are accelerated through the background gas in vacuum, causing the CID breakup of any remaining solvent clusters. Also, the nozzle or capillary exit in these heated systems is usually set off centerline such that the unevaporated droplets of high mass-to-charge ratio are deflected and not accelerated line-of-sight into the mass spectrometer, an event that would cause considerable signal noise to be observed. Alternatively, ES sources have been configured so that the line-of-sight trajectory between the ES needle and the orifice into vacuum is blocked by a heated element. In this case, charged droplets are evaporated as they flow around this heated element aided by concurrent gas flow directed toward the orifice into vacuum. Whatever the method employed in a particular electrospray ion source, efficient drying of charged droplets is essential to the production of a stable ion signal.

[32] J. B. Fenn, *J. Am. Soc. Mass Spectrom.* **4,** 524 (1993).
[33] J. V. Iribarne and B. A. Thomson, *J. Chem. Phys.* **64,** 2287 (1976).
[34] S. K. Chowdhury, V. Katta, and B. T. Chait, *Rapid Commun. Mass Spectrom.* **4,** 81 (1990).
[35] M. H. Allan and M. L. Vestal, *J. Am. Soc. Mass Spectrom.* **3,** 18 (1992).

Operating Range

The stability of the mass spectrometer ion signal with ES ionization is a direct reflection of the stability of the charged droplet formation and the efficiency and consistency of the droplet-drying processes. At atmospheric pressure, unassisted electrospray ionization has a window of stable operation that to a large extent is a function of the range of operating conditions within which uniform charged droplet formation can occur. For consistent unassisted ES, stable charged droplet formation and a stable Taylor cone, filament, and droplet breakoff from the filament must be maintained. Liquid flow rate, solution conductivity, solution surface tension, and operation in the negative ion mode can affect the stability of the charged droplet formation process in unassisted electrospray.

Liquid Flow Rate

If the liquid flow rate fluctuates rapidly, the Taylor cone may become unstable, resulting in an erratic ion signal. When a stable Taylor cone is formed, the diameter of the protruding filament is essentially proportional to the two-thirds power of the liquid flow rate,[36] and the size of the charged droplets produced is proportional to the filament diameter.[37] The net result is a proportional increase in droplet size with liquid flow rate. The production of larger diameter charged droplets has three negative effects on ion evaporation efficiency and hence on overall electrospray performance. First, larger droplets are more difficult to dry in the brief time between droplet formation and entrance into the vacuum system. Second, as the droplets increase in size, the net charge-to-mass ratio of the droplets decreases. Consequently, there is potentially less charge available per solvent molecule. Both effects can cause a fluctuating or reduced ion signal. Third, as the liquid flow rate increases, the filament diameter increases to a point at which the Taylor cone becomes unstable for a given liquid introduction tube tip diameter. If the Taylor cone becomes unstable, then the charge droplet formation is nonuniform, resulting in an unstable ion signal. For an ideal case in which a solution of 100% methanol is electrosprayed at a flow rate of 3 μl/min, the primary charged droplets formed have a mean diameter of 2.9 μm. The implications here for the use of ES/MS as an LC on-line detector are obvious. Because conventional ES has optimum performance at only a few microliters per minute, either some type of

[36] R. C. Willoughby and E. W. Sheenan, "Ion Production in Electrospray: A Thermodynamic Approach." Presented at the 11th Montreux Symposium, Montreux, Switzerland, 1991.
[37] K. Tang and A. Gomez, *Phys. Fluids* **6,** 2317 (1994).

flow splitting or additional technology is required to employ ES/MS as an LC detector.

Solution Conductivity

Direct observation of the electrospray process when spraying solutions with conductivities over 800 micromhos at a few microliters per minute has shown that the filament originating at the Taylor cone can extend for several tens of centimeters without breaking into droplets. The more highly conductive solution counteracts the destabilizing effect of surface tension in the liquid filament. This may be explained as a competition of surface tension and charge repulsion forces. As surface perturbations form along the liquid filament as a result of the surface tension, the diameter of the filament contracts, reducing the radius of curvature of the filament and creating a local region of higher electrostatic field. The net charge rapidly accumulating in the higher conductivity filament region damps out the wave propagation, preventing the surface tension forces from dominating and pinching off a droplet. The result is that charged droplets do not break from the liquid filament extending from an electrospray Taylor cone, as has been observed. If charged droplets are not formed, then no signal can be observed because the ion emission process has been prevented. Higher conductivity solutions can be electrosprayed using lower liquid flow rates in the submicroliter per minute range.[38] Lower liquid flow rates result in smaller filament diameters from which charged droplets can break off with solutions of higher conductivity. The higher the solution conductivity, the lower the liquid flow rate required to achieve stable electrospray. Again, there are serious implications here for the use of ES/MS as an LC or CE detector. The addition of salts, acids, and bases to LC mobile phases and CE buffers is a common and necessary practice. In general, the applied concentrations of these modifiers will increase the conductivity of the solution to a point at which stable unassisted electrospray at higher flow rates is not possible.

Higher Surface Tension Solutions

The higher surface tensions of aqueous solutions can pose a problem for maintaining a stable electrospray ion signal as well. The Taylor cone must pull out sufficiently for a liquid filament to form from its tip. The force applied by the electrostatic potential must overcome the counteracting force applied by the liquid surface tension, which is tending to hold the

[38] J. F. d. l. Mora and I. G. Loscertales, *J. Fluid Mech.* **260,** 155 (1994).

surface of the emerging liquid in a spherical shape. Unfortunately, there is a limit to the electrostatic potential that can be applied in the ES source operated at atmospheric pressure. For unassisted ES operation, if the voltage applied between the needle tip and counterelectrodes exceeds the work function of the electrodes, spontaneous electron emission, leading to a cascading gas phase breakdown and hence a sustained corona discharge, can occur. This results in a loss of or at best an unstable mass spectrometer ion signal. The onset of corona discharge loosely represents a ceiling to the relative voltages that can be applied between electrodes in an ES source operating at atmospheric pressure. Consequently, the applied electrostatic forces have an upper limit. The limited electrostatic force does not compete effectively against the higher surface tension of highly aqueous solutions and results in no Taylor cone or an unstable Taylor cone formation, which in turn results in an unstable charged droplet formation process. Adding an organic solvent to an aqueous solution effectively lowers the surface tension and can allow normal ES operation, but may not be compatible with the solution requirements for optimal LC or CE on-line separation systems. Stable unassisted electrospray can be achieved sometimes with aqueous solutions if low liquid flow rates are used in the submicroliter per minute range in combination with very sharp electrospray needle tips and electron-scavenging sheath gases.

Negative Ion Operating Mode

The sharp liquid introduction needle has a negative potential relative to the surrounding counterelectrodes when the ES source is operated in the negative ionization mode. In this configuration, the onset of corona occurs at a lower potential than when the needle has a relative positive polarity. As the voltage is increased in the negative ion mode, the work function of the metal needle tip is reached and the onset of corona discharge occurs before the field has applied enough force to counteract the liquid surface tension and form a stable Taylor cone. The onset of corona discharge occurring at the needle tip effectively lowers the local field, creating an unstable environment within which the Taylor cone forms. The addition of oxygen at the needle tip or other electron scavenger gas can suppress the onset of corona, thus allowing Taylor cone formation. Also, the use of lower surface tension solvents such as propanol decreases the voltage required for the formation of a stable Taylor cone and hence the onset of electrospray. Once the charged droplets are formed, the ion evaporation mechanisms for positive and negative ionization appear to be quite similar.

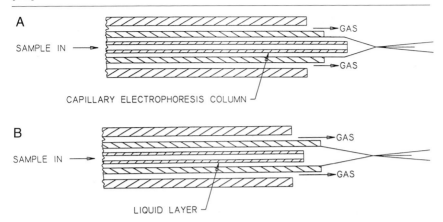

FIG. 2. Electrospray needle tip with sheath liquid flow geometry for (A) CE and (B) LC applications.

Techniques to Extend Range of Electrospray Operation

Sheath Liquid Flow

To extend the range of stable operation, techniques have been applied to electrospray to aid or assist in the charged droplet formation process. For lower flow rate operation, usually below 20 μl/min, the use of sheath liquid as well as sheath gas flows at the needle tip has proved useful in extending the range of stable unassisted electrospray operation. Figure 2 shows a typical geometry used for layering flows at the needle tip for (a) CE/ES/MS and (b) LC/ES/MS applications. The sample introduction tube can be metal or fused silica. A sheath liquid solution is introduced through the second layer tube as illustrated in Fig. 2 and gas can be introduced at the needle tip through an annulus between the second and the third layer tubes.

In this technique, the sheath solvent introduced through a surrounding tube is layered over and mixed with the sample-bearing solution emerging from the innermost tube tip. The sheath solvent can be chosen to reduce the surface tension or conductivity of the sample-bearing solution, thus making ES operation more compatible with solutions used in gradient LC or CE separations. This arrangement has been used for coupling CE to electrospray when the liquid introduction tube as illustrated in Fig. 2A is replaced by a CE column.[39,40] The flow rates in CE applications are usually

[39] R. D. Smith, C. J. Barinaga, and H. R. Udseth, *Anal. Chem.* **60,** 1948 (1988).
[40] R. D. Smith, J. A. Loo, C. G. Edmonds, C. J. Barinaga, and H. R. Udseth, *Anal. Chem.* **62,** 882 (1990).

well below 1 $\mu l/min$, and the buffers used are often aqueous and highly conductive. The addition of a solvent such as methanol or propanol through the second layer tube in this application can serve to increase the total flow rate, decrease the surface tension and conductivity, and make an electrical connection with the fluid exiting the fused silica CE column without adding dead volume. This sheath flow also tends to decouple the charged droplet formation process from the electroosmotic flow process occurring in the capillary electrophoresis column. When the sheath liquid flow technique is used with LC or CE separation systems, correct positioning of the first and second layer tip locations is important to ensure stable performance. Generally, when the first layer tube is a fused silica column, the tip is extended just beyond the second layer tube for optimal performance. When the first layer tube is metal, as may be the case with LC interfacing, the first layer tube is retracted slightly inside the second layer tube for optimal performance. The addition of layered flow to reduce the solution surface tension and conductivity can extend the solution chemistry range of operation of electrospray but may not help to extend the liquid flow rate operating range. The addition of oxygen or other electron scavenger can still be added through the third layer tube during negative ionization mode. There remains much to learn about the use of sheath liquid flow to modify solution chemistry with the goal of achieving higher ion evaporation efficiencies from charged liquid droplets.

Assisted Electrospray

From the realization that ion emission from charged droplets could proceed normally if the charged droplet formation and evaporation could be controlled, two electrospray-assisted techniques, pneumatic nebulization or IonSpray and ultrasonic nebulization or ULTRASPRAY (Analytica of Branford, Inc., Branford, CT), were developed. Both techniques greatly expand the range of solution chemistry and liquid flow rate conditions over which stable electrospray operation can be achieved by mechanically creating the fine charged droplets required for ES and thus eliminating the dependence on Taylor cone formation. In pneumatic nebulization, the addition of a gas flowing concentrically around the liquid introduction needle tip is used to shear off or pneumatically nebulize the liquid as it emerges from the needle tip in the presence of an electrostatic field (Fig. 3A). This technique first reported by Mack et al.[19] and later by Bruins et al.,[41,42] became known as *pneumatic nebulization-assisted electrospray* and

[41] A. P. Bruins, L. O. G. Weidolf, J. D. Henion, and W. L. Budde, *Anal. Chem.* **59,** 2647 (1987).
[42] A. P. Bruins, T. R. Covey, and J. D. Henion, *Anal. Chem.* **59,** 2642 (1987).

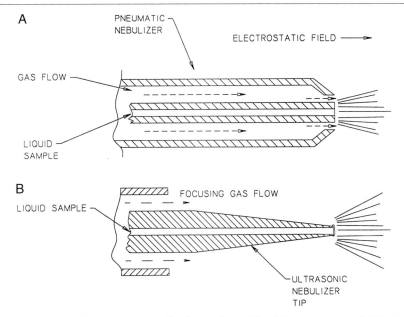

FIG. 3. Assisted electrospray needle tips configured for (A) pneumatic and (B) ultrasonic nebulization.

also as *IonSpray* (Sciex, Thornhill, Ontario, Canada). Pneumatic nebulization tends to create a broader droplet size distribution when compared with unassisted electrospray, and the droplet distribution will change with solution chemistry and liquid flow rate.[43]

The ULTRASPRAY, or ultrasonically assisted electrospray, technique has incorporated a high-frequency ultrasonic nebulizer at the electrospray needle tip to assist in the droplet formation process in the presence of an electrostatic field, as shown in Fig. 3B. This technique creates a narrow range of droplet sizes independent of solution chemistry or liquid flow rates.[44] Both ultrasonic and pneumatic nebulization techniques have been used to extend the liquid flow rate range to over 1000 μl/min. Changes in solution conductivity and surface tension have little effect on the signal stability with assisted electrospray. Both ES-assisted techniques create liquid droplets by mechanical means; however, the addition of a strong electrostatic field at the liquid introduction needle tip is still required to achieve

[43] M. A. Tarr, K. L. Goodner, B. A. Williams, G. Zhu, and R. Browner, "LC-MS Interfacing: Fundamental Studies of Aerosol Formation and Transport." Presented at the Pittsburgh Conference, New Orleans, LA, 1992.

[44] J. F. Banks, J. P. Quinn, and C. M. Whitehouse, *Anal. Chem.* **66,** 3688 (1994).

efficient charging of the droplets created. Negative ionization can be used with assisted ES techniques without the need of an electron-scavenging gas such as oxygen. Sheath flow can also be used with either nebulization-assisted electrospray technique.

Factors Affecting Sensitivity

Electrospray Ionization–Mass Spectrometry
  as a Concentration-Dependent Detector

The behavior of charged droplets and emitted ions during the droplet evaporation step, coupled with transport into vacuum, cause the ES/MS system to respond as a sample concentration-dependent detector. This dependence is analogous to an ultraviolet (UV) detector in LC applications, where the signal response is a function of how much absorbing sample per unit volume of liquid is present. The mass spectrometer ion signal obtained from an ES source for a given compound in solution is, to a first approximation, dependent on the concentration of the sample in a unit volume of solution and is relatively independent of liquid flow rate.

The concentration dependence in ES seems to be the result of droplet repulsion due to space charging in the ES chamber. Work by Gomez[45] has shown that the highly charged droplets produced in electrospray move rapidly away from each other owing to space charge repulsion as they evaporate and drift from the liquid introduction needle toward the counter-electrode or end plate through the atmospheric pressure gas. Because of this space charge repulsion at the droplet level, a limited number of charged droplets can occupy the gas volume in the region of the office into vacuum. The number of ions that are entrained in the gas entering the orifice into vacuum is a function of how much sample is present in each evaporating charged droplet. It is important to note that the space charge limit appears not to occur at the ion level but is established rather at the charged droplet level before the droplets enter the region of ion evaporation.

As an example of the response of an ES/MS system to the concentration of an analyte, Fig. 4 shows the relative ion signal from cytochrome *c* [1 pmol/μl in 1:1 (v/v) methanol–water, 0.1% acetic acid] vs concentration. For this experiment, an HP5989A (Hewlett-Packard, Palo Alto, CA) mass analyzer equipped with an Analytica of Branford ES ion source (Analytica of Branford, Branford, CT) was scanned from 600 to 1300 *m/z* at the rate of 1 sec/scan. The data show that the instrument response is linear over four to five orders of magnitude until the concentration reaches approxi-

[45] A. Gomez, Presented at the LC/MS Workshop of the ASMS, Washington, DC, 1992.

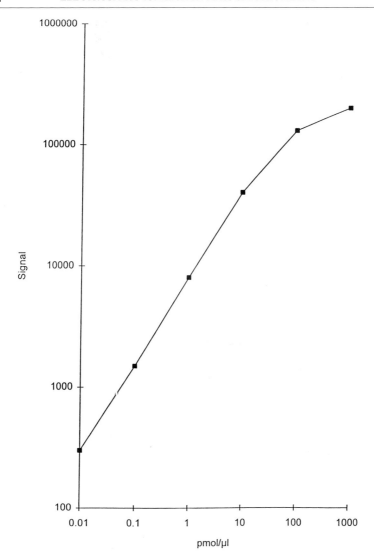

Fig. 4. Cytochrome *c* ion signal vs concentration. Amplitude values are the sum of all multiply charged peak maximums.

mately 100 pmol/μl, at which point a characteristic signal plateau begins. Details relating to this phenomena are discussed below.

A significant consequence of this concentration-dependent behavior is that for a given sample concentration, increasing the solution flow rate increases the amount of sample per time introduced into the ES source but may not result in higher ion currents and hence signal to noise detected by the mass analyzer. In sample-limited applications, lower flow rate separation systems such as microbore LC, capillary LC, or CE will result in higher overall system sensitivity. Running the ES source at higher flow rates with flow injection or higher flow LC applications may simplify the LC coupling and operation but does not significantly increase sensitivity. In reality, for flow rates above a few microliters per minute, the ES source is effectively splitting the sample during the spraying process, sending only a small portion of the total ions produced to the mass spectrometer. Comparisons of on-line LC/ES/MS sensitivity as a function of LC column size used in the LC separation step have been reported.[46] This also suggests that a great portion of LC effluent may be split off for either fraction collection or detection by some other means without loss of ion signal.

Finally, for samples in which the analyte is very low in concentration, preconcentration of the sample on the head of an LC column using gradients or at the beginning of a CE column initially running in the isotachophoretic mode can significantly enhance ES/MS signal to noise.[47]

Sample Handling

An important point to note is that limits in sensitivity when using ES/MS are often imposed by sample-handling techniques. A sample concentration in the low femtogram per microliter range with only a few picograms of total sample present can often be detected by an ES/MS system if the sample is successfully delivered by the liquid flow system to the electrospray needle tip. Such a small quantity of sample can easily be absorbed onto the transfer tube walls or be diluted in dead volume areas of the injector valve or couplings on its way to the electrospray needle tip. Care should be taken in choosing materials that the sample solution will encounter to minimize adsorption effects. Polyetheretherketone (PEEK) tubing is often chosen over stainless steel when analyzing proteins. Capillary electrophoresis columns are in effect specialized transfer lines for which specific coatings can be chosen to minimize sample adsorption to the walls. When mixing, storing, or diluting samples for analysis, losses can occur to the sample container walls. Container materials such as glass can introduce a number of unwanted contaminants into solution, including sodium and potassium. A good grade of polypropylene is preferred, such as those found in Eppendorf-brand microcentrifuge tubes. Samples once in solution should be analyzed

---

[46] T. Covey, "Practicle Principles and Concepts for Analytical Applications of Ion Evaporation Mass Spectrometry" Presented at the ASMS Fall Workshop on Electrospray Ionization, Boston, MA, 1992.

[47] T. J. Thompson, F. Foret, P. Vouros, and B. L. Karger, *Anal. Chem.* **65,** 900 (1993).

as soon as possible to minimize adsorption losses or decomposition in solution.

Solution Chemistry Effects on Ion Emission from Charged Droplets

The sensitivity achievable for a specific analysis may depend heavily on how the solution chemistry chosen affects the efficiency of ion evaporation. The complexity of the ion emission process from charged liquid droplets poses a challenge to the operator to attain optimal performance for a given application but also provides the greatest opportunity for exploring many chemical processes that are difficult to probe using other techniques. A brief attempt is made here to list the solution variables that affect ion emission from droplets and consequently system performance. Positive ions in electrospray are formed by the attachment to the sample molecule of one or more positive cations. H^+ is a common cation for proteins and peptides. Carbohydrates will not show very sensitive response when the charge carrier is hydrogen but will preferentially accept sodium, ammonium, or another cation species as a charge carrier or adduct ion. For negative ions, the charge is usually formed by the removal of a cation, leaving a net negative charge behind. Presumably the most important region on the droplet for these charge transfer processes to take place is at the droplet surface. Hence the species donating or accepting a cation from the sample molecule may play an important role in the charge transfer process. The following is a list of some but certainly not all properties that may influence this charge transfer process: (1) sample concentration, (2) sample pK_a or pK_b, (3) solution conductivity, or a less general but related property, solution pH in such cases when H^+ is the relevant cation, (4) presence of other chemical species that may compete for charge or droplet surface space, (5) sample diffusion rate to the droplet surface, (6) sample molecule shape and size, (7) solution dielectric constant, and (8) droplet charge density. Some of these variables, such as solution pH, conductivity, or sample solubility, can be measured directly in the bulk solution although these values may not hold at the point of ion emission for the evaporated liquid droplet. When these variables are modified, the effect on a given signal level can be quantified.

The body of research work examining these effects has been steadily growing. For example, by examining a series of nucleosides with differing pK_a values it is found that basic analytes with higher pK_a value gave higher H^+ adduct signals in neutral solutions (pH 7.0).[15] This result is as expected after considering a rearrangement of the Henderson–Hasselbalch equation:

$$pK_a = pH + \log \frac{[BH^+]}{[B]} \qquad (1)$$

where [B] is the concentration of the free base and [BH$^+$] is the concentration of the protonated base, respectively. From the relation above, a higher pK_a mandates that a higher concentration of the species will be present in the protonated form, or BH$^+$. Because it is in fact a protonated species that desorbs from the liquid droplet in ES/MS, it is reasonable that compounds with higher pK_a values and thus higher concentrations of BH$^+$ present in solution will give higher ion signals.

The related effect of solution pH due to the presence of dilute acids and bases has also been shown to have substantial effects on the observed ES/MS spectra of proteins and peptides. For example, Fig. 5 shows the ES/MS spectra of cytochrome c taken both without (Fig. 5A) and with (Fig. 5B) 0.1% acetic acid present in the sample solution [10 pmol/μl protein in 1:1 (v/v) methanol–water].[44] Clearly, the signal and number of charges present on each ion are much higher when the acid is present. A similar effect can be seen in the negative ion spectra of insulin, shown in Fig. 6 both without (Fig. 6A) and with (Fig. 6B) 0.01% (v/v) NH$_4$OH present in the sample solution.[48] Again, the signal is greatly improved by the presence of this dilute base. By the same token, the excessive amounts of acid or base can greatly reduce the ion signal observed. To illustrate this point, Fig. 7 shows the relative effect of [NH$_4$OH] on the 4-ion signal of insulin. While signal is greatly improved at the level of 0.01% NH$_4$OH, it is reduced nearly 20-fold at a composition of 1% NH$_4$OH.

Finally, the use of trifluoroacetic acid (TFA) in LC mobile phases calls attention to the interesting effects that solution chemistry can have on the observed ES/MS signal of proteins and peptides. Trifluoroacetic acid inhibits ion signal in ES in two ways. First, TFA is very conductive and very acidic, much more so than acetic acid, for example, and assisted ES must be used when TFA is present even in small amounts. Second, TFA associates with protein and peptide molecules to form ion pairs. This association substantially shifts the acid–base equilibria of proteins and peptides in solution so that attracting H$^+$ becomes unfavorable. In studying this problem, researchers[49] have found that the addition of a solution (75% propionic acid in 2-propanol) postcolumn at one-half of the sample flow rate of the TFA-containing solution prior to introduction into the ES source can greatly improve the ion signal by shifting the acid–base equilibria involved in the ion-pair formation.

[48] J. F. Banks, J. B. Fenn, and C. M. Whitehouse, "Solution Chemistry Effects in Electrospray Mass Spectrometry." Presented at the Pittsburg Conference, Chicago, IL, 1994.

[49] A. Apffel, S. Fischer, P. C. Goodley, and J. A. Sahakian, "Enhanced Sensitivity for Peptide Mapping with Electrospray LC/MS." Presented at the ASMS Conference, Chicago, IL, 1994.

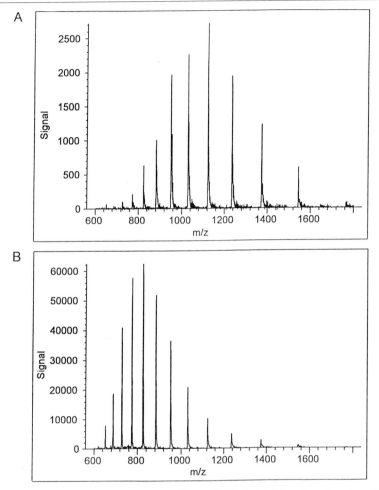

FiG. 5. ES/MS spectra of cytochrome c, 10 pmol/μl in 1 : 1 (v/v) methanol–water with (A) no acid present and (B) 0.1% acetic acid present.

The effect of sample concentration on ion signal has also been studied extensively. As an example, refer again to Fig. 4, in which a signal versus concentration curve for cytochrome c in a solution (1 : 1 methanol and water with 0.1% acetic acid) is shown. The signal response is linear with concentration over four to five orders of magnitude but tapers off at the higher concentrations. Solubility may be a primary factor why the signal-

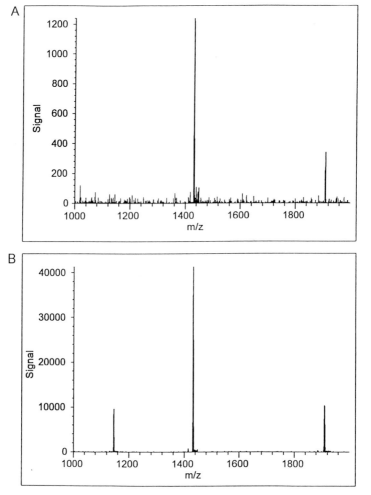

FIG. 6. ES/MS spectra of insulin, 10 pmol/μl in 1:1 (v/v) methanol–water, in the negative ion mode, with (A) no base present and (B) 0.01% NH$_4$OH present.

versus-concentration response shows a negative deviation from linearity at a higher concentration level. The signal plateau beginning at approximately 100 pmol/μl at first glance would not seem to pose a solubility limit. At the point of ion evaporation from the liquid droplet, however, much of the solvent has been evaporated and the nonvolatile cytochrome c sample is considerably more concentrated than it was in the original solution. As the

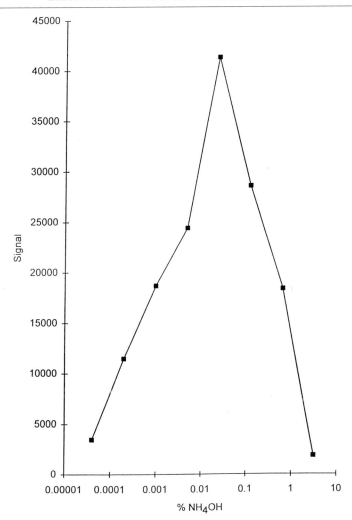

Fig. 7. Insulin −4 ion peak signal vs NH_4OH concentration in 1:1 (v/v) methanol–water.

initial sample concentration is increased beyond the signal plateau, sample clustering through precipitation is often observed in the spectra. These cluster ions appear as dimers, trimers, etc., of the sample species. Compounds with lower solubility are understandably more subject to this effect, and thus the choice of ES solvent can be important for this reason as well. Intriguing studies of system response to solution chemistry variables have been used to study noncovalently bound complexes, including these molecular clusters and even the tertiary structure of large molecules.[15,50-53]

Mass Assignment for Electrospray-Generated Spectra

Deconvolution of Multiply Charged Ions

Electrospray produces multiply charged ions of higher molecular mass compounds that fall typically in an m/z window below 4000. As mentioned in the previous section, in the positive ionization mode a cation is attached to a molecule for every charge added, while in the negative ionization mode a cation is removed for every charge added. Hence for any ion appearing in the mass spectrum, the mass of the molecule can be determined from Eq. (2):

$$K_i = (M/i) + m \qquad (2)$$

where K_i is the measured m/z value, M is the compound molecular weight, i is the charge state, and m is the mass of the charge carrier assuming the electrosprayed ion has only one species of charge carrier. Consequently, a peak appearing in a mass spectrum generated using electrospray ionization has three unknowns: the charge state (i), the charge carrier mass (m), and the compound molecular weight (M). The charge state variable can take on only integer values. There are limited possibilities for the adduct ion identity, for example H^+, Na^+, K^+, or NH_4^+, and these have precise masses associated with them. So, of the three unknowns, the compound molecular weight is the only continuum variable, whereas the charge and adduct ion mass variables have discrete values. To solve for three unknown variables, a minimum of three mass spectral peaks is required. Each related multiply charged peak in a coherent series of multiply charged peaks will satisfy the same value of M in Eq. (2). Solving for the molecular weight M is accomplished in the simplest sense by solving the set of simultaneous linear equations that describes the coherent sequence of multiply charged peaks. Whenever different data points along the m/z scale satisfy or solve for the same values of m and M at different integer values i, the amplitudes of these coherent data points are added and constructively contribute to the amplitude of the deconvoluted parent peak.

[50] M. Hamdan and O. Curcuruto, *Rapid Commun. Mass Spectrom.* **8,** 144 (1993).
[51] B. L. Schwartz, K. J. Light-Wahl, and R. D. Smith, *Rapid Commun. Mass Spectrom.* **5,** 201 (1993).
[52] C. K. Meng and J. B. Fenn, *Org. Mass Spectrom.* **26,** 542 (1991).
[53] S. K. Chowdhury, V. Katta, and B. T. Chait, *J. Am. Chem. Soc.* **112,** 9012 (1990).

Figure 8A shows a mass spectrum containing a series of multiply charged peaks of horse heart cytochrome c, a protein of approximately 12,360 Da in molecular mass. Each multiply charged peak contains the same molecular weight information when Eq. (2) is solved. The charge carrier or adduct ion m in this spectrum is H^+, as is the case with most proteins. It should be noted that the multiply charged peaks observed for cytochrome c in Fig. 8A are actually envelops of a mixture of unresolved isotopic peaks. If the isotopic peaks were well resolved for cytochrome c, as is possible with magnetic sector and Fourier transform mass spectrometry (FTMS) mass analyzers, then the isotopic spacing in m/z would give a direct measure of charge state. A related multiply charged peak with a different charge state than the first would then be required to solve for the value of m in Eq. (2). Computer programs are available for solving for all three variables in Eq. (2), or just M and i, assuming a value for m. When no assumption is made as to the value for the adduct ion mass and all three variables are solved, an indication of mass scale accuracy can be attained and hence an assessment can be made as to the precision of the parent molecular weight determination. An example of this is illustrated in Fig. 8A–C. The multiply charged spectrum of horse heart cytochrome c (10 pmol/μl in water with 1% acetic acid) was acquired with an HP 5989A quadrupole mass analyzer, in which the m/z scale was reasonably well calibrated. All multiply charged peaks in the mass spectrum of Fig. 8A are unresolved isotope profiles. Applying a deconvolution method[54] that simultaneously solves for the three variables in Eq. (2), a three dimensional contour map is generated as illustrated in Fig. 8B. This contour map represents the three-variable solution that yields the deconvoluted peak maximum amplitude based on the measured m/z values and their amplitudes in the cytochrome c spectrum of Fig. 8A. The deconvoluted peak maximum in Fig. 8B ($M = 12,355.19$) falls at an adduct ion mass of 1.251 Da, not precisely the expected 1.008 for hydrogen, thus indicating an error in the m/z locations of the coherent peak maximums. A cross-section taken along the $m = 1.008$ value yields the two-dimensional deconvolution spectrum in Fig. 8C with a maximum at 12,359.13 Da, very close to the accepted value of approximately 12,360. This indicates the presence of random errors in the mass spectra peak maximum and not a mass scale error.

For lower molecular weight ions, for which fewer than three related charge state peaks occur, variables can often be eliminated by assumptions about the adduct ion mass identity and by other spectral features. Assuming H^+ is the adduct ion mass, then two variables remain: charge state and parent molecular weight. As mentioned above, the charge state can be

[54] M. Labowski, C. Whitehouse, and J. Fenn, *Rapid Commun. Mass Spectrom.* **7**, 71 (1993).

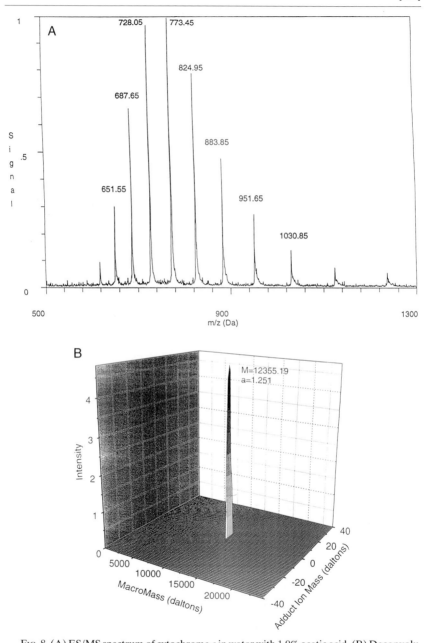

Fig. 8. (A) ES/MS spectrum of cytochrome *c* in water with 1.0% acetic acid. (B) Deconvolution of the cytochrome *c* spectrum, solving simultaneously for charge state, adduct ion mass, and molecular mass as in Eq. (2). (C) Cross-section of the deconvoluted three-dimensional surface in (B), taken along adduct ion mass 1.0008 (H^+).

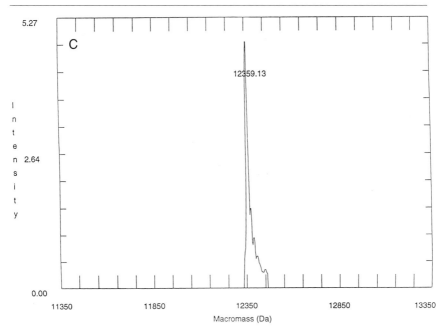

FIG. 8. (*continued*)

determined directly if the resolving power of the mass spectrometer is sufficient to resolve adjacent carbon isotope peaks. The mass-to-charge difference between consecutive carbon isotope peaks yields a direct measure of charge state as the mass difference from ^{12}C to ^{13}C is 1.003355. If the charge state can be determined, the molecular weight can be calculated directly from Eq. (2).

Mass Calibration for Multiply Charged Ions

Consider a tryptic digest from a protein being analyzed using LC/ES/MS with some CID fragmentation. When using a quadrupole mass analyzer, the doubly and triply charged peaks of tryptic fragments may not have resolved isotope profiles. Collision-induced dissociation daughter ion peaks generated from a tryptic digest parent ion may be singly or doubly charged, thus the charge state of a peak appearing in the mass spectrum must first be determined before a mass assignment can be made. Resolution defined as $M/\Delta M$ for a monoisotopic peak decreases with m/z for quadrupole mass spectrometers by the following relation:

$$\text{Resolution} = M/\Delta M = A(m/z) + B \qquad (3)$$

where ΔM is the monoisotopic peak full-width at half the peak maximum amplitude (FWHM) and A and B are variables that can be set with the quadrupole mass spectrometer controls. Often A is called "resolution gain" or "high mass resolution" and B is called "resolution offset" or "low mass resolution." A typical running condition for a quadrupole mass spectrometer would be to set a constant FWHM for singly charged monoisotopic peaks along the m/z scale. For this example using a quadrupole mass spectrometer, assume that the resolution controls on the mass analyzer have been set so that $A = 2$ and $B = 0$. That is, the FWHM of monoisotopic peaks along the m/z scale is set to approximately 0.5 m/z. This resolution setting will yield approximately 12% valley separation between two equal-height, singly charged monoisotopic peaks separated by 1.0 m/z. A doubly charged peak appearing along the m/z scale would have isotopic peaks separated by 0.5 m/z and hence would not be resolved with a quadrupole resolution setting of $A = 2$ and $B = 0$.

As illustrated in Fig. 9A, the isotope profile calculated at a resolution of $M/\Delta M = 2282$ FWHM for gramicidin S + H$^+$ is asymmetric. Figure 9B shows the same isotope profile for gramicidin S + H$^+$ calculated at a resolution of 900 FWHM. Note that from the resolved to the unresolved peak profiles, the isotopic peak maximum has shifted +0.2 Da. Hence, to achieve the correct molecular weight for the gramicidin S most probable monoisotopic peak, a -0.2 Da correction should be applied to the molecular mass assignment obtained from the measured 1140.9 m/z peak, using a rearranged Eq. (2) with the correction factor added, $(K_i \times i) - im - 0.2 = M$.

In the case in which the measured peak has an unknown molecular formula, an estimate of the mass correction can often be made. If proteins or peptides are being mass analyzed, the elemental composition and relative abundance per element can be estimated.[55] Figure 10 is a plot of mass shift in peak maximum from the theoretically most probable resolved isotope (10% valley) m/z value versus resolution for a series of molecular weights. The resolution axis is linearly related to the m/z axis in quadrupoles by Eq. (3), therefore measured peaks of multiply charged ions falling at various points along the m/z scale will require different mass correction factors. The theoretical molecular formulas used for each molecular weight given in Fig. 10 were multiples of relative elemental ratios commonly found in proteins.[55] The formulas used had the elemental ratio 3.667(H):2(C):1(N):1(O). Note that the more symmetric the isotope profile, the less shift in peak maximum occurs for those resolutions typically

---

[55] T. E. Creighton, "Proteins: Structures and Molecular Principles." W. H. Freeman, New York, 1984.

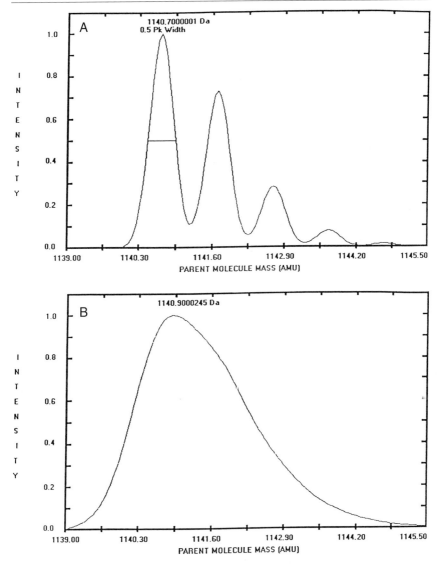

FIG. 9. (A) Calculated isotope distribution of gramicidin S + H$^+$ at a resolution of $M/\Delta M = 2282$. (B) Calculated isotope distribution of gramicidin S + H$^+$ at a resolution of $M/\Delta M = 900$.

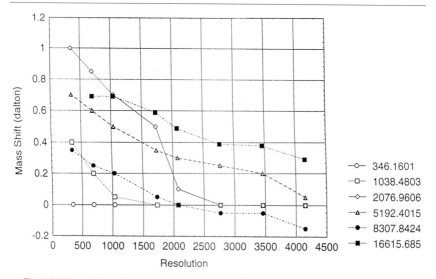

Fig. 10. Mass shift of the peak maximum vs resolution. Molecular formulas are multiples of elemental ratios: $3.667(H) : 2(C) : 1(N) : 1(O)$.

encountered in quadrupole mass spectrometers. Figure 10 serves as an illustration only for proteins. Specific mass correction values versus resolution can be generated for different classes of compounds being studied in a given analytical investigation. Once the charge state of a measured peak is known, the appropriate mass correction can be applied in the case when isotope peaks are not resolved. Often compounds similar to those being analyzed are used to generate calibration files to minimize or eliminate the need to apply mass correction.

Electrospray Ionization–Mass Spectrometry for Protein and Peptide Analysis

For structural elucidation, peptides and proteins are of course often chemically or enzymatically cleaved into smaller fragments that may then be analyzed by LC or CE. The combination of this procedure with MS detection via electrospray ionization results in a highly sensitive and routine method for microsequencing of peptides. As an example of this, Fig. 11 shows a total ion current (TIC) of a tryptic map of myoglobin, generated from 50 pmol of sample injected on-line into an ion trap LC/ES/MS system.[56] An Analytica of Branford ES source operated in the pneumatic nebulization mode interfaced to a Bruker ion trap (ESQUIRE) mass spec-

---

[56] Data were graciously provided by G. H. Kruppa, C. C. Stacey, J. H. Wronka, and F. H. Laukin of Bruker Instruments.

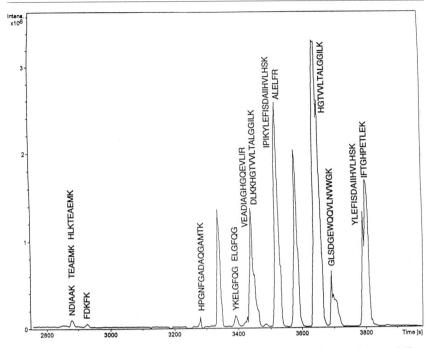

FIG. 11. LC/ES/MS TIC of a 50-pmol digest of myoglobin obtained on an ion trap MS.

trometer (Bruker, Billerica, MA) was used to acquire the data shown in Fig. 11. A 1 × 150 mm 5-μm C_{18} Michrom BioResources LC column was used with a Michrom BioResources LC (Auburn, CA). The separation was performed using a 95% A:5% B to 65% A:35% B gradient in 20 min, where solution A was 0.1% formic acid in water and solution B was 0.1% formic acid in acetonitrile. The liquid flow rate through the microbore LC column directly connected to the ES source was 50 μl/min. All detected peaks in the TIC can be assigned to the known primary sequence of a protein from the molecular weight of the individual peptides. If a mass analyzer capable of multiple stages of analysis is used, such as a triple quadrupole or in this case an ion trap, an individual parent m/z value can be selected and then subjected to MS/MS CID fragmentation. Whereas triple quadrupoles can generate only one stage of MS/MS fragmentation information, ion traps are capable of multiple [or (MS/MS)^n] experiments.

Future Directions

    The future directions in new electrospray mass spectrometric instrumentation that can be applied to applications that complimentarily employ chemical analysis tools are driven directly by more stringent experimental demands imposed by increasingly sophisticated analytical and biological research methods. Several improvements to ES/MS instrumentation and techniques are emerging and will continue to emerge in the near future as the full potential of this analytical tool is realized. Developments have been aimed at improving sensitivity and resolution, decreasing total analysis time, directly studying molecule-folding patterns, investigating noncovalently bound complexes and molecular folding patterns, improving ease of use, and increasing the range of compounds and solutions to which ES/MS can be applied. A few (far from inclusive) examples of new instrumentation and techniques in ES/MS and ES/MS/MS analysis are given below.

Sample-Limited Applications

On-line analysis of complex mixtures of unknown compounds when there are limited sample amounts often requires maximum sensitivity and specificity to solve problems in pharmaceutical drug development and the investigation of diseases. A particularly elegant example of a complex mixture analysis with limited sample available has been reported by Hunt *et al.* in their investigation of the class I and II major histocompatibility complexes (MHCs).[8,57] Mixtures of peptides isolated from different types of carcinoma cells were separated, detected, and sequenced to find those peptides expressed by the MHC proteins that were present in melanoma cells but not found in normal cells. On-line techniques working with individual component amounts in the low femtomolar range were designed to minimize sample loss and allow ES/MS analysis in parallel with fraction collecting. The collected fractions were used to check for biological activity and conduct additional ES/MS/MS analysis off-line. Figure 12 shows a schematic setup similar to what Hunt *et al.* employed, in which the LC analytical column used was a capillary LC column with flow splitting and fraction collecting at column flow rates of only a few microliters per minute. From this type of work, commercially available LC systems will emerge whereby small sample injection and fraction collecting in the 1-μl/min (or lower) flow rate range coupled on-line with ES/MS will become more routine.

Faster Analysis Times

Faster separation systems such as perfusion LC and CE are being employed on-line with ES/MS to decrease overall analysis times without com-

[57] A. L. Cox, J. Skipper, Y. Chen, R. A. Henderson, T. L. Darrow, V. H. Shabanowitz, V. H. Englehard, D. F. Hunt, and C. L. Slingluff, *Science* **264**, 716 (1994).

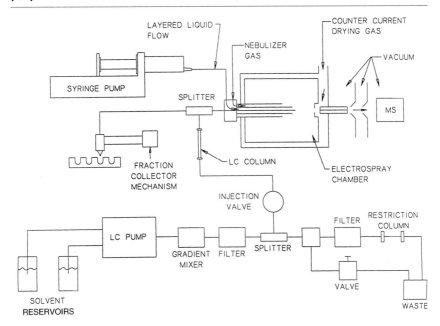

FIG. 12. Diagram of capillary LC/ES/MS system with low flow rate postcolumn fraction collecting.

promising performance. As the separation systems become faster and eluting peaks become narrower, the mass analyzer must be able to acquire data at a rate that does not degrade the chromatographic or CE separation resolution. Scanning instruments when used in scan mode and even in selected ion monitoring mode have difficulty acquiring data at a rapid-enough rate to take full advantage of the faster LC and CE separation techniques. Time-of-flight (TOF) and ion trap mass analyzers can acquire mass spectral scan data at a more rapid rate than the more conventional quadrupole and magnetic sector mass analyzers. Time-of-flight mass spectrometers can generate several mass spectra per second, which are themselves averages of hundreds of individual full mass scans. Figure 13 shows a reconstructed ion chromatogram (RIC) of a CE peptide separation in which the eluting CE peaks are approximately 5 sec wide. In acquiring these data, more than 4000 full scans per second were acquired with 500 scans averaged to produce 8 complete mass spectra per second. With this data acquisition rate, the CE peak profiles were not distorted by the ES/MS detection step. Time-of-flight mass spectrometers have the highest mass range of any mass analyzer and can acquire data at maximum instrument resolution without reducing sensitivity. The TOF analyzer used to generate

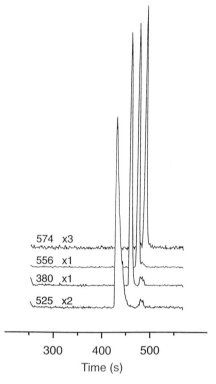

574 x3
556 x1
380 x1
525 x2

300   400   500
Time (s)

FIG. 13. Reconstructed ion currents for angiotensin (525 m/z), Val-Tyr-Val (380 m/z), Leu-enkephalin (556 m/z), and Me-enkephalin (574 m/z) analyzed by CE/ES/TOF/MS.

the TIC in Fig. 13 is capable of acquiring more than 30 mass spectra per second if necessary.[58] Ion traps, although not able to scan as rapidly as TOF, do have the capability of conducting MS/MS and (MS/MS)^n analysis.

Investigation of Molecular Folding Patterns

ES/MS methods are being developed that can shed light on the folding patterns of proteins. For example, in one technique, the supporting solution of a protein sample is varied from 100% aqueous to mixtures of higher organic or acid content while the multiply charged mass spectra are acquired and compared. Relative amplitudes of the m/z peaks shift with the various protein conformation states[3,53] induced by the solution changes and can

[58] J. F. Banks, T. Dresch, P. Haren, and J. Boyle, "Fast Separations with a New ESI-TOF Mass Detector." Presented at the Pittsburgh Conference, New Orleans, LA, 1995.

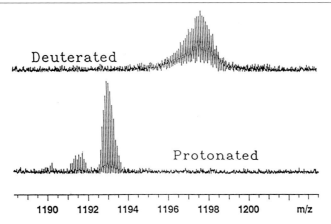

FIG. 14. ES/FTMS of lysozyme +12 charge state peak with isotope peaks resolved; comparison of deuterated and protonated forms.

then be correlated to protein structure and folding. Another technique is to exchange deuterium with exposed hydrogens of a protein and from the rate of exchange and the total percent exchange achieved, infer protein-folding patterns. Deuterium exchange has been conducted in the liquid phase[59] before electrospraying the solution and in the gas phase with electrosprayed ions trapped in FTMS cells.[60] As an example, Fig. 14 shows two mass spectra of the +12 charge state peak of lysozyme, one in its native form and the second after gas phase deuterium exchange in a Bruker ES/FTMS instrument.[61] The rate of deuterium uptake can be measured by the degree of mass shift over time, while the high-resolution capability of FTMS allows the precise measurement accuracy required for these types of experiments.

Lower Liquid Flow Rate Electrospray Ionization–Mass Spectrometry, Higher Sensitivities, Higher Resolution

Sharp electrospray needle tips with bores as small as 3 μm have been used to electrospray liquid samples at flow rates of 25 to 100 nl/min.[62,63]

[59] V. Katta and B. Chait, *Rapid Commun. Mass Spectrom.* **5**, 214 (1991).
[60] D. Suckau, Y. Shi, S. C. Beu, M. W. Senko, J. P. Quinn, F. M. Wampler, and F. W. McLafferty, *Proc. Natl. Acad. Sci. U.S.A.* **90**, 790 (1993).
[61] G. H. Kruppa, C. H. Watson, C. C. Stacey, J. Wronka, F. H. Laukien, C. V. Robinson, S. J. Eyles, R. T. Aplin, and C. M. Dobson, 1996 (submitted).
[62] M. S. Kriger, K. Cook, and R. S. Ramsey, *Anal. Chem.* **67**, 385 (1995).
[63] M. Wilm and M. Mann, "Micro Electrospray Source for Generating Highly Resolved MS/MS Spectra on a 1 μl Sample Volume." Presented at the ASMS Conference, Chicago, IL, 1994.

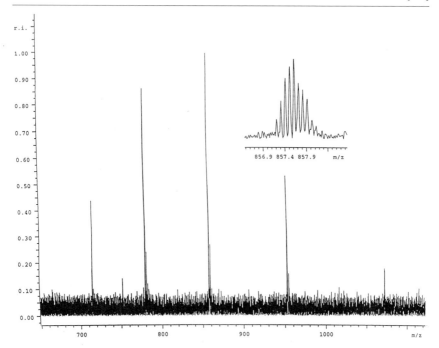

FIG. 15. ES/FTMS analysis of ubiquitin from 1 μl of a 0.1-μg/μl solution loaded into a microtip consuming only 12 amol of sample. *Insert:* +10 peak showing the resolved ^{13}C isotope distribution.

At these liquid flow rates, ES/MS or ES/MS/MS analysis times of over 30 min can be achieved while consuming less than 1 μl of sample. This translates into increased sensitivity as illustrated in Fig. 15, where a microtip was used in an Analytica of Branford ES source interfaced to a Bruker FTMS mass spectrometer. Only 15 amol of ubiquitin was consumed from a 0.1-μg/μl solution to generate the mass spectrum shown[56] with resolution separating the isotopic peaks of the 8.5-kDa protein. Ultimately the ES/MS analysis of just a few molecules may be possible, as has been reported by Bakhtiar *et al.*[64] The refinement of electrospray ion sources as interfaced to TOF, ion trap, and FTMS analyzers will enhance the capability of electrospray mass spectrometry, particularly when complemented by enzymol-

[64] R. Bakhtiar, X. Cheng, S. V. Orden, and R. D. Smith, "Charge State Shifting of Individual Multiply Charged Ions of Bovine Albumin Dimer and Molecular Weight Determination Using an Individual-Ion Approach." Presented at the ASMS Conference, Chicago, IL, 1994.

ogy and on-line separation techniques. Also, increased data analysis capability will greatly improve the range of applications to which ES/MS can be routinely applied and the speed and quantity of information that can be acquired.

[22] Matrix-Assisted Laser Desorption Ionization Mass–Spectrometry of Proteins

By Ronald C. Beavis and Brian T. Chait

Introduction

In this chapter, we provide a practical guide to the application of matrix-assisted laser desorption/ionization–mass spectrometry (MALDI–MS) for the analysis of peptides and proteins. We describe in detail the best methods that are currently available for preparing samples for MALDI–MS, because good sample preparation is the key to successful mass analysis. We consider aspects of the method that are important for obtaining high-quality data. Finally, we describe a selection of strategies for studying proteins with this powerful new technique.

Pulses of laser light have been employed since as early as 1976[1] to produce intact gas phase peptide ions from solid samples. The resulting peptide ions could then be analyzed by mass spectrometry. These early investigations and subsequent measurements over the following decade produced useful mass spectra from only a few short peptides. In addition, the probability for obtaining a useful mass spectrum depended critically on the specific physical properties of the peptide (e.g., photoabsorption spectrum, volatility) under study. This situation changed dramatically with the development by Karas and Hillenkamp.[2] MALDI–MS provides the means to volatilize proteins readily and to make the conditions for volatilization largely independent of the specific physical properties of the protein. This effect is achieved in two steps. The first step involves preparing an appropriate sample by dilutely embedding proteins in a matrix of small organic molecules that strongly absorb ultraviolet wavelength laser light. The second step involves ablation of bulk portions of this solid sample

---

[1] M. A. Posthumus, P. G. Kistemaker, and H. L. C. Meuzelaar, *Anal. Chem.* **50,** 985 (1978).
[2] M. Karas and F. Hillenkamp, *Anal. Chem.* **60,** 2299 (1988).

METHODS IN ENZYMOLOGY, VOL. 270

by intense, short-duration pulses of the laser light. In the ablation step, molecular components of the solid are put into the gas phase and ionized, producing intact protein ions. The molecular masses of these protein ions are easily determined by time-of-flight mass analysis. The marvelous development by Karas and Hillenkamp has been steadily refined and has become a method of choice for characterizing peptides and proteins.[3–8]

MALDI–MS is versatile and effective for the analysis of peptides and proteins because of the special properties and capabilities of the technique. Some of these capabilities and properties are as follows.

> Biological samples can be examined without extensive purification.
> Common biochemical additives such as buffers, salts, glycerol, chelating agents, chaotropic agents, and certain detergents do not interfere with the analysis.
> Most classes of proteins can be examined, provided that the protein can be dissolved in appropriate solvents.
> Posttranslationally modified proteins can be measured.
> Useful mass spectra can be obtained from complex mixtures of peptides and proteins.
> Proteins with masses ranging to greater than 100 kDa can be analyzed.
> The total amount of protein required for an analysis is usually in the range of 1–10 pmol.
> Protein molecular masses can be determined with mass accuracies as high as 1 part in 10,000.
> Complete analyses can be made in a matter of minutes.

It is instructive to compare the properties of MALDI–MS for the analysis of proteins by sodium dodecyl sulfate-polyacrylamide gel electrophoresis (SDS–PAGE), a method widely used in biological research. SDS–PAGE is a universal technique for separating and analyzing proteins because of the effectiveness of the detergent SDS for dissolving proteins and for converting them into entities that migrate on electrophoretic gels with relative velocities that depend on their size. By contrast, there are certain restrictions (see below) on the detergents and additives that can be used in the preparation of samples for MALDI–MS. However, the more complex and expensive instrumentation required by MALDI–MS is justified because MS provides much higher mass accuracy determination (typically three to four orders

[3] F. Hillenkamp, M. Karas, R. C. Beavis, and B. T. Chait, *Anal. Chem.* **63,** 1193A (1991).
[4] B. T. Chait and S. B. H. Kent, *Science* **257,** 1885 (1992).
[5] A. L. Burlingame, R. K. Boyd, and S. J. Gaskell, *Anal. Chem.* **66,** 634R (1994).
[6] R. Aebersold, *Curr. Opin. Biotechnol.* **4,** 412 (1993).
[7] R. Wang and B. T. Chait, *Curr. Opin. Biotechnol.* **5,** 77 (1994).
[8] J. T. Stults, *Curr. Opin. Struct. Biol.* **5,** 691 (1995).

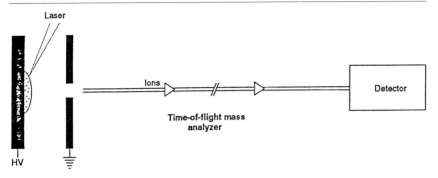

Fig. 1. Schematic diagram of a matrix-assisted laser desorption mass spectrometer with a linear time-of-flight mass analyzer.

of magnitude) and higher resolution than SDS–PAGE. MALDI–MS is also considerably faster than standard SDS–PAGE and can be used to analyze complex peptide mixtures generated by digestion of proteins.

The essential experimental setup for MALDI–MS is illustrated in Fig. 1. Short-duration (nanosecond) pulses of laser light are directed at matrix–protein sample mixtures (protein-doped matrix crystals; see below) that are inserted into the mass spectrometer on a mechanical probe. The laser light causes a portion of the matrix–protein sample to be volatilized and ionized. The resulting gas phase ions are accelerated to a fixed energy in an electrostatic field and directed into a field-free flight tube. After passing through the flight tube, the ions impact on an ion detector, whereupon the intervals (Δt_m) between the pulse of laser light and the ion impacts are measured. The masses of the ions passing through the flight tube can then be determined via the approximate relationship[9,10]

$$m = 2qV\Delta t_m^2/l^2 \qquad (1)$$

where m is the mass of the ion, q is the charge on the ion, V is the potential through which the ion is accelerated, and l is the length of the flight tube.

MALDI–MS is a powerful and versatile new tool for the study of proteins and the required instrumentation is significantly simpler than conventional mass spectrometers. This relative simplicity makes the technique potentially accessible to many biologists. The extent to which MALDI–MS will impact on biological research will depend, to a large degree, both

[9] R. C. Beavis, "Time-of-Flight Mass Spectrometry," American Chemical Society Symposium Series 549 (R. Cotter, ed.), pp. 49–60. American Chemical Society, Washington, DC, 1994.
[10] H. Wolnick, *Anal. Instrument.* **16**, 15 (1987).

on the production of effective commercial instrumentation and on the development of methods for making MS easily accessible to biologists.

Preparing Samples

Sample preparation is crucial for matrix-assisted laser desorption. The surroundings of a protein molecule must be fashioned so that an intense light pulse can transfer the intact molecule into a vacuum. The best method for achieving this effect is to incorporate protein molecules into the crystals of a second material. The protein-containing crystal absorbs the light pulse and uses the absorbed energy to eject material from its surface. If the ejected volume contains protein molecules, they enter the vacuum along with the other ejected materials. Few compounds form crystals that incorporate proteins, eject them intact, and ionize them in the same process. The MALDI literature refers to any material with these properties as a *matrix*.

Several additional properties are important for a matrix compound to be analytically useful. Incorporating the protein into a growing matrix crystal implies that the protein and the matrix must be simultaneously in solution. Therefore, a matrix should dissolve and grow protein-doped crystals in commonly used protein solvent mixtures. The matrix crystal absorbs the light pulse, therefore the compound must contain a stable chromaphore. The protein and matrix must not react to form a stable product. From an instrumental point of view, the matrix crystals must remain in vacuum for extended periods of time (tens of minutes). Therefore, the sublimation rate of a matrix in vacuum should be as slow as possible.

The following sections introduce the matrix compounds commonly used, explain the requirements for protein samples, and give typical recipes for producing protein-doped matrix crystals. These sections do not exhaust all possible combinations of matrix compounds and recipes. Instead, these sections reflect widely used approaches to the problem. Modify a recipe if the requirements of a particular solvent differ from the description— however, keep in mind the general principles outlined below.

Selecting Matrices

Few compounds are good matrix materials for analytical protein MALDI. Many compounds form protein-doped structures that produce ions, but they are disqualified by other factors. The resulting mass spectra may have low signal intensities or poor mass resolution. Other materials work only for certain proteins or in a limited range of solvents. Most matrix compounds produce satellite signals called *adduct peaks* at slightly higher mass than the analyte molecule peaks. These peaks result from the photo-

TABLE I
PROPERTIES OF MALDI MATRICES

| Matrix | Analytes[a] | | Suggested solvent (water : organic) | Ionization[b] | Adduct[c] |
	Peptides	Proteins			
Gentisic acid	+	+/−	9 : 1	+	M + 136
Sinapic acid	+/−	+	2 : 1	+	M + 206
3-Indoleacrylic acid	+	+	2 : 1	++	M + 185
4-HCCA	+	+	2 : 1	+++	—

[a] +, Matrix may be used for most peptides and proteins; +/−, matrix may (or may not) work.
[b] The more + signs, the more intense the signal, and the higher charge state of the most intense peak.
[c] Expected mass of the most intense satellite peak.

chemical breakdown of the matrix into more reactive species, which can add to the polypeptide. The best matrices have low-intensity photochemical adduct peaks.

Table I lists four compounds used for peptide and protein mass measurement and some information about relevant properties. These compounds can be used with 337-nm (nitrogen laser) or 354-nm [neodymium : yttrium/ aluminum/garnet (3)] [Nd : YAG(3) laser] light. Table I gives an idea of the usefulness of a matrix. The utility of a specific matrix for a particular protein or peptide cannot be predicted in advance: trial and error may be necessary to establish a useful recipe. Table I also gives a guide to the most commonly used solvent mixtures. Besides the water and organic solvents (such as acetonitrile, methanol, propanol), the mixture may contain acids (e.g., trifluoroacetic acid), salts, lipids and some types of detergents (see below). An indication of the ion signal intensity and the average number of charges added to the protein in the ejection process are shown in Table I. α-Cyano-4-hydroxycinnamic acid (4-HCCA) does not produce strong photochemical adduct signals, although it seems to encourage copper attachment to some peptides, resulting in [M + Cu] peaks.

Gentisic acid (2,5-dihydroxybenzoic acid, molecular mass 154 Da, CAS 490-79-9) is a useful matrix for a wide variety of peptides and proteins.[11] Gentisic acid forms large dimorphic monoclinic crystals from water, water–alcohol, or water–acetonitrile mixtures. Many researchers use it as their first choice for analyzing peptide mixtures. The protein ions ejected from

[11] K. Strupat, M. Karas, and F. Hillenkamp, *Int. J. Mass Spectrom. Ion Processes* **111**, 89 (1991).

gentisic acid crystals appear to be relatively stable, making them ideal for reflectron mass spectrometers.[12,13]

Sinapic acid (or sinapinic acid, *trans*-3,5-dimethoxy-4-hydroxycinnamic acid, molecular mass 224 Da, CAS 530-59-6) is useful for a wide variety of peptides and proteins.[14] Sinapic acid forms small, thin monoclinic crystals from water–acetonitrile (or alcohol) mixtures. It has a strong affinity for proteins of all types and gives good results with mixtures of proteins. Small peptides (mass <3 kDa) may not produce strong signals.

trans-3-Indoleacrylic acid (molecular mass 203 Da, CAS 29953-71-7) has not found broad application for proteins and peptides,[15] although it has been used for industrial polymers. It produces intense signals and works well for complex mixtures of proteins and peptides.

4-HCCA (α-cyano-4-hydroxycinnamic acid, molecular mass 189 Da, CAS 28166-41-8) produces intense signals from peptides and proteins.[16] It forms polycrystalline clumps made up of twinned monoclinic prisms. Instruments with poor ion transmission efficiency rely on 4-HCCA to overcome their design problems. This matrix produces intense, multiply charged ions in the positive-ion spectra of proteins. The protein ions produced by 4-HCCA frequently undergo metastable decay in the mass spectrometer.[16a]

Preparing Protein Solutions for Analysis

The most important factor in incorporating a protein into matrix crystals is having sufficient protein dissolved in the crystal growing solution. For most proteins, a concentration of 1–10 μM in the crystal growing solution produces useful protein-doped crystals. Lower concentrations of protein may produce results. However, the use of low concentrations of proteins in small volumes means that care must be taken when handling the sample. Protein adsorption to the transferring pipettes, Eppendorf tube walls, etc., may reduce the actual protein concentration to negligible levels.

Protein samples that have been exposed to strong ionic detergents, such as sodium dodecyl sulfate, will not incorporate into matrix crystals. If strong

[12] W. Yu, J. E. Vath, M. C. Huberty, and S. A. Martin, *Anal. Chem.* **65**, 3015 (1993).

[13] R. Kaufmann, D. Kirsch, and B. Spengler, *Int. J. Mass Spectrom. Ion Processes* **131**, 355 (1994).

[14] R. C. Beavis and B. T. Chait, *Rapid Commun. Mass Spectrom.* **3**, 432 (1989).

[15] J. Z. Chou, M. J. Kreek, and B. T. Chait, *J. Am. Soc. Mass Spectrom.* **5**, 10 (1994).

[16] R. C. Beavis, T. Chaudhary, and B. T. Chait, *Org. Mass Spectrom.* **27**, 156 (1992).

[16a] M. Karas, U. Bahr, K. Strupat, F. Hillenkamp, A. Tsarbopoulos, and B. N. Pramanik, *Anal. Chem.* **67**, 675 (1995).

ionic detergents are a necessary part of an isolation procedure, they must be thoroughly removed. A method that successfully removes these detergents from small samples of proteins is a two-phase extraction.[17] Electroblotting from PAGE gels and chromatographic methods may be alternatives for detergent removal.[18,19] Acetone precipitation and dialysis usually do not remove enough detergent for MALDI sample production.

While ionic detergents must be avoided, most other protein solvents do not interfere significantly with sample production. Nonionic detergents (such as Triton X or octylglucoside) can be used[20] (see below for cautions). High concentrations of salts, buffers, urea, glycerol, and formic acid produce useful samples (see below for cautions). The mixture 1:2:3 (v/v) formic acid–2-propanol–water works well for many hydrophobic proteins and peptides.[21] Avoid buffer solutions containing sodium azide: its presence suppresses protein ion formation in the mass spectrometer.

Exposure of proteins to concentrated formic acid should be avoided, if possible, in experiments where accurate mass measurements are required. Formic acid reacts with amino groups (both N-terminal α-amino and lysine ε-amino groups), resulting in a formyl derivative of the protein. Multiple formylation leads to several peaks for each polypeptide, making mass determination ambiguous. If a procedure requires formic acid, exposure should be kept as short as possible. Low temperatures also slow the rate of formylation. Even with this caveat, experience suggests that aqueous 70% formic acid is the best general solvent for the cyanogen bromide peptide cleavage reaction. Dilute HCl (0.1 N) may also be used as a solvent for cyanogen bromide peptide cleavage; however, care must be taken to neutralize the pH of the solution before evaporating the solvent to dryness. (*Note:* Concentrated trifluoroacetic acid also reacts with free amino groups.)

The solvent chosen for the protein must be compatible with the matrix used to form the protein ion-emitting crystals. In most of the recipes that follow, an aliquot of a matrix solution is mixed with an aliquot of a protein solution to make the crystal-forming mother liquor. It is important that neither the matrix nor the protein precipitate when the two solutions mix. Particular care must be taken when the protein solvent does not

[17] L. E. Henderson, S. Oroszlan, and W. Konigsberg, *Anal. Biochem.* **93**, 153 (1979).
[18] W. J. Henzel, T. M. Billegi, J. T. Stults, S. C. Wong, C. Grimley, and C. Watanabe, *Proc. Natl. Acad. Sci. U.S.A.* **90**, 5011 (1993).
[19] W. Zhang, A. J. Czernik, T. Yungwirth, R. Aebersold, and B. T. Chait, *Protein Sci.* **3**, 677 (1994).
[20] O. Vorm, B. T. Chait, and P. Roepstorff, *in* "Proceedings of 41st ASMS Conference in Mass Spectrometry, San Francisco, 1993," p. 621a.
[21] S. L. Cohen and B. T. Chait, *Anal. Chem.* **68**, 31 (1996).

contain any organic solvent, which may lead to precipitation of the matrix on mixing.

Preparing Samples for Mass Spectrometry

Dried Droplet Method

The discovery of the dried droplet method allowed the application of laser desorption to proteins.[2] Drying a droplet of a protein–matrix solution remains the favorite method of most MALDI practitioners. The recipe for producing a dried droplet sample has several simple steps. First, make a saturated solution of the matrix material (see Table I for solvents) and mix in sufficient protein for a final concentration of 1–10 μM. This solution must be thoroughly mixed to ensure reproducible results. A convenient method is to place some saturated matrix solution (5–10 μl) in an Eppendorf tube and then add a smaller volume (1–2 μl) of a protein solution. Agitating the tube with a vortex mixer for a few seconds mixes the solution sufficiently. Second, place a droplet (0.5–2 μl) of the resulting mixture on the sample stage of the mass spectrometer. Dry the droplet at room temperature; blowing room temperature air over the droplet speeds drying. When the liquid has completely evaporated, the sample may be loaded into the mass spectrometer. There is no rush to load the sample. Dried droplets are quite stable: they can be kept in a drawer or in vacuum for days. The deposit may also be washed to etch the surface layer of the deposit crystals; the surface layer is the most heavily contaminated with involatile components of the original solution. Be careful when etching the crystals, as it is easy to wash them off the surface. We recommend thoroughly drying the sample (vacuum dried if possible) followed by a brief immersion in cold water (10–30 sec in 4° water). The excess water should be removed rapidly (e.g., by flicking the sample stage, or by suction).

Some authors suggest placing a drop of saturated matrix solution on the sample stage and then adding a similar volume of protein solution to form the final droplet. This method results in acceptable spectra for samples containing a single analyte. If the protein sample contains more than one protein or peptide component, we recommend thorough mixing of the two solutions in a tube before making the droplet. Prior, thorough mixing increases the reproducibility of the mass spectra obtained.

The simplicity of the dried droplet recipe surprises most people. This method gives good results for many types of protein samples. The recipe has some serious pitfalls that may be encountered.

The protein must be truly dissolved in the solvent. Making a slurry of a peptide powder and solvent is not sufficient.

Be careful of inadvertent changes in solvent composition. Such changes can easily occur when two solutions are mixed, resulting in the precipitation of either the protein or matrix (or both). A similar problem can result from the selective evaporation of organic solvents from aqueous solutions. The latter is a particular problem when small volumes of solution are stored in relatively large containers (e.g., 10 μl of solvent in a 1.5-ml Eppendorf tube).

Do not heat the droplet to speed drying. Changing the temperature of the solution alters matrix crystal formation and protein incorporation, usually in a bad way.

Use fresh matrix solutions whenever possible. Matrix solutions gradually decompose under normal laboratory conditions. Mix small volumes of solution as needed.

Do not use higher protein concentrations than recommended: the final protein concentration should be less than 10 μM. It is tempting to add more protein, in the hope of increasing signal intensity. This approach seldom works. Reducing the protein concentration (not increasing it) frequently increases the signal.

Do not use involatile solvents. The involatile solvents commonly used in protein chemistry are glycerol, polyethylene glycol, 2-mercaptoethanol, Triton X, dimethyl sulfoxide (DMSO), and dimethylformamide (DMF). These solvents interfere with matrix crystallization and coat any crystals that do form with a difficult-to-remove solvent layer. If you cannot avoid using these solvents (or if a droplet does not dry properly), try another method (see below).

Keep the concentration of nonprotein materials to a minimum. The dried droplet method tolerates the presence of salts and buffers well, but this tolerance has limits. Washing the crystals may help, although care must be taken not to wash the crystals off the substrate. If contaminants are suspected of suppressing the signal, try another method (see below).

The pH of the crystal-growing solution must be less than pH 4. The organic acid matrices described above become ions at pH > 4, completely changing their crystallization properties. Using aqueous 0.1% trifluoroacetic acid rather than pure water will usually solve any pH problems.

Do not leave crystals of undissolved matrix in the final protein–matrix solution. Intact matrix crystals act as crystal seeds, leading to inhomogeneous samples. Spinning the matrix solution in a tabletop centrifuge removes the large crystals that cause this problem.

Slow Crystallization (Growing Large Matrix Crystals)

The dried droplet method will fail for samples containing significant concentrations of involatile solvents or high concentrations of salts.[22] The problem seems to be that as the droplet dries, the contaminant concentrations become very high, interfering with matrix crystal growth and with protein incorporation. This concentration effect can be avoided by growing large, protein-doped matrix crystals under more controlled conditions. The technique described below has been shown to be effective for protein solutions containing low protein concentrations (<1 μM) and high concentrations of involatile solvents (e.g., 30% glycerol) or salts (4 M sodium chloride). In mixtures of peptides and proteins, this technique may discriminate in favor of the protein components.[21]

Large matrix crystals can be grown using the following procedure.[22] First, saturate the solvent mixture chosen for the mother liquor with the matrix at room temperature by placing 20–100 μl of the solvent mixture (containing the protein and other solutes) in a 1.5-ml Eppendorf tube with an excess of matrix. Swirl the mixture with a vortex mixer for at least 5 min and then centrifuge the mixture to remove large crystals. Carefully remove the supernatant with a Pasteur pipette and save it in a clean Eppendorf tube as the mother liquor. Seal the Eppendorf tube and heat the solution to approximately 40° to remove any small matrix crystals remaining in suspension. Remove the tube from the heater and allow it to come to room temperature before opening the tube. Place the opened tube into a holder on the rotating stage of a vortex mixer and set the mixer to its lowest speed. Then allow the solvent to evaporate for several hours while swirling the tube at low speed. This procedure results in a collection of protein-doped matrix crystals with long axes of about 0.5–1 mm (for sinapic acid) or small crystals (for 4HCCA).[21] Once the crystals have grown, remove the mother liquor from the Eppendorf tube with a Pasteur pipette. Wash the crystals by adding room temperature water and swirling the tube with a vortex mixer. Pellet the crystals by centrifugation and remove the solvent. Repeat the washing step several times. After the final removal of washing solvent, add a small amount of water, slurry the crystals into the water (taking care not to crush the crystals), and dump the contents onto a piece of filter paper or the sample stage.

There are many possible variations to this method.[21] The saturated matrix solution may be prepared first and cleared of excess crystals prior to the addition of the protein to the solution, taking care not to dilute the matrix solution. Crystallization may also be carried out by lowering the

---

[22] F. Xiang and R. C. Beavis, *Org. Mass Spectrom.* **28,** 1424 (1993).

temperature of the mother liquor by placing the open Eppendorf tube in a cold chamber or a refrigerator.

Crystals can be mounted on a sample holder in a number of different ways. For small crystals (<1 mm), transfer them to the sample holder while they are damp. The crystals usually adhere when as they dry. For larger crystals, it may be necessary to place a thin layer of polystyrene cement on the sample stage and adhere the crystals to the cement. Once they are mounted, the crystals can be examined using the same methods used for dried droplet samples.

Thin Polycrystalline Films

Preparing thin polycrystalline films involves producing a uniform layer of very small crystals on the sample stage of a mass spectrometer that are mechanically well adhered to the substrate.[23] It is a variant of the dried droplet method.[23] The crystals can be thoroughly washed without removing them from the surface. This effect is achieved by creating an activated layer of matrix on the surface of the substrate, which acts as an extended seeding site for growing matrix crystals. A droplet is then dried onto this activated surface, resulting in a thin film of protein-doped matrix crystals attached to the surface. The film grows rapidly, so it is not necessary to wait until the droplet is dry before washing the film, reducing effects caused by increasing contaminant concentrations as the droplet dries. The recipe given below results in reproducible polycrystalline films for most matrices, although it can be difficult to use gentistic acid because of its high solubility in water. The method has been shown to give useful signals from solvents containing as much as 50% (v/v) glycerol or 4 M urea.

First, prepare two solutions containing the matrix at room temperature, referred to below as solution A and solution B. Handle these solutions carefully: it is important to minimize transfer of undissolved crystalline solid from one step to the next. Prepare the solutions fresh each day.

Solution A contains the matrix alone. Prepare it using 0.3 ml of acetonitrile or any appropriate solvent mixture. Saturate the solvent with matrix in a 0.5-ml Eppendorf tube and swirl for at least 1 min with a Vortex mixer. Spin the tube in a benchtop centrifuge to deposit undissolved matrix particles on the bottom of the tube. Remove the supernatant to another tube.

Prepare solution B by saturating 0.3 ml of the selected solvent mixture with matrix and then remove excess matrix particles by centrifugation. Mix an aliquot of this matrix-saturated solution with a protein-containing

[23] F. Xiang and R. C. Beavis, *Rapid Commun. Mass Spectrom.* **8,** 199 (1994).

solution to produce a final protein concentration of approximately 1 μM. It is important to remember that the solvent produced by mixing the protein-containing solution and the matrix-containing solution must keep both the matrix and the protein in solution. Precipitating either solute leads to poor mass spectra.

To make the sample deposit, first place 0.5–1 μl of solution A on the sample stage of the mass spectrometer and allow it to dry. Press firmly on the surface with a clean Kimwipe and stroke the crystalline layer several times, crushing the crystals and rubbing them into the surface. Brush the crushed matrix to remove any loose matrix particles. The layer of crushed matrix serves as the "activated" seeding site for the drying droplet (see below). Alternatively, an activated layer can be made by electrospraying solution A onto the sample stage, being careful that the sprayed layer does not become wet during the spray process. Electrospraying is more difficult than simply crushing crystals, but the results are good and may be worth the trouble for particularly contaminated samples.

Apply 0.5–1 μl of solution B to the spot bearing the smeared matrix material. An opaque film forms over the surface of the substrate below the droplet within a few seconds, covering the metal. After about 1 min, the probe tip can be immersed in room temperature water to remove involatile solvents and other contaminants. It is not necessary to let the droplet dry before washing: the film does not wash off easily. Remove excess water and allow it to dry before loading into the mass spectrometer.

The aliquot of solution B used to create the film must be free of particulate matter. Any particles of matrix present in this solution result in rapid nucleation of matrix crystals throughout the solution, interfering with the formation of the film. If solution B is obviously cloudy, further centrifugation is necessary to clear the liquid.

Seeded Microcrystalline Films

The seeded microcrystalline film method[24] of sample preparation is a direct replacement for the dried droplet technique. It produces more uniform samples than the dried droplet technique and the crystals are better adhered to the surface of the sample stage. The uniformity of the deposit apparently improves the mass resolution and mass accuracy obtained in reflectron-type mass spectrometers. The well-adhered crystals are less likely to be lost during wash–etch cycles. This technique works well with α-cyano-4-hydroxycinnamic acid.

This technique involves first laying down a thin layer of small matrix crystals on the sample stage. A droplet containing the analyte is placed on

[24] O. Vorm, P. Roepstorff, and M. Mann, *Anal. Chem.* **66,** 3281 (1994).

top of the crystalline deposit. The solvent containing the analyte is carefully chosen so that it will redissolve part of the matrix layer. The resulting matrix–analyte solution dries, depositing the matrix on the seed sites provided by the undissolved portion of the original layer.

Two solutions must be prepared (solutions A and B). The solvent for solution A is acetone containing 1–2% water. Matrix is added to form a nearly saturated solution. A droplet (0.5 μl) of solution A is placed on the sample stage and allowed to spread out and dry. The deposit formed should be uniform. The volume of solution A used and the matrix concentration depend on the geometry of the sample stage: the reader is encouraged to experiment with different conditions to suit specific needs.[24]

Solution B contains the peptide/protein of interest at a concentration of approximately 1 μM in an aqueous solution that contains 20–30% organic solvent. Place approximately 0.5 μl of solution B onto the surface prepared from solution A and allow it to dry at room temperature. When the solution is completely dry, the deposit can be washed by placing a drop of pure water [or 0.1% aqueous trifluoroacetic acid (TFA)] onto the deposit for 2–10 sec and then removing excess water.

Measuring Molecular Masses

Mass spectrometers come in a wide variety of configurations. All mass spectrometers can be broken down into three main parts: an ion source, a mass analyzer, and an ion detector. Matrix-assisted laser desorption is carried out in the ion source, producing a current of protein ions whose mass can be measured using a mass analyzer and an ion detector.[3]

Ion Sources

Matrix-assisted laser desorption ion sources all consist of a sample stage (or probe) that is used to carry the protein-doped matrix crystals into the vacuum system of the mass spectrometer. The lasers that produce protein ion ablation are pulsed: intense laser light is emitted for a brief time (<10 nsec). All commercial ion sources use ultraviolet pulsed lasers, either a nitrogen laser ($\lambda = 337$ nm) or a Q-switched neodymium : yttrium aluminum garnet (Nd : YAG) laser, with its fundamental wavelength either tripled ($\lambda = 354$ nm) or quadrupled ($\lambda = 266$ nm). The light emitted by the laser is focused onto the protein-doped matrix crystals in the vacuum system to produce an illumination fluence of approximately 20 mJ/cm^2. Most of the commonly used matrices will absorb sufficient light to produce protein ions at all of these wavelengths. Longer wavelength lasers are currently favored because they reduce the amount of absorption in the analyte protein molecule. Some results have been described using infrared lasers to desorb proteins,[25] but no commercial ion source uses this method.

The relationship between protein ion current and laser fluence is highly nonlinear. At low laser fluences (<10 mJ/cm^2), no protein ions are produced. If the laser fluence is increased, a point will be reached when protein ions suddenly appear ("threshold" fluence, ϕ_{th}). Increasing the laser fluence above the threshold value leads to a rapid increase in protein ion current.[26] It has been found empirically that mass analyzers and detectors work best when the laser fluence is near the threshold value.[27] Different matrices have different threshold fluences and fluence-to-protein ion-current curves [ϕ_{th}(α-cyano-4-hydroxycinnamic acid) $<$ ϕ_{th}(sinapic acid) $<$ ϕ_{th}(gentisic acid)]. For these reasons, the illumination fluence must be easily adjustable so that the operator can set a fluence appropriate for a particular sample.

Mass Analyzers and Detectors

Most currently available mass spectrometers with matrix-assisted laser desorption ion sources use time-of-flight mass analyzers and ion-to-electron conversion detectors. Laser desorption ion sources have also been used successfully with magnetic sector,[28] quadrupole ion trap,[29] and Fourier transform ion cyclotron resonance[30] mass analyzers.

Time-of-flight mass analyzers are easy to understand. Linear time-of-flight analyzers consist of a long, straight, empty flight tube with an ion source on one end and flat detector on the other end. A short-duration pulse of ions emerges from the ion source after acceleration to a common kinetic energy. Light ions move more quickly down the flight tube than heavy ions and therefore strike the detector first. If the signal coming out of the detector is plotted as a function of time, peaks corresponding to ions of different mass appear at different times. The mass-to-charge ratio

[25] K. Strupat, M. Karas, F. Hillenkamp, C. Eckerstkorn, and F. Lottspeich, *Anal. Chem.* **66,** 464 (1994).
[26] W. Ens, Y. Mao, F. Mayer, and K. G. Standing, *Rapid Commun. Mass Spectrom.* **5,** 117 (1991).
[27] R. C. Beavis and B. T. Chait, *Rapid Commun. Mass Spectrom.* **3,** 233 (1989); R. C. Beavis and B. T. Chait, *Anal. Chem.* **62,** 1836 (1990).
[28] J. A. Hill, R. S. Annan, and K. Biemann, *Rapid Commun. Mass Spectrom.* **5,** 395 (1991).
[29] V. M. Doroshenko, T. J. Cornish, and R. J. Cotter, *Rapid Commun. Mass Spectrom.* **6,** 153 (1992); K. Jonscher, G. Currie, A. L. McCormack, and J. R. Yates III, *Rapid Commun. Mass Spectrom.* **7,** 20 (1993); J. C. Schwartz and M. E. Bier, *Rapid Commun. Mass Spectrom.* **7,** 27 (1993); D. M. Chambers, D. E. Goeringer, S. A. McLuckey, and G. L. Glish, *Anal. Chem.* **65,** 14 (1993); J. Qin and B. T. Chait, *J. Am. Chem. Soc.* **117,** 5411 (1995); J. Qin, R. J. J. M. Steenvoorden, and B. T. Chait, *Anal. Chem.* (in press).
[30] See, e.g., Y. Li, R. T. McIver, Jr., and R. L. Hunter, *Anal. Chem.* **66,** 2097 (1994).

(m/z) corresponding to a particular peak can be calculated using simple equations and the time-to-intensity data can be converted into an m/z-to-intensity histogram.

Ions do not emerge from an MALDI ion source with exactly the same kinetic energy. This kinetic energy spread results in relatively low mass resolutions in linear time-of-flight analyzers. This effect can be partially corrected using an ion mirror to reflect the ions back toward a detector positioned near the ion source.[31] An analyzer with an ion mirror is referred to as a "reflectron." The mass resolution and mass accuracy of a reflectron analyzer can be superior to linear analyzers for $m/z < 10,000$. Above this mass-to-charge ratio, the performance of the two types of analyzers is comparable because large ions decay into smaller fragments in flight. This unimolecular decay broadens the signal in reflectrons but has little effect on linear analyzers. Similar improvements in resolution can be obtained using a pulsed extraction ion source[31a,31b] and a linear time-of-flight analyzer.

Ions are converted into signals using ion-to-electron conversion detectors. These detectors have a flat plate surface that is exposed to the incoming protein ions. The ions strike the surface (either a metal or metal oxide coating) and the collision results in the prompt emission of electrons and small negative ions. These electrons and ions are accelerated to another surface that produces electrons exclusively. These electrons are amplified using conventional electron multipliers (microchannel plates or discreet dynode multipliers). *Note:* Multipliers that are easily paralyzed by intense protein or matrix ion signals should not be used.[32]

Mass Calibration

After preparing a sample and obtaining a mass spectrum, one is confronted with the task of assigning masses to the peaks in the spectrum. Time-of-flight mass spectra are recorded as intensity versus time-of-flight tables. The time-of-flight data are converted into mass-to-charge ratio (m/z) data by using a simple equation:

$$(m/z)^{1/2} = at + b$$

The constants a and b can be calculated from the known geometry of an ion source and mass analyzer; however, in practice this calculation is rarely

[31] X. Tang, R. C. Beavis, W. Ens, F. Lafortune, B. Schueler, and K. G. Standing, *Int. J. Mass Spectrom. Ion Processes,* **85,** 43 (1988).

[31a] R. S. Brown and J. J. Lennon, *Anal. Chem.* **67,** 1998 (1995).

[31b] M. L. Vestal, P. Juhasz, and S. A. Martin, *Rapid Commun. Mass Spectrom.* **9,** 1044 (1995).

[32] R. C. Beavis and B. T. Chait, Methods and mechanisms for producing ions from large molecules. NATO ASI Series B. *in* "Physics" (K. G. Standing and W. Ens, eds.), Vol. 269, p. 227. Plenum Press, New York, 1991.

performed. A simpler and more accurate method of determining a and b is to select two peaks of known m and z and derive the constants from their measured flight times. These values of a and b can be used to calibrate other spectra obtained using the same instrument and voltage settings. Applying a set of calibration constants measured in a previous experiment is known as using an "external" calibration.

External calibrations are not always accurate. If experimental conditions change slightly, the use of an external calibration will result in mass assignments that are slightly incorrect, producing a systematic error in mass assignment. This problem can be corrected using known peaks in each spectrum to calculate a calibration for that particular data set, a procedure known as using an "internal" calibration. Internal calibrations are inherently more accurate than external calibrations; however, they require at least two peaks of accurately known mass in every spectrum, which can be difficult to arrange.

The most popular method for the internal calibration of a protein mass spectrum is to mix a known protein into the sample solution. The calibrant protein is added at a concentration such that it produces a signal with an intensity similar to the analyte of interest. Proteins produce ions with different charge states, usually $z = +1$ and $+2$, although there may be signals from ions with $z \gtrsim +3$. These charge states produce signals at predictable m/z values. For a protein of mass M, the series of m/z peaks produced are $z = 1, m/z = M + 1; z = 2, m/z = (M + 2)/2; z = 3, m/z = (M + 3)/3$, etc. Therefore a single protein calibrant usually produces at least two peaks that can be used to calculate the calibration constants. It may be necessary to try several concentrations of calibrant before a spectrum is obtained with both the calibrant and analyte proteins at similar intensities so that a good calibration can be obtained.

Calibrant proteins most commonly used include bovine insulin, horse muscle myoglobin, horse muscle cytochrome c, and bovine trypsinogen. These proteins are used because they normally occur in only one isoform and pure samples can be obtained from commercial suppliers at low cost. Also, they do not oxidize easily and commercial samples do not contain significant protease activity.

It must be pointed out that good mass assignments depend on good signals. If a signal is weak and noisy, the mass assignment will not be as accurate as for intense peaks with large signal-to-noise ratios. The signal-to-noise ratio can be improved by adding together digitized signal transients (spectra) from many individual laser shots taken on the same sample spot. A general rule of thumb is that digitized signal transients from 50 to 100 individual laser shots should be summed to obtain an acceptable signal-to-noise ratio.

Mass Spectrometric Strategies for Studying Proteins

Correlation of Processed Proteins with Their Genes
  by Molecular Mass Measurements

A simple, accurate molecular mass measurement can provide highly useful data concerning a protein sample. The resulting mass spectrum can give information concerning (1) the correctness of a hypothetical structure, (2) the purity of the protein preparation as well as data on the masses of any impurities that are present, (3) protein microheterogeneity, and (4) the presence and likely identities of posttranslational modifications.

Once the cDNA sequence of a gene has been determined, a simple, accurate measurement of the molecular mass of the corresponding protein can provide a host of valuable information. If the measured mass of the protein agrees with that calculated from the gene sequence, it is likely that the deduced sequence is correct, the amino and carboxyl terminals of the mature protein have been correctly assigned, and the protein contains no posttranslationally modified amino residues. However, a difference between the measured and predicted molecular masses implies either an error in the cDNA-deduced sequence or a posttranslational modification of the protein. The power of accurate molecular mass determination in this context is illustrated by consideration of ARPP-16, a cAMP-regulated phospho-protein from the soluble fraction of bovine caudate nuclei.[33] Analysis by SDS–PAGE of the protein yielded an apparent molecular mass of 16,000 Da. The amino terminus was chemically blocked and could not be sequenced by Edman degradation. Inspection of the cDNA sequence suggested a molecular mass considerably lower than that inferred from the SDS–PAGE analysis.[33] The MALDI mass spectrum obtained from the unphosphorylated form of ARPP-16 is shown in Fig. 2. The mass spectrum exhibits two intense peaks corresponding to the singly and doubly protonated intact protein. The average of the two measurements yields a molecular mass of 10,709 Da. This measured mass is much lower than that determined by SDS–PAGE but close to that inferred from the cDNA sequence with an amino-terminal methionine (calculated molecular mass of 10,665 Da). The difference between the measured and calculated masses is 44 ± 2 Da, a value consistent with the presence of an acetyl-blocking group (calculated mass difference of 42 Da) at the amino terminus. Taken together with the cDNA sequence data and partial amino acid sequence data, the simple mass measurement provides compelling confirmation of

---

[33] A. Horiuchi, K. R. Williams, T. Kurihara, A. C. Nairn, and P. Greengard, *J. Biol. Chem.* **265,** 9576 (1990).

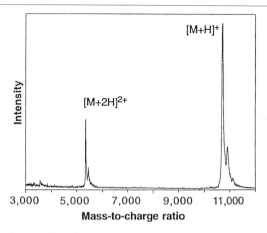

Fɪɢ. 2. Positive ion matrix-assisted laser desorption mass spectrum of the protein ARPP-16, obtained from a matrix of sinapic acid using the dried-droplet sample preparation method (see text). M designates the intact molecule. The small satellite peaks to the right of the main peaks arise from a photochemically produced adduct (see text). [Reproduced from B. T. Chait and S. B. H. Kent, *Science* **257,** 1885 (1992).]

the predicted primary structure of ARPP-16 as well as information concerning a subtle modification of the protein. This detailed information was obtained rapidly (in less than 15 min) from a total quantity of 1 pmol of protein.

Frequently, differences are observed between the measured and calculated molecular masses that are more difficult to interpret than in the example of ARPP-16 given above. The difficulty in defining the source of these mass differences arises from a number of possible sources. Some of these are as follows.

   The mass spectrometric response may be too weak for accurate definition of the mass spectral peak centroids. A weak mass spectrometric response is normally the result of insufficient protein sample, failure to properly dissolve and manipulate the sample, or the presence of impurities.

   The uncertainty in the mass determination accuracy increases as a function of the molecular mass of the protein. For proteins with molecular masses <30,000 Da, the mass accuracy is in the range of 0.01–0.1%. For proteins with molecular masses >30,000 Da, the mass accuracy is in the range of 0.03–0.3%.

   The protein may have a high degree of heterogeneity and the individual components may not be resolved in the mass spectrum. This situation is commonly encountered for glycoproteins.

In all such cases, interpreting the source of mass differences can often be facilitated by an analytical strategy that involves degradation of the protein by chemical or enzymatic means and measurement by MALDI–MS of the total mixture of peptide products (see below).

Identification of Sites of Proteolytic Cleavage of Proteins

An accurate molecular mass determination provides a method of choice for the rapid and reliable identification of sites of proteolytic processing or degradation in proteins. If the protease responsible for the cleavage(s) under study is known and has high specificity, a simple molecular mass determination may be sufficient to yield an unambiguous identification of the portion of the protein produced. If, however, the activity of the protease is unknown or broad, it may be advantageous to define the amino terminus of the cleaved protein by Edman sequencing of a few residues. The carboxyl terminus (and hence the whole portion of the protein of interest) can then be defined without ambiguity by an accurate molecular mass measurement.

The principle of the MS analysis described above is illustrated by consideration of a preparation of streptavidin from *Streptomyces avidinii* that has undergone proteolytic cleavage.[34] The MALDI mass spectrum of this preparation (obtained from Boehringer Mannheim, Indianapolis, IN) is shown in Fig. 3a[35] together with the sequence of the intact precursor subunit (Fig. 3b). Edman degradation of the preparation defines the amino-terminal residues as EAGIT..., showing that the amino terminus begins at residue 14 (Fig. 3b). The molecular mass of the dominant component in the preparation was measured to be 12,969 ± 2 Da. With the knowledge that the amino terminus begins at residue 14, we can calculate (using the known sequence of streptavidin) the molecular masses of the modified protein with all possible C-terminal truncations. The molecule containing residues 14–136 has a calculated molecular mass of 12,971 Da, in good agreement with the measured value. Thus the accurate molecular mass determination has defined unambiguously the C terminus of this modified streptavidin preparation. The sequence of the modified streptavidin is underlined in Fig. 3b. An analogous strategy can be used to define the N terminus if information concerning the C terminus is available. Such a strategy is of particular use for N-terminally blocked proteins.

Assessing Homogeneity and Purity of Protein Preparations

It is often necessary to assess the homogeneity and purity of native, recombinant, or synthetic protein preparations. MALDI–MS provides one

[34] E. A. Bayer, H. BenHur, Y. Hiller, and M. Wilchek, *Biochem. J.* **258,** 369 (1989).
[35] B. T. Chait, *Structure* **2,** 465 (1994).

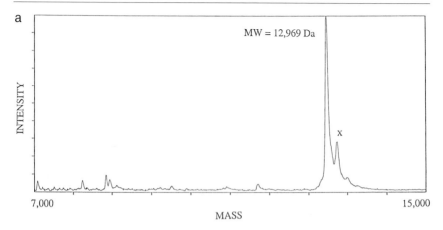

b
```
                           14
DPSKDSKAQVSAAEAGITGTWYNQLGSTFIVTAGADGALTGTYESAV

GNAESRYVLTGRYDSAPATDGSGTALGWTVAWKNNYRNAHSATTW
                                    136
SGQYVGGAEARINTQWLLTSGTTEANAWKSRLVGHDTFTKVKPSAA

SIDAAKKAGVNNGNPLDAVQQ
```

FIG. 3. Mass spectrometry to identify proteolytic cleavage sites. (a) The positive ion MALDI–MS of proteolytically modified streptavidin from *Streptomyces avidinii*, using a sinapic acid matrix and the dried droplet sample preparation method. The peak labeled "x" is an artifact of the MALDI process (see text) and does not arise from an impurity in the sample. However, several small peaks observed in the mass spectrum below mass 12,000 Da do arise from low-abundance impurities in the sample. Interestingly, these impurities were observed to be absent from the mass spectrum of the protein after crystallization (S. E. Darst and B. T. Chait, unpublished data, 1995). (b) The sequence of streptavidin. The sequence of the proteolytically modified streptavidin is underlined. [Reproduced from B. T. Chait, *Structure* **2**, 465 (1994). Copyrighted by Current Science.]

of the most rapid, straightforward, and informative routes for such assessment, which is usually carried out by molecular mass determination followed by more detailed peptide mapping (if necessary). Examples include verification of recombinant[36,37] and synthetic proteins,[38,39] assessment of the

[36] T. M. Billegi and J. T. Stults, *Anal. Chem.* **65**, 1709 (1993).
[37] J. J. M. DeLlano, W. Jones, K. Schneider, B. T. Chait, J. M. Manning, G. Rodgers, L. J. Benjamin, and B. Weksler, *J. Biol. Chem.* **268**, 1 (1993).
[38] R. C. D. Milton, S. C. F. Milton, and S. B. H. Kent, *Science* **256**, 1445 (1992).
[39] B. T. Chait, R. Wang, R. C. Beavis, and S. B. H. Kent, *Science* **261**, 89 (1993).

integrity and purity of proteins used to grow crystals for X-ray diffraction measurements,[35,40] and comparison of different forms of a protein having distinct biological activities.

The utility of MALDI–MS for assessing the integrity and purity of a protein preparation is illustrated by examination of an artificial protein expressed in *Escherichia coli*.[41] A synthetic DNA was constructed by multimerizing oligonucleotide fragments encoding two copies of the repeated undecapeptide of the target protein **I**.

ASMTGGQQMGRDPMFKYSRDPMG-[AGAGAGAGPEG]$_{14}$-ARMHIRPGRYQLDPAANKARKEAELAAATAEQ

I

Amino acid analysis confirmed the composition of purified **I**, and Coomassie blue staining of SDS–PAGE gels revealed no contaminating proteins—the product was observed to migrate as a single tight band. However, the molecular mass reported by SDS–PAGE was found to be 43,000 Da, which is more than twice the expected molecular mass of 17,207 Da. To resolve the apparent molecular weight discrepancy, the protein was analyzed by MALDI–MS.

The mass spectrum of **I** is shown in Fig. 4A. Although the measured m/z (17,264 $\pm$ 2) shows the electrophoretically determined mass to be grossly in error, the observed value remains significantly different from the predicted m/z of 17,208. To determine the origin of the difference, the sequence of a 741-base pair DNA fragment containing the protein-coding region of the expression plasmid was determined. This analysis revealed C $\rightarrow$ T transitions in codons 96 and 101, causing two alanine $\rightarrow$ valine substitutions in the expressed protein. The calculated m/z of the protein with the altered repetitive sequence **II** is 17,264—the experimentally determined value.

-[AGAGAGAGAGPEG]$_6$[AGAGAGVGPEG][VGAGAGAPEG][AGAGAGAGAGPEG]$_6$-

II

Signals arising from low molecular weight polypeptides are also visible in the spectrum shown in Fig. 4A. Analysis of these signals, expanded in Fig. 4B, shows that the peak at m/z 5730 corresponds to fragment **III**, which consists of the intact N-terminal sequence of **I** followed by four copies of the repeating undecapeptide. The calculated mass of the molecular ion derived from **III** is 5733, in agreement with the observed value.

[40] K. L. Clark, E. D. Halay, E. Lai, and S. K. Burley, *Nature* (*London*) **364,** 412 (1993); X. Xie, T. Kokubo, S. L. Cohen, U. A. Mirza, A. Hoffman, B. T. Chait, R. G. Roeder, Y. Nakatani, and S. K. Burley, *Nature* (in press).
[41] R. C. Beavis, B. T. Chait, H. S. Creel, M. J. Fournier, T. L. Mason, and D. A. Tirrell, *J. Am. Chem. Soc.* **114,** 7384 (1992).

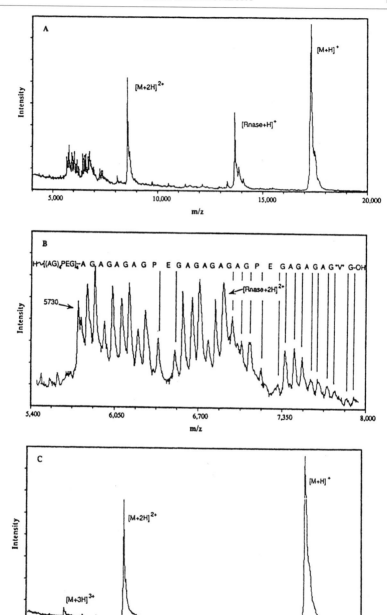

-ASMTGGQQMGRDPMFKYSRDPMG-[AGAGAGAGPEG]$_4$-
III

More striking, however, is the fact that each succeeding signal can be rationalized by addition of a single amino acid residue, proceeding in the N- to C-terminal direction along sequence **I**. Thus, one can read portions of the periodic sequence directly from the mass spectrum, including one of the substituted valines. Although the origin of these fragments has not yet been determined, they probably arise from the action of exo- and endopeptidases, either *in vivo* and *in vitro* or both. The fragments are easily removed by dialysis as shown in Fig. 4C, which shows only the singly, doubly, and triply ionized molecular ions of the intact protein.

Subsequent to the discovery of the contaminating fragments, the sample was rerun on a 25-cm 15% polyacrylamide gel, and the proteins were visualized by Coomassie blue staining. The fragments were not detected by this method, even when the sample was overloaded (up to 50 μg of protein per lane).

Defining Sites and Nature of Posttranslational Modifications by Peptide Mapping

To determine the modifications in a protein, its accurately measured molecular mass is compared with the molecular mass calculated from the cDNA sequence of the corresponding gene or the amino acid sequence of the protein. The difference between the measured and calculated molecular masses provides information concerning the possible nature of the modification(s). If more detailed information concerning the nature and site of a modification is required, it is usually necessary to subject the protein to further analysis, involving enzymatic or chemically induced degradation of the protein followed by further mass spectrometric measurement of the resulting peptide fragments. Partial definition of a posttranslational modification is illustrated by reference to the *a*-chain of the giant extracellular

---

FIG. 4. Analysis of an artificial protein by MALDI–MS. (A) Positive ion spectrum of the target protein **I** with bovine ribonuclease A as an internal calibrant. The dried-droplet sample preparation method was used with a sinapic acid matrix. (B) Spectrum expanded in the region of low molecular mass contaminants. The peaks progress starting with N-terminal fragment **III** (calculated *m/z* 5733, observed *m/z* 5730) with the sequential addition of amino acid residues through three repeats of the target repetitive sequence. The substituted valine is apparent in the seventh repeat. (C) Mass spectrum of the protein sample after removal of the contaminating fragments by dialysis. (Reprinted from R. C. Beavis *et al.*, *J. Am. Chem. Soc.* **114**, 7384. Copyright 1992, American Chemical Society.)

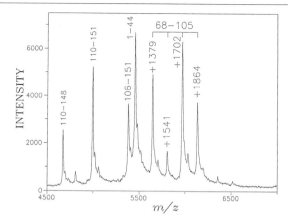

FIG. 5. Positive-ion MALDI–MS spectrum of the *a* chain of the giant extracellular hemoglo-bin from the earthworm *Lumbricus terrestris* after reduction with dithiothreitol and digestion with endoproteinase Lys-C.[43] The dried droplet sample preparation method was used with a matrix of α-cyano-4-hydroxycinnamic acid. Peptide 68–105 was found to be glycosylated.

hemoglobin of the earthworm, *Lumbricus terrestris*.[42] Measurement of the molecular mass of the *a*-chain revealed that it was heterogeneous, having three major components (a_1, 19,386 ± 15 Da; a_2, 19,221 ± 10 Da; and a_3, 18,901 ± 15 Da) with significantly higher molecular masses than that calculated from the known sequence (17,525 Da). The heterogeneity and the differences between these values suggest that the protein is glycosylated and that a_1 has one more hexose than a_2, which has two more hexoses than a_3. More detailed information concerning the site of glycosylation was obtained by reduction of the *a*-chain with dithiothreitol and digestion with Lys-C endoproteinase. The mass spectrum of the unfractionated digest (between *m/z* 4500 and 7000) is given in Fig. 5.[43] The various fragments can be readily assigned because the digesting enzyme is highly specific and because the interpretation of the mass spectrum is simplified by the dominance of singly charged ions. Comparison of the experimentally deter-mined peptide fragment masses with the masses of the unmodified frag-ments calculated from the sequence reveals that the protein is glycosylated on the peptide spanning residues 68–105. This detailed analysis of proteo-lytic fragments reveals the presence of four components that differ from one another by integral numbers of hexoses (hexose residue mass of 162 Da).

[42] D. W. Ownby, H. Zhu, K. Schneider, R. C. Beavis, B. T. Chait, and A. F. Riggs, *J. Biol. Chem.* **268**, 13539 (1993).
[43] K. Schneider, D. W. Ownby, H. Zhu, R. C. Beavis, A. F. Riggs, and B. T. Chait, *in* "Proceedings of 41st ASMS Conference in Mass Spectrometry, San Francisco, 1993," p. 900a.

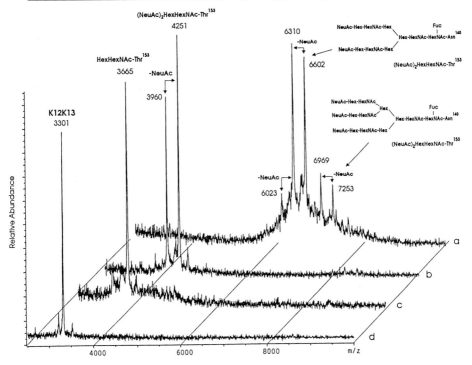

FIG. 6. Positive-ion MALDI–mass spectra of the glycopeptide encompassing residues 138–163 of recombinant human macrophage colony-stimulating factor containing both N- and O-linked oligosaccharides prior to and after treatment with a series of glycosidases. (a) The intact peptide, which was then sequentially treated with (b) PNGase F, (c) neuraminidase, and (d) O-glycanase. The major carbohydrate structures attached to the peptide are shown for each step. The extent of glycosylation and the proposed carbohydrate compositions may be determined on the basis of the observed mass differences from one step to the next and the known specificity of the glycosidase. (Reprinted with permission from M. C. Huberty *et al.*, *Anal. Chem.* **65**, 2791. Copyright 1993, American Chemical Society.)

The difference between these masses and the sequence-derived mass may be accounted for by glycosylation involving 8–12 hexoses. Peptide 68–105 contains three possible sites for O-linked glycosylation. Precise definition of the site of glycosylation requires further analysis, e.g., proteolytic subdigestion or sequencing. Likewise, more detailed definition of the composition and structure of the attached carbohydrate requires further analysis. An example of an analysis that provides such compositional information is given below.

Information concerning the carbohydrate portion of a protein can be obtained by mass determination of component glycopeptides prior to and after treatment with specific glycosidases. This approach is illustrated by analysis of a glycopeptide obtained by digestion of recombinant human macrophage colony-stimulating factor using *Achromobacter* protease I.[44] The MALDI–MS spectrum of the intact glycopeptide (containing both N- and O-linked oligosaccharides) encompassing residues 138–163 of the protein consists of a complex distribution of peaks in the range *m/z* 6000–8000 (Fig. 6a). The two most intense peaks, at *m/z* 6310 and 6602, differ by 292 Da, suggesting the presence of neuraminic acid (291 Da). Treatment of the glycopeptide with PNGase-F (*Flavobacterium meningosepticum*) produces two main products with *m/z* 3960 and 4251 (Fig. 6b), having masses 2351 Da lower than the major components in Fig. 6a. This observation demonstrates that the removed carbohydrate is asparagine linked and provides a measure of the masses of components of the complex carbohydrate. Treatment of the residual glycopeptide with neuraminidase (*Arthrobacter ureafaciens*) removes the neuraminic acid moieties to produce a homogenous product at *m/z* 3665 (Fig. 6c). Finally, treatment of this product with *O*-glycanase (*Streptococcus pneumoniae*) reduces the molecular mass by 364 Da to *m/z* 3301 (Fig. 6d), revealing the presence of a site of O-glycosylation. The molecular mass of the residual peptide is consistent with the calculated mass of the protonated peptide with a deamidated asparagine residue and no further modifications. Therefore, it appears that all of the modifications have been removed by the glycosidase treatments. Compositional analysis of the oligosaccharides can be deduced from these measured mass differences either *ab initio* or by reference to compositional databases.[44]

Peptide Maps of Proteins Eluted from One- and Two-Dimensional Gels

The high sensitivity of MALDI–MS permits the measurement of digestion products generated from proteins separated by one- and two-dimensional (2D) gel electrophoresis.[18,45,48] The combination of electrophoresis

[44] M. C. Huberty, J. E. Voth, W. Yu, and S. A. Martin, *Anal. Chem.* **65,** 2791 (1993).

[45] Y. K. Wang, P.-C. Liao, J. Allison, D. A. Gage, P. C. Andrews, D. M. Lubman, S. M. Hanash, and J. R. Strahler, *J. Biol. Chem.* **268,** 14269 (1993); W. Henzel, T. B. Billeci, J. T. Stults, S. C. Wong, C. Grimley, and C. Watanabe, *in* "Techniques in Protein Chemistry V" (J. W. Crabb, ed.), p. 3. Academic Press, San Diego, CA, 1994; S. Geromanos, P. Casteels, C. Elicone, M. Powell, and P. Tempst, *in* "Techniques in Protein Chemistry V" (J. W. Crabb, ed.), p. 143. Academic Press, San Diego, CA, 1994; S. D. Patterson and R. Aebersold, *Electrophoresis* **16,** 1791 (1995).

[46] Deleted in proof.

[47] Deleted in proof.

[48] W. Zhang, A. J. Czernik, T. Yungwirth, R. Aebersold, and B. T. Chait, *Protein Sci.* **3,** 677 (1994).

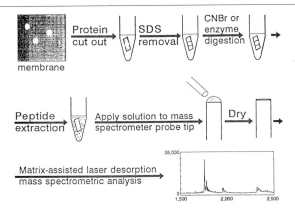

Fig. 7. Strategy employed for the mass spectrometric mapping of peptides generated from proteins electroblotted onto Immobilon CD membranes from two-dimensional electrophoretic gels. [Reprinted from W. Zhang *et al.*, *Protein Sci.* **3**, 677 (1994).]

and mass spectrometry is an effective means for analyzing relatively insoluble proteins and is particularly powerful for assessing the differences between protein isoforms that exhibit, for example, different phosphorylation states. One variation of this approach[48] involves electroblotting of proteins separated by 2D electrophoresis onto a membrane with a cationic surface. The isolated proteins are subjected to chemical and/or enzymatic degradation directly on the membrane, and the resulting unfractionated mixture of peptide fragments is extracted from the membrane into a solution that is compatible with analysis by MALDI–MS. Accurate mass determination of these peptide fragments provide a facile means for detecting the presence of modifications and for correlating such modifications with the differential mobility of different isoforms of a given protein during 2D electrophoresis. The strategy employed for mapping peptides from proteins separated by 2D electrophoresis and electroblotted onto cationic membranes (Immobilon CD; Millipore, Bedford, MA) is outlined in Fig. 7.[48] Protein spots visualized by reverse staining of the blotting membrane are excised, washed, and subjected to chemical or enzymatic digestion. The resulting mixture of peptide fragments is extracted from the membrane, combined with the laser desorption matrix, and analyzed without fractionation.[48a,48b]

Protein Identification from Peptide Maps

A need exists for rapid, sensitive means of identifying proteins that have been isolated on the basis of their biological activity, response to

[48a] A. Sherchenko, M. Wilm, O. Vorm, and M. Mann, *Anal. Chem.* **68**, 850 (1996).
[48b] J. Qin, Doctoral Dissertation, The Rockefeller University, 1986.

stimuli, or specific association with other biomolecules. Such identifications are now frequently made using partial protein sequence information. MALDI–MS provides the basis for a new strategy for protein identification that is fast, sensitive, and accurate.[18,49] This strategy involves mass determination of peptide fragments generated from the protein of interest by an enzyme or chemical reagent with high specificity, followed by screening of these masses against an appropriate database of peptide fragments. The fragment database is calculated from a protein database using the known properties of the cleavage reagent. Proteins can be correctly identified with a relatively limited set of proteolytic peptides, and the approach has sufficient sensitivity for use with proteins separated by 2D electrophoresis (see above). This approach has been improved by using tandem mass spectrometry of the proteolytic peptides to introduce additional constraints for the database search.[51a,52]

Probing Protein–Protein Interactions

The use of MALDI–MS in conjunction with affinity-based biochemical techniques provides potentially powerful new means for probing protein–protein and protein–ligand interactions.[52] The concept is illustrated by reference to a method for rapid mapping of linear epitopes in proteins that are bound by monoclonal antibodies.[53,54] The method consists of three steps (Fig. 8).

The approach is demonstrated through the mapping of a binding epitope in the peptide melittin against a monoclonal antibody that was previously determined to bind to an epitope located at residues 20–26 of melittin.[55,56] Partial digestion of melittin by endoprotease Lys-C yielded four peptide

[49] M. Mann, P. Hojrup, and P. Roepstorff, *Biol. Mass Spectrom.* **22,** 338 (1993); D. J. C. Pappin, P. Hojrup, and A. J. Bleashy, *Curr. Biol.* **3,** 327 (1993); P. James, M. Quadroni, E. Carapoli, and G. Gonnet, *Biochem. Biophys. Res. Commun.* **195,** 58 (1993).
[50] Deleted in proof.
[51] Deleted in proof.
[51a] M. Mann and M. Wilm, *Anal. Chem.* **66,** 4390 (1994); J. R. Yates III, J. K. Eng, A. L. McCormack, and D. Schieltz, *Anal. Chem.* **67,** 1426 (1995); K. R. Clauser, S. C. Hall, D. M. Smith, J. W. Webb, L.-E. Andrews, H. M. Tran, L. B. Epstein, and A. L. Burlingame, *Proc. Natl. Acad. Sci. U.S.A.* **92,** 5072 (1995).
[52] T. W. Hutchens and T-T. Yip, *Rapid Commun. Mass Spectrom.* **7,** 576 (1993).
[53] D. Suckau, J. Kohl, G. Karwath, K. Schneider, M. Casaretto, D. Bitter-Suermann, and M. Przybylski, *Proc. Natl. Acad. Sci. U.S.A.* **87,** 9848 (1990).
[54] Y. Zhao and B. T. Chait, *Anal. Chem.* **66,** 3723 (1994); Y. Zhao, T. W. Muir, S. B. H. Kent, E. Tischer, J. M. Scardina, and B. T. Chait, *Proc. Natl. Acad. Sci. U.S.A.* (in press.)
[55] T. P. King, L. Kochoumian, and A. Joslyn, *J. Immunol.* **133,** 2668 (1984).
[56] P. F. Fehlner, R. H. Berg, J. P. Tam, and T. P. King, *J. Immunol.* **146,** 799 (1991).

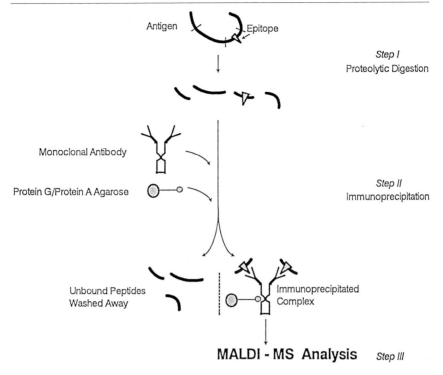

FIG. 8. Three-step strategy for linear epitope mapping. In the first step, an antigen–protein complex is digested by a proteolytic enzyme to produce an appropriate set of peptide fragments. In the second step, peptide fragments containing the linear epitope are selected and separated from the pool of peptide fragments by immunoprecipitation with the monoclonal antibody. In the final step, the immunoprecipitated peptides are identified by MALDI–MS. The method allows the rapid determination of antigenic sites without tedious peptide synthesis or protein mutagenesis. [Reprinted with permission from Y. Zhao and B. T. Chait, *Anal. Chem.* **66,** 3723 (1994).]

fragments that gave intense mass spectral peaks (Fig. 9A). The measured molecular masses of these peptides correspond to the molecular masses predicted for melittin fragments 8–23, 8–26, 1–23, and 1–26 (Fig. 9C). The other two peaks in the spectrum (designated i_1 and i_2) arise from unidentified impurities that were present in the melittin sample. After immunoprecipitation with the antibody, the two impurities and peptide fragments 1–23 and 8–23 were washed away, while fragments 1–26 and 8–26 were identified in the immunoprecipitation complex (Fig. 9B). To further define the antigenic site, an analogous experiment was carried out after chymotrypsin digestion of melittin. The experiment showed that only one of the three chymotryptic peptides was bound by the antibody (Fig. 9C). Inspection of

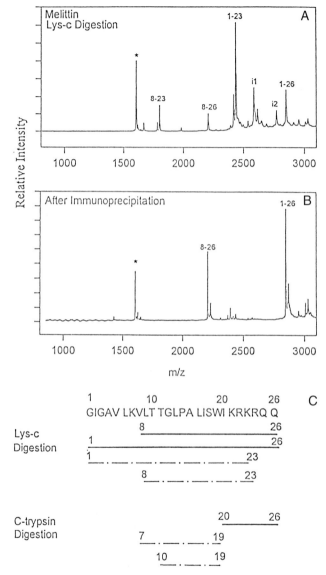

FIG. 9. Positive ion MALDI–MS spectra of (A) peptide fragments produced by endopro-tease Lys-C digestion of melittin and (B) peptide fragments isolated by immunoprecipitation with the anti-melittin monoclonal antibody (see text). The peak labeled with an asterisk (*) corresponds to the internal calibrant dynorphin A 1–13, and peaks i_1 and i_2 correspond to unidentified impurities. (C) Amino acid sequence of melittin. The peptide fragments produced by endoprotease Lys-C and chymotrypsin digestion are indicated by lines. The solid lines represent those peptide fragments that were bound by the anti-melittin monoclonal antibody, and the dashed lines represent those that were not bound. [Reprinted with permission from Y. Zhao and B. T. Chait, *Anal. Chem.* **66,** 3723 (1994).]

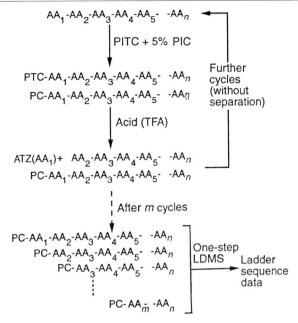

FIG. 10. Protein ladder sequencing principle exemplified by the generation of a set of sequence-determining fragments from an intact peptide chain with controlled ladder-generating chemistry. A stepwise degradation is carried out with a small amount of terminating agent present in the coupling step. In this case, 5% phenylisocyanate (PIC) was added to the phenylisothiocyanate (PITC). The phenylcarbamyl (PC) peptides formed are stable to the trifluoroacetic acid (TFA) used to cyclize and cleave the terminal amino acid (AA) from the phenylthiocarbamyl (PTC) peptide. Successive cycles of ladder-generating chemistry are performed without intermediate isolation or analysis of released amino acid derivatives. Finally, the mixture of PC peptides is read out in one step by MALDI–MS. [Reprinted with permission from B. T. Chait and S. B. H. Kent, *Science* **257,** 1885 (1992). © AAAS.]

the combined data (Fig. 9C) demonstrates that the region encompassing residues 20–26 is sufficient for binding to this monoclonal antibody.

Probing Protein Structure by Wet Chemistry in Combination with MALDI–MS

A powerful set of strategies for probing structural aspects of proteins uses the combination of appropriate wet chemical manipulation of the proteins followed by MALDI–MS measurements of the resulting chemical changes. Examples include strategies for amino acid sequencing; counting acid groups, amino groups, thiols, etc.; determining the solvent accessibility of particular amino acid side groups; chemical cross-linking for determining

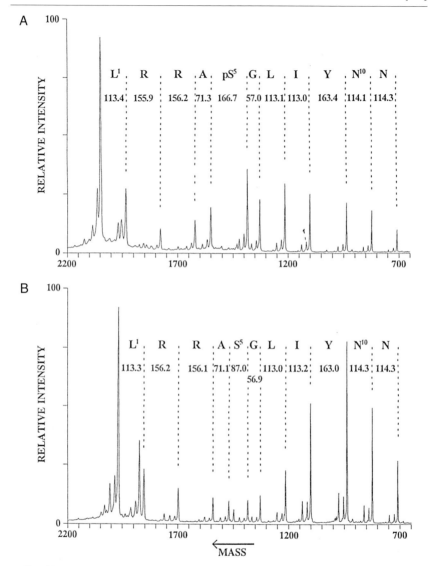

FIG. 11. Protein ladder sequencing of the 16-residue synthetic peptide: Leu-Arg-Arg-Ala-Ser(P$_i$)-Gly-Leu-Ile-Tyr-Asn-Asn-Pro-Leu-Met-Ala-Arg-amide. (A) Phosphorylated peptide. (B) Unphosphorylated peptide. Each peptide sample was subjected to 10 cycles of ladder-generating chemistry. Data defining the 11 N-terminal residues are shown. The Ser(P$_i$) residue is characterized by a mass difference of 166.7 Da observed in positive five (Ser, calculated residue mass 87.1 Da; Ser(P$_i$), calculated residue mass 167.1 Da). There is no evidence for loss of phosphate. [Reprinted with permission from B. T. Chait and S. B. H. Kent, *Science* **257,** 1885 (1992). © AAAS.]

the spatial relationship between proteins; and the footprinting of associated molecules on proteins. We illustrate the utility of such wet chemical manipulation in conjunction with MALDI–MS, using the technique of protein ladder sequencing.[39]

Protein ladder sequencing[39] combines multiple steps of wet degradation chemistry with a final, single-step mass spectrometric readout of the amino acid sequence. First, a sequence-defining concatenated set of peptide fragments, each differing from the next by a single amino acid residue, is chemically generated in a controlled fashion. Second, MALDI–MS is used to read out the complete fragment set in a single operation as a "protein sequencing ladder" data set.

A concatenated set of peptide fragments can be generated in a controlled fashion by carrying out rapid stepwise degradation in the presence of a small amount of terminating reagent, as indicated in Fig. 10. A small proportion of peptide chain blocked at the amino terminus is generated at each cycle. A predetermined number of cycles is performed without intermediate separation or analysis of the released amino acid derivatives. The resulting mixture is read out in a single operation by MALDI–MS. The mass spectrum contains protonated molecule ions corresponding to each terminated polypeptide species present. The mass differences between consecutive peaks each correspond to an amino acid residue, and their order of occurrence in the data set defines the sequence of amino acids in the original peptide chain.

An example of the ladder sequence analysis is shown in Fig. 11 for both phosphorylated and unphosphorylated forms of a 16-residue peptide containing a serine residue that can be phosphorylated by 3′,5′-cyclic AMP-dependent protein kinase. After 10 cycles of ladder-generating chemistry on each form of the peptide, the 2 separate sequence-defining fragment mixtures were each read out by MALDI–MS. The protein ladder-sequencing method directly identified and located a phosphoserine at position 5 in the peptide. Such direct localization and identification of posttranslational modifications are of great potential utility for biological research.

Acknowledgments

The present work was supported in part by grants from the National Institutes of Health (RR00862 and GM38274) and by the Skirball Institute of New York University.

[23] Quadrupole Ion Trap Mass Spectrometry

By Jae C. Schwartz and Ian Jardine

Introduction

The quadrupole ion trap mass spectrometer (QITMS) is now established as a compact, highly sensitive, and extremely specific detector for gas chromatography (GC/MS). QITMS technology has undergone significant developments that now promise to provide similar attractive features for high-performance liquid chromatography (HPLC)/MS and high-performance capillary electrophoresis (HPCE)/MS for the molecular weight and structural analysis of biological macromolecules. These advances include the coupling to the QITMS of new ionization techniques for biological molecules, particularly electrospray ionization (ESI)[1,2] and matrix-assisted laser desorption ionization (MALDI)[3–5]; the extension of the m/z range of the QITMS[6] to accommodate the singly to multiply charged ions of biological macromolecules generated by ESI and MALDI; the discovery of the high mass resolution capability of the QITMS[7,8]; and the demonstration that efficient collision-induced dissociation (CID) in the QITMS with subsequent tandem mass spectrometry (MS/MS) of ions of biological macromolecules[9] provides important structural information such as sequence.[10] A number of reviews describe the history and development of QITMS technology, the operation and theory of QITMS, and numerous applications.[10–17]

[1] J. B. Fenn, M. Mann, C. K. Meng, S. F. Wong, and C. M. Whitehouse, *Science* **246,** 64 (1989).

[2] G. J. Van Berkel, G. L. Glish, and S. A. McLuckey, *Anal. Chem.* **62,** 1284 (1990).

[3] M. Karas and F. Hillenkamp, *Anal. Chem.* **60,** 2299 (1988).

[4] K. Jonscher, G. Currie, A. L. McCormack, and J. R. Yates III, *Rapid Commun. Mass Spectrom.* **7,** 20 (1993).

[5] J. C. Schwartz and M. E. Bier, *Rapid Commun. Mass Spectrom.* **7,** 27 (1993).

[6] R. E. Kaiser, Jr., R. G. Cooks, G. C. Stafford, Jr., J. E. P. Syka, and P. H. Hemberger, *Int. J. Mass Spectrom. Ion Processes* **106,** 79 (1991).

[7] J. C. Schwartz, J. E. P. Syka, and I. Jardine, *J. Am. Soc. Mass Spectrom.* **2,** 198 (1991).

[8] J. D. Williams, K. A. Cox, R. G. Cooks, and J. C. Schwartz, *Rapid Commun. Mass Spectrom.* **5,** 327 (1991).

[9] J. N. Louris, R. G. Cooks, J. E. P. Syka, P. E. Kelly, G. C. Stafford, Jr., and J. F. J. Todd, *Anal. Chem.* **59,** 1677 (1987).

[10] K. A. Cox, J. D. Williams, R. G. Cooks, and R. E. Kaiser, *Biol. Mass Spectrom.* **21,** 226 (1992).

[11] R. G. Cooks, G. L. Glish, S. A. McLuckey, and R. E. Kaiser, *Chem. Eng. News* **69,** 26 (1991).

[12] R. E. March and R. J. Hughes, "Quadrupole Storage Mass Spectrometry." John Wiley & Sons, New York, 1989.

[13] W. Paul, *Angew. Chem. Int. Ed. Engl.* **29,** 739 (1990).

To evaluate the capabilities of the QITMS for the structural analysis of biological macromolecules, we describe a peptide-sequencing study[18] on a large peptide (recombinant human growth hormone-releasing factor or rhGHRF; molecular weight 5040) utilizing a commercial ion trap mass spectrometer (the ITMS from Finnigan MAT (San Jose, CA), which is an MS/MS system) modified with an ESI source, high m/z range, and high resolution capabilities. To evaluate further the potential of the QITMS as an on-line detector for liquid separation systems, this modified ion trap was then coupled to an HPCE system.[19] HPCE was chosen because this separation technology has demanding sensitivity and MS scan speed requirements compared, for example, to HPLC. Analysis of peptides and proteins by MALDI coupled to this QITMS[5] is also briefly described, and other significant advantages of QITMS are discussed to indicate how this technology should play an important role in the high resolution separation and analysis of biological macromolecules.

Experimental Procedures

The extensively modified ITMS system configured for HPCE/ESI/MS/MS has been previously described[5,7,18–20] and is shown in Fig. 1. An ESI source of the Whitehouse/Fenn type[21,22] (see also [21] in this volume[22a]) is coupled to the ITMS using a modified vacuum chamber that allows for differential pumping between the source and the analyzer region. The ions from the ESI source are injected axially into the ion trap at the appropriate time by a gate lens. The apparatus used has modified electronics to apply higher than normal voltages to the end-cap electrodes (0–48 V), to access a range of fundamental radio frequency (RF) trapping frequencies (550–1100 kHz), and to allow computer control of the scan rate from approximately

[14] R. G. Cooks and R. E. Kaiser, Jr., *Acc. Chem. Res.* **23,** 213 (1990).

[15] R. E. March, *Int. J. Mass Spectrom. Ion Processes* **118/119,** 71 (1992).

[16] J. F. J. Todd, *Mass Spectrom. Rev.* **10,** 3 (1991).

[17] S. A. McLuckey, G. J. Van Berkel, D. E. Goeringer, and G. L. Glish, *Anal. Chem.* **6,** 689A and 737A (1994).

[18] J. C. Schwartz, I. Jardine, and C. G. Edmonds, *in* "Proceedings of the 40th ASMS Conference on Mass Spectrometry and Allied Topics, Washington, D.C.," p. 707. 1992.

[19] J. C. Schwartz and I. Jardine, *in* "Proceedings of the 40th ASMS Conference on Mass Spectrometry and Allied Topics, Washington, D.C.," p. 709. 1992.

[20] J. C. Schwartz and I. Jardine, *Rapid Commun. Mass Spectrom.* **6,** 313 (1992).

[21] C. M. Whitehouse, R. N. Dryer, M. Yamashita, and J. B. Fenn, *Anal. Chem.* **57,** 675 (1985).

[22] J. B. Fenn, M. Mann, C. K. Meng, S. F. Wang, and C. M. Whitehouse, *Mass Spectrom. Revs.* **9,** 37 (1990).

[22a] J. F. Banks, Jr. and C. M. Whitehouse, *Methods Enzymol.* **270,** Chap. 21, 1996 (this volume).

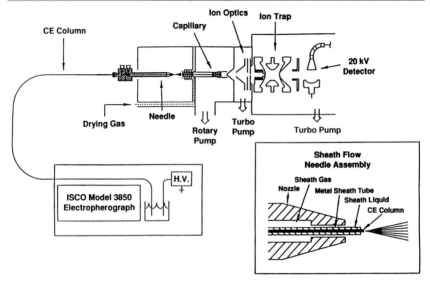

FIG. 1. Electrospray quadrupole ion trap system (ESI/QITMS) configured with a capillary electrophoresis device. The inset shows an enlargement of the electrospray needle tip to indicate the configuration of the CE column with associated liquid sheath needle and gas sheath nozzle.

360 msec/amu to 18 μsec/amu. To aid in the detection of high mass ions, a 20-kV off-axis dynode with a continuous dynode electron multiplier was also added to the ITMS.

In general, the liquid containing the analyte is introduced at a flow rate of 1–5 μl/min (from HPCE or from HPLC or by direct infusion) via a stainless steel needle set at ground potential. A potential of around 3400 V is applied between the needle and the capillary inlet, and this voltage causes the liquid to spray and break up into small, highly charged droplets. The droplets are accelerated toward the capillary, and the solvent in the droplets is completely evaporated by the countercurrent flow of warm nitrogen drying gas, producing desolvated charged analyte ions in the gas phase. This process of electrospray is carried out at atmospheric pressure, which allows convenient coupling of separations systems such as HPLC and HPCE, because a pressure drop across the separation system is not induced by the MS vacuum system.

The resulting dried and charged analyte ions along with the vaporized solvent and drying gas are pulled through the capillary by the first vacuum pump (rotary pump). The capillary skimmer interface (pressure around 1 torr) separates solvent vapor and drying gas from ions. The ions are accelerated through the lensing system, which is pumped by a turbomolecular pump (pressure around 1 millitorr) to an ion optical system including a gated ion trap lens that directs ions through the end cap of the ITMS. The voltage on this lens can be pulsed at an appropriate magnitude and frequency to allow or prevent ions from entering the ion trap, which is pumped by a second turbomolecular pump (pressure around 10^{-5} torr). Positive or negative ions can be analyzed by the appropriate choice of voltage polarities.

Ions are trapped in the ion trap by an RF field created by applying a large RF voltage (up to approximately 7500-V peak) to the center "ring" electrode, of the three-electrode structure of the ion trap. Ions enter and exit the ion trap through holes in the two end-cap electrodes. A cross-section of the ion trap including the hyperbolic surfaces of these electrodes is indicated in Fig. 1. Trapping of ions is assisted and overall performance enhanced by the presence of helium gas at a pressure of approximately 10^{-3} torr. This gas collisionally cools ions and forces them into the center of the ion trap. Ions above a certain minimum m/z remain trapped, cycling in a predominantly sinusoidal motion. The minimum m/z trapped, as well as the amplitude and frequency of the motion of the ions, are determined by the magnitude of the RF voltage. The mathematics used to describe the operation of the ion trap and ion motion[12] utilize the parameters q and β, where $q = 4V/(m/z)\omega^2 r_0^2$, V is the RF amplitude, m/z is the mass-to-charge ratio, ω is the RF frequency, and r_0 is the radius of the ring electrode. Thus, the q value for an ion is inversely proportional to its m/z value. The mathematics also tell us that ions are trapped only if their corresponding q value is less than 0.908, while they are ejected from the trap if their q value is greater than 0.908 ($q_{eject} = 0.908$). Thus, to generate a mass spectrum, the RF voltage is ramped up linearly, causing ions of successive m/z to have a value of 0.908, become unstable, and be ejected axially from the trap. This process describes the "mass selective instability mode" of operating a quadrupole ion trap introduced by Stafford et al.,[23] as is schematically depicted in Scheme 1a. β is used to indicate the frequency f of the predominantly sinusoidal motion of the ions in the trap, i.e., $f = (\omega/2)\beta$ and β ranges from 0 to 1. The particular value of β of an ion is directly related to its q value by an iterative function. It was soon discovered that using an auxiliary RF field, usually created by applying a relatively small voltage (<50-V peak) across the two end-cap electrodes, at a frequency that corresponds to the motion of an ion can also eject ions from the trap in a process called *resonance ejection*. Applying a fixed frequency auxiliary field, and

[23] G. C. Stafford, Jr., P. E. Kelley, J. E. P. Syka, W. E. Reynolds, and J. F. J. Todd, *Int. J. Mass Spectrom. Ion Processes* **60**, 85 (1984).

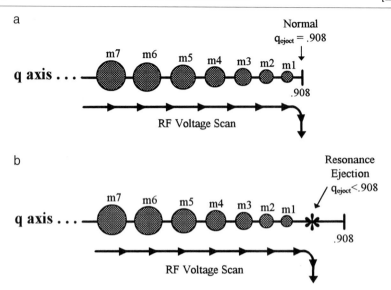

SCHEME 1. Schematic representations of the mass selective instability mode of operating the QITMS (a) and the mass selective instability mode with resonance ejection (b). In the mass selective instability mode, ions of increasing m/z are consecutively moved to the point of ejection (q_{eject}) from the ion trap under the influence of the scanning RF voltage. The addition of an auxiliary resonance ejection voltage lowers the ejection point and results in better sensitivity, resolution, and increased mass range.

again ramping the main RF voltage linearly, causes ions of successive m/z to come into resonance with the auxiliary field, become unstable, and be ejected axially from the trap. This mode of mass selective instability with resonance ejection enhances the resolution and sensitivity of the ion trap mass spectrometer and is schematically depicted in Scheme 1b. Another significant advantage of this technique is that the frequency chosen (expressed as β_{eject}) for the auxiliary field defines a new q_{eject} value of less than 0.908 and the voltage required to eject a particular m/z can be lowered. In effect, this can raise the m/z range of the instrument to large values.[6] Once ions are ejected from the trap, they are efficiently detected by the 20-kV off-axis conversion dynode with an electron multiplier detector. The "off-axis" configuration of the detector reduces background noise from neutral species that are not deflected onto the dynode.

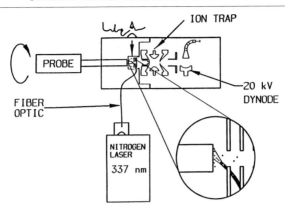

FIG. 2. Matrix-assisted laser desorption quadrupole ion trap system (MALDI/QITMS) configuration.

An HPCE apparatus (Isco model 3850, Isco, Inc., Lincoln, NE) was coupled to the ESI source via the liquid sheath method of Smith *et al.*[24] (Fig. 1), whereby a coaxial flow of water–methanol with 1% (v/v) acetic acid at 1–5 μl/min serves to establish the electrical contact with the HPCE effluent and to provide sufficient liquid flow for stable ESI operation. For HPCE/MS, peptide sample concentrations were generally in the 0.5- to 10-pmol/μl range, electrokinetic injection was performed at 30 kV for variable time periods, and the HPCE buffer was 0.05 M acetic acid. The open tube HPCE capillaries were 370-mm o.d., 50-mm i.d., and 1 meter in length.

For protein and DNA molecular weight studies and for the peptide-sequencing study, the HPCE apparatus was removed and a dilute solution of the analyte, generally around 1–10 pmol/μl, was infused directly into the ESI source at 1 μl/min using a Harvard low-pressure syringe pump.

With a fundamental RF frequency of 880,300 Hz, the ITMS operating parameters used were as follows: for a mass range of 1058 μ, a resonance ejection frequency of 360,760 Hz ($q_{eject} = 0.871$, $\beta_{eject} = 0.820$) and a resonance ejection amplitude of 6.4 Vpp; for a mass range of 2030 μ, a resonance ejection frequency of 147,760 Hz ($q_{eject} = 0.454$, $\beta_{eject} = 0.336$), and a resonance ejection amplitude of 16 V peak-to-peak (pp). Variable ion injection times of 3–500 msec were employed.

The MALDI/QITMS configuration is illustrated in Fig. 2.[5] In brief, a pulsed nitrogen laser beam at 337.1 nm (Laser Science, Cambridge, MA) was focused onto a 200-mm core fiber optic (Newport Corp., Irvine, CA). The fiber transmits the 3-nsec pulse of photons at 10^6–10^7 W/cm^2 to the

[24] R. D. Smith, C. J. Barinaga, and H. R. Udseth, *Anal. Chem.* **60,** 1948 (1988).

external ion source, where a stainless steel probe tip holds a matrix–sample (1000:1, v/v) mixture. A sinapinic acid matrix was used for molecular masses above 20,000 u, and a 2,5-dihydroxybenzoic acid matrix was used for molecular masses below 20,000 u. The desorbed ions were extracted from the external source and transmitted through the ion injection lens system into the ion trap. Both the fundamental RF and the resonance frequency were decreased as the mass range was extended, eventually reaching 130,000 u [fundamental RF = 550,000 Hz; resonance ejection frequency = 3531 Hz (q_{eject} = 0.0182, β_{eject} = 0.0218)].

Results and Discussion

Mass Resolution and Molecular Weight

Molecular weight determination at high sensitivity of complex oligosaccharides, DNA oligomers, peptides, and proteins to over 100,000 Da by ESI/QITMS analysis in both positive and negative ion mode has been demonstrated, mainly in our laboratory[7,25] and at Oak Ridge National Laboratory.[2,26] In general, the data are similar to that obtained by ESI with quadrupole MS analysis (see [21] in this volume)[22a] but mass resolution and sensitivity can be better. Resolution is defined as $m/\Delta m$, where Δm is the full peak width at half-maximum height, or FWHM, of the mass peak. The spectrum of myoglobin (molecular weight 16,950), shown in Fig. 3, is illustrative. In this example, a solution of 3.64 pmol/ml of myoglobin in methanol–water (50:50, v/v) was infused into the ESI/QITMS system at 1 μl/min. Ions were allowed into the trap for 50 msec; the injection lens voltage was then set to prevent any more ions from entering the trap. The trapped ions were then scanned out of the trap to the detector in 120 msec, using a scan speed (5550 u/sec) that is 10 times slower than originally used for ion trap scanning (55,500 u/sec) for this experiment. Slower scanning results in higher resolution.[7] This process was repeated until 26 scans were averaged. The total time taken for this experiment was (50 + 120) × 26 msec, or 4.4 sec. The total amount of myoglobin that was consumed in the ESI source in this time was, therefore, 4.4/60 × 3.64 pmol, or 267 fmol. The maximum quantity of myoglobin that was actually trapped was 50/170 × 267 fmol, or 79 fmol (3 fmol/scan). From the peak width of 0.7 u, and knowing that the minimum peak width possible from the isotopic

[25] I. Jardine, M. E. Hail, S. Lewis, J. Zhou, J. C. Schwartz, and C. M. Whitehouse, in "Proceedings of the 38th Conference on Mass Spectrometry and Allied Topics, Tucson, AZ," p. 16. 1990.

[26] S. A. McLuckey, G. J. Van Berkel, and G. L. Glish, J. Am. Soc. Mass Spectrom. **3**, 60 (1992).

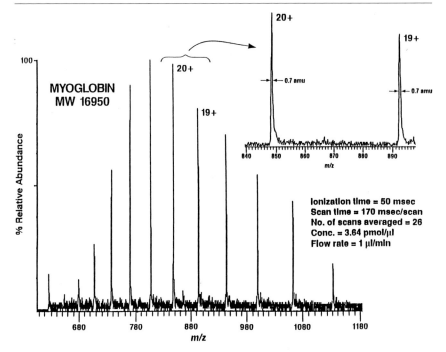

Fig. 3. ESI/QITMS of horse heart myoglobin (molecular weight 16,950) obtained at a scan speed of 5550 u/sec. The inset shows the actual peak widths of the 19+ and 20+ ions, indicating a mass resolution (FWHM) of about 5000.

envelope is about 0.5 u,[27] a resolution of approximately 5000 (FWHM) can be calculated. Such high mass resolution can be useful to "clean up" the spectra of biological macromolecules, as illustrated by the negative-ion ESI/QITMS spectra of polyadenylic acid (Fig. 4), in which sodium adduct ions are more clearly resolved at one-tenth the normal scan rate.

High mass resolutions can be obtained by further slowing of the scan speed, as shown in Fig. 5 for the doubly charged ion of gramicidin (molecular weight 1040; MH_2^{2+} at m/z 571). This illustrates an important use of such high resolution—determining the charge state and thus mass of an unknown multiply charged ion, which under low-resolution conditions (isotopes not resolved) is defined only by an m/z value. High resolution allows clear separation of the isotope clusters of peptide ions, and by measuring the m/z distance between the isotopic ions that differ in actual mass by 1 Da (or by counting the number of peaks contained in 1 m/z unit), the charge

[27] I. Jardine, *Methods Enzymol.* **193,** 441 (1990).

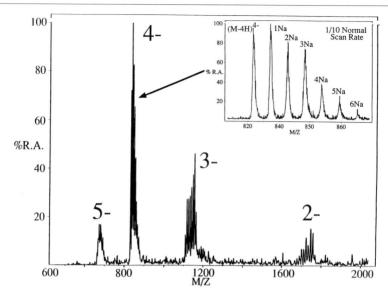

FIG. 4. ESI/QITMS of polyadenylic acid decamer (pd[A]$_{10}$), showing that negative ions can be analyzed and that slower, higher resolution scans can more cleanly resolve sodium adducts (inset).

state, and therefore the mass of the ion, is determined (Fig. 5 and examples to follow).

By drastic reduction of the RF voltage scanning rate from the normal 5555 Da/sec to less than 0.1 Da/sec (i.e., normal scan speed slowed by about 50,000 times), increased mass resolution on ion traps has been demonstrated to well over 1 million using other MS ionization techniques on simple standard calibration compounds [fast atom bombardment (FAB) of cesium iodide[28] and electron ionization (EI) of perfluorotributylamine (PFTBA)[29,30]]. While the usefulness of such ultrahigh mass resolution has been elegantly and clearly demonstrated[31,32] for the ESI/MS analysis of small proteins by high magnetic field Fourier transform ion cyclotron reso-

[28] J. D. Williams, K. Cox, K. L. Morand, R. G. Cooks, R. K. Julian, and R. E. Kaiser, *in* "Proc. 39th Ann. Conf. Am. Soc. Mass Spectrom. Allied Topics, Nashville, TN," p. 1481. 1991.

[29] F. A. Londry, G. J. Wells, and R. E. March, *Rapid Commun. Mass Spectrom.* **7**, 43 (1993).

[30] R. E. March, F. A. Londry, and G. J. Wells, *in* "Proc. 41st ASMS Conf. Mass Spectrom. Allied Topics, San Francisco, CA," p. 790. 1993.

[31] S. C. Beu, M. W. Senko, J. P. Quinn, and F. W. McLafferty, *J. Am. Soc. Mass Spectrom.* **4**, 190 (1993).

[32] B. E. Winger, S. A. Hofstadler, J. E. Bruce, H. R. Udseth, and R. D. Smith, *J. Am. Soc. Mass Spectrom.* **4**, 566 (1993).

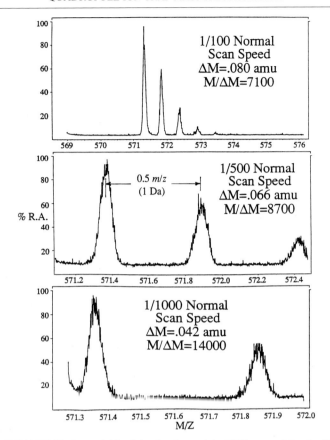

FIG. 5. ESI/QITMS of gramicidin S (molecular weight 1040) around the doubly charged ion (MH_2^{2+}) at m/z 571, illustrating increasing mass resolution with slowing of the scan speed.

nance mass spectrometers (FTICR/MS), equivalent ultrahigh resolution data on real biological macromolecules have not, so far, been generated on QITMS systems. It is not known at present if appropriate experiments have simply not yet been done in this regard, or if this will prove to be a fundamental limitation of QITMS because of "space charge" effects. That is, when high number densities of similarly charged molecules are confined in the QITMS for extended times, adverse effects on resolution can occur by mutual interaction of the ions.

Peptide Sequencing by Multiple Stages of Mass Spectrometry (MS^n)

The positive-ion ESI/QITMS spectrum of recombinant human growth hormone-releasing factor (rhGHRF; molecular weight 5039.7) is shown in

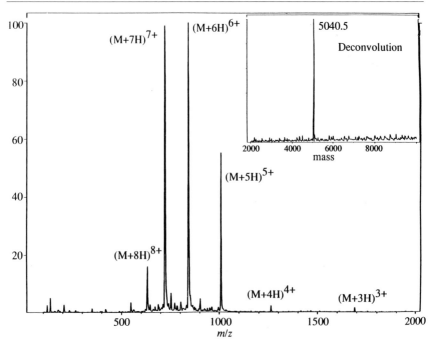

FIG. 6. ESI/QITMS of recombinant human growth hormone-releasing factor (rhGHRF; molecular weight 5039.7) with the multiply charged ion envelope from $(M + 3H)^{3+}$ to $(M + 8H)^{8+}$ on the m/z axis converted or "deconvoluted" to the molecular weight determination of 5040.5 on the mass axis (inset).

Fig. 6.[18] The $(M + 3H)^{3+}$ through $(M + 8H)^{8+}$ ions are clearly identifiable, and deconvolution (i.e., conversion of the relative abundance versus m/z spectrum to a relative abundance versus mass plot)[27] provides a molecular weight determination of 5040.5 ($\Delta = +0.016\%$). Further studies in our laboratory have shown comparable mass accuracies of around ± 0.01–0.05% for a number of peptides and proteins. Using calibration methods involving the use of external standards and a peak-matching procedure, Williams and Cooks[33] have shown that masses can be assigned on current ITMS hardware with accuracies generally better than 50 ppm (parts per million), or 0.005%.

The Finnigan ITMS was originally built in the early 1980s to measure the mass of molecules to 650 Da with only nominal mass accuracy at this mass (usually considered as ± 0.3 Da). The work described here is expanding the original capabilities of the system. New developments in analog (RF)

[33] J. D. Williams and R. G. Cooks, *Rapid Commun. Mass Spectrom.* **6,** 524 (1992).

and digital electronics will provide new generation QITMS systems stable enough to routinely provide more accurate mass measurement. Routine determination of mass accuracy for biological molecules to $\pm 0.05\%$ or better can be expected soon.

One of the important capabilities of the ion trap device is the ability to eject unwanted ions and exclusively retain selected ions. Even the ability to retain selected isotopes from isotopic clusters has been demonstrated.[20] Using this capability, the individual multiply charged ions of rhGHRF were selected and isolated in turn, and then subjected to MS/MS after collision-induced dissociation (CID). The selected $(M + 7H)^{7+}$ ion mass spectrum is shown in Fig. 7a. This selected ion is agitated by resonant excitation and forced to undergo multiple collisions with the residual helium (1 millitorr) in the trap.[9,34] The resulting MS/MS mass spectrum of this ion is shown in Fig. 7b. Notice that the $(M + 7H)^{7+}$ at m/z 720 has been completely fragmented. The labeling of the peptide fragment ions in the MS/MS spectrum follows convention,[35,36] where A, B, and C ions retain the N- or amino-terminal amino acids residues, while the X, Y, and Z ions retain the C- or carboxy-terminal amino acid residues.

The MS/MS spectra of all of the separately isolated multiply charged ions of rhGHRF are shown in Fig. 8. A qualitative examination of the ion trap MS/MS spectra of the various charge states indicates an expected trend of more fragmentation with higher charge state.[37] The higher charge states, however, tend to have many fragments close to the parent m/z (loss of small fragments). Fragments that are distant from the parent m/z are most abundant in MS/MS spectra of the middle charge states, with the MS/MS spectrum of the $(M + 5H)^{5+}$ ion showing almost exclusively these ions. MS/MS spectra of the lower charge states show only losses of ammonia or water (usually the base peak) and very little other fragmentation.

The MS/MS spectrum of the $(M + 7H)^{7+}$ ion shows abundant fragmentation (Fig. 7b); however, identification and sequence ion assignment is difficult owing to the uncertainty in charge state, and therefore mass, assignment. A variety of possibilities for the ions at m/z 266.8, 419.7, and 951.2, for example, are indicated. Higher resolution on the ion trap, obtained by slower scanning across the narrow mass range of interest, can be used to identify the charge state using isotope separation to help identify the correct sequence ion from a set of possibilities. A selection of MS/MS-generated

[34] J. N. Louris, J. S. Brodbelt-Lustig, R. G. Cooks, G. L. Glish, G. J. Van Berkel, and S. A. McLuckey, *Int. J. Mass Spec. Ion Proc.* **96**, 117 (1990).

[35] P. Roepstorff and J. Fohlman, *Biomed. Mass Spectrom.* **11**, 601 (1984).

[36] K. Biemann, *Biomed. Environ. Mass Spectrom.* **16**, 99 (1988).

[37] R. D. Smith, J. A. Loo, C. J. Barinaga, C. G. Edmonds, and H. R. Udseth, *J. Am. Soc. Mass Spectrom.* **1**, 53 (1990).

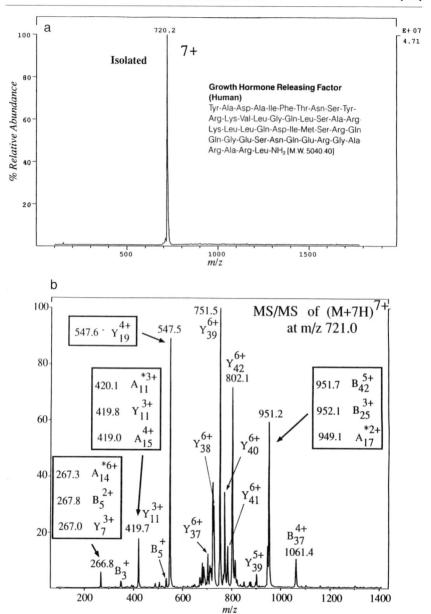

FIG. 7. Isolation of the ESI-generated (M + 7H)^7+ ion at m/z 721 of rhGHRF in the QITMS (a) with subsequent fragmentation and MS/MS analysis of the resulting product ions (b). The extent of parent or precursor ion fragmentation is under the control of the operator, and in this case all of the precursor ions at m/z 721 have been fragmented and none appear in the MS/MS spectrum.

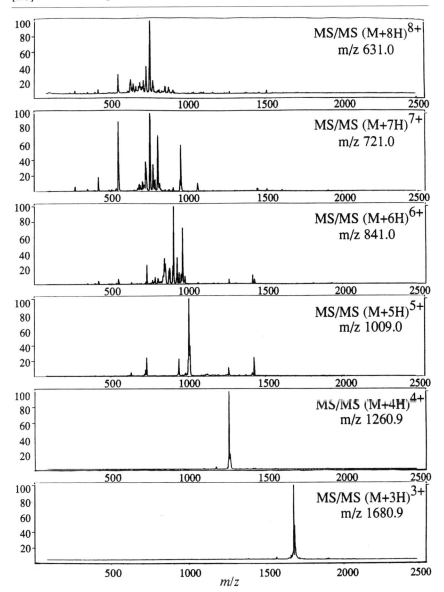

FIG. 8. MS/MS spectra of the $(M + 3H)^{3+}$ through $(M + 8H)^{8+}$ ions of rhGHRF. The ions were isolated in turn and fragmented using approximately the same conditions. The MS/MS spectra indicate that different product ion spectra are obtained from different charge states of the same molecule.

ions whose 3+ charge state, and therefore mass and assignment, was deter-mined in this way, is shown in Fig. 9. Some of the other assignments made in this manner for the MS/MS spectrum of the $(M + 7H)^{7+}$ species are also indicated in Fig. 7b.

On examination of the MS/MS spectrum of the $(M + 7H)^{7+}$ ion (Fig. 7b), it is interesting to note that some of the most prominent ions are complementary fragments such as Y_7^{3+} and B_{37}^{4+}, and Y_{19}^{4+} and B_{25}^{3+}. It is of interest that these ions result from fragmentation occurring on the carboxyl side of glutamate or aspartate residues. The probable mechanism

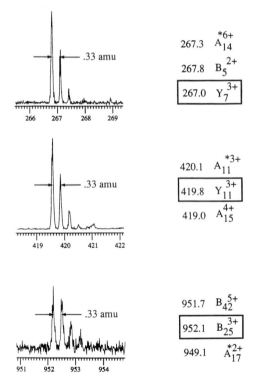

FIG. 9. The product ions at approximately m/z 266–270, 419–422, and 951–954 in the MS/MS spectrum of the $(M + 7H)^{7+}$ ion of rhGHRF (Fig. 7) were reanalyzed under slow-scan, high-resolution conditions (1/100 normal scan rate, 320,760 Hz, 800 mV) to show the m/z separation of the isotopes. The m/z value of 0.33 amu determined in each selected case indicates a 3+ charge state, and therefore a correct assignment of product ion (Y_7^{3+}, Y_{11}^{3+}, and B_{25}^{3+}, respectively) can be made from the choices indicated. The A_{11}^{*3+} choice at m/z 420 is excluded because no associated A_{11}^{3+} ion was present in the MS/MS spectrum.

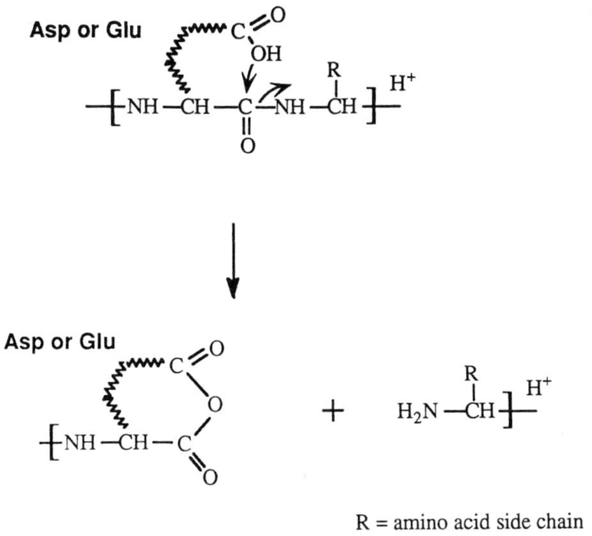

R = amino acid side chain

SCHEME 2. Proposed "remote from the charge site" fragmentation mechanism for the frequently encountered cleavage on the carboxyl side of glutamic acid (Glu) and aspartic acid (Asp) residues in the QITMS/MS spectra of rhGHRF.

of this "remote from the charge site" fragmentation[38] is the cyclization shown in Scheme 2. Results[39] from fragmentation of the ESI-generated multiply charged ions of proline-containing proteins in a triple quadrupole mass spectrometer have suggested that fragmentation due to cleavage of the amide bond to proline is often a dominant reaction, as had been observed earlier for MS/MS of singly protonated peptides[40] obtained by FAB/MS. The protein rhGHRF, however, does not contain a proline residue. Nevertheless, in contrast to MS/MS on other systems, useful complementary fragment ions resulting from cleavage at specific sites in the peptide backbone appear in many of the ion trap MS/MS spectra observed to date.

A major advantage of the ion trap compared, for example, to quadrupole and magnetic sector mass spectrometry systems, is that multiple stages of MS/MS (MS^n) can be induced,[34] with sequential isolation and fragmentation steps, in order to probe more fully molecular structure and sequence. The B$_{25}^{3+}$ MS/MS ion at m/z 951.5, for example, was selected, isolated,

[38] J. Adams, *Mass Spectrom. Rev.* **9,** 141 (1990).
[39] J. A. Loo, C. G. Edmonds, and R. D. Smith, *Anal. Chem.* **65,** 425 (1993).
[40] D. F. Hunt, J. R. Yates III, J. Shabanowitz, S. Winston, and C. R. Haur, *Proc. Natl. Acad. Sci. U.S.A.* **83,** 6233 (1986).

and fragmented to generate the MS/MS/MS (MS³) spectrum shown in Fig. 10. MS/MS/MS spectra of multiply charged intermediate ions lead to the same difficulty in sequence ion assignment because of unknown charge states. Highlighted in Fig. 10 are four regions of this MS³ spectrum scanned under high-resolution conditions (Fig. 11a and b) such that charge states can be determined using isotope separation, and overlapping peaks are resolved. Some sequence ions identified by this process are indicated in Fig. 11a and b for the MS³ m/z regions 908–920, 1086–1094, 1190–1260, and 1290–1330, uniquely identifying, respectively, $B_{24}*^{3+}$, $Y_8{}^+$, $B_{24}{}^{3+}$, $Y_{19}{}^{2+}$, $B_{21}{}^{2+}$, $Y_{21}*^{2+}$, $Y_{11}{}^+$, $Y_{21}{}^{2+}$, $Y_{22}*^{2+}$, $Y_{22}*^{2+}$, $B_{22}{}^{2+}$, $Y_{22}{}^{2+}$, $Y_{23}*^{2+}$, $B_{23}{}^{2+}$, $Y_{23}{}^{2+}$, and $Y_{12}{}^+$ [the asterisk (*) refers to loss of NH_3]. (Note: Labeling of grand product or MS/MS/MS ions is relative to the MS/MS product ion and not the original MS parent or precursor ion.)

Performing MS³ on the complementary ions [e.g., compare MS/MS/MS on $B_{25}{}^{3+}$ (Fig. 10) and $Y_{19}{}^{4+}$ (Fig. 12a)] allows access to completely different parts of the peptide to yield, overall, much more sequence information. Additional sequence information is observed when comparing MS³ spectra of the same fragment ion with different charge states [e.g., compare MS/

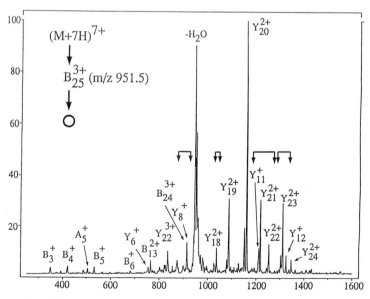

FIG. 10. The $(M + 7H)^{7+}$ ion was isolated and fragmented and then the resulting $B_{25}{}^{3+}$ ion in the MS/MS spectrum was subsequently isolated and fragmented to generate the MS/MS/MS "grand product" ion spectrum shown. The bracketing arrows indicate the regions further analyzed under slow-scan/high-resolution conditions (see Fig. 11).

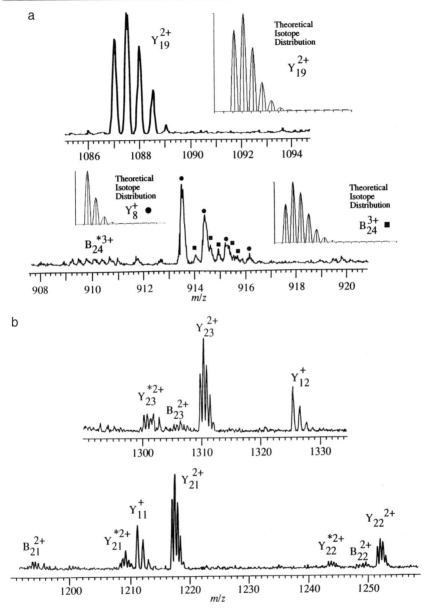

FIG. 11. Slow scans [(a) 1/100 normal scan rate; (b) 1/20 normal scan rate] over selected m/z regions of the $(M + 7H)^{7+} \rightarrow B_{25}^{3+} \rightarrow O$ MS/MS/MS spectrum, showing resolution of overlapping ions and charge state determination from this high-resolution analysis. Some theoretical ion isotope distributions are shown to indicate the fidelity of the measured isotope ratios.

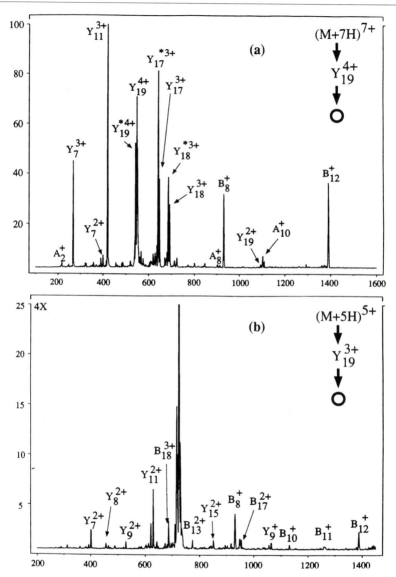

FIG. 12. MS/MS/MS spectra from the $(M + 7H)^{7+} \rightarrow Y_{19}^{4+} \rightarrow O$ (a), and $(M + 5H)^{5+} \rightarrow Y_{19}^{3+} \rightarrow O$ (b) transitions, showing that significantly different fragmentation spectra, which provide different and complementary information, are obtained from the same precursor molecule or fragment (Y_{19}) but with different charge states (4+ and 3+, respectively).

MS/MS of Y_{19}^{4+} obtained from $(M + 7H)^{7+}$ (Fig. 12a) versus Y_{19}^{3+} obtained from $(M + 5H)^{5+}$ (Fig. 12b)].

MS/MS/MS/MS experiments on the hGHRF peptide have been performed in the ion trap, and these experiments do tend to access unique parts of the peptide and therefore give important sequence information. Signals, however, are becoming weak at this point and usually require substantial signal averaging, and therefore more material.

Summarized in Scheme 3 are the cleavages observed by one stage of MS/MS compared to those observed by multiple, and by no means exhaustive, stages of MS/MS (i.e., MS^n). Although no serious attempt was made to test the detection limits in these MS^n experiments, the approximate amount of sample consumed in each experiment was ~100 fmol for MS; ~1 pmol for MS/MS; and ~10–50 pmol for MS^n. The scan times were, in general, about 10 times more than the actual ionization times, and so the total amount of sample actually used was approximately one-tenth these amounts.

Although the practical sequencing of unknowns still presents a complex challenge, these preliminary results suggest that the unique capabilities of the ion trap may be ideally suited for gaining maximum sequence information from peptides. These capabilities include production by efficient fragmentation of unique and complementary sequence information, routine charge state and mass identification, and MS^n experiments. Extension of the application of these capabilities to structural analysis of modified peptides and proteins, such as with natural posttranslational modifications (e.g., phosphorylation, sulfation, glycosylation, and lipidation) as well as other modifications such as alkylation (e.g., with photoactivated probes), acylation, oxidation, and unusual amino acid substitution, is obvious. Indeed, a particular strength of mass spectrometry has always been its ability to allow

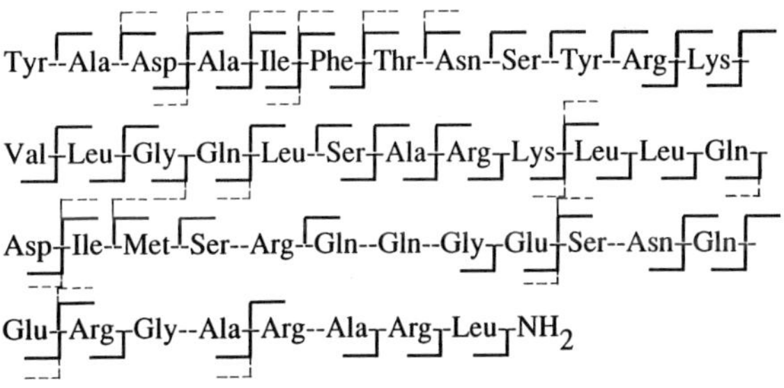

SCHEME 3. Cleavages observed (dashed lines), in the MS/MS spectra of rhGHRF obtained by ESI/QITMS, versus those noted from all the MS^n experiments conducted (solid lines).

probing of unusual structures,[41] and it is anticipated that this capability will reach its apex with MS^n on the QITMS for structures as diverse as synthetic antisense DNA molecules,[42] to complex "hybrid" biological macromolecules such as metalloproteins and protein glycolipid anchor structures.

Peptide Analysis by High Performance Capillary Electrophoresis/ Mass Spectrometry

High-performance capillary electrophoresis (HPCE) is emerging as a powerful technique for the rapid and efficient separation of biological molecules such as peptides and proteins. A review of HPCE/MS by Smith *et al.* has appeared.[43] An HPCE/MS detector that can routinely operate at the low femtomolar level or better, can provide both universal as well as specific detection capabilities, and yet is affordable to the average laboratory is needed. The QITMS has been demonstrated to provide unique and promising capabilities in this regard for the identification and sequencing of peptides and proteins by mass spectrometry. Consequently, an HPCE was coupled to the ESI/QITMS described above to evaluate this combination. The experimental setup is shown in Fig. 1, including (in the inset) the liquid junction interface.[24] The major features of this CE/MS combination that were evaluated included determining whether the m/z range and scan speed capabilities of the QITMS were adequate to handle typical HPCE peak widths of a few seconds; determining the minimum sample loading on the HPCE column that was compatible with CE/MS operation (i.e., detection limits); and determining whether MS/MS experiments could be conducted on-line.

An example of a separation and MS analysis of a 20-sec electrokinetic injection of a 5-ng/μl mixture of 10 closely related angiotensin analogs is shown in Fig. 13, with a resulting sample loading of each peptide of around 0.27–0.49 pmol. In this experiment, full-scan mass spectra were repetitively recorded from m/z 100 to m/z 2000 every 1 sec. The total ion current detected in each scan is plotted in the reconstructed ion chromatogram (RIC) in the lower portion of Fig. 13. Because ESI mass spectra of peptides are generally straightforward, consisting of only a few ions such as $(M + H)^+$, $(M + 2H)^{2+}$, $(M + 3H)^{3+}$, and so on, an alternative MS chromatographic plot is the "base peak" chromatogram, in which only the most intense peak, or base peak, of each spectrum is plotted for every consecutive scan. This plot "picks out" the peptide ions and essentially ignores background ions,

[41] J. A. McCloskey (ed.), *Methods Enzymol.* **193** (1990).
[42] S. A. McLuckey and S. Habibi-Goudarzi, *J. Am. Chem. Soc.* **115,** 12085 (1993).
[43] R. D. Smith, J. H. Wahl, R. D. Goodlett, and S. A. Hofstadler, *Anal. Chem.* **65,** 574A (1993).

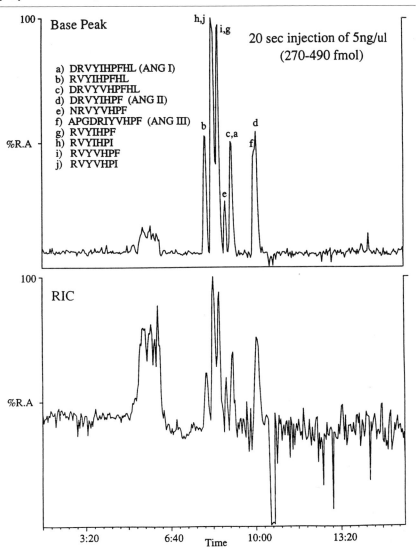

FIG. 13. CE/MS of a mixture of closely related angiotensin peptides recorded by full scan on the ESI/QITMS. The lower trace is the reconstructed ion chromatogram (RIC), which records all of the ion current generated, while the upper trace is a plot of the most abundant ion (base peak) in each consecutive scan.

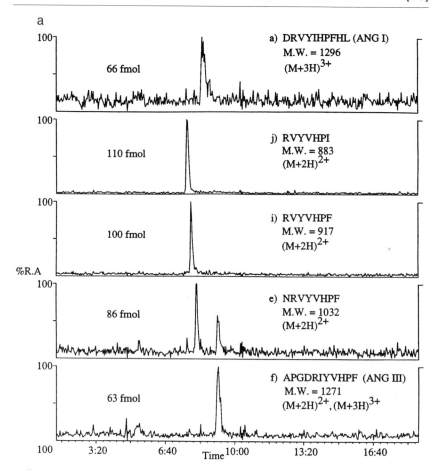

FIG. 14. Selected reconstructed ion chromatograms (SRICs) from full-scan CE/MS spectra for an abundant protonated ion or ions of each of the angiotensin peptides of Fig. 13. In this way all peptides are uniquely identified even if they coelute.

resulting in a much cleaner chromatogram (top portion of Fig. 13). Selected reconstructed mass chromatograms, in which the intensity of selected ions or m/z values are plotted for every consecutive scan, can then be plotted for every peptide in turn. Examples of such plots for the same group of angiotensins run under the same conditions as Fig. 13 but at the 63- to 110-fmol level are shown in Fig. 14a and b. The peptides are uniquely identified in this manner, whether or not they coelute.

These experiments were conducted using bare fused silica HPCE colb

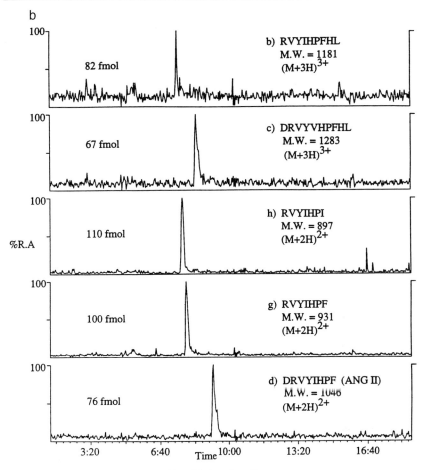

FIG. 14. (*continued*)

umns with a simple acetic acid buffer. These conditions allow optimization of the ESI interface for the analysis of peptides. Under these conditions the HPCE separation efficiency for peptides is not optimized and the peptide bandwidths are on the order of 20 sec or greater at half-height. The separation efficiency for peptides on open columns can be substantially improved by carrying out the separation at basic pH or by the addition of other buffers or modifiers. A major limitation of ESI for CE/MS analysis, however, is the fact that involatile buffers or modifiers will seriously degrade or suppress the ion signals from analytes of interest, such as peptides. Recognizing this,

Moseley *et al.*[44] developed a protocol utilizing aminopropylsilane-coated fused silica columns, to reverse the polarity of the column wall, which then requires reversing the CE polarity, in order to allow separation of peptides at acidic pH and to retain optimized ESI conditions for MS analysis of the peptides. A CE/MS separation and analysis of angiotensin I, renin substrate, and substance P at the 190-fmol sample level loaded on such an aminopropylsilane-coated column is shown in Fig. 15. Separation efficiencies ($N >$ 100,000 plates) are now much more respectable, with bandwidths of 3- to 5-sec FWHM. Nevertheless, the QITMS with a scan speed of one scan per second over an m/z range of 100–2000 is quite capable of high fidelity chromatographic reproduction. Even faster scan speeds may be possible with newer QITMS technology.

Of course, if high mass resolution data are required, as discussed above, slower scan speeds, by as much as 1/10 or more, will be necessary. To provide such high-resolution data, two mechanisms can be considered. The first is to carry out a fast QITMS survey scan, immediately followed by a slow, high-resolution scan, but only over the narrow m/z regions of interest such as over the peptide multiply charged ion isotope clusters. The second method would be simply to lower the HPCE voltage to spread out the band to allow sufficient time for higher resolution, slower scans. The latter method has been convincingly demonstrated by Smith's group[45] for HPCE/MS using FTICR/MS. Indeed, the ability to control the HPCE voltage allows the possibility of other optimization experiments such as reducing to zero or even momentarily reversing the HPCE voltage after an ion packet has been pulsed into the QITMS. The voltage can be reapplied for the next sample packet to be ionized, trapped, and analyzed. In this way, sample is not consumed while the QITMS is scanning.

The result of a 200-fold serial dilution HPCE/MS experiment using triplicate injections of Leu-enkephalin (admittedly a peptide that behaves well in this system with little column sample adsorption, for example) is shown in Fig. 16, and demonstrates full-scan detection limits of below 15 fmol on-column. Figure 17, in fact, shows mass chromatograms for two injections differing by a factor of 10 in sample injected of a mixture containing Leu-enkephalin and oxidized Met-enkephalin. Although the <10-fmol samples are not visible on the RIC trace or even on the base peak trace, both compounds are identified from the reconstructed mass chromatograms at m/z 556 and 590, respectively. The single-scan mass spectra of both peptides are shown in Fig. 18. Background subtraction, a normal

[44] M. A. Moseley, J. W. Jorgenson, J. Shabanowitz, D. F. Hunt, and K. B. Tomer, *J. Am. Soc. Mass Spectrom.* **3**, 289 (1992).

[45] S. A. Hofstadler, J. H. Wahl, J. E. Bruce, and R. D. Smith, *J. Am. Chem. Soc.* **115**, 6983 (1993).

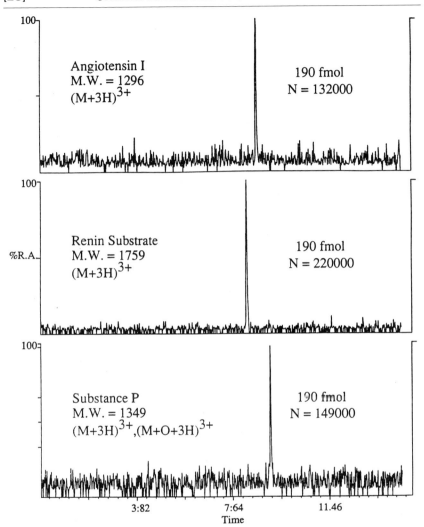

Fig. 15. Selected reconstructed ion chromatograms (SRICs) from full-scan CE/MS spectra for abundant ions of three peptides separated using an aminopropylsilyl-coated fused silica capillary column.

recourse in chromatography/mass spectrometry, provides relatively clean spectra at this level. For qualitative and quantitative target compound analysis, the technique of selected ion storage, in which only the known ions of interest are selectively trapped, may be able to push detection limits even further. Of course, detection limits of analytes in real samples such

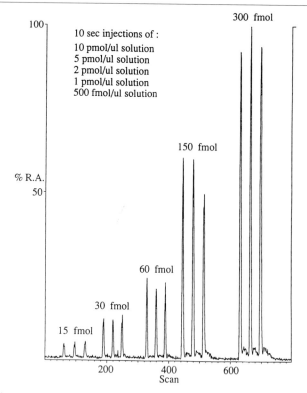

Fig. 16. CE/ME selected ion plots of the protonated ion of Leu-enkephalin from full-scan ESI/QITMS mass spectra with dilution over a 200-fold range down to 15 fmol on-column.

as extracts of biological matrices will be dependent ultimately on the S/N ratio at the masses monitored. To improve further the dynamic range of the QITMS, selected waveform techniques such as stored waveform inverse Fourier transform (SWIFT) excitation are promising.[46-48] Advances in HPCE will also, of course, contribute to the further development of the HPCE/QITMS combination. Three examples are the use of transient iso-tachophoresis to increase the injection volume by 100-fold for HPCE/MS,[49] the implementation of small inner diameter capillaries for subfemtomolar

[46] R. K. Julian, K. Cox, and R. G. Cooks, in "Proc. of the 40th ASMS Conf. Mass Spectrom. Allied Topics, Washington, D.C.," p. 943. 1992.
[47] M. H. Soni and R. G. Cooks, Anal. Chem. 66, 2488 (1994).
[48] S. Guan and A. G. Marshall, in "Proc. 41st ASMS Conf. Mass Spectrom. Allied Topics, San Francisco, CA," p. 451. 1993.
[49] T. J. Thompson, F. Foret, P. Vouros, and B. L. Karger, Anal. Chem. 65, 900 (1993).

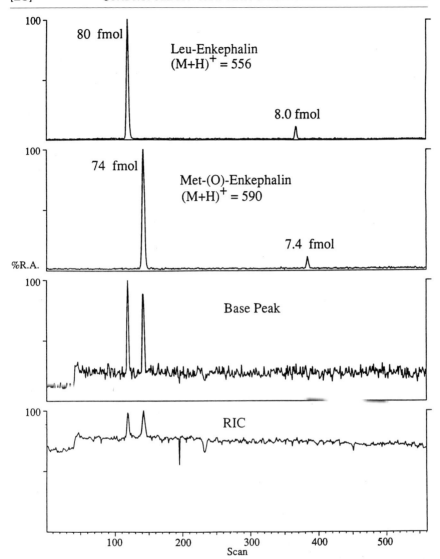

Fig. 17. CE/MS analysis of low femtomolar amounts of Leu-enkephalin and the oxidized form of Met-enkephalin, indicating detection below the 10-fmol level with full-scan MS analysis. The lower amounts are not visible by RIC or base peak analysis, but only by using selected reconstructed ion chromatograms (SRICs).

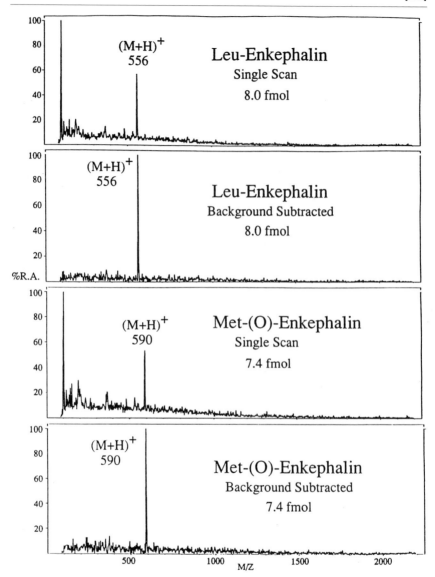

F<small>IG</small>. 18. Full-scan CE/MS data from the experiment of Fig. 17. Single-scan data with and without background subtraction are shown for comparison.

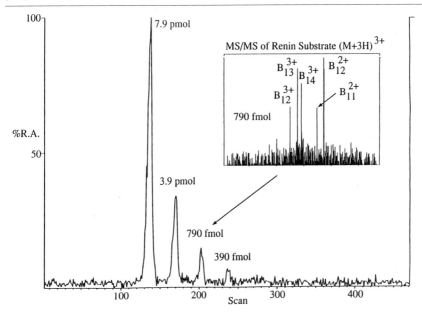

FIG. 19. CE/MS/MS on the ESI/QITMS system, showing selection of the $(M + 3H)^{3+}$ ion of renin substrate with subsequent CID and MS/MS at different dilutions.

analysis by HPCE/MS,[50] and the ability to use complex buffers if electroosmotic flow is minimized.[51]

Only a few preliminary experiments involving HPCE/MS/MS have been carried out to date. An example of the on-line MS/MS analysis of the $(M + 3H)^{3+}$ ion of renin substrate is shown in Fig. 19. Plotted is the sum of the ion currents of the $B_{12}{}^{3+}$, $B_{13}{}^{3+}$, $B_{14}{}^{3+}$, $B_{11}{}^{2+}$, and $B_{12}{}^{2+}$ fragments at four column sample loadings. One reason that so few experiments have been conducted in this regard is that setting up and carrying out such experiments on the currently available QITMS hardware and software are difficult and tedious. A new generation of intelligent computer-controlled QITMS technology should alleviate this problem. In addition, alternative methods are being suggested to make HPCE/MS/MS (and also, therefore, HPLC/MS/MS) easier. Ramsey et al.[52] of the Oak Ridge National Labora-

[50] J. H. Wahl, D. R. Goodlett, H. R. Udseth, and R. D. Smith, *Anal. Chem.* **64,** 3194 (1993).
[51] W. Nashabeh, K. F. Greve, D. Kirby, F. Foret, B. L. Karger, D. H. Reifsnyder, and S. E. Builder, *Anal. Chem.* **66,** 2148 (1994).
[52] R. S. Ramsey, K. G. Asano, K. J. Hart, D. E. Goeringer, G. J. Van Berkel, and S. A. McLuckey, *in* "Proc. 41st ASMS Conf. Mass Spectrom. Allied Topics, San Francisco, CA," p. 749. 1993.

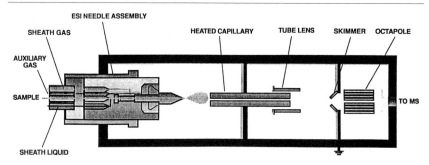

FIG. 20. High-flow/high-sensitivity ESI source showing the electrospray needle assembly, the heated solvent desolvation tube, and high-sensitivity MS injection optics (tube lens, skimmer, RF octapole).

tory group, for example, are developing the use of random noise applied to the end caps of the QITMS[53] to allow trapped ions to be collisionally activated independent of their m/z ratio. This simplifies the MS/MS experiment, assuming reasonable separation of analytes has been obtained before QITMS/MS analysis. The application of low-amplitude noise was also shown by these researchers to improve the S/N ratio in RIC profiles (cf high background in Fig. 13), by reducing the contribution of chemical noise and background derived primarily from ionic buffer components and clusters, including solvent clusters.

Because we have demonstrated that HPCE/MS on an QITMS is feasible, notwithstanding limitations on the use of involatile buffers and other additives, it follows that HPLC/QITMS will also be a powerful combination. For routine LCMS, however, it is important to be able to couple easily any HPLC column diameter (<1 to 4.6 mm) and resulting solvent flow rate (1 μl/min to >1 ml/min) directly to the ESI/MS system. The ESI technology used in the work described above is limited by an inability to handle flow rates above about 5 μl/min. Developments in ESI technology have removed this limitation. In the new ESI system[54] shown in Fig. 20, a different desolvation method using a heated metal tube[55] is used, instead of a countercurrent flow of warm drying gas (cf Fig. 1), and this allows liquid flow rates of from 1 μl/min up to 1 ml/min to be routinely coupled directly to the ESI source. In addition, replacement of the lens focusing system after the skimmer (Fig. 1) with the focusing RF octapole lens (Fig. 20) has provided an increase

[53] S. A. McLuckey, D. E. Goeringer, and G. L. Glish, *Anal. Chem.* **64,** 1455 (1990).

[54] I. Mylchreest, M. Hail, M. Uhrich, and I. Jardine, Paper presented at the 9th Int. Symp. LC/MS, Montreaux, Switzerland, Nov. 4–6, P41 (1992).

[55] S. K. Chowdhury, V. Katta, and B. T. Chait, *Rapid Commun. Mass Spectrom.* **4,** 81 (1990).

in transmission for proteins of 3- to 5-fold, and an even greater increase of >50-fold for small molecules of mass <500 Da.[56] This new ESI source has already proven to be a major advance for the routine analysis of biological macromolecules by HPLC coupled to quadrupole mass spectrometers as well as for magnetic sector MS systems,[57] and preliminary work in our laboratory with this source attached to the QITMS is equally promising.

Other laboratories have also published details of the coupling of HPLC and HPCE to the QITMS. The high-sensitivity analyses of small molecules by the Cornell University (Ithaca, NY) and Oak Ridge National Laboratory groups are particularly notable in this regard.[58–61]

MALDI/QITMS

MALDI has been shown to be extremely powerful for the desorption and ionization of DNA oligomers,[62] complex oligosaccharides,[63] peptides and glycopeptides,[64] proteins and glycoproteins.[65] This important desorption and ionization methodology has now been successfully coupled to the QITMS.[4,5] The MALDI-QITMS combination is particularly promising since the pulsed nature of the ionization technique matches well with the pulsed ion-trap scanning technique.

Two examples of mass spectra generated by MALDI/QITMS in our laboratory on the apparatus shown in Fig. 2 are shown in Figs. 21 and 22. The first spectrum is of α-lactalbumin, a small protein of molecular weight 14,175, and the second is of renin substrate, a peptide of molecular weight 1758, showing a mass resolution of almost 6000 at this mass. Although low picomolar amounts of these materials were placed on the sample target for these experiments, it is certain that only a small proportion of the sample was actually desorbed into the trap, and it is highly likely that much smaller sample amounts will be sufficient for analysis[4] (cf low femtomole amounts of sample used for MALDI/TOFMS[65]).

[56] M. Hail and I. Mylchreest, *in* "Proc. 41st ASMS Conf. Mass Spectrom. Allied Topics, San Francisco, CA," p. 745. 1993.

[57] P. Dobberstein and E. Schroeder, *Rapid Commun. Mass Spectrom.* **7**, 861 (1993).

[58] S. A. McLuckey, G. J. Van Berkel, G. L. Glish, E. C. Huang, and J. D. Henion, *Anal. Chem.* **63**, 375 (1991).

[59] J. D. Henion, A. V. Mordehai, and J. Cai, *Anal. Chem.* **66**, 2103 (1994).

[60] R. S. Ramsey, D. E. Goeringer, and S. A. McLuckey, *Anal. Chem.* **65**, 3521 (1993).

[61] R. S. Ramsey and S. A. McLuckey, *J. Am. Soc. Mass Spectrom.* **5**, 324 (1994).

[62] K. Tang, S. L. Allman, and C. H. Chen, *Rapid Commun. Mass Spectrom.* **7**, 943 (1993).

[63] B. Stahl, M. Setup, M. Karas, and F. Hillenkamp, *Anal. Chem.* **63**, 1463 (1991).

[64] M. C. Huberty, J. E. Vath, W. Yu, and S. A. Martin, *Anal. Chem.* **65**, 2791 (1993).

[65] F. Hillenkamp and M. Karas, *Methods Enzymol.* **193**, 280 (1990).

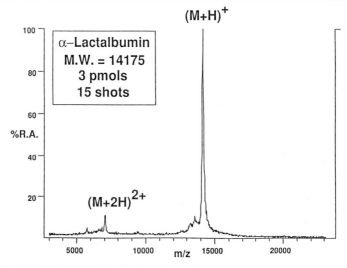

FIG. 21. MALDI/QITMS of α-lactalbumin (molecular weight 14,175; 3 pmol).

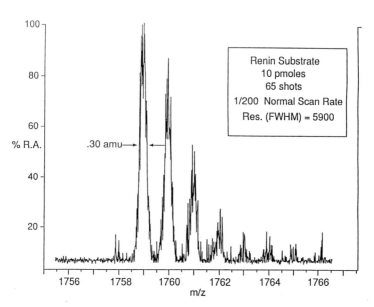

FIG. 22. MALDI/QITMS of renin substrate (molecular weight 1758; 10 pmol) under slow-scan/high-resolution conditions, showing resolved isotopes.

An important advantage of the QITMS for MALDI/MS of biological macromolecules compared, for example, to TOF/MS, is the ready availability of MS^n capabilities, as has been demonstrated.[4,5] It could be argued that MS/MS of the singly protonated ions commonly generated by MALDI of biological macromolecules would be difficult or impossible with the relatively low collision energies available for CID in the QITMS. There are at least three possibilities available to overcome this problem. First, an MALDI matrix that induces the formation of multiply charged ions,[66] which can be more easily fragmented, can be employed; second, surface-induced dissociation (SID) can be implemented in the QITMS[67] to access higher collision energies; and third, laser-induced photofragmentation in the ITMS can be used.[68]

Summary

A number of other features of ITMS systems that will enhance their ability to analyze biological macromolecules are worth mentioning. As has already been demonstrated for ESI/quadrupole, ESI/magnetic sector, and ESI/FTICR systems, the capability of inducing fragmentation of the ESI-generated multiply charged ions of biological macromolecules in the capillary/skimmer region of the ESI source and subsequently selectively analyzing fragments[69] can also be carried out with the QITMS, as we have demonstrated using bovine serum albumin (data not shown). The ability to carry out chemical reactions on biological macromolecules inside the QITMS has been demonstrated by McCluckey et al.[11,70] by showing that the introduction of a pulse of volatile base, such as diethylamine, can result in proton removal from multiply charged protein ions, resulting in species with lower charge states. The application of the technique of deuterium exchange of active hydrogens on peptides[71] to simplify the interpretation of MS/MS sequencing experiments can be implemented for ESI/QITMS. Carrying out such exchange inside the ITMS may also be possible, with resulting analytical advantages. Reports of a hybrid QITMS–TOF system, which was operated with either ESI[72] or MALDI[73] methodology, and which

[66] J. Yates and G. Currie, University of Seattle, Seattle, WA, personal communication, 1995.
[67] S. A. Lammert, Ph.D. thesis, Purdue University, Lafayette, IN, 1992.
[68] W. J. van der Hart, *Mass Spectrom. Revs.* **8,** 237 (1989).
[69] J. L. Loo, C. G. Edmonds, and R. D. Smith, *Science* **248,** 201 (1990).
[70] S. A. McLuckey, G. L. Glish, and G. J. Van Berkel, *Anal. Chem.* **63,** 1971 (1991).
[71] N. F. Sepetov, O. L. Issakova, M. Lebl, K. Swiderek, D. C. Stahl, and T. D. Lee, *Rapid Commun. Mass Spectrom.* **7,** 58 (1993).
[72] S. M. Michael, B. M. Chein, and D. M. Lubman, *Anal. Chem.* **65,** 2614 (1993).
[73] B. M. Chien, S. M. Michael, and D. M. Lubman, *Rapid Commun. Mass Spectrom.* **7,** 837 (1993).

demonstrated low femtomolar sensitivity with higher resolution of the TOF analyzer because of ion injection of essentially monoenergetic ions from the QITMS into the TOF, illustrate additional uses of the QITMS. The reverse combination (e.g., ESI/TOF/QITMS or MALDI/TOF/QITMS) could afford preselection of ions for even higher performance in the QITMS, because space charging (loss of performance such as resolution because of too much charge in close proximity in the ion trap) would be minimized.

Opportunities for the application of QITMS technology for the analysis of biological macromolecules abound, including ultrahigh-sensitivity protein sequencing using specifically derivatized amino acids released by Edman chemistry[74]; rapid sequencing of MHC-associated antigenic peptides of variable length (~nonamers for the MHC I complexes to >dodecamers for the MHC II complexes), which are available in only very low amounts (femtomole/attomole) and in very complex mixtures (5000–10,000 species) of closely related peptide structures[75]; ultrahigh-sensitivity analysis of peptides and proteins directly *in vivo* using microelectrospray[76]; direct analysis of metal ion binding to peptides and proteins and analysis of noncovalent interactions, including conformation[77]; and possible analysis of plasmid DNA, as has been suggested by ESI ionization of a 2-MDa DNA species.[78]

In summary, the ability of the QITMS to interface to key separations systems such as HPLC and HPCE through the critical ionization techniques of ESI and MALDI, coupled with the high mass range, high mass resolution, high sensitivity, high-efficiency CID, and MS^n capabilities of this device, will provide an astonishing array of cost-effective capabilities for the qualitative and quantitative analysis of biological macromolecules.

Acknowledgment

The authors would like to acknowledge Charlie Edmonds for his collaborative efforts on the HGRF studies and to Mark Bier for his contributions to the laser desorption studies.

[74] R. Aebersold, E. J. Bures, M. Namchuk, M. H. Goghari, B. Shushan, and T. Covey, *Protein Sci.* **1,** 494 (1992).

[75] D. F. Hunt, R. A. Henderson, J. Shabanowitz, K. Sakaguchi, N. Sevilir, A. L. Cox, E. Appella, and V. H. Engelhard, *Science* **255,** 1261 (1992).

[76] M. R. Emmett and R. M. Caprioli, *J. Am. Soc. Mass Spectrom.* **5,** 605 (1994).

[77] S. A. McLuckey and R. S. Ramsey, *J. Am. Soc. Mass Spectrom.* **5,** 324 (1994).

[78] C. G. Edmonds, D. L. Springer, B. J. Morris, B. D. Thrall, and D. G. Camp, *in* "Proc. 41st ASMS Conf. Mass Spectrom. Allied Topics, San Francisco, CA," p. 272. 1993.

Author Index

Numbers in parentheses are footnote reference numbers and indicate that an author's work is referred to although the name is not cited in the text.

A

Abad, C., 5, 8(13)
Aberth, W., 454
Abramson, H. A., 298, 355
Ackermann, F., 89
Ackermans, M. T., 317
Ackland, C. E., 225
Adams, J., 567
Aebersold, R., 520, 525, 545, 586
Afeyan, N., 37, 107, 137, 144(6)
Afeyan, N. B., 5(36), 6, 9(36), 78, 87, 92(30), 98(30), 143–144, 145(23), 146(21, 23), 147(24)
Agnello, A., 454, 459(12)
Aguilar, M.-I., 3–4, 12, 13(70), 14, 14(69, 70), 15–18, 20–24, 28, 155, 225
Ahmad, F., 5(47), 6, 10(47)
Ahmad, H., 5(47), 6, 10(47)
Aiken, J. H., 439, 440(55)
Aiuchi, T., 5(35), 6, 9(35)
Akita, S., 519
Alasandro, M., 111, 117(50), 120(49, 50), 121(50), 122, 122(49)
Albert, K., 16
Aldridge, P. K., 273, 281(10)
Alewood, P. F., 5, 5(29), 6, 9(28, 29)
Alexander, J. E., 102
Alexander, L. R., 329
Allan, M. H., 491
Allison, J., 545
Allison, L. A., 187, 191(36)
Allman, S. L., 583
Alpert, A., 30, 75, 137
Altria, K. D., 421
Amankwa, L. N., 445

Anand, R., 259
Anderson, L., 5(43), 6, 10(43)
Andersson, L., 112, 120(59)
Ando, T., 319, 324(2), 325, 328, 328(2), 329–330, 335(24), 340(15)
Andrade, J. D., 84
Andrasko, J., 248
Andrén-Johansson, P., 485
Andreolini, F., 110, 111(33, 48), 117, 120, 122(91), 125(33)
Andrew, P. C., 75
Andrews, P. C., 545
Anflnsen, C. B., 27
Annan, R. S., 532
Anspach, B., 16
Antia, F. D., 53
Apfell, A., 272
Apffel, A., 502
Aplin, R. T., 517
Appella, E., 486(8), 487, 514(8), 586
Arakawa, T., 31, 197
Araki, T., 18
Ardouin, T., 90
Artoni, F., 237
Arvanitidou, E., 277
Asai, K., 5(51), 6, 10(51), 101, 103(3)
Asano, K. G., 581
Astrua-Testori, S., 240, 248
Aubier, M., 98
Avrameas, S., 85
Awasthi, S., 5(47), 6, 10(47)

B

Bachas, L. G., 90
Bailey, A. J., 5(29), 6, 9(29)

Lewis, R. V., 28
Lewis, S., 102, 558
Li, J. P., 139
Li, Q. M., 460, 461(27)
Li, S. F. Y., 320, 338, 420
Li, T. M., 85
Li, W., 423
Li, Y., 532
Li, Y.-C., 5, 16(4)
Li, Y.-F., 66
Li, Y.-M., 311(35), 312
Liao, A. W., 60(20), 65
Liao, J.-L., 301, 309, 309(10), 310(34), 312, 314, 314(36, 37), 317(10, 11, 36), 358, 368(2), 420
Liao, P.-C., 545
Liddel, J. D., 84
Light-Wahl, K. J., 506
Lin, H.-J., 7
Lin, H.-Y., 487, 489(10)
Lin, J.-N., 84
Lin, P. T., 118
Lin, S., 7, 29, 32(31), 39(31), 45(31), 75
Lin, S. N., 420, 485
Lincoln, B., 5(53), 6, 10(53)
Linde, S., 5, 8(15), 16(15)
Linhardt, R. J., 5(41), 6, 10(41), 459
Lin Wang, T.-C., 471, 473
Lipes, B. D., 255
Lipowska, M., 184, 186(23)
Lipsky, S. R., 75, 136
Lisi, D. D., 172
Liu, D., 82, 186(14)
Liu, J., 5(58), 6, 11(58), 26(58), 102, 340–341, 420, 425
Liu, X., 5(35), 6, 9(35)
Loganathan, D., 459
Lohse, D. L., 5(41), 6, 10(41)
Loliger, C., 85
Lommen, D. C., 161
Londry, F. A., 560
Loo, J. A., 5(60), 6, 11(60), 400, 422, 486(3), 487–489, 495, 516(3), 563, 567, 585
Lork, K. D., 16
Loscertales, I. G., 492
Lottspeich, F., 317, 532
Louris, J. N., 552, 563, 563(9), 567(34)
Loyd, L. L., 78
Lu, X. M., 22, 32, 39(42, 43)
Lubman, D. M., 545, 585

Lucarelli, C., 18, 182, 186(19)
Luckey, J. A., 295, 421
Lukacs, K. D., 102, 303, 338
Lumpkin, O. J., 273, 274(2), 277(2), 278(2)
Lunstrum, G. P., 5(52), 6, 10(52)
Lurie, I. S., 220

M

Maa, Y.-F., 53, 142, 149
Maa, Y. H., 75
MacAdoo, D. J., 182, 186(14)
Machiori, F., 5, 8(11)
Mack, L. L., 488, 496(19)
Madajová, V., 317, 399
Madden, T. L., 273, 281(4)
Madurawe, R. D., 84
Maezawa, S., 198
Magda, S., 5(35), 6, 9(35)
Malik, S. K., 330
Malliaros, D. P., 189, 191(48)
Mallis, L. M., 459
Manabe, T., 396–397
Maness, N. O., 184, 186(29)
Maniatis, T., 283
Mann, M., 486, 486(4), 487, 489, 517, 531, 546, 552–553
Manning, G. S., 50
Manning, J. M., 539
Manning, W. B., 90
Mant, C. T., 3, 15–16, 18, 22, 25(3), 151, 155
Manz, A., 118, 421
Manzocchi, C., 239
Mao, Y., 532
Maoka, M., 148
Marák, J., 317, 399
March, R. E., 552, 552(15), 553, 555(12), 560
Marchylo, B. A., 163
Marchylo, B. S., 163, 165(26), 171(26)
Marin, O., 5, 8(11)
Marinkovich, M. P., 5(52), 6, 10(52)
Marino, G., 63
Markey, S. P., 455, 471, 473
Marks, V., 94
Marsden, N. V. B., 27
Marshall, A. G., 578
Martin, F., 89
Martin, G., 199
Martin, J., 118
Martin, M., 105, 125(28), 422, 483(68), 484

van der Hoeven, J. S., 385
van der Wal, S., 114
van der Welle, W., 98
Van der Zee, R., 19
van Dijk, J., 199
van Dijk, L., 199
Van Dorsselaer, A., 5(38), 6, 9(38)
van Nispen, J. W., 394–395
Van Soest, R. E. J., 420
van Zeeland, M. J. M., 394–395
Varady, L., 78, 137, 142–143, 144(6), 146(6, 21)
Varro, R., 85
Vasser, M., 74
Vasta, J. F., 156, 158(17), 159
Vaster, J. F., 12
Vath, J. E., 524, 583
Velander, W. H., 84
Velayudhan, A., 49, 50(3)
Venembre, P., 98
Verheggen, T., 297, 375, 389–390, 421
Vestal, M. L., 491
Vidal-Madjar, C., 84, 138
Videvogel, J., 320
Vilenchik, M., 318
Viovy, J. L., 277, 278(21), 281, 281(21), 282, 282(21, 35), 294(35)
Virtanen, R., 297
Visser, S., 5, 8(14)
Viyayendran, B. R., 199
Vogels, G. D., 385, 387
Volgin, Y. V., 5, 8(6), 16(6), 45
Volk, K. J., 5(58), 6, 11(58), 26(58)
Vollrath, D., 258
Vonderschmitt, D. J., 382–383
Vorhgar, T., 5(33), 6, 9(33)
Vorkink, W. P., 303
Vorm, O., 525, 531
Voth, J. E., 543, 544(44)
Vouros, P., 500, 578
Voyksner, R. D., 487–488, 489(10)
Vratny, P., 184, 186(27)

W

Wachs, T., 420
Wachter, H., 45
Wada, Y., 489
Wahl, J. H., 572, 576, 581
Wahrmann, J. P., 248

Wainwright, A., 340–341
Waki, H., 519
Walbroehl, Y., 420
Waldron, K. C., 433, 434(52)
Wallingford, R. A., 323, 338–339, 421
Walters, R. R., 147, 148(26)
Wampler, F. M., 517
Wampler, F. M. III, 489
Wang, B., 5(55), 6, 10(55)
Wang, D. I. C., 92, 135, 147(2)
Wang, H. M., 459
Wang, P., 423
Wang, R., 520, 525, 539, 539(21), 549(39), 550(39), 551
Wang, S. F., 553
Wang, T., 439, 440(55)
Wang, Y. K., 545
Warburton, P. E., 257
Warner, F. P., 78
Watanabe, C., 525, 545, 545(18), 546(18, 46)
Watanabe, S., 471
Watson, C. H., 517
Watson, E., 5, 8(19)
Wätzig, H., 307
Weast, R. C., 299
Webb, E. C., 46
Weber, I. T., 5, 9(22)
Wegelin, E., 309
Wehr, C. T., 177
Wehr, T., 272, 358, 361, 365(7), 373–374
Weidolf, L. O. G., 496
Weimbs, T., 5(39), 6, 9(39)
Weinberger, R., 178, 186(6)
Weiss, E., 45
Weisweiler, P., 62, 63(16a)
Weksler, B., 539
Welch, L. E., 185, 191(34)
Welinder, B. S., 5, 8(7, 8, 15), 16(15), 18
Welling, G. W., 19
Welling-Wester, S., 19
Wells, G. J., 560
Welply, J. K., 486(5), 487
Wendt, H., 89
Weng, Q. M., 459
Wenisch, E., 251, 252(26), 254
Westermeier, R., 236
Wetlaufer, D. B., 330
Wettenhall, R. E. H., 5, 8(5), 16(5)
Wettlaufer, B., 224
Wheat, T. E., 5, 8(18), 19(18), 176

Z

Zare, R. N., 116, 122, 335, 420–421, 423, 425, 427(40), 428(40)
Zegers, K., 303
Zhang, J. Z., 411, 427
Zhang, R., 312, 314, 314(36, 37), 317(36)
Zhang, W., 525, 545
Zhao, J.-Y., 427
Zhao, Y., 546, 547(54), 548(54)
Zhou, J., 5, 9(27), 558

Zhou, N. E., 5(34), 6, 9(34), 16, 22
Zhu, G., 497
Zhu, H., 542
Zhu, M., 272, 358, 361, 365(7), 369(6), 373–374, 420
Zhu, M.-D., 301, 309(10), 317(10)
Zimm, B. H., 273, 274(2), 277(2, 11), 278(2), 281
Zoccoli, M. A., 90
Zsindely, S., 83

Subject Index

A

Agarose gel electrophoresis, *see* Gel electrophoresis

Alkaline phosphatase, recombinant engineering for immunoassay, 90

Alkaloids, plant, isotachophoresis, 385, 387

Amino acid analysis, microcolumn liquid chromatography, 125–126

Amino acid sequencing
  fast atom bombardment mass spectrometry, 464, 466
  matrix-assisted laser desorption mass spectrometry, 549, 551
  quadrupole ion trap mass spectrometry, 561–563, 566–568, 571–572, 586

Ammonium sulfate, hydrophobic interaction chromatography salt, 38–39

Anion-exchange chromatography, *see* Ion-exchange chromatography

Antibody–antigen complex, *see* Immunocomplex

B

1,2-Bis(4-methoxyphenyl)ethylenediamine, nucleic acid derivatization, 184

Bovine serum albumin, pH titration in free solution zone capillary electrophoresis, 345–350

BSA, *see* Bovine serum albumin

C

Calcium, quantitation by isotachophoresis, 383

Capillary electrophoresis, *see also* Isotachophoresis
  buffer selection
    conductivity, 309–310
    counterions, 309
    optical absorption, 310
    pH, 309
  charge suppression calculation, 356
  coating of capillary
    electroendosmosis suppression, 302, 360–361
    methyl cellulose coating
      allyl methyl cellulose synthesis, 304
      covalent binding step, 304–305
      inner wall activation with methacryl groups, 304
      pH stability, 303
    protein adsorption to fused silica, 302
    types of coating, 302–303, 360–361
  convection suppression, 296–297
  detection systems
    axial-beam detection, 437–439
    design difficulties, 419, 447–448
    electrochemical detection, 421, 443
    indirect fluorometry, 444–446
    laser-induced capillary vibration, 435, 437
    laser-induced fluorescence, 422–423, 425–429, 431–433, 448–449
    mass spectrometry, 421–422, 449
    multipoint detection, 317–319
    multireflection cell, 439–440
    optical absorbance, 420, 433
    radioactivity detection, 421
    refractive index detection
      interference fringe sensing, 441

L

M

ISBN 0-12-182171-4

90038